Foundations of Time Series Analysis and Prediction Theory

A complete list of the titles in this series appears at the end of this volume.

Foundations of Time Series Analysis and Prediction Theory

MOHSEN POURAHMADI
Northern Illinois University

A Wiley-Interscience Publication
JOHN WILEY & SONS, INC.
New York • Chichester • Weinheim • Brisbane • Singapore • Toronto

This text is printed on acid-free paper. ♾

Library of Congress Cataloging-in-Publication Data:

Pourahmadi, Mohsen.
Foundations of time series analysis and prediction theory / Mohsen Pourahmadi.
p. cm. — (Wiley series in probability and statistics)
Includes bibliographical references and index.
ISBN 0-471-39434-3 (cloth : alk. paper)
1. Time-series analysis. 2. Prediction theory. I. Title. II. Series.

QA280 .P68 2001
519.5′5—dc21 2001017918

Printed in the United States of America

10 9 8 7 6 5 4 3 2 1

To
Terri, Nina and Adam

Preface

This book provides a solid *foundation* for time series analysis and prediction theory. The need for a foundation is manifested by the maturity, widespread use and truly interdisciplinary nature of time series analysis lying at the intersection of the mathematical, statistical, computational, physical, engineering and system sciences. The goal is to join the two lines of developments in *time series analysis* centered around the work of Schuster, Yule, Slutsky, Wold, ..., and *prediction theory* based on the work of Szegö, Wold, Cramèr, Kolmogorov, Wiener, ... into a logically sound and pedagogically reasonable theory of modern statistical time series analysis.

We use the language of regression (projection) and the Hilbert space which is powerful, natural, intuitive, spoken widely, and hence capable of facilitating interactions, interdisciplinary efforts to problem solving and communicating results among researchers from diverse and growing fields where time series problems may arise. There is great need for a book emphasizing the fundamental results and structural underpinnings of stationary processes to explain, extend and unify in a mathematically coherent manner the diverse and useful developments of the last few decades. Special effort is made to motivate, present and prove the results on the structure and prediction of stationary processes in the time-domain using the celebrated Wold decomposition. Statistical methods and concepts, however, are mostly motivated using autoregressive (AR) models. In addition due emphasis is given to the spectral-domain results, data analysis and computational issues so as to entice the reader to pursue these areas further.

Our approach does not compete with but rather complements those pursued by Hannan (1970), Anderson (1971), Hannan and Deistler (1988), Brockwell and Davis

(1991) and Fuller (1996), in that we do not present the proofs of consistency and asymptotic normality of the sample covariances, the maximum likelihood estimators of the ARMA parameters and the spectral density estimators. Such classical results, though extremely important, are well established. Instead, we emphasize the mathematical/probabilistic, statistical/data analytic and computational concepts along with some novel and useful results from the research papers that have not appeared in books yet. Proofs of some of these mathematical/probabilistic results are lengthy and tedious or may require advanced background in probability, harmonic and functional analysis. In such cases, we merely motivate, state the results, sketch the proofs and highlight their applications. The books by Robinson (1954), Lambert and Poskitt (1983) and Pollock (1999) are relevant to the topic of developing foundations for time series analysis and prediction theory.

This book is suitable for researchers and advanced students in probability and statistics, mathematics, engineering and system sciences, physical and natural sciences, economics and social sciences who are interested in the deeper aspects of time series analysis. Familiarity with probability and statistics at the level of Casella and Berger's (1990) *Statistical Inference* and knowledge of linear algebra and Hilbert spaces are helpful. Advanced topics including those on Hilbert spaces are used and developed gradually in later chapters as needed. With a careful selection of topics and appropriate supplementation if necessary, the book can be used as a graduate text for either a one- or a two-semester course on time series analysis (Chapters 2–5) and (second-order) stochastic processes (Chapters 5–10). The problems at the end of each chapter and the indicated projects in Chapters 2–4 are useful for gaining skill, deeper appreciation of the covered topics and indicating possible directions of extending the methodology and the theory. However, the book does not attempt to be a traditional manual for time series data analysis and forecasting.

According to the mathematics/statistics level, the book can be divided loosely into three parts comprising of Chapters 1–4, 5–8 and 9–10. At the expense of some duplication, the topics are arranged so that each part can be studied virtually independently of the others. The first part is statistical and covers most ingredients needed for time series data analysis. The second part is probabilistic and develops structural results on stationary processes. Finally, the third part is more mathematical reviewing some background material on Hilbert spaces and function theory and providing proofs of deeper results such as the spectral representation and the Szegö-Kolmogorov prediction theorems. Some of these topics are of independent interest to statisticians due to their widespread use in modern statistics; see Wahba (1990), Small and McLeish (1994) and Vidakovic (1999).

The scope of the subject is wide and the topics covered reflect my interests and the need to control the length of the book. Topics not covered here have received book-length coverage in other sources. Some examples are nonlinear models (Tong, 1990), nonstationary models (Priestley, 1988), long-memory models (Beran, 1994), random fields (Rosenblatt, 1985; Yaglom, 1986), Bayesian models (West and Harrison, 1997), continuous-time processes (Dym and McKean, 1976) and multivariate processes (Rozanov, 1967; Hannan and Deistler, 1988). A broad theory of statistical inference for these diverse cases is provided by Taniguchi and Kakizawa (2000). My hope is that the present book will prepare and motivate the reader for further study and research in these diverse areas centered around the concept of stationarity.

Some of the datasets used in this book are available via my website `http://www.math.niu.edu/~pourahm/book/` at Northern Illinois University. The computations and graphics are done in Splus (MathSoft Inc.). The datasets and the sample programs can also be obtained from the author at the address below. Please feel free to let me know of your questions, comments and constructive criticism.

March 20, 2001

Mohsen Pourahmadi
Division of Statistics
Northern Illinois University
DeKalb, Illinois, 60115 USA
email: pourahm@math.niu.edu

Acknowledgments

Special thanks are due to many people and organizations for contributing directly or indirectly to this work. The project began in 1988 while on sabbatical in the Division of Statistics at University of California, Davis; George Roussas and Bob Shumway were helpful with facilitating the visit and explaining the roles of AIC and BIC, respectively. My understanding of the subject has been shaped and reshaped by many formal and informal discussions with Manny Parzen, P. Brockwell, R. Davis, D. Findley, D. Tjøstheim, A.G. Miamee, H. Salehi, P.R. Masani, S. Cambanis, G. Kallianpur, V. Mandrekar, A.R. Soltani, S. Axler, ..., and my colleagues in the Division of Statistics at Northern Illinois University. Many students have suggested changes and corrections over time, however, the help of B. Tarami and D. Olive in proof-reading various drafts deserves to be singled out. Susan Catron did a superb job of preparing the document in LaTeX and finishing the work of several of her predecessors. Steve Quigley and the editorial staff at John Wiley & Sons were generous with their assistance, timely suggestions and reminders. Research support from the NSF, AFOSR and NSA has contributed greatly to this work. Finally, I would like to thank my family for their patience, understanding, support and good humor while I was working on the project. Leaving a bunch of red pens in my Christmas, 1999, stocking was probably their gentle way of reminding me to bring it to a timely closure.

M. P.

Contents

Acronyms

ACF	Autocorrelation function
AO	Additive outlier
AR	Autoregression
ARIMA	Autoregressive integrated moving average
ARMA	Autoregressive moving average
ARR	Autoregressive representation
BFS	Bayesian Forecasting System
CAPM	Capital asset pricing model
CS	Compound symmetry
DE	Damped Exponential
EWMA	Exponentially weighted moving average
GARCH	Generalized autoregressive conditionally heteroschedastic
GARP	Generalized autoregressive parameters
GLS	Generalized least squares
IACF	Inverse autocorrelation function
IO	Innovation outlier

IV	Innovation variances
MA	Moving average
MLE	Maximum likelihood estimator
MMSE	Minimum mean square error
OSM	Ordinary Scatterplot Matrix
PACF	Partial autocorrelation function
PRISM	Partial Regression-On-Intervenors Scatterplot Matrix
RKHS	Reproducing kernel Hilbert space
SFS	Statistical forecasting system
SUR	Seemingly unrelated regressions
VMO	Vanishing mean oscillation

1

Introduction

Unlike classical statistics, time series analysis is concerned with statistical inference from data that are not necessarily independent and identically distributed (i.i.d.). The ideal framework for time series analysis is stationarity whereas the data encountered in practice are often nonstationary for a variety of reasons. Thus, the challenges facing a time series analyst are to transform the data to stationarity first, and then transform stationary data into an i.i.d. sequence. Some unique features of time series analysis and this book in dealing with these two challenges are described briefly in this chapter.

1.1 NONSTATIONARY DATA AND ORTHOGONALIZATION

A discrete-time stochastic process $\{X_t\}$ representing the evolution of a system over time could be nonstationary because either its *mean* (level) or its *dependence* as gauged by the covariance between two measurements is a function of time. For a given time series data, separating or distinguishing these two aspects of nonstationarity known as mean-nonstationarity and covariance-nonstationarity, respectively, is a difficult task. Traditionally, however, much attention is paid to correcting the mean nonstationarity by relying on simple transformations such as logarithm, differencing and smoothing. These transformations and others from the familiar and popular area of regression analysis are reviewed and illustrated in the early parts of Chapters 2 and 3.

Unfortunately, no simple statistical methodology like regression is known for dealing with covariance-nonstationarity. In addition, the presence of dependence in the data, whether of stationary or nonstationary kind, is problematic for someone who is used to analyzing independent observations. How does one deal with the dependent data? Chapter 2 singles out about eight key results and illustrates their roles in time series data analysis and prediction theory. The first four key results and principles are just enough to encourage a novice to venture into analyzing data and raising questions about the validity and prudence of the commonly used asymptotic procedures in finite sample situations. The balance of the book shows that the idea of *orthogonalization* leads to powerful methodologies for handling stationary and nonstationary data alike. Some of the most important techniques such as the Box-Jenkins methods, the state-space and Kalman filtering algorithm, spectral analysis, the Wold decomposition and the Karhunen-Loeve expansion are centered around the notion of orthogonalizing segments of a stochastic process.

These diverse methods of orthogonalization lead to powerful factorizations of a covariance matrix. In Chapter 3, we show that factorizations and the ensuing graphical methods of time series analysis developed initially to model parsimoniously the covariance matrix of one long stationary series can be generalized to model parsimoniously the common covariance matrix of several short nonstationary series or longitudinal data. The key idea we use is the Cholesky decomposition of the covariance matrix.

1.2 ARMA MODEL FITTING PROCESS

As pointed out earlier there are numerous ways to transform the dependent data $\{X_t\}$ into a sequence of uncorrelated random variables $\{\varepsilon_t\}$ called a *white noise*. Among these a central role is played by the Yule's (1927) pioneering idea of auto-regression and its evolution into the autoregressive and moving average (ARMA) models of orders p and q:

$$X_t = \sum_{i=1}^{p} \alpha_i X_{t-i} + \varepsilon_t + \sum_{j=1}^{q} \beta_j \varepsilon_{t-j}, \tag{1.1}$$

denoted by ARMA(p, q) for short.

The goal of ARMA model fitting process is to use the data $X = (X_1, \ldots, X_n)$ and develop an empirical method for choosing the orders p, q and estimating the parameters $\theta = (\alpha_1, \ldots, \alpha_p, \beta_1, \ldots, \beta_q, \sigma^2)$ so that the fitted residuals are approximately uncorrelated. More precisely, the following three stages are used in the iterative ARMA model fitting process:

(i) Model identification (choosing p and q),

(ii) Parameter estimation,

(iii) Diagnostic checking.

The traditional model identification techniques in time series can be motivated by observing that the very definition of ARMA(p, q) is equivalent to the linear dependence of the random variables

$$X_t, X_{t-1}, \cdots, X_{t-p'}, \varepsilon_t, \varepsilon_{t-1}, \cdots, \varepsilon_{t-q'},$$

for all $p' \geq p$ and $q' \geq q$. To make this observation operational one needs to compute the covariance matrix of the above random variables for various $p', q' \geq 0$ and examine their ranks or eigenvalues. A characterization of ARMA models in terms of singularity of certain covariance matrices lies at the heart of many ARMA model identification techniques often used in times series analysis.

The covariance matrix of the sample $X = (X_1, \ldots, X_n)$, namely,

$$\Gamma_n = (\gamma_{k-\ell})_{k,\ell=1}^{n},$$

where $\gamma_{k-\ell} = \mathrm{Cov}(X_k, X_\ell)$ and its relation with the model parameters θ provide most of the necessary ingredients for the ARMA model fitting process and understanding the structure of such models. This is seen most clearly by focusing on the simple case of auto-regression of order two, AR(2) for short, obtained from (1.1) with $p = 2$ and $q = 0$. Thus, for n consecutive times $t = 1, \ldots, n$, we have

$$\begin{array}{ccccccc}
X_1 & = & \alpha_1 X_0 & + & \alpha_2 X_{-1} & + & \varepsilon_1, \\
X_2 & - & \alpha_1 X_1 & = & \alpha_2 X_0 & + & \varepsilon_2, \\
X_3 & - & \alpha_1 X_2 & - & \alpha_2 X_1 & = & \varepsilon_3, \\
\vdots & & & & & & \vdots \\
X_n & - & \alpha_1 X_{n-1} & - & \alpha_2 X_{n-2} & = & \varepsilon_n.
\end{array}$$

Setting $\varepsilon = (\varepsilon_1, \ldots, \varepsilon_n)$ and $e = (X_{-1}, X_0)$, the above can be written as the regression-like model:

$$AX = Be + \varepsilon, \tag{1.2}$$

where

$$A = \begin{bmatrix} 1 & 0 & 0 & \cdots & \cdots & 0 \\ -\alpha_1 & 1 & 0 & \cdots & \cdots & 0 \\ -\alpha_2 & -\alpha_1 & 1 & \cdots & \cdots & 0 \\ 0 & \ddots & \ddots & \ddots & \ddots & \vdots \\ \vdots & \ddots & \ddots & \ddots & \ddots & 0 \\ 0 & \cdots & 0 & -\alpha_2 & -\alpha_1 & 1 \end{bmatrix}, B = \begin{bmatrix} \alpha_2 & \alpha_1 \\ 0 & \alpha_2 \\ \cdots & \cdots \\ 0 & \cdots & 0 \\ \vdots & & \vdots \\ 0 & \cdots & 0 \end{bmatrix} = \begin{bmatrix} B_1 \\ \cdots \\ 0 \end{bmatrix}.$$

This analogy with the regression models is exploited extensively and systematically to motivate and develop the theory and methods for analysis of general ARMA models. To elaborate further, for the moment assuming the ε and e are independent (the causality assumption), it follows from (1.2) that

$$A\Gamma_n A' = \begin{bmatrix} B_1 \Gamma_2 B_1' & 0 \\ 0 & 0 \end{bmatrix} + \sigma^2 I_n. \tag{1.3}$$

This identity clearly reveals the nonlinear relationship between the covariance matrix Γ_n of the data and the model parameters $\theta = (\alpha_1, \alpha_2, \sigma^2)$; more surprisingly, it shows that the unit lower triangular banded matrix A nearly diagonalizes the covariance matrix Γ_n. In fact, the only obstacle in the way of perfect diagonalization is the rank-2 matrix associated with the two *initial values* needed to start an AR(2) model. The identity (1.3) is extremely useful in parameter estimation where one needs to evaluate the determinant and inverse of Γ_n for various values of n and the parameter θ.

Since the length of a time series grows over time, one has to work with a sequence of positive definite matrices $\Gamma_n, n = 1, 2, \ldots$. Heeding to the positive-definiteness constraint is one of the most awkward stumbling block in modeling covariances. Fortunately, Yule's (1927) ingenious auto-regression idea, leading to (1.3), has evolved into a successful statistical methodology for approximating the entries of Γ_n by the auto-covariance of ARMA(p, q) models involving only $p + q + 1$ parameters which does not grow with n. In this framework, the operationally difficult positive-definiteness constraint is replaced by the seemingly simpler requirement that the roots of the characteristic polynomials $\alpha(z) = 1 - \alpha_1 z - \ldots - \alpha_p z^p$ and $\beta(z) = 1 + \beta_1 z + \ldots + \beta_q z^q$ lie outside the unit circle in the complex plane.

Most classical results about ARMA models including conditions for stationarity, causality, invertibility, predictors, state-space representation, use of the Kalman filter in evaluating the Gaussian likelihood function are discussed in Chapter 4. In addition, particular attention is paid to the functional form of the likelihood function of Gaussian ARMA models by writing it explicitly in terms of the model parameters. This requires explicit computation of the inverse of the covariance matrix of the data. For AR processes, the fact that the entries of the inverse covariance matrix are quadratic functions of the model parameters, see (1.3), allows one to introduce sufficient statistics and to write exact likelihood equations for these parameters.

How "dense" is the class of stationary ARMA models within the class of all stationary processes? Are there other "finite-dimensional" covariance modeling methods that can achieve the positive-definiteness of $\Gamma_n, n = 1, 2, \ldots$, or the covariance sequence $\{\gamma_k\}$? The answers to these and other questions in times series analysis and prediction theory can be sought in the unconstrained or mildly constrained reparameterization of a general covariance function $\{\gamma_k\}$. These are reviewed next and covered in more details in Chapters 5–8.

1.3 REPARAMETERIZATIONS

Time series analysis is rich in offering statistically meaningful and (mildly) unconstrained reparameterizations of the covariance function $\{\gamma_k\}$. These are consequences of various ways of diagonalizing Γ_n as in (1.3) or orthogonalizing a segment of the stationary process $\{X_t\}$ giving rise to two broad categories of parameterization corresponding to *spectral-* and *time-domain* approaches to times series analysis.

Spectral Parameters: These are nonnegative functions and consequence of writing the covariance function $\{\gamma_k\}$ as the Fourier transform of a distribution function

F (Herglotz Theorem). More precisely,

$$\gamma_k = \int_{-\pi}^{\pi} e^{ik\lambda} dF(\lambda), \; k = 0, \pm 1, \cdots ,$$

the function F called the *spectral distribution* is the new "infinite-dimensional parameter" of the process. Note that the non-negativity requirements on F is quite mild compared to the positive-definiteness constraint on the covariance. When $\sum_{k=0}^{\infty} |\gamma_k| < \infty$, then F is absolutely continuous with the derivative f given by

$$f(\lambda) = \frac{1}{2\pi} \sum_{k=-\infty}^{\infty} \gamma_k e^{-ik\lambda},$$

called the *spectral density* of the process. Fourier coefficients of and the functions $f^{-1}(\lambda)$ and $\log f(\lambda)$ are also useful in time series analysis and prediction theory; these appear in Chapters 2, 8 and 10. Spectral representations of a stationary process and its covariance function, mean-convergence of the natural estimators of the covariance and spectral density functions, and basic results of prediction theory are discussed in Chapter 5.

Predictor Parameters: These are sequences of numbers appearing as coefficients of certain predictors, and are consequence of the Wold decomposition (Wold, 1938) and other similar representations of the process or its predictors. It follows from the Wold decomposition that the covariance function $\{\gamma_k\}$ can be decomposed as

$$\gamma_k = \sigma^2 \sum_{i=0}^{\infty} b_i b_{i+k},$$

where $\{b_i\}$ satisfies $\sum_{i=0}^{\infty} b_i^2 < \infty$ and the zeros of the function $\varphi(z) = \sum_{k=0}^{\infty} b_k z^k$ lies on or outside the unit disc $(|z| < 1)$ in the complex plane. The sequence $\{b_i\}$ is called the *moving average* (MA) parameters of the process $\{X_t\}$. Note that the requirements on $\{b_i\}$ are quite mild compared to those on $\{\gamma_k\}$. Since $\varphi(z)$ does not have zeros in the unit disc, the Taylor expansion of its reciprocal

$$\frac{1}{\varphi(z)} = 1 - a_1 z - a_2 z^2 - \cdots,$$

gives rise to $\{a_i\}$ the *autoregressive (AR) parameters* of the process. A fairly elementary and versatile method of parameterizing stationary processes, even when the covariance or the spectrum is not defined, is discussed in Chapter 6. This method, based on the simple idea of regression, is used to introduce AR and MA representations and parameters for second-order stationary as well as infinite variance stationary processes. The approach has a finite-dimensional character and reveals a sort of duality between the AR and MA representations and parameters of a stochastic process.

Closely related to the AR parameters is the coefficients of the linear least squares predictor of X_t based on the finite past $X_{t-1}, \ldots, X_{t-n}$. This predictor is of the form $\sum_{k=1}^{n} a_{k,n} X_{t-k}$ where $\boldsymbol{a}_n = (a_{1,n}, \ldots, a_{n,n})$ satisfies the *Yule-Walker equations*

$$\Gamma_n \boldsymbol{a}_n = \boldsymbol{\gamma}_n, \ n = 1, 2, \cdots, \tag{1.4}$$

with $\boldsymbol{\gamma}_n = (\gamma_1, \ldots, \gamma_n)$, which can be solved recursively in n using the Durbin-Levinson algorithm. Surprisingly, the sequence of scalars $\{a_{n,n}\}_{n=1}^{\infty}$ called the *partial autocorrelation function* (PACF) provides a reparameterization of $\{\gamma_k\}$ which plays a distinguished role in linking results from diverse areas of applications such as time series analysis, orthogonal polynomials, digital signal processing, inverse scattering and numerical linear algebra. For a fixed k, the finite predictor coefficient $a_{k,n}$ converges to the kth AR parameter a_k as $n \to \infty$. It turns out that many important problems in time series analysis such as consistency of ARMA order selection using AIC and BIC, and asymptotic properties of autoregressive spectral density estimators are closely tied to the rate and mode of convergence of $(a_{1,n}, \ldots, a_{n,n}, 0, \ldots)$ to $(a_1, a_2, \ldots)$. Some of these problems and their connections with the theory of orthogonal polynomials on the unit circle and maximum entropy estimation are discussed in Chapter 7.

Canonical correlations between past and future of a stationary process provides another reparameterization of the process. These are of interest since a stationary process with only a finite number of non-zero canonical correlations is essentially "finite-dimensional" and turns out to be an ARMA process. In Chapter 8, these concepts are introduced and the apparently distinct problems of interpolation of missing values, prediction based on incomplete past, model identification and estimation of the impacts of interventions on a given response are solved using the set of multi-step predictors of a stationary process. Most of the topics of Chapters 6–8 are fairly new and this is probably the first coherent presentation of these diverse topics in a book.

1.4 MATHEMATICS OF PREDICTION THEORY

Since the pioneering work of Kolmogorov (1941), the Hilbert space framework has been used routinely in solving linear least squares prediction problems of stochastic processes. Some background topics such as inner product spaces, projection and bounded operators, the theory of positive-definite functions and reproducing kernel Hilbert spaces are covered in Chapter 9. These are then used to establish the spectral theorem for unitary operators by noting that the function

$$\gamma_k = (U^k x, x)$$

is positive-definite on the set of integers, for any fixed x from a Hilbert space and a unitary operator U. From this the spectral representation of stationary processes, the Kolmogorov's isomorphism between time-domain and spectral-domain and solution of a generic linear filtering problem are immediate.

For a stationary process $\{X_t\}$ with known spectral density function $f(\lambda)$, the isomorph $g_1(\lambda)$ of the linear least squares interpolator of X_0 based on the *rest* $\{X_t;\ t \neq 0\}$ is found easily to be

$$g_1(\lambda) = 1 - c_1 f^{-1}(\lambda), \tag{1.5}$$

where c_1 is a constant. In sharp contrast, the isomorph $g_2(\lambda)$ of the linear least squares predictor of X_0 based on the *past* $\{X_t; t < 0\}$ is much harder to arrive at and turns out to be

$$g_2(\lambda) = 1 - c_2 \varphi^{-1}(\lambda), \tag{1.6}$$

where c_2 is a constant and $\varphi(\lambda)$ is the optimal factor of the spectral density, that is,

$$f(\lambda) = |\varphi(\lambda)|^2, \tag{1.7}$$

where $\varphi(\lambda)$ is an outer function in the Hardy space H^2. The class H^2 of analytic functions on the unit circle and some relevant function-theoretic results like the Beurling theorem characterizing outer functions and the Kolmogorov-Szegö distance theorem expressing the prediction error variance c_2 in (1.6) in terms of the spectral density function are covered in Chapter 10. Though the isomorphs g_1 and g_2 are similar, computing the latter in the spectral-domain requires much more elaborate mathematical machinery. However, the results in Chapters 5 and 8 clearly reveal that in the time-domain the Wold decomposition can be used to solve the prediction, interpolation and many other problems in a unified manner.

2

Time Series Analysis: One Long Series

An overview of time series data analysis, its tools, concepts, history and connection with statistics, prediction theory and orthogonal polynomials on the unit circle are presented in this chapter. The goal is to single out a few (about eight) key results and illustrate their roles in data analysis and statistical time series analysis. The first four key results and principles are just enough to encourage a novice to venture into analyzing time series data and raising questions about the validity and prudence of the commonly used asymptotic procedures in the finite sample situations.

The chapter begins with the basic notion of stationarity. Since most time series data in practice are nonstationary, simple tools such as differencing and linear regression models are used in Sections 2.2 and 2.3 to transform the data to stationarity. In Section 2.4, dependent time series data are made nearly uncorrelated using discrete Fourier transform, simple moving average and autoregressive transforms. Periodograms and correlograms are used in the search for periodicity and model identification. The Box-Jenkins iterative three-stage model fitting process is reviewed. Various statistical forecasting systems are introduced in Section 2.5; these are motivated using autoregressive models and the Yule-Walker estimate of its parameters. Here, we naturally encounter numerous conceptual, statistical, probabilistic and mathematical questions and challenges. The central question is that of defining and establishing the efficiency of statistical forecasting systems using the data at hand. The maximum likelihood estimation of parameters of ARMA models is reviewed in Section 2.6. The basics of the Kolmogorov-Wiener theory of prediction of a stationary time series is presented in Section A2.7 as an appendix to the chapter.

2.1 TIME SERIES DATA

Measurements $x_1, x_2, \cdots, x_n$ recorded over time is called a *time series dataset.* Examples of time series datasets are the daily price of the IBM stock, daily temperature in a certain city, monthly unemployment rate, etc. The numbers $x_1, x_2, \cdots, x_n$ are usually viewed as the observed values of a segment of a discrete-time *stochastic process* $\{X_t\}$, or a collection of random variables whose index set $\mathbb{Z} = \{t = 0, \pm 1, \cdots\}$ consists of all integers.

Historically, time series data were collected and analyzed first in commerce and banking. In the 19th and 20th century, time series analysis flourished under the stimulus of applications in almost every area of human endeavors including meteorology, biology, astronomy, insurance, business and economics, demography, genetics, traffic flow, clinical trials, neurophysiology, medical imaging, earthquakes, signal processing and telecommunications, just to name a few. The engineering applications referred to under the broad heading of 'signal processing' were inspired by Norbert Wiener in the early 1940's, who insisted on viewing a signal, say, the path of an airplane, as a stochastic process.

Datasets: The following three datasets, in addition to others, are used throughout to illustrate various methodologies. The first two have been around for quite a while, in fact, the lynx data has achieved a benchmark status in the literature of time series analysis. However, the third dataset is fairly new and unexplored.

Airline data: The data consist of the number of international airline monthly total passengers (in thousands) during Jan. 1949–Dec. 1960 with $n = 144$ months (Box, Jenkins and Reinsel, 1994, Series G). In the sequel we refer to this dataset simply as the airline data; see Figure 2.1.

Lynx data: The data consist of the annual record of the number of Canadian lynx trapped in the Mackenzie River district of Northwest Canada for the period 1821–1934 with $n = 114$ observations. It is known that these are a fairly noisy set of field data and have been used as a benchmark to gauge the performance of various new time series methods and models; see Moran (1953), Tong (1990), Lin and Pourahmadi (1998) and Stenseth et al. (1998).

Snow data: Snow water equivalent (SWE) is a measure in inches of the amount of snow water contained in snow (Isaacson and Zimmerman, 2000). It is important to cities and agriculture in the western United States, because most of the summer water supply comes from melting snow in the Rocky Mountains and nearby mountain ranges. The SWE data we consider here are from one particular watershed near the Yellowstone National Park in Wyoming measured on April 1 of each year. There are two methods to measure SWE. The older method is a manual snow survey, called the snow course measurements (SNOCOM), and involves a person hiking to specified sites measuring the depth of the snow using a measuring stick. After the depth of the snow is measured it is converted to SWE based on an equation involving some other variables such as temperature. SNOCOM data are available for the years 1935–1996 for a total of 62 years. Since this system of measurement is dangerous, a newer method called snow telemetry (SNOTEL) consists of electronic devices set up at designated sites and has been in use since 1967. We have SNOTEL SWE data for the period of 1967–1996, for a total of 30 years. The general objectives here are to determine: (a) if there is significant evidence of a trend in SWE in this watershed over the years 1935–1996; and (b) which method is more reliable; see Figure 2.4. In addition to these objectives, the dataset is ideal for fostering interactions among linear models, times series analysis, Kalman filtering and serves as a great link to the repeated measure data discussed in Chapter 3.

2.1.1 Stationarity

A major goal of *time series analysis* or *inference for stochastic processes* is to infer the unknown structure (distribution) of the process $\{X_t\}$ using the available data $x_1, x_2, \cdots, x_n$. This structure (model) is then used to forecast and control the future of the process. Of course, this is a nearly impossible task if the system is unstable or chaotic over time. Thus, some kind of *stability* or *stationarity* is needed. There are many definitions of stationarity, the most useful for the purpose of prediction is the notion of *covariance stationarity* due to Khinchin (1932, 1934) where the first two moments are time-invariant in the sense defined below.

A stochastic process $\{X_t\}$ is said to be *stationary* if for all integers t, k,

$$\begin{aligned} E(X_t) &= \mu, \\ \text{Cov}\,(X_{t+k}, X_t) &= \gamma_k. \end{aligned} \tag{2.1}$$

The first identity in (2.1) suggests that $\{X_t\}$ fluctuates around a fixed level μ (mean of the process) and the second suggests the variation around this level does not change over time; since setting $k = 0$ one obtains

$$\text{Var}\,(X_t) = \gamma_0, \text{ for all } t.$$

The function $\{\gamma_k\}$ defined on the set of all integers, by the second equation in (2.1), is called the *covariance function of* the stationary process $\{X_t\}$. Its *autocorrelation function* (ACF), denoted by $\{\rho_k\}$, is given by

$$\rho_k = \frac{\text{Cov}(X_{t+k},\ X_t)}{\sqrt{\text{Var}(X_{t+k})\text{Var}(X_t)}} = \frac{\gamma_k}{\gamma_0}, k = 0, \pm 1, \cdots.$$

Note that due to stationarity the covariance and correlation between X_{t+k} and X_t depends only k, their *time-separation* or *lag*, and not on the actual values of time or the time origin. Moreover, the Chebyshev's inequality,

$$P(|X_t - \mu| \geq c\sqrt{\gamma_0}) \leq \frac{1}{c^2},$$

suggests that (for c large) a small percentage of the measurements $x_1, \cdots, x_n$ could fall outside the interval $(\mu - c\sqrt{\gamma_0}, \mu + c\sqrt{\gamma_0})$. Graphically speaking, this suggests that the time series plot or the scatterplot of x_t versus $t, t = 1, 2, \cdots, n$, should be contained in a relatively narrow horizontal band regardless of n. We shall rely on this simplified, but *ideal picture of stationarity* to assess such property in a time series data.

Stationary time series have been used successfully to model a wide range of fluctuating phenomena ranging from business and economics to natural, physical and engineering sciences. In these areas, one usually estimates the covariance function $\{\gamma_k\}$ using the data $x_1, \cdots, x_n$, then the problem is to select an appropriate probabilistic description of the process on the basis of this estimated covariances. Often in practice, but not always, the phrase 'an appropriate probabilistic description of the process' amounts to expressing a future value in terms of its predecessors.

For a stationary process, its mean, covariance and correlation functions are estimated by $\bar{x} = \frac{1}{n}\sum_{t=1}^{n} x_t$ and for $k = 0, 1, \ldots, n-1$,

$$\begin{cases} \hat{\gamma}_k = \frac{1}{n}\sum_{t=1}^{n-k}(x_{t+k} - \bar{x})(x_t - \bar{x}), \\ \hat{\rho}_k = \hat{\gamma}_k / \hat{\gamma}_0. \end{cases} \tag{2.2}$$

These estimators, particularly the plot of $\hat{\rho}_k$ vs. k called the *sample correlogram*, play crucial roles in selecting an appropriate probabilistic description of the process generating the data. Some statistical properties of these estimators are discussed in Sections 2.4.3, 2.6 and Chapter 5. Meanwhile, it is instructive to note that for a fixed k, $\hat{\rho}_k$ in (2.2) is essentially the Pearson product-moment correlation between $X = X_{t+k}$ and $Y = X_t$ using the lagged pairs of observations (x_{t+k}, x_t), $t = 1, \ldots, n-k$. Thus, in the context of exploratory time series data analysis, it is possible to exploit effectively the intuition gained from the delicate interplay between correlation and prediction in the regression analysis framework. We elaborate on this point further in the next few sections.

2.1.2 Nonstationarity

Time series plots of many data sets encountered in applications do not resemble the ideal picture of stationarity; they are *nonstationary*, because of *trend, seasonality* and other *time dependent patterns*. Usually the subject matter does provide some valuable information about the existence and nature of such time dependent patterns and components. For example, in Figure 2.1 (a) the plot of the airline data has a clearly nonstationary appearance. How does one deal with such nonstationary and dependent data?

There are two relatively distinct techniques in *time series analysis* for dealing with nonstationarity and dependence, characteristics that make this subject quite distinct from the classical theory of statistics that deals mostly with samples from populations or observations that are independent and identically distributed (i.i.d.). These are (i)

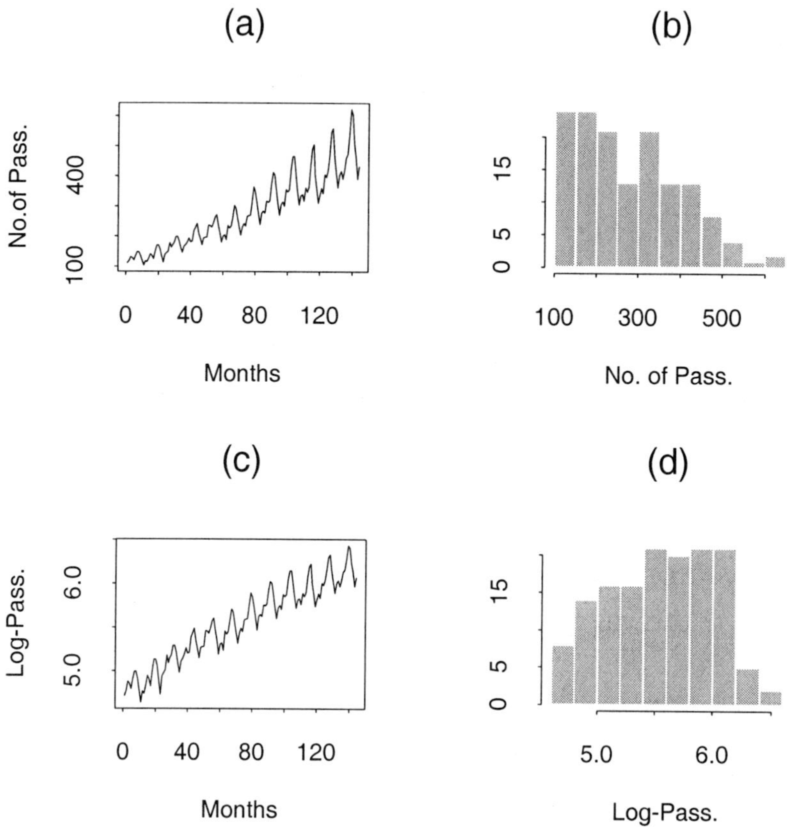

Fig. 2.1 (a) Time series plot of the airline data, (b) The histogram of the data, (c) Plot of natural log of the data and (d) Histogram of log of the data.

transformations to stationarity, and (ii) *transformations to orthogonality* which are reviewed briefly in this chapter and discussed in greater detail in later chapters.

2.2 TRANSFORMATIONS TO STATIONARITY: TREND ELIMINATION

When confronted with nonstationarity, it is prudent to find simple transformations of the data to reduce the degree of nonstationarity. Since most time series of growth phenomena have monotone trend and variability, see Figure 2.1 (a), taking logarithm, successive *differences* and some other linear operations are the simplest and most useful transformations available to time series data analysts for eliminating trends, seasonality, variability and reducing data to stationarity.

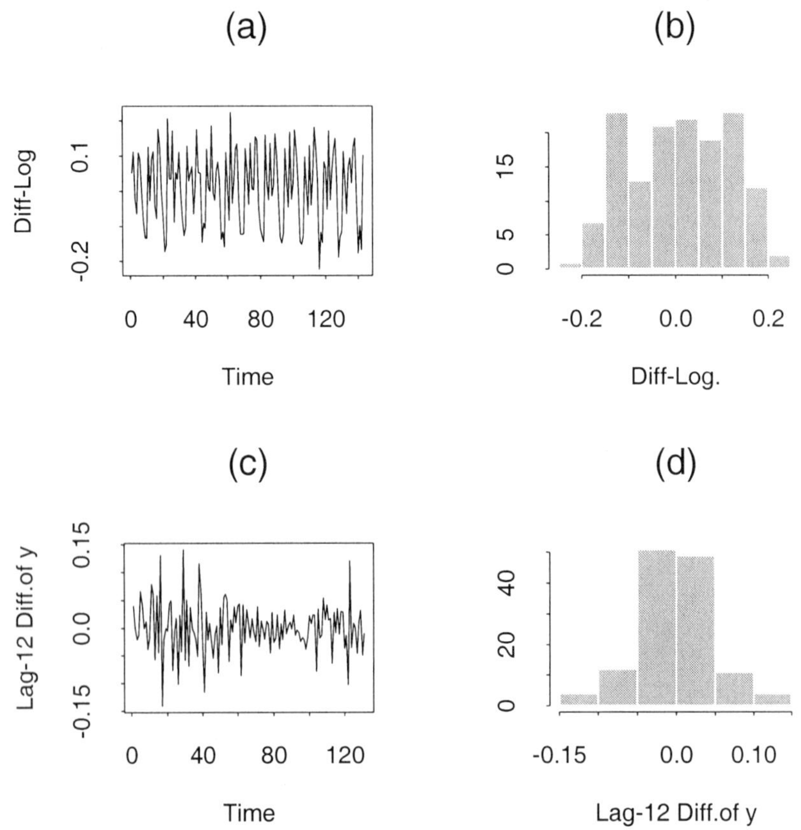

Fig. 2.2 (a) and (b) The time series plot and histogram of the first difference of the natural logarithm of the airline data. (c) and (d) The time series plot and histogram of the seasonally differenced (lag-12) series in (a).

Applying first the natural logarithmic transformation to the airline data and then differencing, one obtains two time series $\{log\ x_t\}_1^{144}$ and

$$y_t = \log\ x_t - \log\ x_{t-1}, t = 2, \cdots, 144.$$

Figures 2.1 (c) and 2.2 (a) show that these simple transformations substantially reduce the variability and nonstationarity of the airline data. The scatterplot of $\{y_t\}$ vs. t, the *logarithmically transformed* and *differenced series*, indicates the presence of a strong periodic or seasonal component with a trough around January and peak around August. Since the month of August is the peak season for international travel, a 12-month *seasonal differencing* of $\{y_t\}$ seems appropriate:

$$z_t = y_t - y_{t-12}, t = 13, 14, \cdots, 144.$$

Scatterplot of $\{z_t\}$ in Figure 2.2 (c) shows that the resulting series is, indeed, more compatible with the ideal picture of a stationary time series. The histograms in these two figures show that the logarithmic and differencing operations have the unintended but pleasant consequence of reducing the original series to a time series with symmetric or nearly Gaussian marginal distribution.

2.2.1 Smoothing

Smoothing is another useful tool for emphasizing certain major features of the data such as *seasonal patterns* while de-emphasizing other features like the *random fluctuations*. Given a time series data, a *3-term smoother* replaces x_t by $s_t = \frac{x_{t-1}+x_t+x_{t+1}}{3}$, a running average of the neighboring terms. The new series $s_1, \ldots, s_n$ is called the *smoothed* series and the *residuals*

$$y_t = x_t - s_t, t = 1, 2, \cdots, n, \tag{2.3}$$

is usually *rough* and re-emphasizes the random aspects of the data without the distraction of the seasonal pattern, Diggle (1990, p. 22). At times the phrase moving average is also used for the above smoothing operations. Figure 2.3 (a), (b) depicts the effect of a 3-term smoother on the logarithm of the airline passenger data. It turns out that differencing and smoothing, though apparently different concepts, are closely related to each other; see Problem 2.1.

Similar to (2.3), one could compute the residuals from a 5-term smoother or more general linear smoothers of the form

$$s_t = \sum_{j=-m}^{m} w_j x_{t+j}, \tag{2.4}$$

obtained as a general weighted average of the terms involved, where m and the preferably symmetric weights w_j's are to be chosen by the data analyst. Note that for $m = 1$ and $w_{-1} = w_0 = w_1 = \frac{1}{3}$, (2.4) reduces to the 3-term smoother used earlier. Other examples of (2.4) are discussed in the sequel. Hastie and Tibshirani (1990, Chapter 2) provide an excellent review of smoothing methods and their relationships.

2.2.2 Linear Regression

The idea of linear regression and the ordinary least square (OLS) estimation of its parameters provides an alternative and viable method for selecting the weights in (2.4). For example, the upward trend in Figure 2.1 (a) and (c) suggests the simple linear regression model,

$$\log x_t = \beta_0 + \beta_1 t + e_t \ , \ t = 1, \cdots, 144, \tag{2.5}$$

for the data, where $\{e_t\}$ is the residual series. Let $\hat{\beta}_0, \hat{\beta}_1$ be the OLS estimates of the parameters. Figure 2.3 (c) and (d) provide time series plots of the estimated residuals $\{\log x_t - \hat{\beta}_0 - \hat{\beta}_1 t\}$ and its lag-12 differences. Although there are some similarities between Figure 2.3 (d) and Figure 2.2 (c), comparing these with Figure 2.3 (b), it

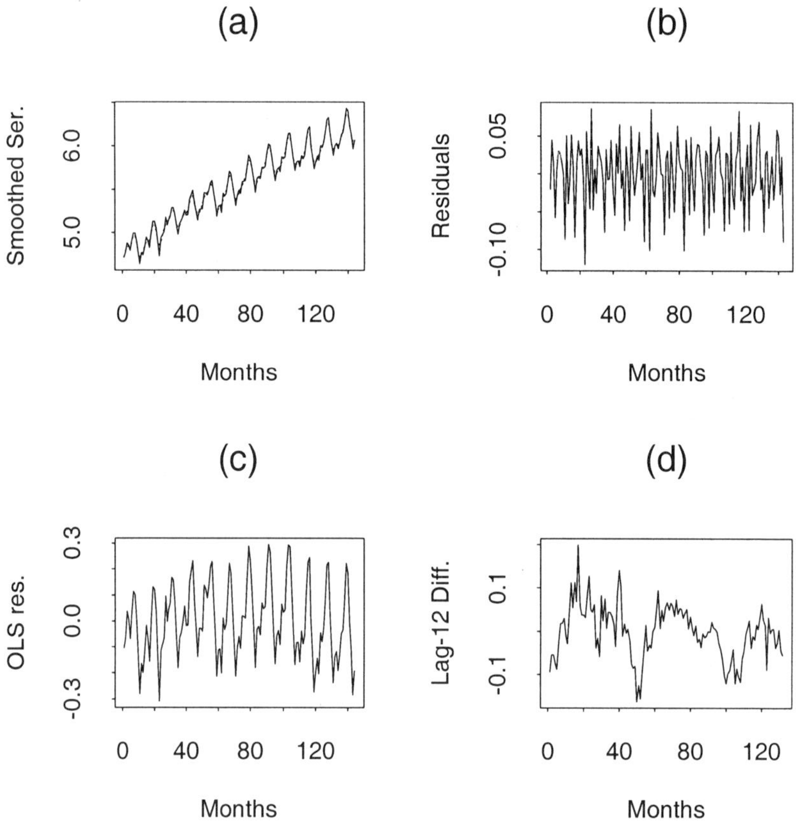

Fig. 2.3 (a) 3-term smoothed series of the airline data, (b) Residuals from a 3-term smooth, (c) OLS residuals of the simple linear model (2.5), (d) Lag-12 difference of the OLS residuals.

is apparent that the 3-term smoother does a better job of eliminating the trend than the methods corresponding to the earlier two figures. In a way this performance is expected of a *smoothing* (nonparametric) method since, unlike the previous two methods, it does not assume a form or model for the underlying mechanism other than assuming its smoothness.

In practice, appropriate transformations are found by carefully examining the plot of the data and comparing it with the ideal scatterplot of a stationary time series. This stage of time series analysis is purely empirical and art-like, it may remain so for the times to come. We know of NO simple and powerful statistical test of stationarity. Finding such a test is nearly impossible since the class of stationary processes is so broad. The alternative hypothesis must deal with the class of nonstationary processes which is even broader! Fortunately, the popularity and availability of interactive, easy-to-use software with powerful graphic features have made the first stage of exploratory time series data analysis accessible to everyone.

The history of massaging time series data and search for stationarity is long, its early development were inspired by the activities of merchants, speculators and individuals associated with business and financial institutions. An excellent account of the history from this perspective is presented in Klein (1997). She reveals the key role of applications in leading to the mathematical concept of stationarity formulated in the early 1930's by A. Khinchin. According to Klein (1997, p.290), "We readily accept that in our economic system time is money; it stands to reason that commercial arithmetic could be a source of algorithms for time series analysis. Merchants, financier, and speculators were among the earliest manipulators of time series data, and it is from their practice that many of the tools associated with the analysis of stationary time series derive. Although the tools have been appropriated to different contexts, an understanding of their earlier uses sheds light on their present capacities. Differencing time series is a key technique for getting from fluctuations to deviations by freezing history. It is no coincidence that a financier with a short-term profit perspective should find first differences as useful as the statistician anxious to make her or his data "stationary". Similarly, studying the veiling/persuading history of truth-seeking tools such as index numbers and moving average enables us to understand their potential and shortcoming for revelation."

In the terminology and practice of modern time series analysis, the above activities are related to *eliminating the trend* in the data. In areas like econometrics and environmetrics the setup of *regression with time series errors* is commonly used to *estimate the trend* or understand the nature of mean-nonstationarities in the data, since the mean is of intrinsic value in such fields.

2.3 TIME SERIES REGRESSION: TREND ESTIMATION

Differencing and smoothing are two powerful and closely related methods for *eliminating the trend* in the data. However, at times it is preferred to estimate the non-constant mean (trend) of the series while studying its dependence or covariance structure. This happens, for example, in economics and environmental sciences where estimation and testing for trends in presence of correlated data sometimes takes the center stage. Verification of the global warming hypothesis and testing whether the national economy is in the growth (expansion) phase are some of the most prominent cases in these areas.

Fortunately, this and the more general problem of relating explanatory (exogenous) variables $\{x_t\}$ to a time series of interest $\{Y_t\}$ can be solved within the familiar linear regression framework.

2.3.1 Simple Linear Regression

Consider the simple linear regression model

$$Y_t = \beta_0 + \beta_1 x_t + e_t, \tag{2.6}$$

where $\{x_t\}$ denotes a fixed explanatory (exogenous, input) variable, β_0 and β_1 are the parameters and hence $\beta_0+\beta_1 x_t$ represents the average effect of $\{x_t\}$ on the (response, output) series $\{Y_t\}$. Here, $\{e_t\}$ is an unobservable zero-mean time series referred to as the noise or the error term in the model. If $\{e_t\}$ were an i.i.d. sequence of random variables, then (2.6) is simply the standard simple linear regression. However, in the applications we have in mind, $\{e_t\}$ is almost always a dependent sequence of random variables (time series). Hence, we refer to (2.6) as the *time series regression* or more precisely as *regression with time series errors*.

Having observed $Y = (Y_1, \ldots, Y_n)$, the goals of time series regression are:

(i) to estimate β_0, β_1,

(ii) to identify the nature of dependence of $\{e_t\}$ and

(iii) to predict the next future value Y_{n+1}.

The dependence structure of $\{e_t\}$ is not that important for the first task, but is crucial for the third. Let $\Sigma = \sigma^2 V$ be the covariance matrix of the error vector $e = (e_1, \ldots, e_n)$. For the moment, we assume that $V = I_n$ is the $n \times n$ identity matrix. The problem of estimating β_0, β_1 usually proceeds by minimizing the residual sum of squares $Q(\beta) = \sum_{t=1}^{n}(Y_t - \beta_0 - \beta_1 x_t)^2$ over the range of $\beta = (\beta_0, \beta_1)$. Differentiating with respect to β_0 and β_1 and setting to zero we obtain the ordinary least squares (OLS) estimates

$$\hat{\beta}_1 = \sum_{t=1}^{n} c_t Y_t \ , \ \hat{\beta}_0 = \overline{Y} - \hat{\beta}_1 \overline{x}, \tag{2.7}$$

where $c_t = \dfrac{(x_t - \overline{x})}{\sum_{t=1}^{n}(x_t - \overline{x})^2}$ and $\overline{x}$ and $\overline{Y}$ are the sample averages of the observed x_t's and Y_t's, respectively.

These estimators of β_0 and β_1 are simple linear functions of the data. It can be shown that they are unbiased using only the assumption that $\{e_t\}$ is mean zero or that the model $E(Y_t) = \beta_0 + \beta_1 x_t$ is correct for the mean. Surprisingly, this unbiasedness property does not depend on the knowledge of the true covariance matrix Σ of e or Y. However, the variance (efficiency) of the above estimators do depend on the knowledge of the covariance.

From (2.6) the natural predictor for Y_{n+1} and its prediction error are given by

$$\begin{cases} \hat{Y}_{n+1} & = & \hat{\beta}_0 + \hat{\beta}_1 x_{n+1}, \\ Y_{n+1} - \hat{Y}_{n+1} & = & e_{n+1} + (\beta_0 - \hat{\beta}_0) + (\beta_1 - \hat{\beta}_1) x_{n+1}. \end{cases} \tag{2.8}$$

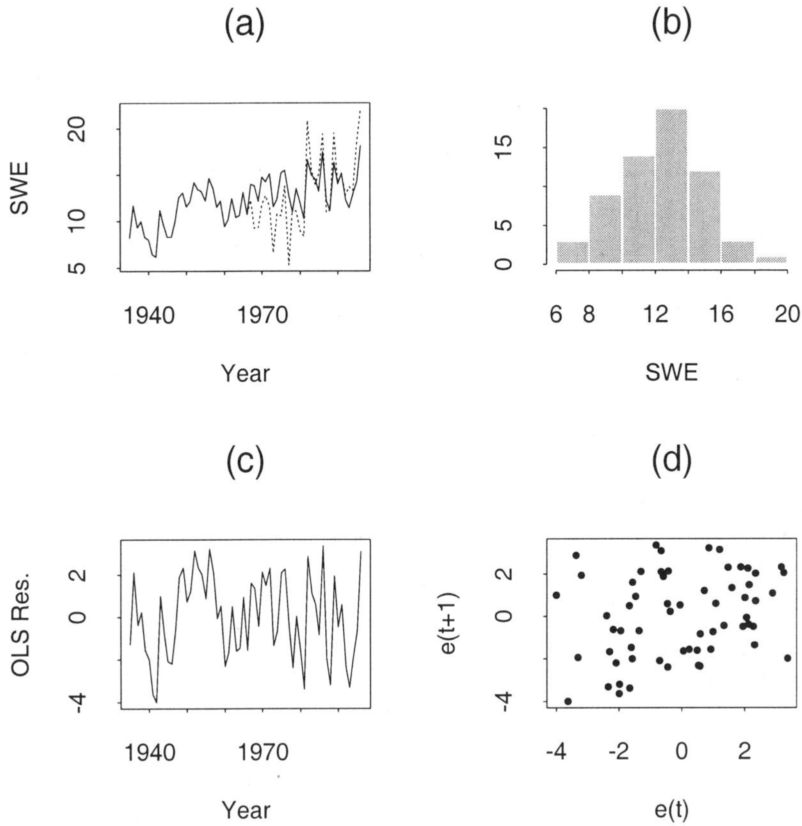

Fig. 2.4 (a) Time series plots of SWE from SNOCOM (—) and SNOTEL (– – –), (b) The histogram of SWE from SNOCOM, (c) OLS residuals of model (2.6) and (d) The scatterplot of $(\hat{e}_t, \hat{e}_{t+1})$.

This is an *unbiased predictor* for Y_{n+1} in the sense that $E(Y_{n+1} - \hat{Y}_{n+1}) = 0$. The variance of the prediction error using (2.8) and (2.7) is given by

$$\begin{aligned} \text{Var}(Y_{n+1} - \hat{Y}_{n+1}) &= \sigma^2 + \text{Var}(\hat{\beta}_0) + x_{n+1}^2 \text{Var}(\hat{\beta}_1) + 2x_{n+1}\text{Cov}(\hat{\beta}_0, \hat{\beta}_1) \\ &= \sigma^2 \left(1 + \frac{1}{n} + \frac{(x_{n+1} - \overline{x})^2}{\Sigma(x_t - \overline{x})^2}\right). \end{aligned} \tag{2.9}$$

If the formula in (2.9) is to be used in practice, one needs to estimate σ^2. To this end, we use the estimated regression residuals

$$\hat{e}_t = Y_t - \hat{\beta}_0 - \hat{\beta}_1 x_t \ , \ t = 1, \cdots, n, \tag{2.10}$$

to form the following unbiased estimator for σ^2:

$$\hat{\sigma}^2 = \frac{\sum_{t=1}^{n} \hat{e}_t^2}{n-2}. \tag{2.11}$$

The estimated residuals $\{\hat{e}_t\}$ is, indeed, a time series on its own right. Thus, it can be plotted and viewed either as a time series plot or regression residuals plot and exploited for the following diagnostic purposes:

(1) assessing stationarity of $\{e_t\}$,
(2) assessing the goodness-of-fit of the model (2.6),
(3) postulating a model for the unknown covariance matrix Σ of e.

Snow data (cont.) These features and use of (2.6)–(2.11) are illustrated next using the snow data. We focus on the SWE data obtained from the SNOCOM method. Its time series plot in Figure 2.4 (a) reveals a monotone trend over this period. Thus, model (2.6) with $x_t = t,\ t = 1, \ldots, 62$, seems plausible for the data. The OLS estimates of β_0, β_1 and σ are:

$$\hat{\beta}_0 = 9.40(0.51),\ \hat{\beta}_1 = 0.09(0.014),\ \hat{\sigma} = 1.98,$$

where the numbers in the parenthesis are the standard errors of the estimates.

The significance of the trend in SWE over the years 1935–1996, noted visually in Figure 2.4 (a), can be tested formally, for example, by computing the t-statistic for testing $H_0 : \beta_1 = 0$ versus $H_1 : \beta_1 \neq 0$. The value of the t-statistic, $\frac{0.09 - 0}{0.014} = 6.48$, is quite large and provides ample evidence to confirm the existence of a trend over this period at the given location. Is it appropriate to use the t-test here? The histogram of SWE in Figure 2.4 (b) supports the required approximate normality of the data. How about the required independence of the data or the error terms in (2.6)? The time series plot of the OLS residuals $\{\hat{e}_t\}$ in Figure 2.4 (c) appears to be trend-free and fairly homogeneous. However, the scatterplot of $(\hat{e}_t, \hat{e}_{t+1}),\ t = 1, \ldots, 61$, in Figure 2.4 (d) indicates presence of mild temporal dependence in the error terms $\{e_t\}$. Overall, the fit of the model (2.6) to the SWE data is quite satisfactory.

For purely illustrative reasons at this time, we use the fitted model to forecast the SWE for the year 1997 with $x_{1997} = 63$. We obtain

$$\hat{Y}_{1997} = 9.40 + 0.09 \times 63 = 15.07,$$

and a 95% forecast interval for Y_{1997} in view of (2.9) is given by

$$15.07 \pm 1.96 \times \hat{\sigma} \times 1.03 = 15.07 \pm 4.01.$$

Similarly, one can issue point forecasts and forecast intervals for later years.

The OLS estimates of the parameters of model (2.6) for the SNOTEL data are:

$$\hat{\beta}_0 = 8.93(1.26),\ \hat{\beta}_1 = 0.26(0.071),\ \hat{\sigma} = 3.36.$$

Note that the estimate of $\hat{\beta}_1$ is about three times of that obtained using the old method of measuring SWE and the estimate of $\hat{\sigma}$ is about two times higher. Of course, the latter is expected in view of the higher voltality of SNOTEL measurements depicted in Figure 2.4 (a). In the sequel we shall have several opportunities to revisit the snow data.

2.3.2 Multiple Linear Regression

When there are more than one explanatory (exogenous) variables potentially related to our time series of interest $\{Y_t\}$, as a direct generalization of (2.6) one can use multiple linear regression models to investigate this relationship.

For example, for p explanatory variables $X_1, \ldots, X_p$, the model takes the form

$$\begin{aligned} Y_t &= \beta_1 x_{t1} + \beta_2 x_{t2} + \cdots + \beta_p x_{tp} + e_t \\ &= x_t'\beta + e_t,\ t = 1, \cdots, n, \end{aligned} \tag{2.12}$$

where $x_t = (x_{t1}, \ldots, x_{tp})$ is the vector of known values of the explanatory variables at time t, $\beta = (\beta_1, \ldots, \beta_p)$ is the vector of unknown regression parameters and $\{e_t\}$ is an unobservable zero-mean time series referred to as the noise or the error term in the model. With $Y = (Y_1, \ldots, Y_n)$, $e = (e_1, \ldots, e_n)$ and $X = (x_1', \ldots, x_n')'$, the model in (2.12) rewritten in matrix form becomes

$$Y = X\beta + e. \tag{2.13}$$

The basic assumptions are that

$$E(e) = 0,\ \text{Cov}(e) = \Sigma = \sigma^2 V, \tag{2.14}$$

and that the design matrix X is fixed (nonrandom) and of full rank p.

In the rest of this section, as before, we take $V = I_n$ and derive the OLS estimate of β and some of its properties. The more general case of V is discussed in the next section. The least squares criterion for estimation of β, namely,

$$Q(\beta) = \sum_{t=1}^{n} (Y_t - x_t'\beta)^2 = (Y - X\beta)'(Y - X\beta),$$

when minimized leads to the normal equations for $\hat{\beta}$:

$$X'X\hat{\beta} = X'Y, \tag{2.15}$$

with solution

$$\hat{\beta} = (X'X)^{-1}X'Y. \tag{2.16}$$

Some noteworthy properties of the OLS estimate $\hat{\beta}$ and the associated forecast of the next future value are given below. For the last two properties we assume that the observations are independent and normally distributed, that is, $Y \sim N_n(X\beta, \sigma^2 I_n)$; for information on multivariate normal distribution, see Appendix A.

1. **Unbiasedness:** $E(\hat{\beta}) = \beta$.

2. **Normality:** $\hat{\beta} \sim N_p[\beta, \sigma^2(X'X)^{-1}]$.

3. **Forecast Distribution:** A forecast of the next observation y_{n+1} is

$$\hat{Y}_{n+1} = \hat{\beta}_1 x_{n+1,1} + \cdots + \hat{\beta}_p x_{n+1,p} = x'_{n+1}\hat{\beta} \sim N(x'_{n+1}\beta, \sigma^2_{pe}), \tag{2.17}$$

where σ^2_{pe}, the variance of the prediction error, is given by

$$\sigma^2_{pe} = \sigma^2[1 + x'_{n+1}(X'X)^{-1}x_{n+1}], \tag{2.18}$$

with $x_{n+1} = (x_{n+1,1}, \ldots, x_{n+1,p})$. A $100(1-\alpha)\%$ forecast interval for Y_{n+1} is given by

$$\hat{Y}_{n+1}, \ \pm t_{\frac{\alpha}{2}}(n-p)\hat{\sigma}[1 + x'_{n+1}(X'X)^{-1}x_{n+1}]^{\frac{1}{2}},$$

where $t_{\frac{\alpha}{2}}$ is the ordinate corresponding to the upper $\frac{\alpha}{2}$ probability of the t-distribution with $n-p$ degrees of freedom, and with $\hat{e} = Y - X\hat{\beta}$ the OLS residuals,

$$\hat{\sigma}^2 = \frac{\hat{e}'\hat{e}}{n-p}$$

is an unbiased estimator of σ^2.

Snow data (cont.) As a concrete example of (2.12)–(2.18) we revisit the snow data with the hope of combining $\mathbf{y}_1$, the 30 measurements from SNOTEL, and $\mathbf{y}_2$, the 62 measurements from SNOCOM, into a longer vector of observations $Y = (\mathbf{y}'_1, \mathbf{y}'_2)$. However, since the data from the newer SNOTEL device are visibly (Figure 2.4 (a)) more volatile than those from the SNOCOM, one needs to be concerned about the assumption of equality of variances or heterogeneity of the error terms. This problem is a bit harder to handle here and is discussed in detail in the next section. In addition to the differences in the variances (precisions), the two devices could have different measurement biases. As an indication of this, note the difference in the estimated $\hat{\beta}_0$ for SNOTEL and SNOCOM reported in the previous section. Let b_1 and b_2 be the true measurement biases associated with these two methods. The expanded model (2.6) for each device incorporating the potential measurement biases are:

$$\begin{cases} Y_{1,t} &= \beta_0 + b_1 + \beta_1 t + e_{1,t}, \ t = 1, \cdots, 30, \\ Y_{2,t} &= \beta_0 + b_2 + \beta_1 t + e_{2,t}, \ t = 1, \cdots, 62. \end{cases} \tag{2.19}$$

For the combined data, this can be rewritten as

$$Y = \beta_0 x_0 + b_1 x_1 + b_2 x_2 + \beta_1 x_3 + e = X\beta + e, \tag{2.20}$$

with $\beta = (\beta_0, b_1, b_2, \beta_1)$ and x_i's, $i = 0, 1, 2, 3$, are the appropriate columns of the design matrix

$$X = \begin{bmatrix} 1 & 1 & 0 & 1 \\ \vdots & \vdots & \vdots & \vdots \\ 1 & 1 & 0 & 30 \\ 1 & 0 & 1 & 1 \\ \vdots & \vdots & \vdots & \vdots \\ 1 & 0 & 1 & 62 \end{bmatrix}, \tag{2.21}$$

and $\text{Cov}(e) = \sigma^2 I_{92}$.

Note that the matrix X is not of full rank, so that β_0, b_1, b_2 and β_1 are not identifiable. This difficulty is resolved by removing the first column of X or introducing the new parameters $\alpha_1 = \beta_0 + b_1, \alpha_2 = \beta_0 + b_2$. Now (2.20) holds with a modified 92×3 matrix X and $\beta = (\alpha_1, \alpha_2, \beta_1)$. The OLS estimate of the regression parameters $(\alpha_1, \alpha_2, \beta_1)$ and their standard errors (SE) are $(11.32, 8.85, 0.11)$, $SE(\hat{\beta}) = (0.55, 0.65, 0.018)$ and $\hat{\sigma} = 2.62$. These numerical results are dramatically different from their counterparts in Section 2.3.1 based on simple linear regression.

For a more detailed coverage of regression methods and their use in time series data analysis, see Abraham and Ledolter (1983) and Shumway and Stoffer (2000).

2.3.3 Generalized Least Squares (GLS) and Unbiased Prediction

Now, we return to the general linear regression setup in (2.13)–(2.14) where V is not I_n, but assumed to be a general covariance matrix. Instead of estimating β or forecasting Y_{n+1}, we focus on estimating a generic unknown scalar Z that could be either a parameter, a random variable or a function of these. For this task, however, specification or knowledge of $E(Z)$ and $\text{Cov}(Z, Y)$ is needed when Z is a random variable. This general setup allows a unified discussion of problems of estimation of parameters and prediction of random variables. As such it is applicable to models in regression analysis, times series analysis, spatial data analysis and kriging as well as to the linear mixed models. For more applications and discussions see Hayes and Haslett (1999).

Consider the problem of finding that linear combination of the available data Y which is best for the estimation of the unknown Z among those linear estimators which are constrained to be *unbiased*. In other words, we need to find the n-vector $a = (a_1, \ldots, a_n)$ such that $\hat{Z} = a'Y$ has the *minimum mean square error* (MMSE) $E(Z - \hat{Z})^2$, subject to the *unbiasedness* constraint $E(Z - \hat{Z}) = 0$. Note that when Z is not a random variable, then $E(Z) = Z$ and $\text{Cov}(Z, Y) = 0$. Otherwise, it is convenient to specify $E(Z) = b'\beta$ for a known p-vector b. The unbiasedness constraint on $\hat{Z}$, now reduces to $b' - a'X = 0$. Thus, our problem is reduced to that of minimizing

$$\begin{aligned} Q(a, c) &= \text{Var}(Z - \hat{Z}) + c'(X'a - b) \\ &= \text{Var}(Z) + a'\Sigma a - 2a'\text{Cov}(Y, Z) + c'(X'a - b), \end{aligned} \tag{2.22}$$

where c is a p-vector of Lagrange multipliers. Setting to zero the derivatives with respect to a and c yields the system of normal equations

$$\begin{pmatrix} \Sigma & X \\ X' & 0 \end{pmatrix} \begin{pmatrix} a \\ c \end{pmatrix} = \begin{pmatrix} \mathrm{Cov}(Y, Z) \\ b \end{pmatrix}. \tag{2.23}$$

It is important to note that the matrix on the left-hand side of (2.23) concerns only the "past" or what we know, namely, the data and the constraint; it does not depend on Z and stays the same no matter what Z being estimated. However, the right-hand side contains the term $\mathrm{Cov}(Z, Y)$ concerning the "past and future" that can be computed depending on what Z is being considered. It is this exceptional and intuitively appealing feature of the system (2.23) that lies at the heart of our generic approach to the unified problem of statistical estimation and forecasting arising from linear models, time series analysis, spatial data and kriging, etc.

When more than one unknown is to be estimated, like the next k future values, we may stack them as a vector Z. In analogy with $k = 1$, set $\hat{Z} = a'Y$ where a is now an $n \times k$ matrix of coefficients, and specify $E(Z) = b\beta$ where b is a $k \times p$ known matrix. Finding a requires elementwise minimization of $\mathrm{diag}[\mathrm{Var}(Z - a'Y) + c'(X'a - b')]$, where c is a $p \times k$ matrix of Lagrange multipliers. The system of equations for a k-vector Z are of the exact same forms as (2.23), except now the vectors a, b, c are replaced by their matrix counterparts.

Let Γ be the coefficient matrix of the system (2.23) and let us partition its inverse as

$$\Gamma^{-1} = \begin{pmatrix} A & B' \\ B & C \end{pmatrix}. \tag{2.24}$$

Then, $\hat{Z}$ can be computed using any of the following forms

$$\begin{aligned} \hat{Z} &= a'Y = (a', c') \begin{pmatrix} Y \\ 0 \end{pmatrix} = (\mathrm{Cov}(Z, Y), b')\Gamma^{-1} \begin{pmatrix} Y \\ 0 \end{pmatrix} \\ &= (\mathrm{Cov}(Z, Y), b') \begin{pmatrix} AY \\ BY \end{pmatrix}. \end{aligned} \tag{2.25}$$

The MMSE $E(Z - \hat{Z})^2$ in this case can be expressed as

$$\mathrm{Var}(Z - \hat{Z}) = \mathrm{Var}(Z) - \begin{pmatrix} \mathrm{Cov}(Y, Z) \\ b \end{pmatrix}' \begin{pmatrix} A & B' \\ B & C \end{pmatrix} \begin{pmatrix} \mathrm{Cov}(Y, Z) \\ b \end{pmatrix}. \tag{2.26}$$

The identification of the elements A, B and C of Γ^{-1} in terms of Σ and X can be done using standard identities involving inverse of partitioned matrices. However, many of these results are unintuitive, clumsy and do not provide much conceptual insights into the problems and their solutions (Problem 2.5). The formula (2.25) and the following simple examples shed more light on the roles and actions of the submatrices A, B and C of Γ^{-1}.

Example 2.1 Special Cases of GLS

(a) Consider estimating the fixed unknown vector β in (2.13). Since $\text{Cov}(\beta, Y) = 0$ and $b = I_p$, from (2.23) it follows that $a = B$ so that

$$\hat{Z} = \hat{\beta} = B'Y, \tag{2.27}$$

and from (2.26) we obtain

$$\text{Var}(\hat{\beta}) = \text{Var}(Z - \hat{Z}) = -C.$$

It can be shown that (Problem 2.5), $B = (X'\Sigma^{-1}X)^{-1}X'\Sigma^{-1}$ and $C = -(X'\Sigma^{-1}X)^{-1}$. The $\hat{\beta}$ above is called the *generalized least squares* (GLS) estimate of β with the weight Σ. To emphasize this dependence on Σ, sometimes it is denoted by $\hat{\beta}_{\Sigma}$.

(b) $Z = \mu = E(Y)$ is a fixed unknown vector, so that $\text{Cov}(\mu, Y) = 0$ and $b = X$. Consequently,

$$\hat{Z} = \hat{\mu} = XB'Y = X\hat{\beta},$$

as expected.

(c) $Z = Y$ is a random vector, so that $\text{Cov}(Z, Y) = \text{Var}(Y) = \Sigma$ and since $E(Z) = \mu = X\beta$ with $b = X$, it follows from (2.23) that $a = I_n$ and hence $\hat{Z} = Y$ is the vector of observed values.

(d) $Z = Y - \mu$ is a linear combination of a fixed and a random vector, so that $\text{Cov}(Y - \mu, Y) = \text{Var}(Y) = \Sigma$ and $E(Z) = 0$ with $b = 0$. From (b) and (c) above, we obtain

$$\hat{Z} = \hat{Y} - \hat{\mu} = Y - \hat{\mu},$$

where from here on, $\hat{e} = Y - \hat{\mu}$ is referred to as the vector of GLS residuals of the general linear model (2.13)–(2.14). Thus far, the submatrix A from (2.24) has not made a debut. To see its role, note that an alternative and direct derivation of $\hat{Z}$ from (2.23)–(2.24) gives

$$\hat{Z} = \Sigma AY \text{ or } \Sigma^{-1}\hat{e} = AY. \tag{2.28}$$

Consequently, A determines $\hat{e}$ or the *standardized GLS residuals* $\Sigma^{-1}\hat{e}$ from Y. It can be shown that (Problem 2.5) $A = \Sigma^{-1}X(X'\Sigma^{-1}X)^{-1}X'\Sigma^{-1}$.

(e) Finally, the most important example for our purpose is $Z = Y_{n+1}$, the prediction of the next future observation corresponding to the covariate x_{n+1}. Since $EZ = x'_{n+1}\beta = \mu_{n+1}, \text{Cov}(Z, Y) = \text{Cov}(Y_{n+1}, Y)$ and $Y_{n+1} = \mu_{n+1} + (Y_{n+1} - \mu_{n+1})$ from (d) and (2.25) we obtain

$$\hat{Y}_{n+1} = x'_{n+1}\hat{\beta} + \text{Cov}(Y_{n+1}, Y)\Sigma^{-1}\hat{e}. \tag{2.29}$$

Referring to the first term as the *regression predictor* and the second term as the *dependence dividend*, since it vanishes when Y_{n+1} and Y are uncorrelated, it is seen in (2.29) that the optimal predictor of Y_{n+1} is decomposed into two uncorrelated components. The appearance of $\hat{e}$ in the latter also suggests to possibly interpret it as updating the former. This view is more common in the literature of state-space and dynamic models (West and Harrison, 1997).

The two terms in (2.29), which depend on Σ, are orthogonal so long as β and Σ are assumed known. What is the situation for unknown β and Σ? What are the optimal properties of the *plug-in predictors*? Namely, the one obtained from (2.29) replacing β and Σ by their estimators. These fundamental questions which are ubiquitous in regression, time series analysis and other areas of statistics dealing with correlated data are dealt with in Section 2.5.

2.3.4 Iterative Estimation of β and Σ

Simultaneous modeling of the mean and dependence structures is a real challenge in statistics. The difficulty lies in separating the mean and dependence components of the data. A compromise is usually reached based on substantive knowledge or exploratory data analysis or simply assuming parametric mean and covariance structures for the data that are known up to the values of few unknown parameters. In what follows we work within the framework of the linear model (2.13) for the mean and assume that the covariance matrix $\Sigma = \Sigma(\theta)$ is parameterized with θ as its unknown parameters. Details of parameterizing or formulating models for covariances are discussed at the end of this section and Chapters 3–7.

The following iterative method is usually used for simultaneous estimation of the regression and covariance parameters β and θ:

Stage 1. Select an initial value $\tilde{\beta}$ for β.

Stage 2. Estimate the covariance parameters θ using the fitted residuals $\hat{e} = Y - X\tilde{\beta}$, to obtain $\hat{\Sigma} = \Sigma(\hat{\theta})$ to be used in the next stage.

Stage 3. Update the value $\tilde{\beta}$ using the current estimate $\hat{\Sigma}$ via

$$\hat{\beta} = (X'\hat{\Sigma}^{-1}X)^{-1}X'\hat{\Sigma}^{-1}Y.$$

Stage 4. Stop the process if $\hat{\beta} \simeq \tilde{\beta}$ and take $\hat{\beta}$ as an estimate of β and $\hat{\Sigma} = \Sigma(\hat{\theta})$ as an estimate of $\Sigma(\theta)$. Otherwise, repeat Stages 2–3.

A convenient initial value for β is its OLS estimate given in (2.16). Its updated value in Stage 3 is the same as the GLS estimate (2.27) using $\Sigma = \hat{\Sigma}$. For estimating the covariance parameter θ in Stage 2, one could either minimize $(Y - X\tilde{\beta})'\Sigma(\theta)^{-1}(Y - X\tilde{\beta})$ or the quasi-Gaussian likelihood (Whittle, 1961; Rosenblatt, 2000),

$$GL(\theta; Y) = \log|\Sigma(\theta)| + (Y - X\tilde{\beta})'\Sigma(\theta)^{-1}(Y - X\tilde{\beta}), \tag{2.30}$$

for a fixed $\tilde{\beta}$, even if Y is not normally distributed. For a matrix Σ, $|\Sigma|$ stands for its determinant.

We illustrate this iterative procedure using the snow data.

Table 2.1 Parameter estimates for the combined snow data

Iterations	$\hat{\alpha}_1$	$\hat{\alpha}_2$	$\hat{\beta}_1$	$\hat{\sigma}_1$	$\hat{\sigma}_2$
1	11.32	8.85	0.11	3.57	1.99
2	10.88	7.95	0.14	3.48	2.13
3	10.97	8.14	0.13	3.50	2.09
4	10.95	8.10	0.13	3.49	2.10
5	10.96	8.11	0.13	3.50	2.10
6	10.96	8.10	0.13	3.50	2.10
7	10.96	8.11	0.13	3.50	2.10
8	10.96	8.10	0.13	3.50	2.10
9	10.96	8.11	0.13	3.50	2.10
10	10.96	8.11	0.13	3.50	2.10

Snow data (cont.) It was noted earlier that the precisions (variances) of the two devices for arriving at SWE were different. To account for these differences, let σ_i^2 be the variance of $\mathbf{y}_i, i = 1, 2$. Then, $\text{Cov}(Y) = \Sigma = \text{diag}(\sigma_1^2 I_{30}, \sigma_2^2 I_{62}) \neq \sigma^2 I_{92}$, and β in (2.20) is usually estimated by minimizing the GLS criterion

$$\begin{aligned}(Y - X\beta)'\Sigma^{-1}(Y - X\beta) &= \sigma_1^{-2}(\mathbf{y}_1 - X_1\beta)'(\mathbf{y}_1 - X_1\beta) \\ &\quad + \sigma_2^{-1}(\mathbf{y}_2 - X_2\beta)'(\mathbf{y}_1 - X_2\beta),\end{aligned}$$

where $\beta = (\alpha_1, \alpha_2, \beta_1)$ and X_1, X_2 are appropriate submatrices of X. The advantages of using GLS in situations of this kind is evident from above where the low-quality data from SNOTEL with larger variance σ_1^2 is given a much lower weight (σ_1^{-2}) than the weight (σ_2^{-1}) for the high quality data from SNOCOM.

For the snow data, the above iterative procedure stabilizes essentially after four iterations, except for $\hat{\alpha}_2$ that converges after ten iterations. The resulting parameter estimates for the first ten iterations are reported in Table 2.1. The standard error of the final regression parameter estimates are

$$SE(\hat{\beta}) = (0.55, 0.87, 0.02).$$

We note that these estimates are generally in agreement with the OLS estimates obtained at the end of Section 2.3.2.

Note that the unstructured covariance matrix Σ in (2.30) has $\frac{n(n+1)}{2}$ parameters constrained so that it is positive-definite. When Y is stationary the number of parameters in Σ is reduced to n which is still too large compared to the length of the data. Specific dynamic models, with stationary solutions introduced in the next section, could have as few as two parameters; one accommodating the variation and the other for dependence. The following simple example indicates how dependence can be introduced in a meaningful manner among observations so that Σ in (2.30) is not necessarily diagonal. For more examples of this kind, see Chapter 3.

Example 2.2 Compound Symmetry (CS) Covariance Let U and $\varepsilon_1, \ldots, \varepsilon_n$ be independent mean-zero random variables with $\text{Var}(U) = \sigma_u^2$ and $\text{Var}(\varepsilon_t) = \sigma_\varepsilon^2$. Then, the random variables $Y_t = U + \varepsilon_t$, $1 \leq t \leq n$, sharing (the trait) U are dependent with covariance

$$\begin{aligned} \text{Cov}(Y_{t+k}, Y_t) &= \text{Cov}(U + \varepsilon_{t+k},\ U + \varepsilon_t) \\ &= \text{Var}(U) + \text{Cov}(\varepsilon_{t+k}, \varepsilon_t) \\ &= \begin{cases} \sigma_u^2 & , \quad k = 0 \\ \sigma_u^2 + \sigma_e^2 & , \quad k \neq 0. \end{cases} \end{aligned}$$

The covariance matrix Σ of $Y = (Y_1, \ldots, Y_n)$ has the simple form

$$\Sigma = \sigma^2(I_n + \rho 1_n 1_n'), \tag{2.31}$$

where $\sigma^2 = \sigma_u^2 + \sigma_\varepsilon^2$, $\rho = \dfrac{\sigma_u^2}{\sigma^2}$, I_n is the $n \times n$ identify matrix and $1_n = (1, \ldots, 1)$. This matrix with two restricted parameters ρ and σ^2 is known as the *compound symmetry* (CS), *exchangeable* or the *intra-block* covariance matrix. In genetic applications the parameter ρ is meaningful and is known as the *heretability index* (Searle, Casella and McCulloch, 1992). Note that as $\sigma_u^2 \to 0$, then $\rho \to 0$ and larger σ_u^2 leads to bigger ρ.

2.4 TRANSFORMATIONS TO ORTHOGONALITY

Thus far, we have focused on handling trends using differencing, smoothing and regression analysis. Handling the remaining two components of a time series, namely, the *seasonality* (cycle, period) and *random variation* (fluctuation) in a systematic manner requires more sophisticated techniques of Gram-Schmidt type to orthogonalize a sequence of dependent observations and parameterize its covariance matrix *parsimoniously*. Namely, with as few parameters as possible.

Most ideas and tools of time series analysis are means to orthogonalize dependent random variables. Two important and widely used tools for this purpose are, (1) *the Fourier transform* (spectral decomposition) involving the exponentials $\{e^{it\lambda}\}$ and (2) *the Gram-Schmidt procedure* (Wold decomposition) involving uncorrelated random variables (white noise). Although orthogonalization was not the primary motivation behind Schuster *periodogram* in the search for *hidden periodicity* and Yule's (1937) use of *autoregression* (AR) in the search for *flexible periodicity*, but they were quickly seen as the prototypes of distinct and powerful techniques for *eliminating* dependence among time series observations. For a thorough discussion on predecessors of periodogram and autoregressions, see Klein (1997, Chaps. 9, 10).

2.4.1 Spectral Analysis and Periodogram

Spectral analysis of time series is widely used in physical, natural, engineering and medical sciences where usually problems of interest lend themselves to formulation in terms of frequency, cycle, period, rhythm,..., and large data sets are available due to using on-line digital devices for collecting data. However, in economics and social sciences where not enough data are available and due to the nature of the questions of interest a time-domain approach in the spirit of correlation and regression analysis is desirable. Bartlett's (1978) paper on "Correlation or spectral analysis?" provides insightful suggestions on using one or the other approach.

The *(finite) Fourier transform* and the *periodogram* of a time series $x_1, \cdots, x_n$, given by

$$\begin{cases} d_n(\lambda) = n^{-\frac{1}{2}} \sum_{t=1}^{n} x_t e^{-it\lambda}, \lambda \in R, \\ \\ I_n(\lambda) = | \, d_n(\lambda) \, |^2, \end{cases} \tag{2.32}$$

where $i = \sqrt{-1}$, are the key ingredients in the spectral analysis of a stationary time series. Observe that $d_n(\lambda)$ is a complex-valued random variable. For simplicity, often we suppress their dependence on n. Since $d(\lambda + 2\pi) = d(\lambda)$ and $\overline{d(\lambda)} = d(-\lambda)$ in the case of real-valued time series, from here on the principal domain of $d(\lambda)$ and $I(\lambda)$ is taken to be $0 \leq \lambda \leq \pi$. The concept of periodogram was introduced and its statistical properties were developed by Schuster (1898, 1906) to find the exact but unknown period p (hidden periodicity) of data generated, for example, from models of the form

$$X_t = \alpha \cos \omega t + \beta \sin \omega t + \varepsilon_t, t = 1, \cdots, n, \tag{2.33}$$

where $\omega = \frac{2\pi}{p}$ is the frequency cycle of the fluctuations in the data, α and β are the parameters and $\{\varepsilon_t\}$ is the measurement error. For this model, it can be seen that the function $I(\lambda)$ attains its maximum at $\lambda = \frac{2\pi}{p}$, so that scanning the plot of $I(\lambda_k)$ vs. the *Fourier frequencies* $\lambda_k = \frac{2k\pi}{n}, k = 0, 1, \cdots, n-1$, can reveal the period p. For more general models of this type, see Section 5.2.

Figure 2.5 (a) gives for $\alpha = 2$, $\beta = 1, p = 6$ and $n = 200$, the plot of $\alpha \cos \omega t + \beta \sin \omega t$ versus t, superimposed on the plot of $\{X_t\}$. Here $\{\varepsilon_t\}$ is a sequence of i.i.d. normal random variables with mean zero and variance 2. Periodograms of these series are given in Figure 2.5 (b). Both of these attain their maximum at about λ_{34}, thus from $\frac{2\pi \cdot 34}{200} \approx \frac{2\pi}{p}$, one estimates the period to be $p \approx \frac{200}{34} = 5.88$, which is not too far from the true value 6.

Two important properties of $d(\lambda)$ are, (1) it is completely determined by its values at the frequencies $\lambda_k = \frac{2\pi k}{n}$, $k = 0, 1, \cdots, n-1$ and (2) the vector $d = (d(\lambda_0), \cdots, d(\lambda_{n-1}))$ of these values is obtained from the data $x = (x_1, \cdots, x_n)$ via a nearly orthogonal transformation:

$$d = U_n x, \tag{2.34}$$

where $U_n = (u_{k,t})$ and $u_{k,t} = n^{-\frac{1}{2}} \exp\{-it\lambda_k\}$; see Problem 2.7.

The spectral decomposition (representation) of stationary processes, is a more natural and ideal source of motivation for using the periodogram. This powerful result was proved by Cramer (1940) using stochastic integrals and Kolmogorov (1941) using the spectral theorem of unitary operators, see Masani (1995) for the history and

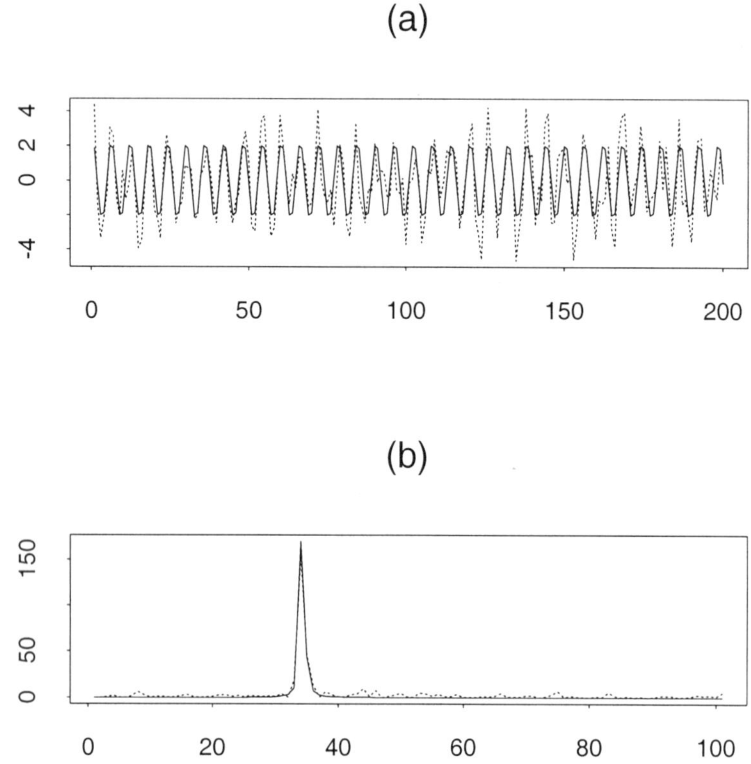

Fig. 2.5 (a) Plot of $2\cos\frac{2\pi}{6}t + \sin\frac{2\pi}{6}t$ (solid line) with added noise (dotted line). (b) Periodograms of the series in (a).

ramifications. This result is stated below, the unfamiliar terms and concepts are defined in Chapter 5; the proof is given in Chapter 9. We use Roman numerals for the eight key theorems in this chapter so that they can be distinguished from other results and theorems.

Theorem I: Let $\{X_t\}$ be a mean-zero stationary process. Then,

(a) $\{X_t\}$ is the Fourier transform of a process:

$$X_t = \int_{-\pi}^{\pi} e^{it\lambda} dZ(\lambda), t = 0, \pm 1, \cdots, \tag{2.35}$$

where $\{Z(\lambda); -\pi < \lambda \leq \pi\}$ has *orthogonal increments*, that is, Cov $(dZ(\lambda_1), dZ(\lambda_2)) = 0$ for $\lambda_1 \neq \lambda_2$.

(b) The *spectral decomposition* (2.35) of the process leads to the spectral representation of its *covariance function* $\{\gamma_k\}$:

$$\begin{cases} \gamma_k &= \text{Cov}(X_{t+k}, X_t) = \int_{-\pi}^{\pi} e^{ik\lambda} dF(\lambda), k = 0, \pm 1, \cdots, \\ dF(\lambda) &= E|dZ(\lambda)|^2. \end{cases} \tag{2.36}$$

The function $F(\lambda)$, being nondecreasing on $(-\pi, \pi]$ and $f(\lambda)$ the derivative of its absolutely continuous part are called the *spectral distribution* and the *spectral density* of the stationary process $\{X_t\}$, respectively. Note that $F(\lambda)$ and $f(\lambda)$ are merely required to be nonnegative, this restriction is a lot less sever than the *positive-definiteness* requirement on the covariance function $\{\gamma_k\}$; see (5.1). For this reason often it is more convenient to work in the spectral-domain dealing with $f(\lambda)$, than in the time-domain and working with the covariance $\{\gamma_k\}$.

It is instructive to know how $\{X_t\}, \{Z(\lambda)\}, F(\lambda)$ and the periodogram $I(\lambda)$ relate to each other or better yet to find ways to interpret these quantities. To this end, setting $k = 0$ in (2.36) one obtains

$$\gamma_0 = \text{Var}(X_t) = \int_{-\pi}^{\pi} dF(\lambda), \tag{2.37}$$

This shows the role of the spectral distribution in dividing up the *total power*, $\text{Var}(X_t)$, among the frequencies λ in $(-\pi, \pi]$. But, such distribution of power is also controlled by the random process $\{Z(\lambda); -\pi < \lambda \leq \pi\}$. So that one would like to express $Z(\lambda)$ in terms of the original process $\{X_t\}$. Informal use of the inversion formula of the Fourier analysis expresses $dZ(\lambda)$ as the divergent infinite series $\sum_{t=-\infty}^{\infty} X_t e^{-it\lambda}$.

Truncating this series or replacing by zero the potentially unobservable segment of the process, one obtains

$$\begin{cases} dZ(\lambda) \approx \sum_{t=1}^{n} X_t e^{-it\lambda}, \\ dF(\lambda) \approx \frac{1}{n} \left| \sum_{t=1}^{n} X_t e^{it\lambda} \right|^2 = I(\lambda), \end{cases}$$

which are closely related to the finite Fourier transform and periodogram defined in (2.32).

The building block of the class of stationary process is the sequence of uncorrelated random variables $\{\varepsilon_t\}$ with mean zero and variance σ^2, called a *white noise*, we use the symbol $WN(0, \sigma^2)$ to denote such a sequence. It is the simplest of all stationary processes, with covariance function $\gamma_0 = \sigma^2, \gamma_k = 0, k \neq 0$. It follows from (2.36) that its spectral density function is given by $f(\lambda) = \frac{\sigma^2}{2\pi}$, which is flat like the spectrum of white light, this is the main reason for christening it as white noise. For more information on white noise, see Section 5.5.

The statistical importance of (2.32) is mostly due to the fact that for large n and when the stationary process $\{X_t\}$ is weakly dependent or satisfies the conditions of a

Central Limit Theorem (CLT), then the random variables $d(\lambda_0), d(\lambda_1), \ldots, d(\lambda_{n-1})$ are approximately independent with normal and $I(\lambda_0), \ldots, I(\lambda_{n-1})$ are approximately independent with a multiple of chi-squared distributions, respectively. The periodogram is a naive and notoriously bad estimator of the spectral density function $f(\lambda)$ of a stationary process because it is highly irregular and erratic; see Figure 1.6 and Section 5.6. But, since periodogram ordinates evaluated at the Fourier frequencies are approximately independent, a much better and smoother estimator $\hat{f}(\lambda)$ is obtained by *averaging (smoothing)* periodogram ordinates in a neighborhood of λ. Thus, motivated by (2.4) for a non-negative integer m we introduce the *smooth density estimator*

$$\hat{f}(\lambda_j) = \frac{1}{2m+1} \sum_{k=-m}^{m} I(\lambda_{j+k}). \tag{2.38}$$

The size of this neighborhood determined by m is called the *window length* or *bandwidth* or *smoothing parameter*. It is important in controlling the level of smoothness of the estimator and in the trade-off between bias and variance of the estimator; see Koopmans (1974), Bloomfield (1976) and Diggle (1990). We summarize the key results on the nonparametric estimator of the spectral density of a stationary time series in the following theorem.

Theorem II: Let $\{X_t\}$ be a mean-zero stationary process with the spectral density $f(\lambda)$ which satisfies the conditions of a CLT. Then, as $n \to \infty$,

(a) The discrete Fourier transforms are asymptotically normal and independent. That is, for $0 < \lambda < \pi$,

$$d_n(\lambda) \Rightarrow N^c(0, f(\lambda)),$$

where N^c stands for a complex normal distribution. Furthermore, for $j \neq k$, $d_n(\lambda_j)$ and $d_n(\lambda_k)$ are asymptotically independent. For a definition of complex normal random vector, see Appendix A.

(b) $|d_n(\lambda)|^2$ converges in distribution to $\frac{1}{2} f(\lambda)\chi_2^2$ and hence from (2.38),

$$\hat{f}(\lambda_j) \Rightarrow f(\lambda_j)\frac{\chi^2_{2(2m+1)}}{2(2m+1)},$$

where χ_v^2 stands for a chi-square random variable with v degrees of freedom. Furthermore, $\hat{f}(\lambda_j)$ and $\hat{f}(\lambda_k)$ are independent for all j, k such that $j - k \geq 2m + 1$.

(c) When $\{X_t\}$ is Gaussian, the set of discrete Fourier transforms $\{d_n(\lambda_0), \cdots, d_n(\lambda_{n-1})\}$ is sufficient for the original data $\{X_1, \cdots, X_n\}$; see Anderson (1971).

When $\{X_t\}$ is i.i.d. normal, proofs of the above assertions are not too difficult and follows the standard approach. For more general linear processes introduced in the next section and Chapter 5, one could reduce the proofs to the i.i.d. case using the technique of Olshen (1967).

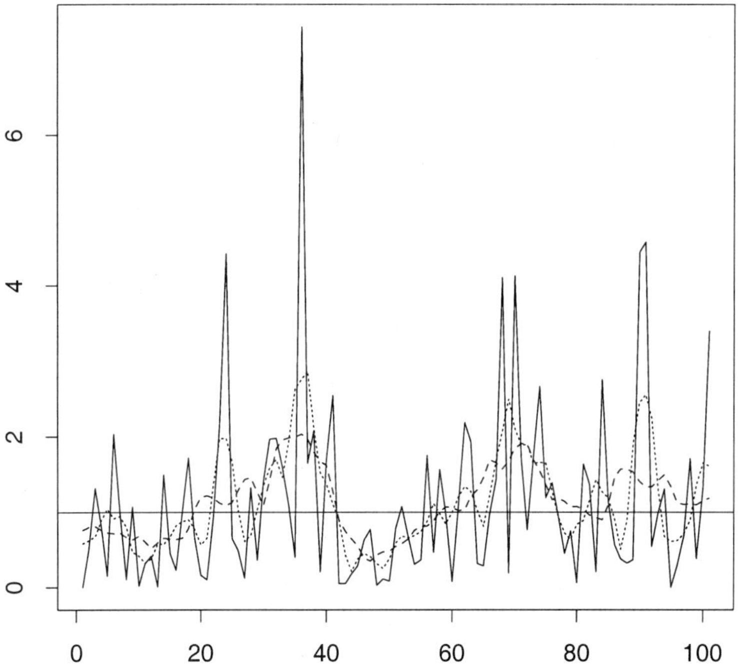

Fig. 2.6 Periodogram of $n = 200$ simulated normally distributed white noise (—), smoothed periodograms or spectral density estimators corresponding to $m = 2$ $(\cdots)$ and $m = 5$ $(- \ - \ -)$. The horizontal solid line is the true spectral density function of a white noise.

The independence and normality of ordinates of $d(\lambda)$ at the Fourier frequencies have been used effectively to formulate spectral-domain analogues of the classical statistical methods such as the analysis of variance, see Tukey (1962); regression, principal components and canonical correlation analysis, see Brillinger (1981) and Shumway and Stoffer (2000). Priestley (1981) provides an encyclopedic coverage of the spectral analysis of time series and its interplay with the time-domain approach.

2.4.2 Time-Domain Analysis: MA and AR Models

The predominant view of time series data up to the 1920's was that the observations were functions of time and could be represented by sums of sines and cosines as in (2.33) with exact periods (hidden periodicity). This view and Schuster's periodogram had proved successful in some areas of applications. At the same time, it was known that the assumption of exact periods could not be realistic in all applications. For example, in the theory of business cycles the assumed model must be flexible to allow change in periods (flexible periodicity) and amplitudes. How can one write models with such characteristics?

Starting from a white noise $\{\varepsilon_t\}$, Yule (1921, 1926) and Slutsky (1927), respectively, had shown that differences of a fixed order and certain general linear operations could generate time series data similar to sinusoidal waves with slowly changing amplitude and phase mimicking the cycles in economic time series. These efforts gave rise to the important notion of *moving average* (MA) models

$$X_t = \varepsilon_t + b_1\varepsilon_{t-1} + \cdots + b_q\varepsilon_{t-q}, \tag{2.39}$$

where $b_1, \cdots, b_q$ are scalars and q, a nonnegative integer, is the *order* of the MA model.

The paradigm shift from viewing "time series as a deterministic function-of-time" to viewing "time series as a collection of random variables indexed-by-time" was a monumental achievement credited mostly to Yule's (1927) concept of *autoregression* (AR) expressing a future value in terms of the previous two values and a *relationship error*. Applying this concept to the Wolfer's sunspot data, he arrived at the model

$$X_t = 1.34X_{t-1} - 0.65X_{t-2} + \varepsilon_t,$$

where the term ε_t represents the relationship error rather than the traditional *measurement error*. Only four years later, Walker (1931) generalized the model of autoregression to allow for dependence on more than two previous values:

$$X_t = a_1X_{t-1} + a_2X_{t-2} + \cdots + a_pX_{t-p} + \varepsilon_t, \tag{2.40}$$

where p is an integer and $a_1, \cdots, a_p$ are the parameters of the model. Following the footsteps of Yule, Walker also used the graph of the correlations versus the lag (the correlogram) as an aid to selecting the order p.

In honor of these pioneers, the set of equations for estimating the parameters $a_1, \cdots, a_p$ and $\sigma^2 = \mathrm{Var}(\varepsilon_t)$ in (2.40) for *known covariance* $\{\gamma_k\}$, namely,

$$\left\{\begin{array}{l} \begin{bmatrix} \gamma_0 & \gamma_1 & \cdots & \gamma_{p-1} \\ & \ddots & & \vdots \\ & & \ddots & \gamma_1 \\ & & & \gamma_0 \end{bmatrix} \begin{bmatrix} a_1 \\ \vdots \\ a_p \end{bmatrix} = \begin{bmatrix} \gamma_1 \\ \vdots \\ \gamma_p \end{bmatrix}, \\ \sigma^2 = \gamma_0 - \sum_{k=1}^{p} a_k\gamma_k, \end{array}\right. \tag{2.41}$$

is called the *Yule-Walker equations*. These are obtained from minimizing $Q(a, 0) = E\left(X_t - \sum_{k=1}^{p} a_kX_{t-k}\right)^2$ as in (2.22)–(2.26) by taking $Z = X_t$ and $Y_t = X_{t-k}$, $k = 1, \ldots, p$. Solutions of several diverse least squares problems in statistics and engineering also lead to variants of Yule-Walker equations; see Section 7.1. This set of equations has attained a prominent position in the theory and practice of time series analysis and signal processing. Very often, one collects data sequentially over time so that (2.41) must be solved for increasing orders $p = 1, 2, \cdots$.

Solving the linear system (2.41) without inverting the *Toeplitz matrices* $\Gamma_n = (\gamma_{i-j})_{i,j=1}^{n}, n = 1, 2, \cdots$, of increasing orders is a fascinating task linking apparently diverse areas of time series analysis, digital signal processing, computational linear algebra, inverse scattering, function and operator theory, orthogonal polynomials, etc. The sequence of Toeplitz matrices $\{\Gamma_n\}$ had appeared earlier in the non-statistical work of Szegö and Schur around 1917 (Gohberg, 1986) eventually giving rise to the modern notion of *Schur parameters* in the theory of orthogonal polynomials and computational linear algebra, *reflection coefficients* in inverse scattering and geophysics, and *partial autocorrelation function* (PACF) $\{\pi_k\}$ in time series analysis (Kailath, 1987; Landau, 1987a).

Levinson's (1947) algorithm for solving (2.41) with increasing p was rediscovered by Durbin (1960) in recursive fitting of autoregressive models of increasing orders to time series data. The influential Durbin-Levinson algorithm, its theoretical and practical ramifications, are discussed in Chapter 7. Many extensions of this algorithm, its great impact on communication and electrical engineering and other fields of applications, and the historical development of the linear prediction of stationary processes in the context of communication engineering are presented brilliantly by Kailath (1974, 1987).

2.4.3 Model Fitting Process

The iterative three-stage statistical model-fitting process initially popularized by Box and Jenkins in the context of time series analysis (Box, Jenkins and Reinsel, 1994) has penetrated essentially every area of applied statistics. The three stages are:

(i) Model identification or formulation,

(ii) Model estimation,

(iii) Model checking or diagnostics.

Our focus here is on (i) and (iii), however, (ii) is discussed in some detail in Section 2.6 and Chapter 4. Strictly speaking, the techniques behind these three stages fall into two distinct groups, since most diagnostic checks rely on testing whether or not the fitted residuals behave like a white noise sequence.

The term *correlogram* was coined by Wold (1938, p. 7) as a substitute to the Schuster's periodogram and has been used effectively ever since as means of identifying model or appropriate probabilistic description of time series data. Wold followed the graphical practice of Yule and Walker, which was simply based on the correlation coefficients plotted against their lags. He showed how correlograms could be used to distinguish the then three available schemes of hidden periodicities, moving average and autoregressive. For this, Wold (1938, p.147) generated ideal, hypothetical correlograms for these three key models, and then compared these with sample correlograms to identify appropriate model for the data.

To illustrate the Yule, Walker and Wold's idea on the use of correlogram, note that for the MA(q) model (2.39), its covariance function for $k \geq 0$ is

$$\begin{aligned}\gamma_k = \text{Cov}\,(X_{t+k}, X_t\,) &= \text{Cov}\left(\sum_{i=0}^{q} b_i \varepsilon_{t+k-i}, \sum_{j=0}^{q} b_j \varepsilon_{t-j}\right) \\ &= \sum_{i=0}^{q}\sum_{j=0}^{q} b_i b_j \text{Cov}(\varepsilon_{t+k-i}, \varepsilon_{t-j}) \\ &= \sigma^2 \sum_{i=0}^{q}\sum_{j=0}^{q} b_i b_j \delta_{k-i,j} \\ &= \sigma^2 \sum_{i=0}^{q} b_i b_{i+k},\end{aligned} \tag{2.42}$$

where $b_0 = 1, b_i = 0$ for $i > q$ and $\delta_{i,j}$ is the Kronecker's delta. Thus, its autocorrelation function (ACF) $\{\rho_k\}$ *drops off to zero* after the lag q, the order of the MA(q) model. The hope is to see the same pattern in the sample ACF of data generated by (2.39). Thus, examining the sample *correlogram* was the key idea used by Yule, Walker and Wold to identify the type and orders p, q of models (2.39) and (2.40), respectively.

For $q = 1$ and $b_1 = b$, (2.42) reduces to

$$\gamma_0 = \sigma^2(1 + b^2), \gamma_{\pm 1} = \sigma^2 b, \gamma_{\pm k} = 0, k \geq 2, \tag{2.43}$$

so that the covariance matrix Γ_n is a simple tri-diagonal $n \times n$ matrix. On the other hand, for the AR(1) model (2.40) with $p = 1$ and $a_1 = a, |a| < 1$, it follows from (2.41) that

$$\gamma_k = \sigma^2(1 - a^2)^{-1} a^{|k|},\ k = 0, \pm 1, \cdots, \tag{2.44}$$

so that all entries of the covariance matrix Γ_n are nonzero. Fortunately, for AR(p) models the PACF drops off to zero after the lag p as indicated next.

For a stationary process with covariance function $\{\gamma_k\}$, its lag k *partial autocorrelation*, denoted by π_k, is the correlation coefficient between X_1 and X_{1+k} after eliminating the effect of the intervening values $X_2, \cdots, X_k$. We show in Chapter 7 that π_k is, indeed, the last entry of the solution of (2.41) for $p = k$. Note that $\pi_1 = \rho_1$.

The use of the PACF $\{\pi_k\}$ in identifying the order p in (2.40) was initiated by Quenouille (1949) exploiting the fact that $\pi_k \equiv 0$ for autoregressive models of order p when $k > p$ (Chapter 7). Plot of π_k versus k gives a variant of the correlogram, called the *partial correlogram* (Diggle, 1990, p.168); see Figure 2.7 (c)). A fairly complete strategy for identification of model type and orders p and q based on comparing patterns of theoretical and sample correlograms is developed and popularized by Box et al. (1994).

Figure 2.7 (a) provides time series plot of $n = 200$ simulated values of a Gaussian MA(2) process with $b_1 = 1.34, b_2 = -0.65$. Its sample correlograms computed

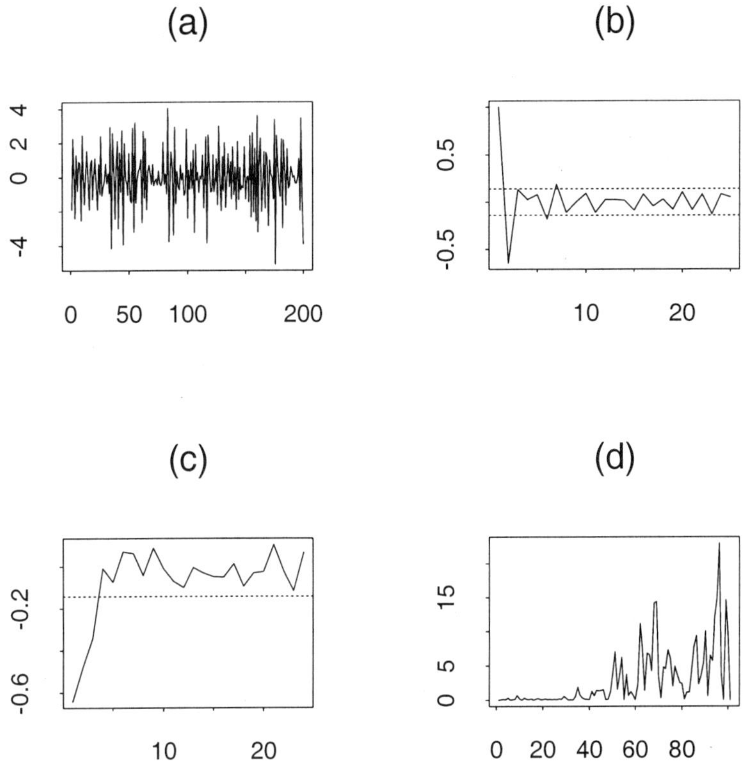

Fig. 2.7 (a) Time series plot of $n = 200$ simulated values of a Gaussian MA(2), (b) Sample correlogram, (c) Sample partial correlogram and (d) Periodogram.

using (2.2), partial correlogram and periodogram are also given for the purpose of comparison. In this case, it is evident that the sample correlogram mimics closely the behavior of the theoretical correlogram. The dotted lines provide a 95% pointwise confidence band for the sample ACF as described in Theorem III below.

The key statistical result or justification behind graphic model identification strategy is the consistency and asymptotic normality of the estimated (partial) correlations for a purely random series and more general linear processes (Brockwell and Davis, 1991, Chapter 7).

Theorem III: If $\{X_t\}$ is an i.i.d. sequence with mean zero and finite variance, then

(a) For n large and $K < n$, the estimated correlations $\hat{\rho}_1, \cdots, \hat{\rho}_K$ are approximately i.i.d. normal random variables with mean zero and variance $1/n$. Moreover, the random variable $Q_K = n \sum_{k=1}^{K} \hat{\rho}_k^2$ is approximately chi-square with K degrees of freedom.

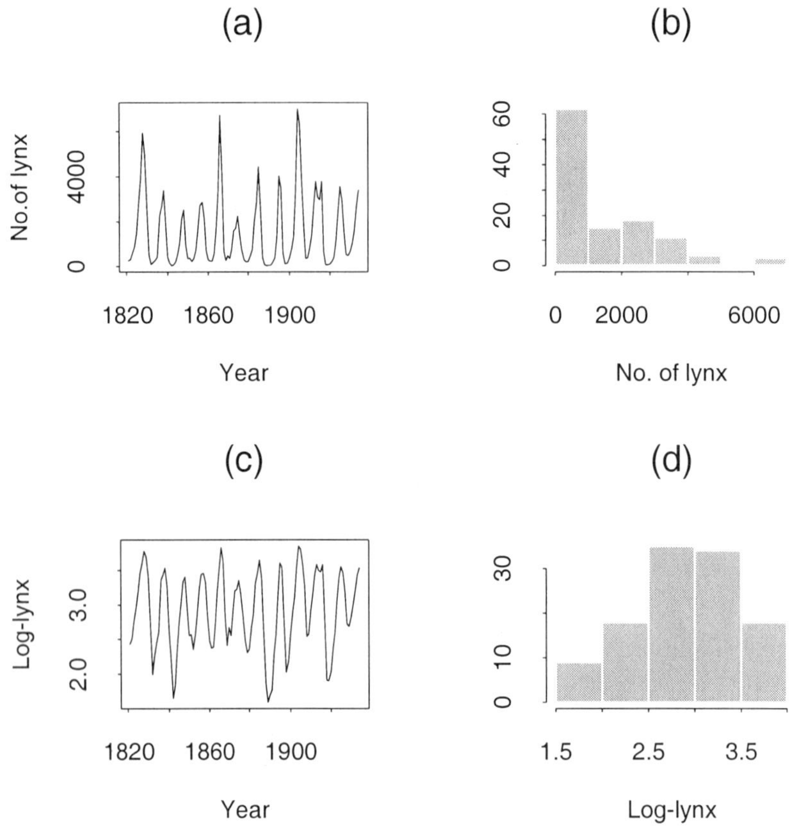

Fig. 2.8 (a) and (b) time series plot and histogram of the raw lynx data, (c) and (d) time series plot and histogram of logarithm of the lynx data.

(b) The above statement also holds for the estimated partial correlations $\hat{\pi}_1, \cdots, \hat{\pi}_K$.

The asymptotic normality of the individual $\hat{\rho}_k$ and $\hat{\pi}_k$ can be used to construct pointwise confidence intervals for ρ_k and π_k; see Figure 2.7 (b) and (c). These are in turn useful in checking whether the observations are from a truly i.i.d. process. The statistic Q_K is often used to check the overall adequacy of a fitted model to a time series data. These are illustrated next using an actual data set.

Lynx data (cont.) We analyze the Canadian lynx data and use it to illustrate various ideas and techniques developed so far. Moran (1953) was probably the first to analyze the lynx data. Because the cycle in the raw data is very asymmetrical with a sharp and large peak and a relatively smooth and small trough, see Figure 2.8 (a), he used a base-10 logarithmic transformation of the data, plotted in Figure 2.8 (c), and studied the series $x_t = \log_{10}$ (No. of lynx trapped in year $1820 + t$), $t = 1, \cdots, 114$. From here on by the lynx data we shall mean the log-transformed series $\{X_t\}$. For

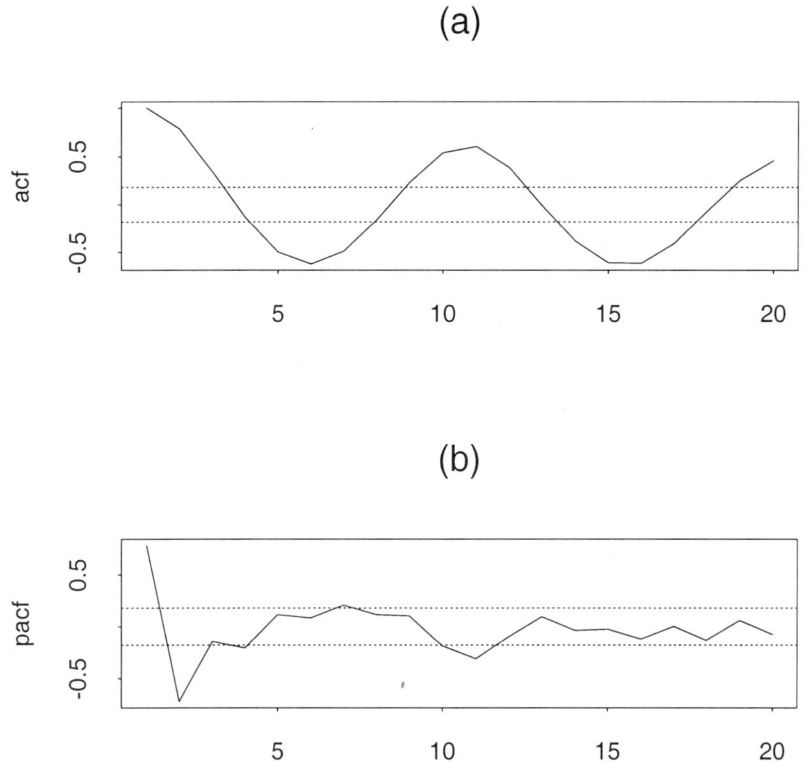

Fig. 2.9 (a) The sample correlogram of the lynx data, (b) The sample partial correlogram of the lynx data.

this series we have

$$\bar{x} = 2.904\ ,\ \hat{\gamma}_0 = 0.309,\ \hat{\gamma}_1 = 0.243,\ \hat{\gamma}_2 = 0.105,$$

with its sample correlogram and partial correlogram plotted in Figure 2.9. Scanning the latter, we note that $\hat{\pi}_k \approx 0$ for $k > 2$, thus an AR(2) model seems appropriate for the centered data (Moran, 1953). For $p = 2$, replacing γ_k's in (2.41) by their sample counterparts given above the fitted AR(2) model is

$$X_t - 2.904 = 1.30(X_{t-1} - 2.904) - 0.72(X_{t-2} - 2.904) + e_t,$$

or

$$X_t = 1.03 + 1.30X_{t-1} - 0.72X_{t-2} + e_t, \tag{2.45}$$

where $e_t \sim WN(0, 0.059)$; recall that $WN(0, \sigma^2)$ stands for a white noise process with mean 0 and variance σ^2.

The presence of cycle is evident both in the plots of data and the sample correlogram in Figure 2.9 (a). One major role of the periodogram, as mentioned earlier, is the

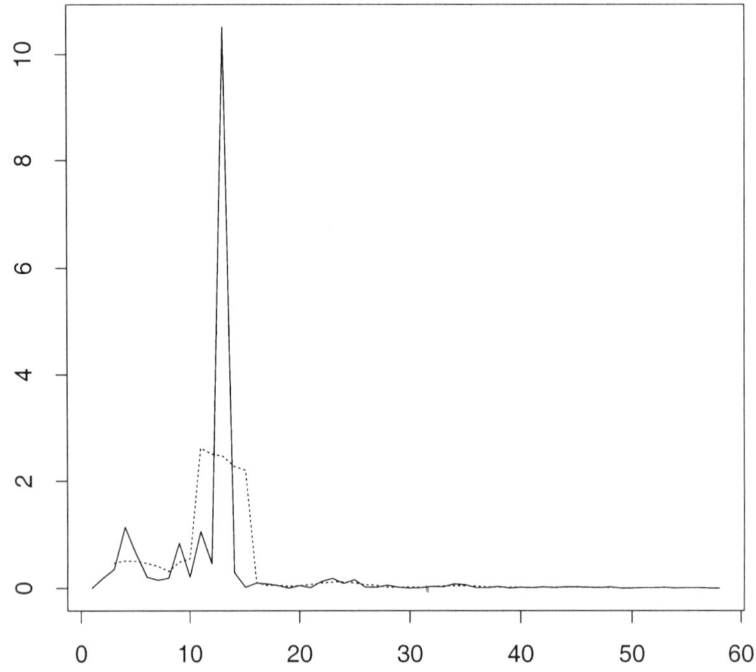

Fig. 2.10 The periodogram (—) and an estimated spectral density (- - -) of the lynx data.

detection and estimation of cycles or periods in the data. The periodogram and an estimated spectral density of the lynx data are given in Figure 2.10. The peak occurs near $k = 13$, so that from $\frac{2\pi \times 13}{114} \approx \frac{2\pi}{p}$, we obtain a very rough estimate of $p \approx \frac{114}{13} = 8.76$. It was observed by Campbell and Walker (1977) that the time plot of $\{X_t\}$ is an approximately sinusoidal curve (Figure 2.8 (c)). Among other models, they selected and fitted the following AR(2) model with an additive harmonic component with period 9.63:

$$X_t = 0.85 - 0.24 \cos \frac{2\pi}{9.63}(t - 1.33) + 0.97X_{t-1} - 0.27X_{t-2} + e_t, \qquad (2.46)$$

where $e_t \sim WN(0, 0.0420)$. Note that the estimated residual variance of this model is less than that in model (2.45). The broad features of the sample and fitted correlograms do provide qualitative methods for assessing the goodness-of-fit of models in time series. For the lynx data, while there is good agreements between sample and fitted partial correlograms, the same cannot be said about the sample and fitted correlograms.

An alternative and more common method of assessing the goodness of a fitted model is based on an analysis of its residuals. If a selected model fits the data well, one expects its residuals to mimic a white noise, or a purely random series. Thus, by examining the time series plot of the residuals one could qualitatively assess the goodness of a fitted model. However, a test of white noise can be developed using Theorem III(a). In fact, Box and Pierce (1970), had developed a *portmanteau test* using the fact that for n large,

$$Q_K = n \sum_{k=1}^{K} \hat{\rho}_k^2 \sim \chi^2_{K-r}, \tag{2.47}$$

where r is the number of estimated parameters of the fitted model. Larger values of Q_K supports departures from white noise, the sensitivity of the Q_K-statistic to such departures depends on the choice of K.

The computed test statistic for the residuals from the AR(2) model fitted to the lynx data with $r = 2$ estimated parameters and $K = 10$ turns out to be

$$Q_{10} = 15.25,$$

on 8 degrees of freedom which is on the borderline of being significant at 0.05 level. Goodness-of-fit of this model and others based on their predictive power is discussed in Section 2.5.4.

2.4.4 Lagged Scatterplot Matrix

Graphical tools for visualizing correlations in the raw data prior to fitting models and transformations to remove nonstationarity are of great interest in exploratory data analysis. The most common tool for this purpose is the *scatterplot matrix*, a two-dimensional array of pairwise scatterplots of multivariate standardized responses, which is graphically equivalent to the sample correlation matrix. Fortunately, for stationary data, it suffices to focus on the first row of a certain scatterplot matrix, called the *lagged scatterplot*. However, the use of this powerful tool is not as widespread as it should be in time series data analysis.

To highlight the importance and origin of the scatterplot matrix, Cleveland (1993, p. 274) states: "An award should be given for the invention of the *scatterplot matrix*, but the inventor is unknown—an anonymous donor to the world's collection of visualization tools .... The inventor may have not appreciated the significance of the method or may have thought the idea too trivial to bring it forward, but its simple, elegant solution to a difficult problem is one of the best visualization ideas around."

The *lagged scatterplot matrix* for a time series $\{x_t\}$ consists of scatterplots of pairs of values $(x_t, x_{t+\ell})$, separated by ℓ time units for $\ell = 1, \ldots, L$. Figure 2.11 provides the lagged scatterplot matrix of the lynx data (Series 1) and its logarithm (Series 2) for $L = 3$. A circular shape for a scatterplot at lag ℓ indicates little correlation at that lag. However, an elliptical shape in the 45 degree direction indicates positive correlation. The lagged scatterplot matrix of the logarithmically transformed lynx

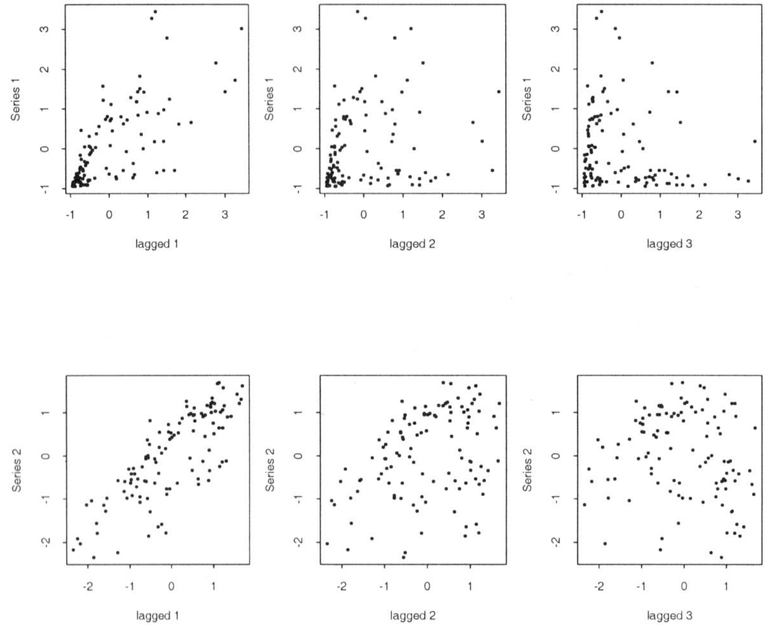

Fig. 2.11 Lagged scatterplots of the lynx data (Series 1) and its logarithm (Series 2).

data reveals stronger correlations for the transformed data for the indicated lags. This provides additional support for the log-transformed lynx data and the fitted AR(2) model (2.45).

Snow data (cont.) For the OLS residuals of the model (2.6) fitted to SNOCOM data, its lagged scatterplot (L=6) is given in Figure 2.12. Since the scatterplots at larger lags are more circular and in view of a large partial correlation at lag one in Table 2.2, it seems reasonable to assume that the OLS residuals $\{e_t\}$ follow an AR(1) model,

$$e_t = ae_{t-1} + \varepsilon_t,\ t = 2, 3, \cdots .$$

The Yule-Walker estimates of the parameters a and $\sigma_\varepsilon^2 = \text{var}(\varepsilon_t)$ are obtained using (2.41), the ACF in Table 2.2 and noting that for the OLS residuals $\hat{\gamma}_0 = 3.85$. Thus,

$$\begin{aligned}
\hat{a} &= \frac{\hat{\gamma}_1}{\hat{\gamma}_0} = \hat{\rho}_1 = 0.27 \\
\hat{\sigma}_\varepsilon^2 &= \hat{\gamma}_0 - \hat{a}\hat{\gamma}_1 = \hat{\gamma}_0(1 - \hat{a}\hat{\rho}_1) \\
&= \hat{\gamma}_0(1 - \hat{a}^2) = 3.85(1 - 0.27^2) = 3.57.
\end{aligned}$$

Table 2.2 The sample correlations and partial correlations of the OLS residuals for model (2.6) of SWE data (1935–1996).

Lag	ACF	PACF
1	0.27	0.27
2	0.07	0.00
3	0.14	0.13
4	0.08	0.01
5	0.06	0.04
6	-0.09	-0.14
7	0.01	0.07
8	-0.24	-0.31
9	-0.25	-0.09
10	-0.18	-0.13
11	-0.24	-0.11
12	-0.16	-0.05

A more careful scrutiny of Table 2.2 reveals large values for ACF and PACF at lag eight. Therefore, models more general than the AR(1) model might be appropriate. However, for simplicity we recommend this AR(1) model, Isaacson and Zimmerman (2000) provide a thorough analysis of this data set based on such AR(1) model for the errors.

2.5 FORECASTING SYSTEMS

It might be surprising to know that Schuster, Slutsky, Yule and Walker, the founders of modern time series analysis, were primarily interested in determining and/or generating periods or cycle in the data. In fact, the AR and MA models were merely viewed as powerful alternatives to (2.33) and Schuster's periodogram in their search for *flexible periodicity*. Nevertheless, the work on periodicity had the genesis of spectral analysis of time series as well as the time-domain approach. Forecasting future was not the main item on their agenda at that time. For an extensive history of forecasting, see McClosky (1992).

Wold (1938) integrated the mathematical developments in probability theory due to Kolmogorov, Khinchin's notion of stationarity and the empirical investigation of Schuster, Yule, Slutsky and Walker and showed that the models of hidden periodicities, autoregression and moving average are special cases of the class of stationary processes. He unlocked the structure of stationary processes using the concept of regression (prediction) as the key tool in approximating a future value in terms of

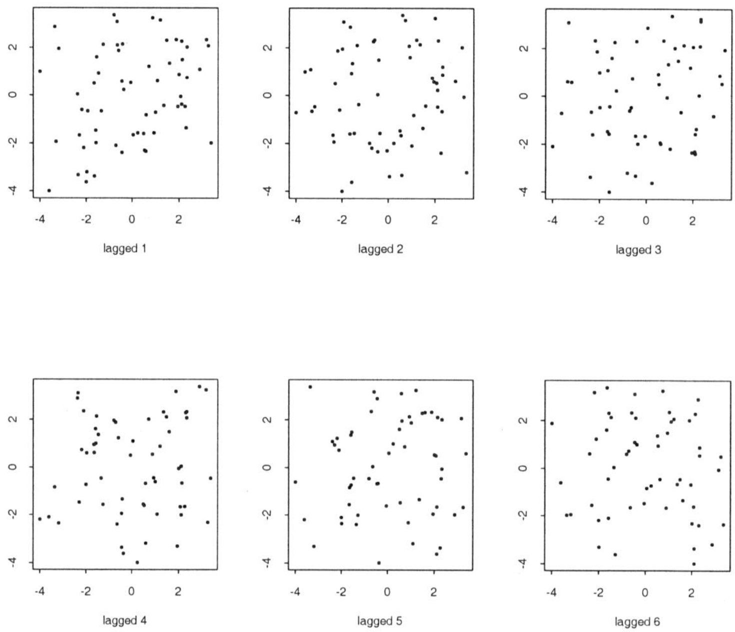

Fig. 2.12 Lagged scatterplots of the OLS residuals of the model (2.6) fitted to the SNOCOM data.

its predecessors. The celebrated Wold decomposition theorem or the infinite order moving average representation of stationary processes was the most revealing and satisfying result about such processes that he obtained in 1938. Wold employed it effectively to obtain either rigorously or formally most forecast formulas, their error variances, relationships between the AR and MA parameters, ..., that are used in modern time series analysis texts. Rigorous derivation of some of these results had to wait for the spectral representation of stationary processes; see Cramer (1940), and the more abstract formulation of the prediction problem and its solution by Kolmogorov (1939, 1941); see Section A2.7 and Chapter 10.

2.5.1 Prediction Theory and Time Series Analysis

Given the covariance function and past of a stationary process, *prediction theory* is concerned with deducing stochastic relationship among its variables. The most important of such relationships is the *autoregressive representation* expressing X_t in terms of its predecessors; see Section A2.7 for more information. By contrast, in *time series analysis* given the data $x_1, \ldots, x_n$, one estimates the covariance function and fits models, such as (2.33), (2.39)–(2.40) to the data with the ultimate goal of issuing forecasts for future values.

From the standpoint of forecasting future, it is instructive to divide the body of knowledge on stationary processes into three disjoint, but increasingly unrealistic categories:

I. *Statistical Time Series Analysis* is concerned with obtaining statistically optimal forecast of future using only the data $x_1, \ldots, x_n$.

II. *Finite Prediction* is concerned with prediction of future based on the *finite past* $X_1, \ldots, X_n$ and the *knowledge* of $\gamma_0, \ldots, \gamma_n$.

III. *Prediction Theory* is concerned with prediction of future based on the *infinite past* and the *knowledge* of the whole covariance function $\{\gamma_k\}$.

Clear demarcation of these categories is not essential, but to minimize confusion and for pedagogical reasons it is helpful to distinguish between the words *forecast(ing)* and *prediction*. Though literally identical, they are different technically in the sense that the former relies on the data alone (category I) and the latter does not (categories II and III). The somewhat artificial category II serves as a bridge linking the extreme practical and theoretical problems.

Practically unrealistic results and formulas obtained in II and III are put to practical use in I, upon simply replacing the *unknown covariances* by their estimated values and *truncating the infinite series* whenever necessary. This *plug-in forecasting system* is used routinely in time series analysis to issue forecast of future values. The study of the adverse effects of such substitutions and truncations is an important part of statistical time series analysis that relies heavily on prediction theory and limit theorems of probability theory. In other words, theories of prediction and probability are vital for justifying the methodologies of statistical times series analysis. According to Hannan (1987), "The classical Wiener-Kolmogoroff prediction theory is not fully relevant to practice because it uses the infinite past. However, it brings an understanding that is central to the statistical theory of stationary time series."

To make Hannan's remark clearer and render prediction theory relevant, next we replace the assumptions in III by more realistic ones and introduce ideas and notation that are needed in practice and in the study of errors introduced by truncating the infinite series for the predictor (bias) and estimating the unknown covariance (variability).

2.5.2 The Yule-Walker Forecasting System

Autoregressive models and the corresponding Yule-Walker equations are the source of one of the simplest, most transparent and widely used forecasting systems. The basic concepts and some relevant statistical issues are introduced in this section.

When $X_1, \cdots, X_n$ have zero means and the covariances $\gamma_0, \cdots, \gamma_n$ are known, the linear least squares predictor of X_{n+1}, based on the m immediate past observations

$X_n, \cdots, X_{n+1-m}$, in symbols $X_{n+1,m} = \sum_{k=1}^{m} a_{k,m} X_{n+1-k}$, is determined by the scalars $a_{k,m}$'s minimizing $E\left(X_{n+1} - \sum_{k=1}^{m} a_{k,m} X_{n+1-k}\right)^2$ as in (2.22). The *finite predictor coefficients* $a_{k,m}$ and the *finite prediction error variance*, in symbols σ_m^2, satisfy the *Yule-Walker equations*:

$$\begin{bmatrix} \gamma_0 & \cdots & \gamma_{m-1} \\ \ddots & & \vdots \\ & \ddots & \gamma_0 \end{bmatrix} \begin{bmatrix} a_{1,m} \\ \vdots \\ a_{m,m} \end{bmatrix} = \begin{bmatrix} \gamma_1 \\ \vdots \\ \gamma_m \end{bmatrix}, \tag{2.48}$$

$$\sigma_m^2 = E(X_{n+1} - X_{n+1,m})^2 = \gamma_0 - \sum_{k=1}^{m} a_{k,m} \gamma_k. \tag{2.49}$$

Computationally this prediction problem is the same as fitting AR(m) models to data; see (2.41).

Having observed $x_1, \cdots, x_n$, how does one issue forecasts for the future values $X_{n+1}, X_{n+2}, \cdots$? There are two natural methods that come to mind. First, the *linear least squares forecast* of X_{n+1} based on the m immediate past values given by

$$\hat{X}_{n+1,m} = \sum_{j=1}^{m} \hat{a}_{j,m} x_{n+1-j}, \tag{2.50}$$

where $\hat{a}_{1,m}, \cdots, \hat{a}_{m,m}$ are chosen to minimize the sum of squared forecast errors:

$$\sum_{t=1}^{n+1-m} (x_{m+t} - \hat{X}_{m+t,m})^2. \tag{2.51}$$

This usually gives rise to somewhat complicated normal equations. As a simpler alternative, we focus on finding the coefficient vector $\hat{\mathbf{a}}_m$ in (2.50) that satisfies the *sample Yule-Walker equations*:

$$\begin{cases} \hat{\mathbf{\Gamma}}_m \hat{\mathbf{a}}_m = \hat{\boldsymbol{\gamma}}_m, \\ \hat{\sigma}_m^2 = \hat{\gamma}_0 - \sum_{k=1}^{m} \hat{a}_{k,m} \hat{\gamma}_k \end{cases}, \tag{2.52}$$

where $\hat{\mathbf{\Gamma}}_m = (\hat{\gamma}_{i-j})_{i,j=1}^m, \hat{\boldsymbol{\gamma}}_{\mathbf{m}} = (\hat{\gamma}_1, \cdots, \hat{\gamma}_m)$ and $\hat{\gamma}_k$ is given in (2.2). Once the coefficients $\hat{a}_{j,m}$ are found, $\hat{X}_{n+1,m}$ provides a prediction rule for issuing forecast of X_{n+1}. We refer to this as the *Yule-Walker forecasting system*.

The integer m, often depending on n, is a critical entity so far as performance of the forecast $\hat{X}_{n+1,m}$ is concerned. Thus, the following questions are natural to ask: What is the optimal choice of m? Does $\hat{X}_{n+1,m}$ converge as $m \to \infty$? Does $\hat{a}_{j,m}$ converge to some a_j as $m \to \infty$? If yes, what is the statistical meaning or relevance of a_j? Does $\hat{\sigma}_m^2$ converge to σ_m^2 or σ^2 in (2.49) or (2.52)? These questions beg meaningful parameterization of stationary processes studied in Chapter 6. Their answers, which rely on techniques from statistical time series analysis, prediction theory and harmonic analysis, are addressed in Chapters 5–10. The choice of m, however, is closely related to the model identification problem discussed next.

2.5.3 AIC and BIC: Autoregressive Approximation

Traditionally, *model identification*, that is, selecting the order p of an AR model was based on comparing the sample correlogram and partial correlogram of the data with their theoretical counterparts for AR processes; see Section 2.4.3. By and large such graphical techniques rely on pattern recognition and are subjective. However, there has been great progress starting with the work of Akaike (1969) in formulating order selection as a decision-theoretic problem involving penalized likelihood functions. For the Gaussian data this amounts to scrutinizing the prediction error variances $\{\hat{\sigma}_m^2\}$ in (2.52).

Consider the more general problem of approximating the true model for a stationary process $\{X_t\}$ by an AR(p) model using the data $x_1, \cdots, x_n$. A plausible approach to selecting the order p is fitting AR models of successively increasing order $m = 0, 1, \cdots, M$, where M is a preassigned bound on the order. For each fitted model compute the residual variance $\hat{\sigma}_m^2$ using (2.52). Scanning the plot of $\hat{\sigma}_m^2$ vs. m could provide useful information about the "optimal" value of p. However, since $\hat{\sigma}_m^2$ is decreasing in m this approach always pick M, the largest value allowed. Figure 2.13 presents such a plot for the lynx data. A way around this difficulty is to impose a penalty for choosing p too large. This might harness the unabated decline of $\hat{\sigma}_m^2$ and provide a balance between fidelity to the data and parsimony of the model. The AIC (Akaike Information Criterion), Akaike (1969), and BIC (Bayesian Information Criterion), Schwarz (1978), two of the most popular suggestions coming out of such considerations, are defined for $m = 0, 1, \cdots, M$, by

$$\begin{cases} AIC(m) = log\, \hat{\sigma}_m^2 + 2mn^{-1}, \\ BIC(m) = log\, \hat{\sigma}_m^2 + 2(log\, n)mn^{-1}. \end{cases} \tag{2.53}$$

Note that the penalty for each additional parameter is $2n^{-1}$ in AIC and $2(log\, n)n^{-1}$ in BIC, where the latter levies a harsher penalty for larger models. Smaller values of AIC and BIC indicate better fitting models, so that scanning plots of AIC(m), BIC(m) vs. $m = 0, 1, \cdots, M$ suggests a viable method for selecting the unknown order p of an approximating AR model. For more information on these and other order selection criteria, see Choi (1992), Hannan and Deistler (1988, Chap. 5).

Figure 2.13 also presents AIC and BIC plots for the lynx data where the AIC $(\cdot)$ and BIC $(\cdot)$ have minima at $p = 11$ and 2, respectively. An approximating AR (11) model for the lynx data is also supported by the large value of $\hat{\pi}_{11}$ in the partial correlogram; see Figure 2.9 (b). The huge discrepancy between the orders of models selected using AIC and BIC is not incidental and should be of concern so far as stability of forecast is concerned. Shibata (1976) has shown that AIC tends to pick larger orders and is

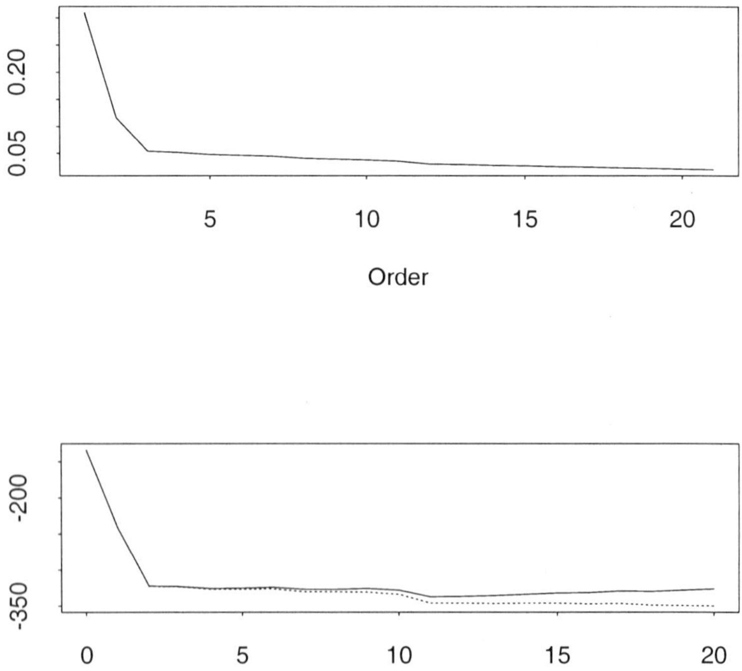

Fig. 2.13 Plots of residual variances (top), AIC (—) BIC ($\cdots$) (bottom) of AR(m) models fitted to the lynx data for $m = 0, \ldots, 20$.

not consistent. This might be because of its low penalty of $2n^{-1}$ per parameter as compared to the $2(\log n)n^{-1}$ for BIC. Using the idea of subset autoregression and a bias-corrected version of AIC which levies a slightly heavier penalty, Brockwell and Davis (1991, p.550) fit an AR(12) model to the lynx data with 7 parameters:

$$\begin{aligned} X_t = 1.123 \quad &+ \quad 1.084X_{t-1} - .477X_{t-2} + 0.265X_{t-2} + 0.265X_{t-3} \\ &- \quad 0.218X_{t-4} + 0.180X_{t-1} - 0.224X_{t-12} + e_t, \end{aligned} \tag{2.54}$$

where $e_t \sim WN(0, 0.0396)$. We refer to this model as the SAR(12) in the sequel.

2.5.4 Forecasting the Lynx Data

In a sense the ultimate purpose of time series analysis is the prediction of certain future values or events. Hence, it is conceivable that the goodness of a model should be assessed in terms of its ability to predict future values accurately. Implementation of this simple idea, however, turns out to be a daunting task in practice, as it requires choosing a criterion for measuring accuracy and more importantly having access (knowing) the future values so that the prediction errors and hence the accuracy measure can be computed. The future data are seldom available, so that one must resort to some alternatives and deal with less genuine prediction setups.

As a compromise in this introductory section, we illustrate how the two models (2.45) and (2.54) fitted to the lynx data can be compared so far as their empirical predictive powers are concerned. To this end, let us partition the $n = 114$ observations into two groups of 100 (past) and 14 (future) observations. The first group $X_1, \cdots, X_{100}$ will be used to refit the models and compute the forecasts of the next 14 future values. In computing these forecasts, we "pretend" that the remaining 14 observations $\{X_t; t = 101, \cdots, 114\}$ are unknown and use them only later when computing the out-of-sample prediction errors. The empirical performance of these models will be assessed by computing the standard deviation of the 14 prediction errors, that is, $s = \sqrt{S/14}$, where S is the sum of squares of the 14 multi-step ahead prediction errors.

To be more precise, let $\hat{X}_{100+\ell}$ be the linear least-squares forecast of $X_{100} + \ell$ based on the past $\{X_1, \cdots, X_{100}\}$. It follows from (2.45) that

$$X_{100+\ell} = 1.03 + 1.30X_{99+\ell} - 0.72X_{98+\ell} + \varepsilon_{100+\ell},$$

and hence the forecasts satisfy the simple recursion:

$$\hat{X}_{100+\ell} = 1.03 + 1.30\hat{X}_{99+\ell} - 0.72\hat{X}_{98+\ell}, \ell = 1, 2, \cdots, 14.$$

The desired forecasts are computed as follows:

$$\begin{array}{rcl} \hat{X}_{101} & = & 1.03 + 1.30X_{100} - 0.72X_{99}, \\ \hat{X}_{102} & = & 1.03 + 1.30\hat{X}_{101} - 0.72X_{100}, \\ \hat{X}_{103} & = & 1.03 + 1.30\hat{X}_{102} - 0.72\hat{X}_{101}, \\ & \vdots & \vdots \\ \hat{X}_{114} & = & 1.03 + 1.30\hat{X}_{113} - 0.72\hat{X}_{112}. \end{array}$$

Note that when the (unknown) $X_{101}, X_{102}, \cdots$, are needed they are replaced by their forecasts $\hat{X}_{101}, \hat{X}_{102}, \cdots$. Similar ideas are used in computing forecasts for the SAR(12) model (2.54) and the more general ARMA models. The forecasts and values of s are given in Table 2.3. Comparing s, the measure of empirical performance of the models, indicates that the AR(1) model is inferior to the SAR(2) model.

Computing the ℓ-step ahead prediction error variance, that is Var$(X_{100+\ell} - \hat{X}_{100+\ell})^2$, is essential for setting up prediction intervals for the forecasts. For this, the Wold decomposition of a stationary process and its MA parameters, introduced in Section A2.7, and in Chapters 4 and 5, play crucial roles.

2.5.5 Efficient Statistical Forecasting Systems

How does one come up with a forecasting method and assess its optimality? In this section, we introduce and compare methods of forecasting based on their *empirical*

Table 2.3 One-step ahead forecasts for the lynx data and values of s for the AR(2) and SAR(12) models

ℓ	Observed	AR(2) forecasts	SAR(12) forecasts
1	2.36	2.30	2.35
2	2.60	2.56	2.79
3	3.05	2.70	2.87
4	3.39	2.70	3.23
5	3.55	2.59	3.35
6	3.47	2.46	3.32
7	3.19	2.36	2.98
8	2.72	2.33	2.67
9	2.69	2.36	2.43
10	2.82	2.42	2.82
11	3.00	2.48	2.97
12	3.20	2.51	3.24
13	3.42	2.51	3.41
14	3.53	2.48	3.55
s		0.339	0.138

predictive performance using the data at hand. We avoid as much as possible the traditional and unrealistic references to hypothetical replications of the data generating mechanism, alternative data sequences or events that did not occur. In particular, the relevance of *expected performance* will be discussed as the measure of merit of a *statistical forecasting system* (SFS). The latter criterion computed by averaging over all possible realizations involves data never observed and forecasts never issued.

We follow the recent work of Skouras and Dawid (1998, 1999) and focus primarily on the task of issuing optimal or efficient forecasts, rather than searching for the best model or efficient parameter estimates. It is conceivable that the model selection and estimation problems can be embedded within the more general problem of issuing efficient forecasts.

Consider a forecaster who observes a sequence $X = \{X_t,\ t \geq 1\}$, of random variables. His task at each time t is to issue a point forecast for the next potential observation X_{t+1} by using the past data $x^t = (x_1, \ldots, x_t)$. The forecaster uses a rule which, for every set of outcomes for x^t and any other relevant information, specifies a point forecast for X_{t+1}. Such a rule is called a *point prediction system.* Of course, for any forecasting task there are many prediction rules that a forecaster can choose from as the next simple example reveals.

Example 2.3. As a concrete example, the forecaster A may use the average of the past observations $A_{t+1} = \frac{1}{t}\sum_{j=1}^{t} x_j$, $t \geq 1$ as a prediction rule to predict X_{t+1}. Another forecaster B, can use a *discounted average* of the past observations of the form $A_{t+1}(\lambda) = c_t \sum_{j=0}^{t-1} \lambda^j x_{t-j}, t \geq 1$, for a known or unknown *discount factor* $\lambda (0 \leq \lambda \leq 1)$, where the constant c_t satisfies $c_t \sum_{j=0}^{t-1} \lambda^j = 1$. Note that $A_{t+1}(1) = A_{t+1} = \bar{x}_t$ is the average of all the past values up to and including t, and $A_{t+1}(0) = x_t$ picks the latest observation as the forecast of the next future value. The first prediction rule is said to have long memory and the second is Markovian of order one or is said to have short memory. It is evident that values of λ closer to one produce prediction rules of the first type and those closer to zero give rise to prediction rules with shorter memory. For a given dataset it is desirable to develop concrete methodology for selecting λ that is optimal in the sense defined below.

The family $\{A_t(\lambda);\ 0 \leq \lambda \leq 1\}$ generates a host of point prediction rules. How does one compare two point prediction systems using only the available data? To answer this question, we need a measure of the empirical performance of a SFS. Suppose A is a SFS and $(A_1, A_2, \ldots, A_n)$ are the point forecasts issued for the first n observations $(x_1, \ldots, x_n)$. Then, we may assess the *empirical performance* of A up to time n by the running *sum of squared prediction errors*

$$S_n(A) = \sum_{t=1}^{n} (x_t - A_t)^2.$$

Note that for any SFS, $S_n(A)$ is increasing in n and may converge to $+\infty$.

Having defined the above measure of empirical performance for SFSs using only the available data of length n, we can compare any two SFSs, say A and B, using the difference between the cumulative losses for A and B. Evidently, negative values of $D_n(A, B) = S_n(A) - S_n(B)$ indicates A is performing better than B and the opposite for positive values. Such verdicts may depend on the data length n, so that for some n's A may perform better than B and the reverse may hold for others. A way around this problem is to take an asymptotic view and consider a specific infinite sequence $(x_1, \ldots, x_n, \ldots)$ of outcomes. If the difference $D_n(A, B)$ tends to $-\infty$, then we say the SFS A is *better* than B, and the opposite if $D_n(A, B)$ tends to $+\infty$. When the difference $D_n(A, B)$ stays bounded above and below, the two SFSs cannot be distinguished and would be considered as *equivalent*. This heuristic and asymptotic notion of empirical predictive performance can be made more precise by assuming some knowledge of the data generating mechanism.

Figure 2.14 provides plots of the running sum of squared prediction errors $S_t(\lambda) = S_t(A_{t+1}(\lambda))$ for the Canadian lynx data, where $A_{t+1}(\lambda)$ is the discounted average prediction rules introduced in Example 2.3. These plots correspond to the five distinct values of the discount factors $\lambda = 0.0(0.25)$; it is evident that smaller values of λ

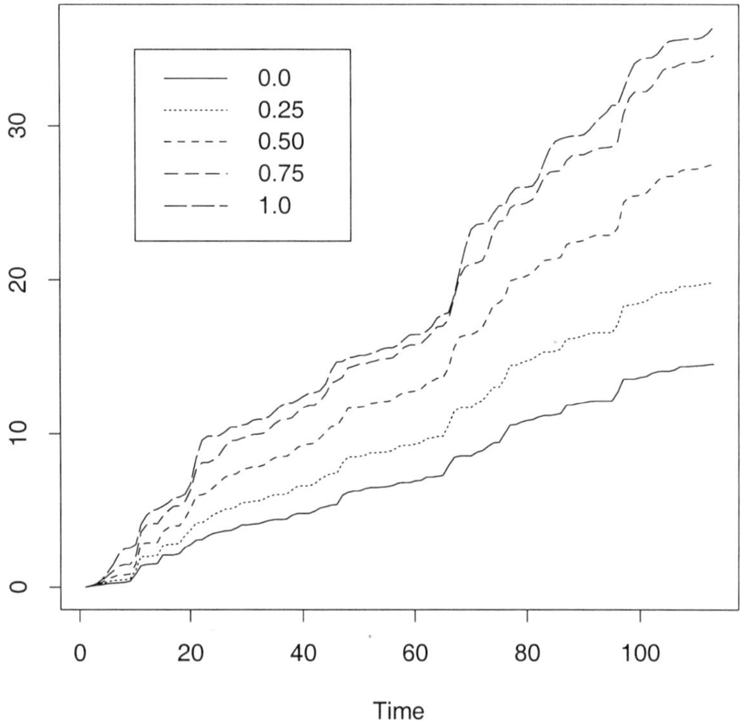

Fig. 2.14 Plots of $S_t(\lambda)$ for the lynx data corresponding to five distinct discount factors $\lambda = 0.0(0.25)$ in Example 2.3.

produce better prediction rules as judged by data-based criterion $S_n(A)$. This suggests using Markovian or short memory models for prediction purposes for the lynx data.

Thus far, it seems one can introduce point prediction systems and compare them without knowing the distribution of the sequence $X = \{X_t;\ t \geq 1\}$. While this is true, lack of knowledge of the distribution of X makes it impossible to define or characterize *efficient* SFSs. In what follows, we assume that the distribution of the sequence X is known to be a member of a given parametric family $\mathcal{F} = \{f_{X|\theta}; \theta \in \Theta\}$ of distributions. As a concrete example, one may assume the sequence X is Gaussian whose mean and covariance functions are parameterized using only p unknown parameters. More specifically, if the mean of X is known and, say, is zero and its covariance structure is introduced via AR or MA models, then θ is simply the AR or MA coefficients of the model. Additional and common examples of distributions for the data are introduced in Appendix A.

Given a parametric family $\{f_{X|\theta}\}$ of distributions for the sequence X, a SFS A is said to be *efficient* for $\{f_{X|\theta}\}$, if for any other SFS B, we have

$$\lim_{n\to\infty} \sup D_n(A,B) < \infty,$$

for "almost all" data and $\theta \in \Theta$. That is, A is efficient if with probability one and for almost all θ, its asymptotic empirical predictive performance will be at least as good as any other SFS. This is quite a strong property and without the "almost all" qualification, no SFS may be efficient. This heuristically appealing notion of efficiency is based on the principle that prediction rules should be assessed using their empirical performance for the actual data. An efficient SFS is guaranteed, up to the " almost all" qualification, to produce optimal performance so long as the data sequence is generated from the assumed distribution.

We note that in these developments and definitions, the traditional concept of *expected performance* has not made a debut yet. The next example introduces a simple, popular SFS from the 1950's (Abraham and Ledolter , 1983, Chapter 3) which is closely related to the discounted average introduced in Example 2.3, but is optimal for the *expected performance* criterion.

Example 2.4: Exponentially Weighted Moving Average (EWMA):
A forecaster, using EWMA issues a point forecast of the form

$$\hat{x}_{t+1} = (1-\lambda)\sum_{j=0}^{\infty} \lambda^j x_{t-j},\ 0 < \lambda < 1,$$

for X_{t+1} given a time series $x_t, x_{t-1}, \ldots$. The point forecast satisfies the recursion

$$\hat{x}_{t+1} = (1-\lambda)x_t + \lambda \hat{x}_t,$$

which is ideal for updating the current forecast in the face of a new datum x_t, without storing the past values; only the current data point is needed. This prediction rule is popular for forecasting sales of many items simultaneously and in the construction of control charts in statistical process control in engineering applications.

Is the EWMA forecasting system optional in some sense? Muth (1960) has shown that the EWMA forecasts are the *minimum mean square error* forecasts in the sense that they minimize the *expected performance* $E(\hat{X}_{t+1} - X_{t+1})^2$, for observations $x_t, x_{t-1}, \ldots$ generated by the pair of models

$$\begin{cases} X_t &= \mu_t + e_t, \\ \mu_t &= \mu_{t-1} + \eta_t, \end{cases}$$

where $\{e_t\}$ and $\{\eta_t\}$ are independent zero-mean and constant-variance white noises. For more on the history, developments and applications of such forecasting methods based on more general dynamic models for $\{\mu_t\}$, see Section 2.5.7; Durbin (2000); West and Harrison (1997) and Abraham and Ledolter (1983).

Note that the optimality of the EWMA forecasting system as well as the Yule-Walker forecasting systems are based on the *expected performance criterion.* For their implementation one generally needs to know the *first two moments* of the distribution $f_{X|\theta}$.

2.5.6 Plug-in Statistical Forecasting Systems

A widely used method of issuing point forecast is that of replacing the unknown parameters of an optimal predictor by suitable estimators based on the past data. This so-called plug-in SFS and the situations where it is efficient are discussed in this section.

Suppose we observe the past X^t and want to predict the next future value X_{t+1} under the squared error loss. That is, we are looking for a predictor $g(X^t)$ minimizing $E\left[X_{t+1} - g(X^t)\right]^2 = E\left\{E\left[X_{t+1} - g(X^t\right]^2 |X^t\right\}$. The optimal predictor is $g(X^t)$ $= E(X_{t+1}|X^t)$, the conditional mean of X_{t+1} given the past X^t. Since the joint distribution of (X^t, X_{t+1}) can be derived from the parametric family of distributions $\{f_{X|\theta}\}$ indexed by θ, we will have a different optimal predictor for every θ, which we denote by $m_{t+1}(\theta) = E(X_{t+1}|X^t, \theta)$. Thus, there are two sources of uncertainty in such point forecasts, the first due to the randomness of X_{t+1} is beyond our control, but the second from not knowing the parameter θ is handled by using either a plug-in method described below or the Bayesian paradigm discussed briefly in Appendix B.

It is common to form forecast rules by replacing the unknown parameter θ in the optimal predictor $m_{t+1}(\theta)$ with a suitable estimate $\hat{\theta}_t$ based on the past data. Thus, the point forecast for X_{t+1} is given by $m_{t+1}(\hat{\theta}_t)$. We illustrate this method using the lynx data. Figure 2.15 provides plots of S_t for $t = 11, 12, \ldots, 113$ for the Canadian lynx data corresponding to AR(1) and AR(2) models fitted to the lynx data sequentially, that is, fitted to $x^t = (x_1, \ldots, x_t)$ for $t = 10, \ldots, 113$ by solving the relevant sample Yule-Walker equations (2.52). Then, the forecast $\hat{x}_{t+1}$ is computed for each model and compared to the actual datum x_{t+1} in order to compute S_{t+1}. We have started from $t = 10$ so that reasonable estimates of the autoregressive parameters are obtained. The simple forecast rule based on the average of the immediate past values $\frac{1}{t}\sum_{j=1}^{t} x_j$ is included for the sake of comparison. From the plot it is evident that the plug-in SFS based on the AR(2) model does much better than the AR(1) and the average predictors. Computing S_{t+1}, for $t = 10, \ldots, 113$ requires fitting 103 AR(2) models to x^t of increasing lengths. This shows how computationally intensive the proposed procedure could become when comparing more complicated models. For general ARMA models it seems the use of state-space representation and the Kalman filter introduced in the next section and Chapter 4 could reduce the computational burden considerably.

How good are plug-in SFSs? Does use of optimal estimators $\hat{\theta}_t$ guarantee the efficiency of the resulting plug-in SFS? The following important example sheds light on these questions in some of the most practically useful situations; see Skouras and Dawid (1998, Section 8).

Example 2.5: Suppose $Y = (Y_t)$ are generated from the regression model (2.12), where e_t's are i.i.d. $N(0, \sigma^2)$, $x_t = (x_{t1}, \ldots, x_{tp})$, $X_t = \begin{pmatrix} x_1' \\ \vdots \\ x_t' \end{pmatrix}$ and the regres-

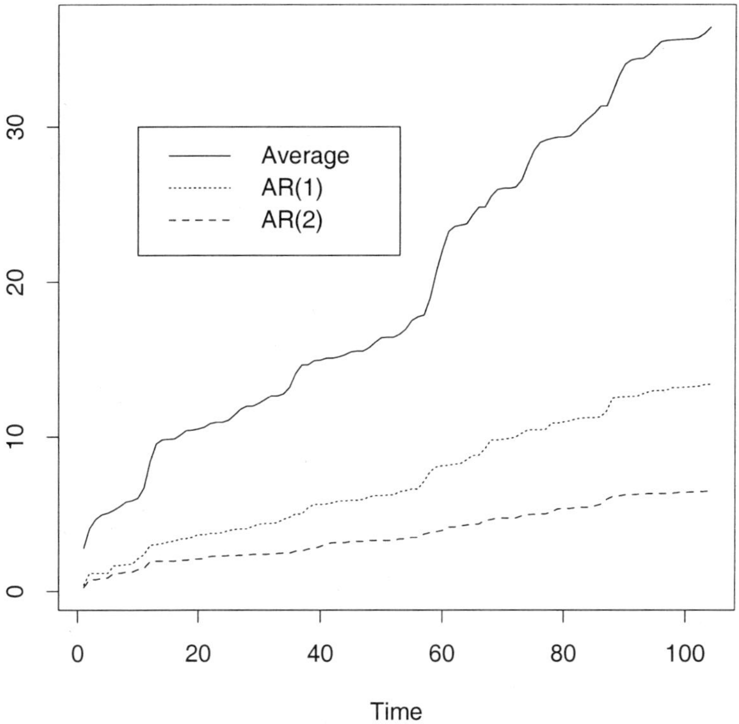

Fig. 2.15 Plots of S_t for the lynx data corresponding to plug-in SFS based on AR(1) and AR(2) models. The simple prediction rule based on the average of past values is included for comparison.

sor x_{t+1} may depend on the previous responses and regressors, $x_1, Y_1, \ldots, x_t, Y_t$. Then, the plug-in SFS based on the least squares estimator (LSE) of β is efficient. Similarly, when $Y = (Y_t)$ follows the AR(p) model (2.40) with $\{e_t\}$ as above and $\alpha = (\alpha_1, \ldots, \alpha_p)$ so that the roots of the characteristic polynomial $\alpha(z) = 1 - \alpha_1 z - \ldots - \alpha_p z^p$ are outside or on the unit circle (nonexplosive case), then the plug-in SFS based on the LSE of α is efficient.

Efficiency of more general plug-in SFSs can be established by comparing their performance with an efficient Bayesian forecasting system where our initial uncertainty about θ is quantified by means of a *prior distribution* for θ. The basic principles of Bayesian forecasting along with a few examples are presented in Appendix B.

2.5.7 ARMA Models and the State-Space Approach to Prediction

The unrealistic requirements of stationarity and knowing the infinite past and the phenomenal evolution of computational technology from analog to digital were some of the reasons for a major paradigm shift in time series analysis and digital signal processing in the 1950's and 1960's. The new era based on "finite-dimensional" nonstationary models was strikingly different from the Wold-Kolmogorov-Wiener's "infinite-dimensional" formulation of prediction problems based on general stationary processes, but in spirit were much closer to Yule's AR and Slutsky's MA models.

It was noted by Doob (1944) and others that a stationary process with a rational spectral density can be written as a dynamic linear model driven by a white noise. This simple observation had the impetus of a new emerging era in time series analysis centered around the class of mixed *autoregressive moving average* (ARMA) models:

$$X_t = \sum_{i=1}^{p} \alpha_i X_{t-i} + \varepsilon_t + \sum_{j=1}^{q} \beta_j \varepsilon_{t-j}, \tag{2.55}$$

where $\alpha_1, \cdots, \alpha_p, \beta_1, \cdots, \beta_q, p, q$ are its AR, MA coefficients and orders, and $\{\varepsilon_t\}$ is a white noise with mean zero and variance σ^2. This and Kalman's (1960) state-space models, changed the conventional Wold-Kolmogorov-Wiener formulation of the prediction problem by giving, not the covariance of the process, but a dynamic model for its evolution containing a finite number of parameters which allowed recursive computation of the predictors (Kalman and Bucy, 1961).

Depending on the values of the AR coefficients, the stochastic difference equation (2.55) could have stationary or nonstationary solutions. Whereas the MA(q) process $\{X_t\}$ in (2.39) is always stationary, solutions $\{X_t\}$ of (2.55) are not always stationary. For it to have a stationary solution, no roots of the AR-polynomial

$$\alpha(z) = 1 - \alpha_1 z - \cdots - \alpha_p z^p,$$

must lie on the unit circle in the complex plane; see Chapters 4 and 9. Similarly, whereas the AR(p) process $\{X_t\}$ in (2.40) is always *invertible*, namely, ε_t is expressible in terms of present and past observations, $\varepsilon_t = X_t - \sum_{i=1}^{p} \alpha_i X_{t-i}$, this is not always the case for (2.55). For it to be *invertible*, all roots of the MA-polynomial

$$\beta(z) = 1 + \beta_1 z + \cdots + \beta_q z^q,$$

must lie outside the unit circle. Under these conditions of *stationarity* and *invertibility*, the ACF and PACF of a stationary solution of (2.55) decay to zero exponentially fast for large lags; see Chapter 4. This property of ACF and PACF is useful in identifying the orders p and q of ARMA models using sample correlogram and partial correlogram introduced earlier in Section 2.4.3. ARMA model selection using suitable extensions of AIC and BIC is discussed in Section 2.6.3.

As noted in Section 2.2, sometimes a nonstationary series $\{X_t\}$ can be reduced to stationarity by differencing. Consider the formal *backward shift operator* B defined by $BX_t = X_{t-1}$ and the *difference operator* $\Delta = I - B$ so that $\Delta X_t = (I - B)X_t = X_t - X_{t-1}$ is the series of the first-order differences. For an integer $d \geq 0$, the *dth-order difference* $\Delta^d X_t$ is defined by repeated applications of Δ. Now, we define a very specific class of nonstationary process called *autoregressive integrated moving average* (ARIMA) models. We say a series $\{X_t\}$ satisfies an ARIMA(p, d, q) model if its dth-order difference $\{\Delta^d X_t\}$ follows a stationary ARMA(p, q) model. Thus, using the shift B an ARIMA(p, d, q) can be written compactly as

$$(I - B)^d \alpha(B) X_t = \beta(B) \varepsilon_t, \tag{2.56}$$

where $\alpha(B) = I - \alpha_1 B - \ldots - \alpha_p B^p$ and $\beta(B)$ is a similar polynomial in B. Note that for $d = 0$ the new class reduces to the standard ARMA models. In some genuine applications, differencing may not produce the required degree of stationarity. For these and other situations the flexible class of state-space models introduced next might be the ideal setup.

The *state-space approach* is appreciated better through the meaning of the word "state". The state of a physical system is the minimal set of variables needed for predicting its future course. For example, to track a satellite in the space, it is crucial to know its position, direction and velocity, so that the state at time t is at least a 5-dimensional vector. The use of vector-valued processes is quite natural in state-space modeling. Thus, let $\{\mathbf{X}_t; t \in \mathbb{Z}\}$ be a d-dimensional process standing for the state of a system. Often, the state is not directly observable and the recorded data $\{\mathbf{Y}_t\}$ is a noisy version of $\{\mathbf{X}_t\}$. The state-space approach consists of two models, the first describing the evolution of the state is Markovian in nature and the second is of the form signal (state) plus noise:

$$\begin{cases} \mathbf{X}_t = A\mathbf{X}_{t-1} + \boldsymbol{V}_t, & \textit{state equation}, \\ \mathbf{Y}_t = G\mathbf{X}_t + \mathbf{W}_t, & \textit{observation equation}, \end{cases} \tag{2.57}$$

where A and G are known $d \times d$ and $p \times d$ matrices and $\{\boldsymbol{V}_t\}$ and $\{\mathbf{W}_t\}$ are vector-valued white noises orthogonal to each other with known covariances $\Sigma = E(\boldsymbol{V}_t \boldsymbol{V}_t')$ and $R = E(\mathbf{W}_t \mathbf{W}_t')$. Conditions for stationarity of $\{\mathbf{X}_t\}, \{\mathbf{Y}_t\}$, and state-space representation of ARMA models can be found in Chapter 4.

Example 2.6 Let s_t be a random variable standing for the unobservable SWE for the year t in the snow data. From the analysis in Sections 2.3 and 2.4, it is plausible to assume that $\{s_t\}$ follows an AR(1) model, so that the *state equation*, with $d = 1$, is

$$s_t = a s_{t-1} + \boldsymbol{V}_t, \; t = 1, 2, \cdots .$$

Since the SNOTEL and SNOCOM devices provide two potentially correlated readings $y_{1,t}$ and $y_{2,t}$ of s_t with measurement errors, the *observation equation* in this situation, with $p = 2$, is given by

$$\boldsymbol{Y}_t = \begin{bmatrix} 1 \\ 1 \end{bmatrix} s_t + \boldsymbol{W}_t, \; t = 1, 2, \cdots, 62,$$

where $\boldsymbol{Y}_t = (y_{1,t}, y_{2,t})$ and $R = E(\boldsymbol{W}_t \boldsymbol{W}_t')$ is a 2×2 covariance matrix, which is not necessarily diagonal. Note that since SNOTEL was employed starting in 1967, the first component of $\boldsymbol{Y}_t$ is missing for the first thirty-two years.

The state-space models are quite versatile and subsume virtually all ARIMA models and "finite-dimensional" forecasting techniques in the literature. By now it is a mature and well-developed topic on its own right and covered in many excellent books such as Brockwell and Davis (1991, Chap.12); Priestley (1981); Harvey (1989); Hannan and Deistler (1988); Shumway and Stoffer (2000). Its relation to the Bayesian forecasting is discussed in West and Harrison (1997). Unfortunately, compared to ARMA models the widespread use of state-space models is hampered by lack of software on the general statistical packages. In spite of this, state-space models have been used increasingly in the last three decades to handle time series problems because of its key structural advantages relative to the ARIMA models. These advantages are discussed in detail in Durbin (2000), a summary is given below:

(1) **Structural advantages:**
State-space models for a time series problem is usually arrived at through a structural analysis of its components that make up the series. These components may include trend, seasonal, cycle, together with the effects of explanatory variables, interventions, outliers and missing values. In contrast, the ARIMA modeling is a passive 'black box' approach in which model identification relies solely on the data without prior analysis of the system that generated the data. The main tools for ARIMA model identification are the sample ACF and PACF which are notoriously imprecise because of their high sampling variability. Handling missing values, outliers, explanatory variables and multivariate time series is a lot more complicated within the ARIMA modeling framework.

(2) **Computational advantages:**
State-space models are Markovian in nature and their dynamic or evolution through time is quite transparent. Consequently, the calculations needed to implement such models are recursive and can handle increasingly large models and datasets without disproportionate increases in the computational burden. No extra theory is required to compute the forecasts and their standard errors. In contrast, the computational burden for ARIMA model usually increases with the size of the datasets and order of the MA part.

More details on estimation and implementation of these two classes of models can be found in Chapter 4.

2.6 STATISTICAL TIME SERIES ANALYSIS

In this section we discuss briefly some key ideas and results on the sample covariances, maximum likelihood estimation of parameters of ARMA models, AR and ARMA spectral density estimators.

Statistical analysis of data from a stationary process requires development of new techniques capable of accounting for dependence among the observations $x_1, \cdots, x_n$. For example, the familiar estimator of the mean μ, that is $\bar{x} = \frac{1}{n}\sum_{t=1}^{n} x_t$, has the variance,

$$E(\bar{x}-\mu)^2 = \frac{1}{n^2}\sum_{i=1}^{n}\sum_{j=1}^{n}\gamma_{i-j} = \frac{\gamma_0}{n}\left[1 + 2\sum_{k=1}^{n-1}(1-k/n)\rho_k\right], \tag{2.58}$$

which is quite different from the more familiar formula σ^2/n when the observations are i.i.d. Thus, the (asymptotic) distribution and statistical properties of $\bar{x}$ depend primarily on the correlation function $\{\rho_k\}$ and the rate at which it goes to zero. This is in sharp contrast with the classical theory of statistics where the asymptotic properties of the estimators depend on the shape, smoothness and other properties of a single density function (Casella and Berger, 1990).

2.6.1 Uniform Convergence of Sample Covariances

Using the *ergodic theorem* (law of large numbers) one can show that the estimators $\bar{x}$ and $\hat{\gamma}_k$'s converge as $n \to \infty$, to their population counterparts. Stronger and uniform rate of convergence of $\hat{\gamma}_k$ to $\gamma_k, 0 \le k \le m(n)$ are needed to show that an *estimated model* converges in some sense to the underlying theoretical model, if there is such a model at all. Here, $m = m(n)$ the number of estimated covariances must be substantially smaller than n (the number of observations). There has been considerable progress in proving such results in the last three decades employing limit theorems of probability theory; see Hannan (1974), Berk (1974), An, Chen and Hannan (1982), Hannan and Deistler (1988).

Usually existence of moments of up to order four is essential in proving such uniform convergence results. For simplicity, in the rest of this section $\{X_t\}$ is assumed to be a zero-mean Gaussian stationary process. The following theorem due to An, Chen and Hannan (1982) gives a flavor of such results.

Theorem IV Let $\{X_t\}$ be a zero-mean, purely nondeterministic Gaussian stationary process with AR parameters $\{a_k\}$, see (2.79), satisfying

$$\sum_{k=1}^{\infty}\sqrt{k}\,|a_k| < \infty, \ \sup_k k|a_k| < \infty.$$

Then,

(a) $\max_{0\le k\le n-1}|\hat{\gamma}_k - \gamma_k| \le c_n\frac{log\ n}{n}$,
where $\{c_n\}$ is a sequence of scalars (nonrandom) converging to zero.

(b) $\sup_m \lambda_1(\Gamma_m^{-1}) \le c < \infty$,
where for a square matrix A, $\lambda_1(A)$ stands for its smallest eigenvalue.

(c) $\lambda_1(\hat{\Gamma}_m^{-1}) \leq c < \infty$, for $0 < m \leq c_n \sqrt{\frac{n}{log\ n}}$.

Proofs of these results are rather lengthy and tedious relying on the law of iterated logarithm for martingales. They are not discussed here. Hannan and Deistler (1988, Chap. 7) provide a complete coverage.

2.6.2 Maximum Likelihood Estimation

Consider a zero-mean Gaussian stationary process $\{X_t\}$ with a covariance function $\{\gamma_k\}$ parameterized by a vector θ. Given the data $x_1, \cdots, x_n$, the likelihood function of θ is given by

$$\ell(\theta) = \frac{1}{(2\pi)^{\frac{n}{2}} |\Gamma_n|^{\frac{1}{2}}} e^{-\frac{1}{2}x' \Gamma_n^{-1} x}. \tag{2.59}$$

This is simply the probability density of the data $x = (x_1, \cdots, x_n)$ viewed as a function of the parameters, here $\Gamma_n = (\gamma_{k-\ell})_{k,\ \ell=1}^n$, $|\Gamma_n|$ stands for the determinant of Γ_n and the dependence of Γ_n on θ is suppressed. The maximum likelihood estimation of θ amounts to finding a $\hat{\theta}$ maximizing the function in (2.59). Such a $\hat{\theta}$ is called the *maximum likelihood estimator* (MLE) of θ. Closed-form expression for the MLE of $\hat{\theta}$ is hard to come by mostly because $|\Gamma_n|$, Γ_n^{-1} and consequently the partial derivatives of the loglikelihood function are intractable. Even when an exact expression is found for $\hat{\theta}$, it may not be as revealing and useful for theoretical and computational purposes as the following result for the simple AR(1) models indicates.

Example 2.7: Let $\{X_t\}$ be the zero-mean AR(1) process

$$X_t = \alpha X_{t-1} + \varepsilon_t,\ t = 0, \pm 1, \cdots,$$

where $|\alpha| < 1$ and $\{\varepsilon_t\}$ is a sequence of i.i.d. $N(0, \sigma^2)$ random variables. To write the likelihood function, recall that the joint distribution of $X = (X_1, \ldots, X_n)$ is the product of the conditional distribution of $(X_2, \ldots, X_n)$ given X_1 and the marginal distribution of X_1. The latter can be shown to be $N\left(0, \dfrac{\sigma^2}{1-\alpha^2}\right)$ and the former is

$$\prod_{t=2}^{n} \left(\frac{1}{\sqrt{2\pi}\sigma}\right)^{n-1} e^{-(1/2\sigma^2)(X_t - \alpha X_{t-1})^2},$$

so that the loglikelihood function of $\theta = (\alpha, \sigma^2)$ simplifies to

$$\log \ell(\theta) = -\frac{n}{2} \log \sigma^2 + \frac{1}{2} \log(1-\alpha^2) - \frac{1}{2\sigma^2}(d_{22}\alpha^2 - 2d_{12}\alpha + d_{11}),$$

where

$$d_{rs} = \sum_{t=s}^{n-r+1} X_t X_{t+r-s},\ r, s, \geq 1.$$

Thus, the MLE $(\hat{\alpha}, \hat{\sigma}^2)$ of (α, σ^2) must satisfy the equations

$$\sigma^2 = \frac{1}{n}(d_{22}\alpha^2 - 2d_{12}\alpha + d_{11}),$$

$$\frac{n-1}{n}d_{22}\alpha^3 - \frac{n-2}{n}d_{12}\alpha^2 - \left(d_{22} + \frac{d_{11}}{n}\right)\alpha + d_{12} = 0.$$

Let us denote the left-hand side of the last equation by $g(\alpha)$. White (1961) examined the signs of $g(\alpha)$ at $-\infty, -1, +1, +\infty$ and, using the fact that $d_{22} > 0$ he concluded that the cubic equation $g(\alpha) = 0$ has three real roots, one for each of the open intervals $(-\infty, -1)$, $(-1, 1)$ and $(1, +\infty)$, and hence that it has exactly one solution in the feasible domain $(-1, 1)$. Moreover, this solution can be shown to be a local maximum and therefore it is the MLE of (α, σ^2). The exact formula for the MLE of α given by Minozzo and Azzalini (1993) is

$$\hat{\alpha} = \frac{2}{3}\sqrt{c_2^2 - 3c_1}\cos\left(\frac{\theta}{3} + \frac{4\pi}{3}\right) - \frac{c_2}{3},$$

where

$$\theta = \arccos\left\{\frac{9c_2c_1 - 27c_0 - 2c_2^3}{2(c_2^2 - 3c_1)^{3/2}}\right\},$$

and

$$c_0 = \frac{n}{n-1}\frac{d_{12}}{d_{22}},\ c_1 = -\frac{n}{n-1}\left(1 + \frac{d_{11}}{nd_{22}}\right),\ c_2 = -\frac{n-2}{n-1}\frac{d_{12}}{d_{22}}.$$

Note that the formula for $\hat{\alpha}$ is a transcendental function of the data $x = (x_1, \ldots, x_n)$ and is far more complicated than the Yule-Walker estimate $\frac{\hat{\gamma}_1}{\hat{\gamma}_0} = \hat{\rho}_1$ in this case.

For Gaussian ARMA models with known orders p and q, the Γ_n has $p+q+1$ parameters and computing the MLE of the parameter vector $\theta = (\alpha_1, \cdots, \alpha_p, \beta_1, \cdots, \beta_q, \sigma^2)$ and establishing its asymptotic normality are studied in detail in the literature, see Brockwell and Davis (1991, Chap. 8). We provide a brief review of various methods for computing the estimator here. The statement of asymptotic normality of $\hat{\theta}$ is given at the end of this section.

One may start the search for the MLE of θ by choosing a grid for the parameter space and computing $\ell(\theta)$ for each and all grid points. The most expensive parts of evaluating the likelihood function at a grid point are those of $|\Gamma_n|$, and Γ_n^{-1}. Unfortunately, this cost grows with n, the length of the series. However, orthogonalizing the data or diagonalizing Γ_n using either the finite Fourier transform of the data or the Kalman filter (Gram-Schmid procedure), could make the cost bearable and shed significant light on the likelihood function itself as indicated next.

It was pointed out first by Whittle (1951) that to the first order of magnitude for large n, maximizing $\ell(\theta)$ is the same as minimizing

$$\begin{aligned} L(\theta) &= n^{-1}\log|\Gamma_n| + n^{-1}x'\Gamma_n^{-1}x \\ &\approx \log\sigma^2 + n^{-1}\sum_{k=0}^{n-1}\frac{I(\lambda j)}{2\pi f(\lambda_j)}, \end{aligned} \tag{2.60}$$

where $I(\lambda), f(\lambda)$ and σ^2 are defined in (2.32), (2.65) and (2.74). The approximation on the right-hand side follows from:

(a) $\frac{1}{n} \log |\Gamma_n| \to log\, \sigma^2$, as $n \to \infty$,

(b) The matrix U_n in (2.34) approximately diagonalizes Γ_n with the diagonal elements being $2\pi f(\lambda_j)$.

We give a proof of (a) in Lemma 7.1 (e) and (b). The expression on the right side of (2.60) is known as the *Whittle likelihood* and has been used extensively to obtain approximate MLE of θ with much less computation than using (2.59). Its performance, however, is not that great when n is small and the roots of the characteristic polynomials of (2.55) are near the boundary of the unit circle.

Box et al. (1994) made one of the first attempts to provide complete inferential procedure for ARMA models. They replaced (2.60) by

$$\log \sigma^2 + \sigma^{-2} n^{-1} \sum_{t=1}^{n} \varepsilon_t^2(\theta), \tag{2.61}$$

where for a given $\theta, \varepsilon_t(\theta)$ is obtained from (2.55) replacing $\varepsilon_t(\theta)$ and X_t by zero for $t \leq 0$. Then, an estimate of θ is used to "backforecast" X_t for $t \leq 0$; for example, for the AR(2) model (1.2) this amounts to forecasting the "initial values" X_{-1} and X_0 using the observations $X_1, \ldots, X_n$. In fact, if ε_t and X_t were known and treated as constants for $t \leq 0$, then (2.61) is the true loglikelihood for θ. In this approach, it is instructive to view the ARMA model (2.55) as a transformation or filter to "whiten" or orthogonalize $x_1, \cdots, x_n$; see (1.2)–(1.3). The Kalman filter or the "innovation algorithm" also can be applied to the ARMA models to calculate the likelihood function more expeditiously. This is discussed in detail in Chapter 4.

In recent years there has been considerable progress in developing solid statistical theory for estimating the parameters and orders of ARMA(p, q) models simultaneously; see Hannan and Deistler (1988). The methodology and proofs often do rely on the spectral density and the periodogram of the data; see (2.60). Suggesting that even in the time domain, the mathematics of the subject leads one to the arena of spectral domain. On the other hand, progress in the time domain methodologies has in turn produced alternative methods for estimating quantities of interest in the spectral analysis, the most notable example is estimation of the spectral density by fitting higher order ARMA models. We note that this interaction and synergy between time and spectral domain methodologies are quite essential both in the theory and practice of time series analysis. Another example of this synergy occurs in the proof of the following useful and deep theorem; see Brockwell and Davis (1991, Sec. 10.8).

Theorem V: Let $\{X_t\}$ be the zero-mean, Gaussian, stationary and invertible ARMA (p,q) process (2.55) where $\{\varepsilon_t\}$ is an i.i.d. sequence with mean zero and variance $\sigma^2, \alpha(z)$ and $\beta(z)$ have no common zeros and (p, q) is known. Then, with $\theta = (\alpha_1, \cdots, \alpha_p, \beta_1, \cdots, \beta_q)$ and $\hat{\theta}$ its MLE estimator, we have

$$\sqrt{n}(\theta - \hat{\theta}) \Rightarrow N(0, \boldsymbol{V}_\theta).$$

The asymptotic covariance matrix $\boldsymbol{V}_\theta$ is computed from

$$\boldsymbol{V}_\theta = \sigma^2 \begin{bmatrix} E\boldsymbol{Y}_t\boldsymbol{Y}_t' & E\boldsymbol{Y}_t\boldsymbol{Z}_t' \\ E\boldsymbol{Z}_t\boldsymbol{Y}_t' & E\boldsymbol{Z}_t\boldsymbol{Z}_t' \end{bmatrix}^{-1}, \tag{2.62}$$

where $\boldsymbol{Y}_t = (Y_t, \cdots, Y_{t+1-p}), \boldsymbol{Z}_t = (Z_t, \cdots, Z_{t+1-q})$, $\{Y_t\}$ and $\{Z_t\}$ are the autoregressive processes

$$\begin{aligned} Y_t &= \alpha_1 Y_{t-1} + \cdots + \alpha_p Y_{t-p} + \varepsilon_t, \\ Z_t &= -\beta_1 Z_{t-1} - \cdots - \beta_q Z_{t-q} + \varepsilon_t. \end{aligned}$$

Example 2.8 We illustrate the computation of $\boldsymbol{V}_\theta$ for few important cases.

(a) AR(p) process: Since $q = 0$, we obtain from (2.62) that

$$\boldsymbol{V}_\alpha = \sigma^2[E\boldsymbol{Y}_t\mathbf{Y}_t']^{-1} = \sigma^2\Gamma_p^{-1}, \tag{2.63}$$

where $\Gamma_p = (\gamma_{i-j})_{i,j=1}^p$ and $\{\gamma_k\}$ is the covariance function of the AR(p) process $\{Y_t\}$. The situation is similar for MA(q) processes.

(b) ARMA(1,1) process: Here $\boldsymbol{V}_\theta$ is given by

$$\begin{aligned} \boldsymbol{V}_{(\alpha,\beta)} &= \sigma^2 \begin{bmatrix} (1-\alpha^2)^{-1} & (1+\alpha\beta)^{-1} \\ (1+\alpha\beta)^{-1} & (1-\beta^2)^{-1} \end{bmatrix}^{-1} \\ &= \frac{1+\alpha\beta}{(\alpha+\beta)^2} \begin{bmatrix} (1-\alpha^2)(1+\alpha\beta) & -(1-\beta^2)(1-\alpha^2) \\ -(1-\beta^2)(1-\alpha^2) & (1-\beta^2)(1+\alpha\beta) \end{bmatrix}. \end{aligned} \tag{2.64}$$

This follows from calculating

$$\begin{aligned} EY_tZ_t &= E(\varepsilon_t + \alpha\varepsilon_{t-1} + \alpha^2\varepsilon_{t-2} + \cdots)(\varepsilon_t - \beta\varepsilon_{t-1} + \beta^2\varepsilon_{t-2} - \cdots) \\ &= \sigma^2[1 - \alpha\beta + (d\beta)^2 + \cdots] = \sigma^2(1-\alpha\beta)^{-1}. \end{aligned}$$

For more general ARMA(p, q) processes, calculation of $E\boldsymbol{Y}_t\boldsymbol{Y}_t'$ and $E\boldsymbol{Y}_t\mathbf{Z}_t'$ proceeds as above using the MA representations of $\{Y_t\}$ and $\{Z_t\}$. The above computation and the asymptotic normality of $\hat{\theta}$ in Theorem V are the key ingredients for setting up asymptotic confidence regions for α and β.

2.6.3 ARMA Spectral Density Estimator

The class of spectral density estimator discussed briefly in Section 2.4.1 is based on smoothing the periodograms and is nonparametric in the sense that no functional form is assumed for the unknown density function. Such nonparametric estimators are quadratic function of the data so long as the bandwidth selection is not data-based.

However, if we happen to know that the series conforms to the stationary ARMA model (2.55), then the functional form of its spectral density is known (see Corollary 9.31) and given by

$$f(\lambda) = \frac{\sigma^2}{2\pi} \left| \frac{1 + \beta_1 e^{i\lambda} + \cdots + \beta_q e^{iq\lambda}}{1 - \alpha_1 e^{i\lambda} - \cdots - \alpha_p e^{ip\lambda}} \right|^2 . \tag{2.65}$$

This suggests estimating (2.65) by plugging in an estimate of $\theta = (\alpha_1, \cdots, \alpha_p, \beta_1, \cdots, \beta_q, \sigma^2)$, thus providing a parametric, non-quadratic spectral density estimator using solely time-domain techniques, this is closer to and intertwined neatly with the task of forecasting the future.

To implement this method, one must first select the orders p and q of the true or approximating ARMA model. The AIC or BIC in (2.53) can be modified to this more general setup as follows. Let $\hat{\sigma}^2_{k,\ell}$ be the residual variance of an ARMA(k, ℓ) model fitted to the data, then define the bivariate functions

$$\begin{cases} AIC(k,\ell) &= \log \hat{\sigma}^2_{k,\ell} + 2(k+\ell)n^{-1}, \\ BIC(k,\ell) &= \log \hat{\sigma}^2_{k,\ell} + 2(\log n)(k+\ell)n^{-1}, \end{cases} \tag{2.66}$$

for $k = 0, \cdots, K, \ell = 0, \cdots, L$, where K and L are the preassigned bounds on the AR and MA orders, respectively. As before smaller values of AIC or BIC indicate better fitting models.

For $q = 0$, the above method of estimating the spectral density function is known as *autoregressive spectral estimation* and was introduced by Akaike (1969) and Parzen (1974). In the Parzen's formulation the time series may not conform to a finite AR model, but is assumed to have an infinite order autoregressive representation. Here, one expects a finite order AR model to provide an adequate approximation to a truly infinite order AR process analogous to a polynomial approximating a smooth function. The more general ARMA type spectral estimation was introduced by Akaike.

A2.7 PREDICTION THEORY

Some of the key concepts and results from the prediction theory of stationary processes relevant to time series analysis are summarized in this section as an appendix. They are used informally in this and the next two chapters. More complete discussions with proofs can be found in Chapters 5–10.

Prediction theory is concerned with approximating future in terms of the observed past when the covariance function and the past of a stationary process are known. Its goal is to express the predictors in terms of the known past values and the second-order moments of the process. In the spectral domain this amounts to solving some trigonometric polynomial approximation problems and studying convergence of certain Fourier series.

A2.7.1 Wold Decomposition and Prediction

A discrete-time stationary process $\{X_t; t = 0, \pm 1, \cdots\}$ consists of an infinite number of random variables. The question of linear independence of these variables is important in solving the least squares problems. A stationary process is said to be *deterministic* (singular) if X_t can be predicted correctly (with zero error) using the entire past $X_{t-1}, X_{t-2}, \ldots$; otherwise, it is said to be *nondeterministic*. For more discussions about these concepts, see Section 5.5.

With the elegance and simplicity of (2.40) in mind, Wold's search for relations among values and parameterization of a stationary process started with approximating X_t in terms of its predecessors $X_{t-1}, \cdots, X_{t-m}$:

$$X_t = a_{1,m} X_{t-1} + \cdots + a_{m,m} X_{t-m} + \varepsilon_{t,m}, \tag{2.67}$$

where $a_{k,m}$'s are some scalars and $\varepsilon_{t,m}$ is the approximation error; see Wold (1938, Sec. 19). The next natural step was to let $m \to \infty$. Here, convergence of $a_{k,m}$ and $\sum_{k=1}^{m} a_{k,m} X_{t-k}$ were problematic due to dependence of $\{X_t\}$. Thus, Wold aborted the earliest attempt to define the AR parameters and obtain the *autoregressive representation* (ARR) of a stationary process. Instead, he showed that $\{\varepsilon_{t,m}\}$ converges in the mean to a white noise $\{\varepsilon_t\}$, called the *innovation process* of $\{X_t\}$, and shifted attention to approximating X_t in terms of the past innovations $\varepsilon_t, \varepsilon_{t-1}, \cdots, \varepsilon_{t-m}$:

$$X_t = \varepsilon_t + b_{1,m}\varepsilon_{t-1} + \cdots + b_{m,m}\varepsilon_{t-m} + V_{t,m}, \tag{2.68}$$

for some scalars $b_{1,m}, \cdots, b_{m,m}$ and a random variable $V_{t,m}$. The orthogonality of $\{\varepsilon_t\}$ implies that $b_{k,m}$ does not depend on m, so that $b_{k,m} = b_k$ and furthermore $\{b_k\}$ is square-summable. Letting $m \to \infty$ in (2.68) one obtains the Wold decomposition of a nondeterministic stationary process.

Theorem VI: (Wold decomposition). Any nondeterministic, real-valued stationary process can be decomposed as

$$X_t = \sum_{k=0}^{\infty} b_k \varepsilon_{t-k} + V_t = U_t + V_t, \ \sum_{k=0}^{\infty} b_k^2 < \infty, \ \text{Var}\,(\varepsilon_t) \ = \ \sigma^2, \tag{2.69}$$

where $b_0 = 1$, $\{U_t\}$ and $\{V_t\}$ are *purely nondeterministic* and *deterministic* stationary processes, respectively, and orthogonal to each other.

This fundamental result shows that stationary processes are, indeed, constructed from two distinct orthogonal components: the deterministic component $\{V_t\}$ generalizing processes of hidden periodicities and the purely nondeterministic component $\{U_t\}$ subsuming finite AR and MA models. Its far-reaching consequences for prediction and other areas (see Chapter 10) were noticed immediately by Kolmogorov, Wiener and Krein among others. The situation is best summarized by Wold himself, in Gani (1982, p.192), "My main incentive for the decomposition of a general stationary time series was to cover as special cases the three type of stationary time series that were known at the time. These were (i) the scheme of hidden periodicities, (ii) the finite moving average model, (iii) the finite autoregression model. When establishing the decomposition I was only vaguely aware of its close affinity to the spectral analysis of time series, and even less aware of its close affinity to operator theory. It so happened that the decomposition gave a clue to overcoming an earlier impasse in spectral analysis and operator theory, as was shown by Andre Kolmogorov in 1939, and Norbert Wiener in 1942. Thanks to this profound aspect, my decomposition is referred to in the Encyclopedia Britanica, 13th ed., in the article on "Automata"."

Writing a stationary process as the sum of two orthogonal stationary components, splits virtually every relevant characteristics of $\{X_t\}$ into "orthogonal" sums. From (2.69), for all $t,\ k \in \mathbb{Z}$, we have

$$\begin{cases} \mathrm{Cov}(X_t, \varepsilon_{t-k}) = \sigma^2 b_k, \\ \mathrm{Cov}(X_k, X_0) = \mathrm{Cov}(U_k, U_0) + \mathrm{Cov}(V_k, V_0). \end{cases} \tag{2.70}$$

Using the spectral representation of covariances, see (2.36), with obvious notation, it follows from the second identify in (2.70) that

$$dF_X(\lambda) = dF_U(\lambda) + dF_V(\lambda) = f(\lambda)d\lambda + dF_V(\lambda). \tag{2.71}$$

This is precisely the Lebesgue decomposition of dF_X into its absolutely continuous and singular parts with respect to the Lebesgue measure on $(-\pi, \pi]$. Also from (2.69) for any integer $\ell \geq 1$, we have

$$X_0 = \sum_{k=0}^{\infty} b_k \varepsilon_{-k} + V_0 = \sum_{k=0}^{\ell-1} b_k \varepsilon_{-k} + \sum_{k=\ell}^{\infty} b_k \varepsilon_{-k} + V_0, \tag{2.72}$$

where the three components on the right are orthogonal to each other. Thus, $\hat{X}_{0,\ell}$ the ℓ-step ahead predictor of X_0 based on the infinite past (the orthogonal projection of X_0 onto the closed linear span of $X_{-1}, X_{-2}, \cdots$), its error and error variance are given, respectively, by

$$\begin{cases} \hat{X}_{0,\ell} = \sum_{k=\ell}^{\infty} b_k \varepsilon_{-k} + V_0, \\ X_0 - \hat{X}_{0,\ell} = \sum_{k=0}^{\ell-1} b_k \varepsilon_{-k}, \\ \mathrm{Var}(X_0 - \hat{X}_{0,\ell}) = \sigma^2 \sum_{k=0}^{\ell-1} b_k^2. \end{cases} \tag{2.73}$$

For $\ell = 1$, we simply use $\hat{X}_0$ to denote the one-step ahead predictor of X_0 based on the entire past. Then, the last two equations in (2.73) rewritten as

$$\begin{cases} X_0 = \hat{X}_0 + \varepsilon_0, \\ \mathrm{Var}\,(X_0 - \hat{X}_0) = \sigma^2, \end{cases} \tag{2.74}$$

provides opportunities to interpret $\{\varepsilon_t\}$ and its variance σ^2. The suggestive phrase *innovation process* for $\{\varepsilon_t\}$ was used by Wiener and Masani (1957) to emphasize writing X_0 in (2.74) as the sum of $\hat{X}_0$ (something from past) and ε_0 (something *new* from the future). The quantity σ^2 is usually called the *one-step ahead prediction error variance* or *innovation variance* or *intrinsic variance* of the process $\{X_t\}$. Both $\{\varepsilon_t\}$ and σ^2 play crucial roles in prediction theory and time series analysis.

In Chapter 8, we show that solutions of diverse and seemingly unrelated problems such as interpolation of missing values, prediction based on incomplete past, computation of canonical correlations between past and future, estimation of the impacts of interventions on a given response and detection of outliers are closely related and can be solved using the multi-step predictors and their error variances given in (2.73).

A2.7.2 Kolmogorov's Isomorphism and Contributions

Kolmogorov (1939, 1941), inspired by the geometric nature of Wold's decomposition, developed a comprehensive theory of prediction for stationary processes by abandoning the original probabilistic character of such processes and viewing them as sequences of vectors taking values in a Hilbert space (Chapters 9–10). Using the spectral representation of stationary processes (Theorem I) along with Wold's work he revealed the deeper aspects of the Wold decomposition through a similar decomposition of the spectral distribution, given in (2.71).

Kolmogorov's prediction-theoretic results were obtained by linking the study of stationary processes to certain approximation problems in the *spectral (frequency) domain* $L^2(dF)$, namely, the space of functions square-integrable with respect to the measure dF; see Chapter 9. Let $H(X)$ be the linear subspace generated by the values $X_t,\ t = 0, \pm 1, \ldots$, of the process, called its *time-domain*. Rewriting (2.35) as

$$\gamma_{k-\ell} = \int_{-\pi}^{\pi} e^{i(k-\ell)\lambda} dF(\lambda) = (e^{ik\lambda}, e^{i\ell\lambda})_{L^2(dF)} = \mathrm{Cov}(X_k, X_\ell), \tag{2.75}$$

allows that the covariance between X_k and X_ℓ can be viewed as the inner product of $e^{ik\lambda}$ and $e^{i\ell\lambda}$ in $L^2(dF)$. Thus, geometrically the process $\{X_t\}$ is the same as the sequence of exponentials $\{e^{ik\lambda}\}$. The norm-preserving map $T : X_t \to e^{i\lambda t}$ is called the *Kolmogorov's isomorphism* between the *time-domain* $H(X)$ and the spectral-domain $L^2(dF)$. In its wake, computing $a_{k,m}$'s and the finite prediction error variance σ_m^2 in (2.48) is equivalent to choosing $\{c_k\}$ to minimize

$$E\left(X_{n+1} - \sum_{k=1}^{m} c_k X_{n+1-k}\right)^2 = \int_{-\pi}^{\pi} \left|1 - \sum_{k=1}^{m} c_k e^{ik\lambda}\right|^2 dF(\lambda). \tag{2.76}$$

In words, the *(finite) linear prediction problem* in the time-domain is isomorphic to the *polynomial approximation* on the unit circle. The polynomial approximation problem, however, was solved around 1917 by Szegö for an absolutely continuous F. Kolmogorov (1941) independently solved it for a general spectral distribution. The connection between the finite prediction problem and Szegö's work was noted by

Krein (1944) who extended some of Kolmogorov's result to continuous-time stochastic processes, a direction that was also pursued vigorously and independently by Wiener around 1942.

For prediction-theoretic purposes implementation of (2.73) requires the knowledge of $\{b_k\}$, $\{\varepsilon_t\}$ and $\{V_t\}$. By exploiting the above isomorphism and results from the Hardy class of analytic functions on the unit circle, Kolmogorov characterized nondeterministic stationary processes and obtained explicit formulas for σ^2 and $\{b_k\}$ in terms of the Fourier coefficients of $\log f(\lambda)$ as summarized next. For the proofs of these results, see Chapter 10.

Theorem VII: Let $\{X_t\}$ be a stationary process with the spectral distribution function as in (2.71). Then,

(a) $\{X_t\}$ is a nondeterministic process, if and only if

$$\int_{-\pi}^{\pi} \log f(\lambda) d\lambda > -\infty. \tag{2.77}$$

It is purely nondeterministic, if and only if (2.77) holds and $F_V \equiv 0$.

(b) (Szegö-Kolmogorov). For a nondeterministic process, its one-step ahead prediction error variance is given by

$$\sigma^2 = \exp\left\{\int_{-\pi}^{\pi} \log f(\lambda) d\lambda / 2\pi\right\}.$$

(c) For a nondeterministic process with $c_k = \int_{-\pi}^{\pi} e^{ik\lambda} \log f(\lambda) \frac{d\lambda}{2\pi}, k = 0, \pm 1, \cdots$ its spectral density and MA parameters $\{b_k\}$ satisfy

$$\begin{cases} f(\lambda) = |\varphi(\lambda)|^2, \\ \varphi(z) = \sigma(1 + b_1 z + b_2 z^2 + \cdots) = \exp\left\{\dfrac{c_0}{2} + \sum_{j=1}^{\infty} c_j z^j\right\}, |z| < 1. \end{cases} \tag{2.78}$$

Note that if the spectral density function $f(\lambda)$ is zero on a set of positive Lebesgue measure, then $\int_{-\pi}^{\pi} \log f(\lambda) d\lambda = -\infty$ and (2.77) does not hold. Thus, for a stationary process to be nondeterministic, its spectral density function cannot be zero too often. It can be shown that the spectral density of a stationary ARMA model in (2.65) always satisfies (2.77). However, a stationary process with the spectrum $f(\lambda) = \exp\{-|\lambda|^{-\alpha}\}$ is nondeterministic if $\alpha < 1$ and deterministic if $\alpha \geq 1$, that is, when the spectral density has a high order zero at $\lambda = 0$. The last part of the theorem gives the so called *spectral factorization* of its density, it clearly indicates how the MA parameters of a stationary process are related to the Fourier coefficients of $\log f(\lambda)$. In fact, differentiating both sides of the equation in (2.78) we obtain

$$b_1 + 2b_2 z + 3b_3 z^2 = (1 + b_1 z + b_2 z^2 + \cdots) \sum_{j=1}^{\infty} j c_j z^{j-1}, |z| < 1.$$

Now, matching coefficients of the like powers of z one arrives at the following useful recursive formula for computing the MA parameters b_k's in terms of the Fourier coefficients of the logarithm of the spectral density function:

$$b_{k+1} = \sum_{j=0}^{k} \left(1 - \frac{j}{k+1}\right) c_{k+1-j} b_j, \, k = 0, 1, 2, \cdots, b_0 = 1.$$

For more details and applications, see Pourahmadi (1984).

A2.7.3 Wiener's Explicit Predictor Formula

The work of Wold, Kolmogorov and Krein did not provide explicit formulas for the predictor in terms of the observed past. In fact, the Wold decomposition is not suited for forecasting, since it expresses the predictor (2.73) in terms of the unobservable innovations $\{\varepsilon_t\}$.

During World War II, engineering problems related to anti-aircraft fire-control led Wiener (1949) to formulate a continuous-time linear prediction problem and derive an explicit formula for the predictor in terms of the past data. The work of Kolmogorov and Krein mentioned earlier were independent of Wiener's with some overlap. Due to war efforts Wiener's work though completed around early 1942 was classified, and published later in 1949. There seems to be some uncertainty on the chronology and priority, Wiener (1949, p.59) sheds some light on these issues.

The autoregression (2.67) is in line with the Wiener's approach where the predictor is expressed explicitly in terms of the observed quantities. Due to its practical importance, it is desirable to have spectral necessary and sufficient conditions for a stationary process to have an autoregressive representation of the form

$$X_t = \sum_{k=1}^{\infty} a_k X_{t-k} + \varepsilon_t, \tag{2.79}$$

where the infinite series is mean-convergent. Wiener and Masani (1958) showed that *boundedness* of the spectral density function, that is,

$$0 < c \leq f(\lambda) \leq d < \infty, \tag{2.80}$$

is sufficient for the mean-convergence of the series in (2.79). This condition holds, for example, for all invertible stationary ARMA models. The sequence $\{a_k\}$ is called the *autoregressive parameters* of $\{X_t\}$. When (2.79) holds, the AR parameters $\{a_k\}$ and the prediction error variance σ^2 satisfy the *discrete Wiener-Hopf or the infinite-order*

Yule-Walker equations:

$$\begin{cases} \begin{bmatrix} \gamma_0 & \gamma_1 & \ddots & \cdot \\ & \gamma_0 & \gamma_1 & \\ & & \ddots & \ddots \\ & & & \ddots \end{bmatrix} \begin{bmatrix} a_1 \\ a_2 \\ \vdots \end{bmatrix} = \begin{bmatrix} \gamma_1 \\ \gamma_2 \\ \vdots \end{bmatrix}, \\ \sigma^2 = \gamma_0 - \sum_{k=1}^{\infty} a_k \gamma_k. \end{cases} \tag{2.81}$$

Its solution requires the *spectral factorization* of the density $f(\lambda)$ of the process as in (2.78). This fundamental factorization problem was solved by Wiener and Hopf in 1931 in a different context. Wiener (1949) also noted that for a rational spectral density (2.65) one can obtain the factorization by locating its zeros and poles. Later, Doob (1944) pointed out the simple and (Markovian) dynamic structure of stationary processes with rational spectral densities.

A2.7.4 Finite-Segment Wiener-Hopf Equations

A truncation of (2.79) or (2.81) offers a simple method of approximating the solution of the latter via the Yule-Walker equations (2.48), which is usually used for solving finite prediction problems. How does such truncation affect the accuracy of prediction? Interestingly, this question is related to many asymptotic results in the statistical time series analysis and particularly to the accuracy of the autoregressive spectral density estimator introduced in Section 2.6.3; see Berk (1974) and Hannan and Deistler (1988).

It turns out that the rates of convergence of $a_{k,m}, \delta_m = \sigma_m^2 - \sigma^2$ and others of interest in statistical time series analysis are dominated by the truncation error as measured by the tail of the autoregression series in (2.79):

$$t_m = E \left| \sum_{k=m+1}^{\infty} a_k X_{t-k} \right|^2 .$$

Moreover, the estimation errors admit decompositions as sum of truncation error (bias) and sampling error (variance) when we use $\hat{\gamma}_k$ in place of γ_k:

$$\begin{cases} \hat{\sigma}_m^2 - \sigma^2 = (\hat{\sigma}_m^2 - \sigma_m^2) + (\sigma_m^2 - \sigma^2), \\ \hat{a}_{k,m} - a_k = (\hat{a}_{k,m} - a_{k,m}) + (a_{k,m} - a_k). \end{cases} \tag{2.82}$$

Under additional smoothness conditions than (2.80) on the spectral density of the process, one can show that the truncation error δ_m and $a_{k,m} - a_k$ converge to zero, as $m \to \infty$. However, obtaining fast and uniform rates of convergence for these errors are crucial in many problems in statistical time series analysis. A key result for controlling the truncation error (bias) is the *Baxter's inequality* (1962):

Theorem VIII: Let $\{X_t\}$ be a stationary process with a positive and continuous spectral density function, autoregressive parameters $\{a_k\}$ and finite predictor coefficients $a_{k,m}$ as in (2.48). Then, there exists an integer N and a constant K so that for all $m \geq N$,

$$\sum_{k=1}^{m} |a_{k,m} - a_k| \leq K \sum_{k=m+1}^{\infty} |a_k|. \tag{2.83}$$

Baxter's inequality also plays a major role in studying the sampling error (variance) or the first terms on the right-hand sides of (2.82). It follows from (2.48) and (2.52) and some algebraic manipulations that for $j = 1, 2, \cdots, m$,

$$\begin{aligned}\sum_{k=1}^{m}(\hat{a}_{k,m} - a_{k,m})\hat{\gamma}_{k-j} &= \sum_{k=0}^{m} a_k(\gamma_{k-j} - \hat{\gamma}_{k-j}) \\ &\quad + \sum_{k=1}^{m}(a_k - a_{k,m})(\hat{\gamma}_{k-j} - \gamma_{k-j}) \\ &= A_j + B_j.\end{aligned} \tag{2.84}$$

Thus, for example, the second term is bounded by

$$\begin{aligned}|B_j| &= \left|\sum_{k=1}^{m}(a_k - a_{k,m})(\hat{\gamma}_{k-j} - \gamma_{k-j})\right| \\ &\leq (\max_{0\leq k\leq m} |\hat{\gamma}_k - \gamma_k|) \sum_{k=1}^{m} |a_k - a_{k,m}|,\end{aligned} \tag{2.85}$$

so that in addition to Baxter's inequality, fast and uniform convergence of $\hat{\gamma}_k - \gamma_k, 0 \leq k \leq m = m(n)$ are crucial for controlling the sampling errors. Furthermore, viewing (2.84) as a system of linear equations in $\hat{a}_{k,m} - a_{k,m}, 1 \leq k \leq m$, with the $m \times m$ coefficient matrix $\hat{\Gamma}_m = (\hat{\gamma}_{k-j})$, its solution can be shown to be bounded by the right-hand side of (2.85), provided that the sequence $\{\lambda_1(\hat{\Gamma}_m^{-1})\}$ is bounded. Results on the rate of convergence and the nature of estimated covariances are stated in Theorem IV.

The calculation in (2.84) is also useful in studying $\hat{\sigma}_m^2 - \sigma_m^2$, since

$$\hat{\sigma}_m^2 - \sigma_m^2 = \sum_{k=0}^{m} a_{k,m}(\gamma_k - \hat{\gamma}_k) + \sum_{k=1}^{m} \hat{a}_{k,m}(\gamma_k - \hat{\gamma}_k), \tag{2.86}$$

where $a_{0,m} = -1$. Extensive applications of (2.83) and the subsequent results to the problems of system identification and selecting the orders of ARMA(p, q) models are discussed in Hannan and Deistler (1988).

Problems

2.1 The first difference of a time series $\{x_t\}$, denoted by $\{\Delta x_t\}$, is defined by the transformation

$$\Delta x_t = x_t - x_{t-1}.$$

Higher-order differences are defined by repeated applications. For example, the second difference $\{\Delta^2 x_t\}$ is defined by

$$\Delta^2 x_t = \Delta(\Delta x_t) = \Delta x_t - \Delta x_{t-1} = x_t - 2x_{t-1} + x_{t-2}.$$

(a) If $\{x_t\}$ is a linear function of t, that is, $x_t = \alpha + \beta t$, show that $\{\Delta x_t\}$ is a constant time series.

(b) If $\{x_t\}$ is a quadratic polynomial in t plus a stationary random component $\{z_t\}$, show that $\{\Delta^2 x_t\}$ is a stationary time series.

(c) (*Relationship between differencing and smoothing*). For a time series $\{x_t\}$ show that

$$x_t - \frac{1}{3}(x_{t-1} + x_t + x_{t+1}) = -\frac{1}{3}\Delta^2 x_{t-1}.$$

2.2 Let $\{X_t\}$ be a mean-zero stationary process with covariance and correlation functions $\{\gamma_k\}$ and $\{\rho_k\}$, respectively.

(a) For any scalar a and integer k set $Q(a, k) = E(X_{t+k} - aX_t)^2$, show that

$$Q(a, k) = \gamma_0(1 + a^2) - 2a\gamma_k.$$

(b) In the literature of geostatistics $Q(1, k)$, viewed as a function of k, is called the *variogram* of the process $\{X_t\}$. Show that

$$Q(1, k) = \gamma_0(1 - \rho_k).$$

(c) For fixed k, show that $Q(\cdot, k)$ is minimized at

$$a = \frac{\gamma_k}{\gamma_0} = \rho_k,$$

and the minimum is given by

$$Q(\rho_k, k) = \gamma_0(1 - \rho_k^2).$$

Discuss the statistical relevance of minimizing $Q(\cdot, k)$ and the value of a when predicting the future value X_{t+k} using the single past observation X_t.

(d) Consider predicting X_t using the past two observations X_{t-1} and X_{t-2} by minimizing the expected prediction error

$$Q(a_1, a_2) = E(X_t - a_1 X_{t-1} - a_2 X_{t-2})^2$$

over the scalars a_1 and a_2. Show that the minimizers satisfy

$$\begin{bmatrix} \gamma_0 & \gamma_1 \\ \gamma_1 & \gamma_0 \end{bmatrix} \begin{bmatrix} a_1 \\ a_2 \end{bmatrix} = \begin{bmatrix} \gamma_1 \\ \gamma_2 \end{bmatrix},$$

and express the minimum of $Q(\cdot, \cdot)$ in terms of γ_0, γ_1 and γ_2.

2.3 Let $x_1, \ldots, x_n$ be a time series data from the stationary process $\{X_t\}$ in Problem 2.2 and $S(a, k)$ be the sample analogue of $Q(a, k)$ given by

$$S(a,k) = \sum_{t=1}^{n-k} (x_{t+k} - a x_t)^2, \; k = 1, 2, \cdots, n-1.$$

(a) For a fixed k, find the minimizer $\hat{a}$ and the minimum of $S(\cdot, k)$. Compare the results with those in Problem 2.2 (c) and the formulas for $\hat{a}$ and $\hat{\rho}_1$ in (2.2).

(b) Compute and plot $S(1, k)$ versus k, for the lynx data.

(c) A sample analogue of $Q(a_1, a_2)$ in Problem 2.2 (d) is

$$S(a_1, a_2) = \sum_{t=2}^{n} (x_t - a_1 x_{t-1} - a_2 x_{t-2})^2.$$

Find the minimizer $\hat{a} = (\hat{a}_1, \hat{a}_2)$ and the minimum of $S(\cdot, \cdot)$. Compare the results with their theoretical counterparts in Problem 2.2 (d).

2.4 **(Projects)** Let $y_1, \ldots, y_n$ be a time series data from a nonstationary process $\{Y_t\}$.

(a) Suppose $E(Y_t) = \beta_0 + \beta_1 x_t$ for some known x_t's and unknown parameters β_0 and β_1. Consider predicting the next future value Y_{n+1} using

$$\hat{y}_{n+1} = \hat{\beta}_0 + \hat{\beta}_1 x_{n+1} + \hat{\beta}_2 y_{n-1},$$

where $\hat{\beta}_i$'s are the minimizers of

$$S(\beta_0, \beta_1, \beta_2) = \sum_{t=2}^{n} (y_t - \beta_0 - \beta_1 x_t - \beta_2 y_{t-1})^2.$$

Note that $\hat{\beta}_i$'s are simply the OLS estimators of the parameters of the following linear regression model

$$y_t = \beta_0 + \beta_1 x_t + \beta_2 y_{t-1} + e_t, \; t = 2, \cdots, n,$$

which when rewritten in matrix form looks just like (2.13), that is,

$$y = X\beta + e,$$

but with

$$y = \begin{bmatrix} y_2 \\ \vdots \\ y_n \end{bmatrix}, X = \begin{bmatrix} 1 & x_2 & y_1 \\ \vdots & \vdots & \vdots \\ 1 & x_n & y_{n-1} \end{bmatrix}, \beta = \begin{bmatrix} \beta_0 \\ \beta_1 \\ \beta_2 \end{bmatrix}, e = \begin{bmatrix} e_2 \\ \vdots \\ e_n \end{bmatrix},$$

where y, X and e have lost their first row (element) due to the appearance of the lagged term y_{t-1} as an explanatory variable in the model. Nevertheless, (2.13) and the subsequent results can be used to compute the OLS estimate $\hat{\beta}$ and draw inferences about the model. Apply this model and procedure to the SNOCOM data and compare your numerical results with those reported at the end of Section 2.3.1. Also, plot and compare the OLS residuals of the two models. Which one is closer to the ideal picture of stationarity? Which one is closer to a white noise?

(b) For the logarithm of the lynx data $y_1 \ldots, y_{114}$, the model

$$y_t = \beta_0 + \beta_1 \cos \frac{2\pi}{9.63}(t - 1.33) + \beta_2 y_{t-1} + \beta_3 y_{t-2} + e_t,$$

was suggested by Campbell and Walker (1977). Write this model in the matrix form $y = X\beta + e$, and

(i) identify the vector y and matrix X, compute $X'y$ and note the appearance of the cosine-transform of the data.

(ii) Compute the OLS estimate $\hat{\beta}$ and compare them with those for the models (2.45) and (2.46).

(iii) Plot and compare the OLS residuals for models (2.46) and the above AR(2) model with an additive harmonic component. Which one is closer to the ideal picture of stationarity? Which one is closer to a white noise?

(iv) Use the above model and the procedure in Section 2.5.4 to predict the last fourteen observations of the lynx data. Compare your numerical results with those in Table 2.3.

(c) For the logarithm of the airline data, denoted by y_t, the first difference Δy_t followed by the lag-12 difference Δ_{12} amounts to the following linear transformation of y_t:

$$\begin{aligned} \Delta_{12}\Delta(y_t) &= \Delta_{12}(y_t - y_{t-1}) = y_t - y_{t-1} - (y_{t-12} - y_{t-13}) \\ &= y_t - y_{t-1} - y_{t-12} + y_{t-13}. \end{aligned}$$

If this were to reduce the nonstationarity and orthogonalize the data, then the model

$$y_t = \beta_0 + \beta_1 y_{t-1} + \beta_2 y_{t-12} + \beta_3 y_{t-13} + e_t,$$

where $\{e_t\}$ is either stationary or a white noise, might provide a suitable model for the data. Verify this claim using the following steps:

(i) Write this 4-parameter model in the matrix form $y = X\beta + e$, identify y and X. How many elements of y are lost due to using lagged values as explanatory variables?

(ii) Compute the OLS estimate $\hat{\beta}$. Plot the OLS residuals and compare it with Figures 2.2 (c) and 2.3 (c). Which one is closer to the ideal picture of stationarity? Which one is closer to a white noise?

(iii) If the OLS residuals are not close to white noise but look stationary, compute and plot its ACF and PACF. Identify a model for the OLS residuals and modify the above 4-parameter model to a possibly larger one. Repeat (i) and (ii).

This iterative process is roughly the Box-Jenkins' three-stage statistical model-fitting process; see Section 2.4.3.

2.5 Show that if $\begin{pmatrix} \Sigma & X \\ X' & 0 \end{pmatrix}^{-1} = \begin{pmatrix} A & B' \\ B & C \end{pmatrix}$, then

$$\begin{aligned} A &= \Sigma^{-1} - \Sigma^{-1}X(X'\Sigma^{-1}X)^{-1}X'\Sigma^{-1}, \\ B &= (X'\Sigma^{-1}X)^{-1}X'\Sigma^{-1}, \\ C &= -(X'\Sigma^{-1}X)^{-1}, \end{aligned}$$

provided that the indicated matrices are invertible.

2.6 For the simple linear regression model (2.6), derive the OLS estimates $\hat{\beta}_0$ and $\hat{\beta}_1$ and the prediction error variance (2.9).

2.7 For the matrix U_n in (2.34) compute $U_n U_n^*$ and $U_n^* U_n$ and compare them with the $n \times n$ identity matrix I_n. Recall that for a matrix $U_n = (u_{k,t})$ with complex-valued entries, U_n^* is defined by $U_n^* = (\bar{u}_{t,k})$ where the overbar stands for the complex conjugate.

2.8 Show that the periodogram in (2.32) can be written as

$$I(\lambda) = n^{-1}x'A(\lambda)x,$$

for some matrix $A(\lambda)$. Find the matrix A and express it in terms of powers of the simple matrix

$$U = \begin{bmatrix} 0 & 1 & 0 & \cdots & 0 \\ 0 & 0 & 1 & \cdots & 0 \\ \vdots & \vdots & \vdots & & \vdots \\ 0 & 0 & 0 & \cdots & 1 \\ 1 & 0 & 0 & \cdots & 0 \end{bmatrix}.$$

Does $A(\lambda)$ have any recognizable structure? How about U?

2.9 Simulate $n = 200$ observations from the Yule's AR(2) model

$$X_t = 1.34X_{t-1} - 0.65X_{t-2} + \varepsilon_t,$$

where $\{\varepsilon_t\}$ is a Gaussian $WN(0, 1)$. Plot this series and examine the stationarity of the simulated data. Compute the sample correlations using (2.2) and plot the counterparts of the Figure 2.7 for this time series data set.

2.10 Repeat Problem 2.9 for the model (2.45), but with $e_t \sim WN(0, 0.059)$ and compare your plots with those in Figures 2.8–2.9.

2.11 Develop an analogue of Figure 2.13 for the SNOCOM component of the snow data.

2.12 Compute and compare the empirical predictive powers of models (2.46) and (2.54) for the lynx data.

Hint: Try to produce an analogue of Figure 2.14 for these models.

2.13 Given $x_1, \ldots, x_t$, the prediction rule for issuing a point prediction of X_{t+1} can be obtained by minimizing the discounted sum of squares

$$Q_t(p) = \sum_{j=0}^{t-1} \lambda^j (x_{t-j} - p)^2.$$

(a) Show that the minimizer of $Q_t(\cdot)$ is of the form

$$p_{t+1} = c_t \sum_{j=0}^{t-1} \lambda^j x_{t-j},$$

find the c_t in terms of λ.

(b) Show that when $t \to \infty$, the p_t satisfies the recursion

$$p_{t+1} = (1 - \lambda) x_{t+1} + \lambda p_t.$$

2.14 For a time series data $x_1, \ldots, x_n$, one can devise many prediction rules to issue forecast of the next future value X_{n+1}. The *least square forecasting system* is based on minimizing (2.51).

(a) Derive the normal equations for minimizing (2.51) and compare them with (2.52).

(b) For $m = 2$, write and solve these equations for the lynx data. Compare the estimated coefficients and variances with those in (2.45).

(c) Compare the least square and Yule-Walker forecasting systems based on their empirical performance in issuing forecast for the last 14 observations of the lynx data.

2.15 The CAT (criterion autoregression transfer function) due to Parzen (1974) is defined by

$$\mathrm{CAT}(m) = \begin{cases} -(1 + n^{-1}), & m = 0 \\ \frac{1}{n} \sum_{j=1}^{m} \frac{1}{\tilde{\sigma}_j^2} - \frac{1}{\tilde{\sigma}_m^2}, & m = 1, 2, \cdots, M, \end{cases}$$

where $\tilde{\sigma}_j^2 = (1 - \frac{j}{n})^{-1}\hat{\sigma}_j^2$.

(a) Show that

$$\text{CAT}(m) - \text{CAT}(m-1) = \frac{1}{\tilde{\sigma}_{m-1}^2}\{\text{AIC}(m) - \text{AIC}(m-1)\} + r_n,$$

where r_n is a very small quantity for n large. This implies that, ignoring the last term, CAT(m) and AIC(m) rise and fall together and would be minimized at the same value.

(b) Compute and plot CAT$(\cdot)$ for the lynx data. Compare the plot and the minimum value with that for the AIC$(\cdot)$ in Figure 2.13 (b).

3

Time Series Analysis: Many Short Series

Much of the statistical time series analysis is concerned with the statistical inference from a single long realization of a stochastic process where in-built replication is usually provided by invoking stationarity and some kind of asymptotic independence (ergodicity). By contrast, in this chapter we deal with a number of short series of fixed lengths where instead the number of replications may tend to infinity.

Such data may arise in studies in social, environmental, epidemiological, medical, biological and physical sciences where many experimental units (human and animal subjects, households, firms, schools, ...) are measured repeatedly over time giving rise to many short time series. These studies, often referred to as *repeated measures, longitudinal or panel studies*, have the broad goal of understanding and detecting changes over time within subjects and relating them to treatments, interventions, group conditions and other relevant covariates using the idea of regression. They have been the subject of intense study in the last two decades; Diggle, Liang and Zeger (1994) and Lindsey (1999) provide excellent review of the literature and models developed for this emerging area of statistics that lies at the intersection of time series analysis and multivariate statistics.

Since repeated measures on the same subject are correlated, it is prudent to take account of this dependence properly when fitting regression models so that valid scientific inferences can be drawn. To appreciate the nature and direction of dependence (across units and/or over time), let us denote by y_{it} the measurement made on the ith unit at the tth occasion as displayed in Table 3.1. The designs in the longitudinal studies are so that the units or the rows in Table 3.1 can be viewed as realizations of independent random vectors with different means and a common $n \times n$ covariance matrix. The dichotomy of a small number of long series versus a large number of

Table 3.1 The shape of a rectangular repeated measure data

		Occasion					
		1	2	$\cdots$	t	$\cdots$	n
	1	y_{11}	y_{12}	$\cdots$	y_{1t}	$\cdots$	y_{1n}
	2	y_{21}	y_{22}	$\cdots$	y_{2t}	$\cdots$	y_{2n}
Unit	$\vdots$	$\vdots$	$\vdots$		$\vdots$		$\vdots$
	i	y_{i1}	y_{i2}	$\cdots$	y_{it}	$\cdots$	y_{in}
	$\vdots$	$\vdots$	$\vdots$		$\vdots$		$\vdots$
	m	y_{m1}	y_{m2}	$\cdots$	y_{mt}	$\cdots$	y_{mn}

short series provides a useful framework for bringing time series methodology to bear on the longitudinal data, spatial data and other similar situations dealing with correlated data.

As in Chapter 2 our focus is on modeling the dependence. The first challenge in identifying models for the covariance matrix of data from longitudinal and panel studies is the apparent nonstationarity of the repeated measurements for each subject. This challenge is usually met by assuming that the subjects are independent and homogeneous; thus, inferences about their common covariance matrix can be made by borrowing strength across many subjects (pooled time series analysis). The second challenge is the awkward positive-definiteness constraint on a covariance matrix, this can be dealt with by introducing an unconstrained reparameterization of the covariance matrix. The key technical idea is to decompose a nonstationary covariance matrix into two components capturing its "variance" and "dependence" structures. For the first component we follow the successful methodology for modeling the mean (first moments) using the regression techniques. For modeling the dependence (second moments), however, we develop some new graphic methods based on special triangular factorization of the covariance matrix which guarantees the positive-definiteness of the estimated covariance matrix. We provide details of identifying models for the nonstationary covariance matrix using graphical techniques which are generalizations and enhancements of the correlogram and partial correlogram developed in Chapter 2. We show that these graphical tools though initially developed (Yule, 1927; Quenouille, 1949) for the parsimonious modeling of one long stationary series are applicable to analogous modeling of several independent homogeneous short nonstationary series.

The notation and basic parametric framework for modeling the mean vectors and covariance matrices and a real dataset are introduced in Section 3.1. Specific parametric covariance structures such as AR(1), compound symmetry, linear covariance models and linear mixed models are introduced in Section 3.2. The sample correlation matrix and its relation with the scatterplot matrix are discussed in Section 3.3. The natural generalizations of the correlogram, partial and inverse correlograms to the nonstationary setup and their roles in identifying parsimonious models for covariances are discussed in Section 3.4. Unfortunately, a drawback of covariance models identified and estimated without heeding to the positive-definiteness constraint is that the estimated covariance matrix is not guaranteed to be positive-definite. This difficulty is resolved in Section 3.5 by introducing an unconstrained and statistically meaningful reparameterization of a covariance matrix using its modified Cholesky decomposition and a genuine enhancement of the generalized partial correlograms. Certain nonstationary analogue of AR models known as antedependence models (Gabriel, 1962) and their roles in modeling the covariance of longitudinal data are discussed in Section 3.6 and applied to a real dataset.

3.1 THE MODELING FRAMEWORK, DATA AND EXAMPLES

In this section we introduce the notation and basic parametric framework for modeling the mean vectors and covariance matrices along with a data set and some concrete examples. Our focus, however, is on developing parsimonious models for the covariance matrices.

3.1.1 The Parametric Modeling Framework

Consider a longitudinal study with m subjects. Let $\boldsymbol{y}_i = (y_{i1}, \ldots, y_{in_i})$ be the vector of n_i measurements on the ith subject collected at the measurement times $\boldsymbol{t}_i = (t_{i1}, \ldots, t_{in_i})$. The set of measurement times in the study is usually called the *measurement schedule*. Measurement times may be unequally spaced within a subject and may differ across subjects. If measurement times are common across subjects, we call the measurement schedule *rectangular*. Although experiments are designed to have rectangular schedules, dropouts and other forms of incomplete data can ruin the rectangularity of the schedules and hence have important implications for modeling the covariance structure. Associated with each measurement y_{it} we have a p-vector of covariates x_{it} so that the design matrix corresponding to the n_i measurements $\boldsymbol{y}_i$ is the $n_i \times p$ matrix $X_i = (x_{i1}, \ldots, x_{in_i})'$.

For modeling the mean we work within the framework of linear regression models discussed in Chapter 2. More precisely, for the subject i and a generic measurement time t we assume that the response y_{it} and the covariates x_{it} are related linearly, that is,

$$y_{it} = x'_{it}\beta_i + \varepsilon_{it}. \tag{3.1}$$

Consequently, for each subject we have a multiple linear regression model,

$$\boldsymbol{y}_i = X_i\beta_i + \varepsilon_i, \; i = 1, \cdots, m, \tag{3.2}$$

where β_i is a p-vector of regression coefficients and $\varepsilon_i = (\varepsilon_{i1}, \ldots, \varepsilon_{ini})$ is the vector of possibly dependent error terms. For the time being, it is convenient to think of $\boldsymbol{y}_i$'s as the realizations of normal random vectors

$$\boldsymbol{Y}_i \sim N_{n_i}(X_i\beta_i, V_i(\theta)), \; i = 1, \cdots, m, \tag{3.3}$$

where $V_i(\theta)$ is a $n_i \times n_i$ covariance matrix and θ is a q-vector of unknown covariance parameters constrained so that $V_i(\theta)$ is a positive-definite matrix. It is convenient to think of y_{ij} as a realization of a random variable $Y_i(t_{ij})$ where $\{Y_i(t);\ t \in R\}$, $i = 1, \ldots, m$, are a set of mutually independent Gaussian stochastic processes with the generic covariance function $\gamma_{s,t} = \text{Cov}(Y_i(s),\ Y_i(t))$ determined by the q-vector of parameters θ.

Let $\boldsymbol{y} = (\boldsymbol{y}_1', \ldots, \boldsymbol{y}_m')'$ be the complete set of $N = \sum_{i=1}^{m} n_i$ measurements from m subjects, $X = (X_1', \ldots, X_m')'$ and Σ be a block-diagonal matrix with non-zero $n_i \times n_i$ blocks $V_i(\theta)$ corresponding to the covariance matrix of the vector of measurements on the ith subject. The β_i's are not usually the same, but when they are the same and equal to β, one may pool the available data from all subjects to form the model

$$\boldsymbol{y} = X\beta + \varepsilon. \tag{3.4}$$

In this case, the OLS estimator of β is $\hat{\beta} = (X'X)^{-1}X'\boldsymbol{y}$ and its GLS estimator for known Σ is $\hat{\beta}_{GLS} = (X'\Sigma^{-1}X)^{-1}X'\Sigma^{-1}\boldsymbol{y}$. Even if the normality assumption (3.3) is violated, still one can use twice the negative of the pseudo-loglikelihood for the observed data $\boldsymbol{y}$,

$$L(\beta, \theta|\boldsymbol{y}) = \sum_{i=1}^{m} \left\{\log|V_i| + (\boldsymbol{y}_i - X_i\beta)'V_i^{-1}(\boldsymbol{y}_i - X_i\beta)\right\},$$

as the objective function for estimation of β and θ. Note that this is just like the likelihood in (2.30), but added over the contributions of m independent subjects, so that the details of iterative estimation of β and θ generally proceeds as in Section 2.3.4.

In what follows, we assume that β is known and focus on modeling the dependence or the covariance function of the data. We note that when the measurement schedule is rectangular then $V_i = V$ and $\Sigma = I_m \otimes V$, where $\otimes$ stands for the Kronecker product of two matrices. Specifically, for $A = (a_{ij})$ and $B = (b_{ij})$ their Kronecker product $A \otimes B = (a_{ij}B)$.

3.1.2 The Cattle Data

The cattle data will be used in this chapter to illustrate various methods for modeling a covariance matrix and the nature of nonstationarity prevalent in practice. Kenward (1987) provides the data and describes an experiment for controlling intestinal parasites in cattle. The experiment was conducted by randomly assigning 60 animals to two treatments A and B with $m = 30$ animals in each group. The weights of the animals was recorded every two weeks during the grazing season, except the last observation that was taken after one week interval. The total number of periods considered was $n = 11$. The measurement times were common across animals and are rescaled so that $\boldsymbol{t}_i = \boldsymbol{t} = (1, 2, \ldots 10, 10.5)$. Note that the measurement times are not equally-spaced, yet the measurement schedule is rectangular.

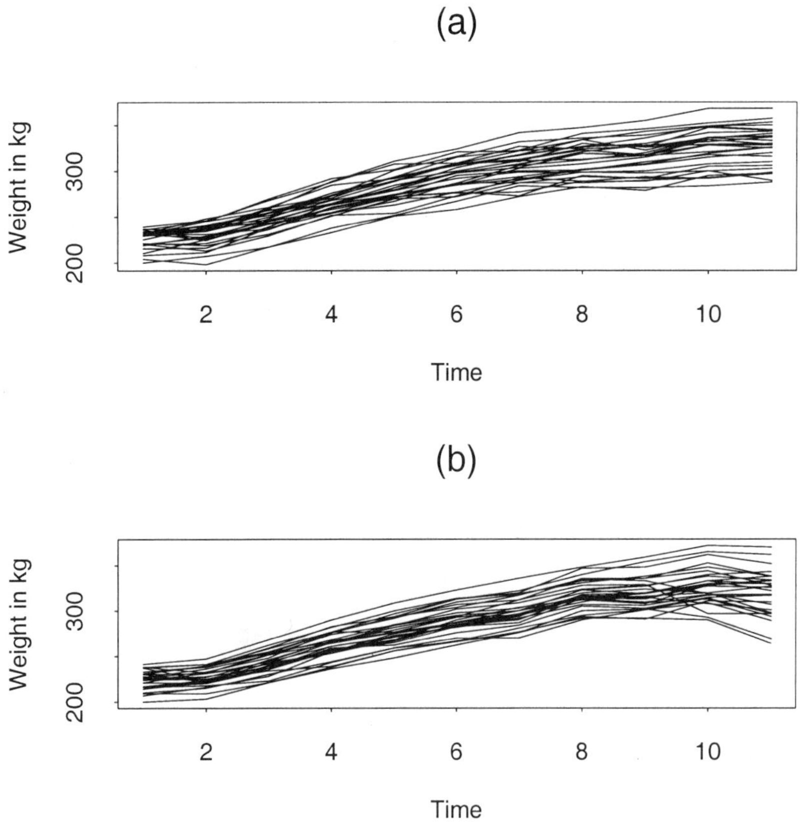

Fig. 3.1 Profile plots of the raw cattle data, (a) treatment A, (b) treatment B.

Similar to the role of scatterplot in regression analysis, for fixed i the plot of y_{ij}'s versus t_{ij}'s called the *profile plot* of the ith subject could reveal a great deal about the time-evolution of the response variable. Figures 3.1 and 3.2 are the profile plots of the raw and standardized data, respectively. It is evident from the first that the variability is increasing over time for both treatments and it is more so for group B at and beyond the 10th measurement. However, it is difficult to discern the nature of the correlation in the data. To resolve this, one could adjust for the means and variances over time by standardizing the measurements and examining their profile plots or equivalently the associated *parallel axis system* (Dawson et al., 1997) in Figure 3.2. Here presence of many nearly parallel profiles suggests the existence of positive correlation among the measurements.

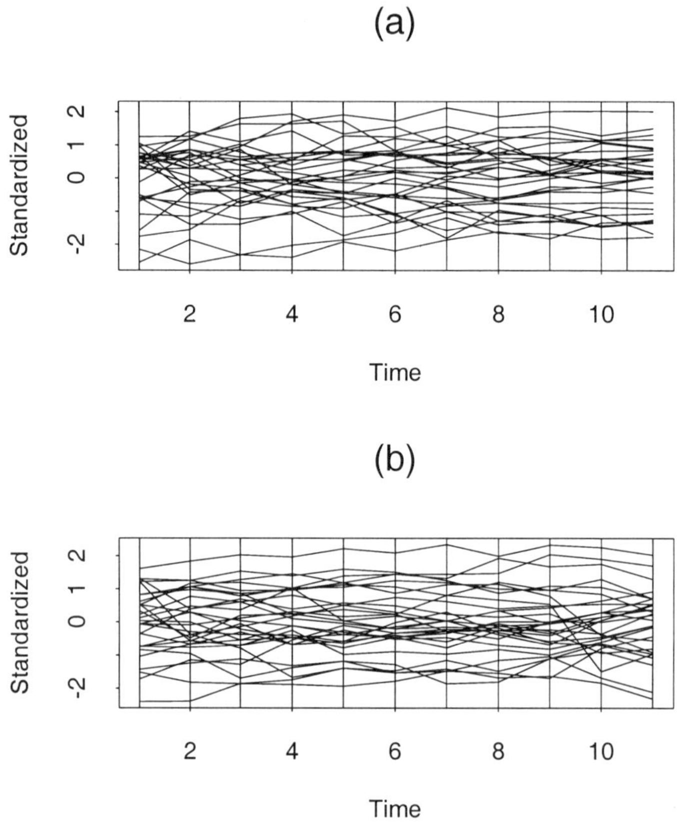

Fig. 3.2 Profile plots (parallel axis systems) of the standardized cattle data, (a) treatment A, (b) treatment B.

3.1.3 Additional Examples of Repeated Measures Data

Repeated measures data from business, economics and social sciences are usually called panel data where it is common to have $n_1 = \ldots = n_m = n$ and $t_{ij} = j$, $j = 1, \ldots, n$, see Table 3.1. In the most basic form of panel studies the columns of the table are viewed as realizations of independent random vectors with different means and a common $m \times m$ covariance matrix, however, lately it is becoming more common in such studies to allow intercorrelation in each row as well. This more general dependence scheme provides feasible models for studies in environment, ecology and biology as well where neighboring sites and countries could have impact on each other contemporaneously and temporally.

When β_i's are different by stacking up the data as before or models (3.2) for the m units we arrive at

$$y = X^* \beta_a + \varepsilon, \tag{3.5}$$

where $\boldsymbol{y} = (y_1', y_2', \ldots, y_m')', X^* = \text{diag}(X_1, X_2, \ldots, X_m), \beta_a = (\beta_1', \beta_2', \ldots, \beta_m')'$, $\varepsilon = (\varepsilon_1', \varepsilon_2', \ldots, \varepsilon_m')'$ where $\boldsymbol{y}$ and ε are of the same dimension $N = mn$. In the panel studies the errors ε_{it}'s are assumed to be normally distributed with zero mean, serially uncorrelated and possibly contemporaneously correlated, so that $\text{Cov}(y) = \text{Cov}(\varepsilon) = \Sigma \otimes I_n$, where now $\Sigma = (\sigma_{st})$ stands for the $m \times m$ contemporaneous covariance matrix of the tth column in the Table 3.1, hereafter denoted by $y_{(t)}$, $t = 1, \ldots, n$. This contemporaneous dependence is usually justified because of the existence of common market forces influencing all firms or some of them in certain industry sectors. The model (3.5) also known as the *seemingly unrelated regressions* (SUR) has been used extensively in the econometrics literature since its introduction by Zellner (1962). The interest in SUR models lies in the fact that the m regression models in (3.2) are correlated. Consequently, testing a relationship between the various β_i's will force one to treat the models as a multivariate regression model (3.5). Three celebrated examples of SUR models fitted to panel data are discussed in the rest of this section.

Capital Asset Pricing Model: The capital asset pricing model (CAPM) states that the excess stock returns from individual firms are linearly related to the excess market return without an intercept (Sharpe, 1964). The excess stock return is defined as the nominal return minus the risk free rate, usually the one month Treasury bill rate is used as a proxy for the risk free rate. The multivariate CAPM is a simpler form of (3.2):

$$y_{(t)} = \beta x_t + \varepsilon_{(t)} = X_t \beta + \varepsilon_{(t)},$$

where $y_{(t)}$ is the $m \times 1$ vector of monthly excess returns at time t for m firms, $\beta = (\beta_1, \ldots, \beta_m)$ is the $m \times 1$ vector of regression coefficients and β_i is called the *systematic risk* of the ith firm, x_t is the scalar excess market return at time t, $X_t = x_t I_m$ is a $m \times m$ diagonal matrix and $\varepsilon_{(t)}$ is the $m \times 1$ error vector at time t with $\text{Cov}(\varepsilon_{(t)}) = \Sigma$. In this case the computation of the OLS and GLS estimators of β are simpler since

$$X'(\Sigma^{-1} \otimes I)X = \Sigma^{-1} \sum_{t=1}^{n} x_t^2,$$
$$X'(\Sigma^{-1} \otimes I)Y = \Sigma^{-1} \sum_{t=1}^{n} x_t y_{(t)}.$$

Smith and Kohn (2000) have analyzed the data for $m = 89$ firms on S & P 100 that traded continuously between November 1986 and November 1996 with $n = 120$ observations per firm.

Dynamic Panel-Data Models: Dynamic models are often employed in econometrics to generate forecasts of the growth rate of real gross national product (GNP) for a group of countries such as the Organization for Economic Cooperation and Development (OECD) over various time periods, Min and Zellner (1993); Hoogstrate,

Palm and Pfann (2000) have used the following model for the OECD countries:

$$\begin{aligned} y_{it} &= \beta_{0i} + \beta_{1i} y_{it-1} + \beta_{2i} y_{it-2} + \beta_{3i} y_{it-3} + \beta_{4i} x_{1,it-1} \\ &\quad + \beta_{5i} x_{1,it-2} + \beta_{6i} x_{2,it-1} + \beta_{7i} x_{3,it-1} + \varepsilon_{it}, \end{aligned} \tag{3.6}$$

where

y_{it} is the first difference of the logarithm of the real output of country i in period t;
$x_{1,it}$ is the first difference of the logarithm of a stock price index divided by a general price index;
$x_{2,it}$ is the first difference of the logarithm of the nominal money supply divided by a general price index;
and
$x_{3,it}$ is the world return, which equals the median of countries real stock return in period t.

Note that the above model reduces to the standard AR(3) model when the $\beta_{4i} = \beta_{5i} = \beta_{6i} = \beta_{7i} = 0$. The 8×1 vector of covariates $x_{it} = (1, y_{it-1}, y_{it-2}, y_{it-3}, x_{1,it-1}, x_{1,it-2}, x_{2,it-1}, x_{3,it-1})$ in this example is allowed to contain lagged values of the response y_{it}. Hoogstrate et al. illustrate the gains from pooling the models (3.6) when n is small by performing the actual one-step-ahead forecasts for the period 1983–1990. Chib and Greenberg (1995) have carried out a hierarchical analysis of more general SUR models with serially correlated errors and time-varying parameters. These models could be employed for the analysis of the next case.

Intra-day Electricity Demand: Utility companies use various regression models to forecast demand at an intra-day level to guide their day-to-day operational decisions and determine any pricing policies. One popular method is to estimate separate regression equations for each intra-day period and allow the errors to be correlated across equations. Smith (2000) has developed a SUR model for the total New South Wales (NSW) electricity demand observed during the $n = 15$ working days of the three week period of March 8–28, 1993. Longer lengths of data are not viable because the parameters in the model vary over season and are only effectively static over periods of two to three weeks. Using half-hourly average demand data, we have $m = 48$ regression equations with an unknown 48×48 error covariance matrix Σ. Each regression features a constant μ_i along with a temperature effect based on the average half-hourly temperature T_{it} measured at the Sydney central business district, which is the central point of NSW electricity demand. The temperature effect is modeled as an inverted V with the point of inversion at 18.3° C (65° F), which is a commonly used nonlinear functional form for the relationship between electricity demand and temperature (Engle, Granger, Rice and Weiss 1986). Thus, the model is

$$y_{it} = \mu_i + \beta_i |T_{it} - 18.3| + \varepsilon_{it}, \; i = 1, \cdots, 48, \; t = 1, \cdots, 15,$$

where y_{it} is the NSW electricity demand at half hour i on day t. Using $n = 15$ vector observations to estimate a 48×48 covariance matrix or $\dfrac{48 \times 49}{2} = 1176$ unknown parameters is difficult unless a parsimonious parametric structure for Σ is postulated or formulated using the data at hand. The regression models above can be stacked and rewritten as in (3.5) with $\boldsymbol{y} = (y_1', \ldots, y_m')'$, $\beta = (\mu_1, \beta_1, \mu_2, \beta_2, \ldots, \mu_m, \beta_m)'$, X the appropriate $nm \times 2m$ sparse design matrix and $\text{Cov}(\varepsilon) = I_m \otimes \Sigma$. For details of the computation and model fitting process, see Smith (2000) and Smith and Kohn (2000).

3.2 STRUCTURED COVARIANCE MATRICES

As pointed out earlier, in longitudinal studies measurements made on the same subject over time are correlated and the goal is to relate the means of the responses to a set of covariates such as time and treatments, and making inferences about the regression parameters. An ongoing debate in the formulation of models for repeated measure data is the extent to which one should impose a specific parametric structure on the generic covariance function $\gamma_{s,t}$ and hence on the V_i's in (3.3). Our ultimate goal is to use the data and provide graphic tools that are useful in formulating parsimonious models for the covariance matrices. A very tight specification of the covariance matrix risks invalidating the inference about the mean structure, which is usually of primary interest, whereas too loose a specification leads to a loss of efficiency, since then some of the information in the data is wasted by estimating redundant nuisance parameters. An appropriate balance between these two extremes is certainly desirable in practice, but reaching such balance depends among other things on the sizes of m and n_i's. For large m and small n_i's, an unstructured V_i is feasible as advocated, for example, in the generalized estimating equations (GEE) method (Liang and Zeger, 1986). However, when m is small and the n_i's are relatively large it is essential to impose structural restrictions on V_i's to reduce the dimension of θ. Most parametric models in the literature is based on the assumption that the variation about the mean response is stationary. When nonstationary variation is allowed, it is of a highly structured form, as in the random effect models of Laird and Ware (1982) and the antedependence models of Gabriel (1962) and Kenward (1987) and the first difference models of Cullis and McGilchrist (1990).

A popular way to reduce the number of parameters of the common covariance matrix is to use specific parametric covariance models. The following covariance matrices with 2 or 3 parameters are used frequently in the literature of longitudinal data and other areas of statistics dealing with correlated data:

Compound Symmetry (CS):

$$\Sigma = \sigma^2(I_n + \rho \mathbf{1}_n \mathbf{1}_n'), \text{ where } \sigma^2 > 0, \ -\frac{1}{n-1} < \rho < 1 \text{ and } \mathbf{1}_n = (1, \ldots, 1).$$

This is also known as the exchangeable or the intra-block covariance matrix with two restricted parameters $\theta = (\rho, \sigma^2)$, see Example 2.2.

$AR(1)$:

$\Sigma = \sigma^2(\rho^{|i-j|})$, $\sigma^2 > 0$, $-1 < \rho < 1$, contains two parameters $\theta = (\rho, \sigma^2)$. This corresponds to the covariance matrix of an AR(1) model introduced in Chapter 2.

Damped Exponential (DE):

For $t_1 < t_2 < \ldots < t_i < \ldots < t_n$ denoting the measurement times on a typical subject, the $n \times n$ covariance matrix given by

$$\Sigma = \sigma^2 \left(\rho^{(t_i - t_{i-j})^\gamma}\right)$$

with $\theta = (\rho, \gamma, \sigma^2)$ having three parameters, is known as the *damped exponential* covariances (Munoz et al., 1992). Evidently this is more general than the AR(1) covariance structure.

Note that these covariances correspond to assuming stationarity of the repeated measures. However, stationarity forces the variances to be constant over time and equidistant measurements to be equicorrelated, features that are seldom present in the longitudinal and growth studies. Consequently, as partial remedy heterogeneous versions of most stationary models (Wolfinger, 1996) can be introduced by allowing distinct variances along the main diagonal of the covariance matrix. Thus, if $\boldsymbol{\rho} = (\rho_{|i-j|})$ is a stationary correlation matrix and $\boldsymbol{\sigma} = \text{diag}(\sqrt{\sigma_{11}}, \ldots, \sqrt{\sigma_{nn}})$ is a diagonal matrix of standard deviations, a *heterogeneous covariance matrix* is of the form

$$\Sigma = \boldsymbol{\sigma}\boldsymbol{\rho}\boldsymbol{\sigma}. \tag{3.7}$$

In general, it contains $n + n - 1 = 2n - 1$ parameters, a considerable reduction compared to the $\dfrac{n(n+1)}{2}$ parameters of an unstructured covariance matrix. Although one may reduce the number of parameters of $\boldsymbol{\rho}$ to just one, $\boldsymbol{\sigma}$ is usually left unstructured in the current implementation of SAS (SAS Institute, Inc., 1997).

A more flexible class of nonstationary models than the class of heterogeneous covariance matrices introduced next deserves to be studied further and used more often as candidates to model dependence in longitudinal data.

Linear Covariance Models:

Consider the class of *linear covariance models* (Anderson, 1973) defined by

$$\Sigma = \theta_1 A_1 + \cdots + \theta_q A_q, \tag{3.8}$$

for some integer q, known symmetric matrices $A_1, \ldots, A_q$ and the vector of unknown parameters $\theta = (\theta_1, \ldots, \theta_q)$ is constrained so that Σ is positive-definite. Except for the awkward positive-definiteness constraint, (3.8) is the analogue of the linear models for the mean. This class referred to as LIN(q) in the MIXED procedures (SAS

Institute, Inc., 1997, p.626) is general enough to include many specific covariance structures provided in SAS; see Problem 3.2. In fact, any $n \times n$ covariance matrix can be written as LIN(q) with $q = \dfrac{n(n+1)}{2}$. To this end, note that for $n = 2$, we have

$$\begin{pmatrix} \sigma_{11} & \sigma_{12} \\ \sigma_{12} & \sigma_{22} \end{pmatrix} = \sigma_{11} \begin{pmatrix} 1 & 0 \\ 0 & 0 \end{pmatrix} + \sigma_{22} \begin{pmatrix} 0 & 0 \\ 0 & 1 \end{pmatrix} + \sigma_{12} \begin{pmatrix} 0 & 1 \\ 1 & 0 \end{pmatrix},$$

which is of the form (3.8), with $\theta = (\sigma_{11}, \sigma_{22}, \sigma_{12})$ simply the vector of the entries of Σ. We shall develop graphical techniques analogous to the stationary correlograms to identify q and the matrices $A_1, \ldots, A_q$ for a given dataset. The procedure is similar to ARMA model formulation in time series analysis where the sample covariance matrix is the main input. However, one must pay due attention so that the estimated covariance matrix is positive-definite.

Linear Mixed Models:

Consider the *linear mixed (random effects) model*

$$\boldsymbol{y}_i = X_i\beta + Z_i\boldsymbol{b}_i + \varepsilon_i, \; i = 1, \cdots, m, \tag{3.9}$$

where X_i and Z_i are known design matrices of orders $n_i \times p$ and $n_i \times q$, respectively, β is the fixed $p \times 1$ vector of unknown (nonrandom) regression parameters and the $q \times 1$ vectors $\boldsymbol{b}_i$ are random effects with $E(\boldsymbol{b}_i) = 0$ and $\text{Cov}(\boldsymbol{b}_i) = \sigma^2 D$. Finally, the ε_i are $n_i \times 1$ vectors of random errors whose elements are correlated with $\text{Cov}(\varepsilon_i) = \sigma^2 R_i$, and for $i \neq i'$, we have $\text{Cov}(\varepsilon_i, \varepsilon_{i'}) = 0$, $\text{Cov}(\boldsymbol{b}_i, \boldsymbol{b}_{i'}) = 0$, and $\text{Cov}(\boldsymbol{b}_i, \varepsilon_j) = 0$ for all i and j. These assumptions on $\boldsymbol{b}_i$ and ε_i seem reasonable where subjects are assumed to be independent, but the repeated measures on a given subject are correlated; see Laird and Ware (1982); Searle et al. (1992). From these assumptions and (3.9) we obtain

$$\text{Cov}(\boldsymbol{y}_i) = \sigma^2 Z_i D Z_i' + \sigma^2 R_i = \sigma^2 (Z_i D Z_i' + R_i), \tag{3.10}$$

which gives rise to a variety of covariance structures by varying the triplet (Z_i, D, R_i).

For example, the choice $(\mathbf{1}_{n_i}, 1, \boldsymbol{I}_{n_i})$ gives rise to the compound symmetry covariance structure introduced earlier; the choices $([\mathbf{1}_{n_i}, t_i], D, \boldsymbol{I}_{n_i})$ and $([\mathbf{1}_{n_i}, t_i, (t_{i1}^2, \ldots, t_{in_i}^2)'], D, \boldsymbol{I}_{n_i})$ give rise to the linear and quadratic random coefficient models for the means, respectively, and parsimonious covariance structures with nonconstant variances and nonstationary correlations.

The m submodels in (3.9) can be stacked as before to give a single model

$$\begin{bmatrix} \boldsymbol{y}_1 \\ \boldsymbol{y}_2 \\ \vdots \\ \boldsymbol{y}_m \end{bmatrix} = \begin{bmatrix} X_1 \\ X_2 \\ \vdots \\ X_m \end{bmatrix} \beta + \begin{pmatrix} Z_1 & 0 & \cdots & 0 \\ 0 & Z_2 & \cdots & 0 \\ \vdots & \vdots & \ddots & \vdots \\ 0 & 0 & \cdots & Z_m \end{pmatrix} \begin{bmatrix} \boldsymbol{b}_1 \\ \boldsymbol{b}_2 \\ \vdots \\ \boldsymbol{b}_m \end{bmatrix} + \begin{bmatrix} \varepsilon_1 \\ \varepsilon_2 \\ \vdots \\ \varepsilon_m \end{bmatrix},$$

or with obvious notation,

$$\boldsymbol{y} = X\beta + Z\boldsymbol{b} + \varepsilon, \tag{3.11}$$

where

$$\text{Cov}(\boldsymbol{b}) = \sigma^2 I_m \otimes D = \sigma^2 G,$$
$$\text{Cov}(\varepsilon) = \text{diag}(\sigma^2 R_1, \cdots, \sigma^2 R_m) = \sigma^2 R,$$

and hence

$$\text{Cov}(\boldsymbol{y}) = \sigma^2 (ZGZ' + R).$$

For the moment, assuming that G and R are known, the GLS estimate of $(\beta, \boldsymbol{b})$ in (3.11) can be found by solving the system of *mixed model equations*:

$$\begin{bmatrix} X'R^{-1}X & X'R^{-1}Z \\ Z'R^{-1}X & Z'R^{-1}Z + G^{-1} \end{bmatrix} \begin{bmatrix} \hat{\beta} \\ \hat{\boldsymbol{b}} \end{bmatrix} = \begin{bmatrix} X'R^{-1}y \\ Z'R^{-1}y \end{bmatrix}. \tag{3.12}$$

These normal equations can be obtained using (2.23) and (3.11). If in addition, multivariate normality is assumed for $\boldsymbol{b}_i$ and ε_i $i = 1, \ldots, m$, the $\hat{\beta}$ and $\hat{\boldsymbol{b}}$ are also the MLE and maximum likelihood predictor of β and $\boldsymbol{b}$, respectively. For more information on (3.12) and MLE of β and $\boldsymbol{b}$, see Searle et al. (1992).

3.3 SAMPLE COVARIANCE AND CORRELATION MATRICES

The estimated nonstationary correlation and covariance matrices described below are the key ingredients for subsequent analysis. We assume that the measurement schedule is rectangular and that there is a relatively small portion of missing data occurring toward the end of the study (dropouts) which are ignorable (Diggle et al., 1994, Chapter 11).

In estimating the correlation matrix, it is prudent to use saturated models for the means and variances. Let $\bar{y}_j$ and s_j be the sample mean and sample standard deviation, respectively, of the nonmissing responses at time t_j. Then, the standardized responses $\frac{y_{ij} - \bar{y}_j}{s_j}$ can be used to estimate and suggest structures for the underlying correlation matrix $\boldsymbol{\rho} = (\rho_{jk})$; see Figure 3.2. Its (j, k)th element is estimated using the Pearson's product-moment formula

$$\hat{\rho}_{jk} = \frac{\sum_{i=1}^{m} (y_{ij} - \bar{y}_j)(y_{ik} - \bar{y}_k)}{s_j s_k}, \tag{3.13}$$

where the summation is over the nonmissing pairs of observations. The sample covariance matrix $\hat{\Sigma}$ of the data is obtained either directly or by using the standard factorization formula (3.7),

$$\hat{\Sigma} = SRS, \tag{3.14}$$

where $R = (\hat{\rho}_{jk})$ is the sample correlation matrix and $S = \text{diag}(s_1, \ldots, s_n)$ is a diagonal matrix of sample standard deviations. The sample variances and correlations of the Group A cattle data are given in Table 3.2.

Table 3.2 Sample variances (along the main diagonal), correlations (below the main diagonal), fitted correlations using model (3.16)) (above the main diagonal) and fitted variances (last row) using OLS for the group A cattle

106	0.94	0.89	0.84	0.79	0.74	0.68	0.63	0.58	0.53	0.48
0.82	155	0.94	0.89	0.84	0.79	0.74	0.68	0.63	0.58	0.53
0.76	0.91	165	0.94	0.89	0.84	0.79	0.74	0.68	0.63	0.58
0.66	0.84	0.93	185	0.94	0.89	0.84	0.79	0.74	0.68	0.63
0.64	0.80	0.88	0.94	243	0.94	0.89	0.84	0.79	0.74	0.68
0.59	0.74	0.85	0.91	0.94	284	0.94	0.89	0.84	0.79	0.74
0.52	0.63	0.75	0.82	0.87	0.93	307	0.94	0.89	0.84	0.79
0.53	0.67	0.77	0.84	0.89	0.94	0.93	341	0.94	0.89	0.84
0.52	0.60	0.71	0.77	0.84	0.90	0.93	0.97	389	0.94	0.89
0.48	0.58	0.70	0.73	0.80	0.86	0.88	0.94	0.96	470	0.94
0.48	0.55	0.68	0.71	0.77	0.83	0.86	0.92	0.96	0.98	445
124	143	165	191	222	256	297	343	397	460	495

Formal inferences regarding entries of a covariance matrix are conducted using the following asymptotic results which are analogue of Theorem III. For a sample of size m from a $N(\mu, \Sigma)$ population, let $\hat{\rho}_{st}(\hat{\pi}_{st})$ stand for the sample correlation (partial correlation) between y_s and y_t. It is known (Anderson, 1984, p.123 and p.133) that for m large, $\hat{\rho}_{st} \sim N(0, 1/m)[\hat{\pi}_{st} \sim N(0, \frac{1}{m})]$ if $\rho_{st} = 0(\pi_{st} = 0)$. These can be used to construct approximate pointwise confidence intervals and test the significance of the population (partial) correlations. For example, the horizontal dotted lines in Figure 3.4 (a) and (c) correspond to the approximate 95% confidence intervals $(\pm\frac{2}{\sqrt{m}})$ for π_{st}'s.

All computations here are based on sample correlation and covariance matrices computed as above. However, there are many alternatives and improvements of these estimators that one could use instead of R and $\hat{\Sigma}$ from (3.14). These alternatives include the nonparametric methods (Diggle & Verbyla, 1998), Bayesian and shrinkage methods (Yang and Berger, 1994; Brown, Le and Zidek, 1994; Barnard, McCulloch and Meng, 2000). Diggle and Verbyla (1998) provide an excellent review of the literature and a viable nonparametric method for estimating covariance matrices of severely unbalanced longitudinal datasets using variogram of squared residuals from a saturated model for the mean response. Barnard et al. (2000) provide ample evidence in favor of separating out the standard deviations and correlations as in (3.14) in the Bayesian analysis of covariance matrices.

3.3.1 The Scatterplot Matrix and PRISM

Graphical tools for visualizing correlations in the raw data prior to fitting a mean model is of great interest in the longitudinal data analysis (Diggle et al., 1994, Chs. 4,5; Dawson et al., 1997; Zimmerman , 2000). The most common tool is the scatterplot matrix (draftman's display), a two-dimensional array of pairwise scatterplots of standardized responses, which is graphically equivalent to the sample correlation matrix. This graphical display facilitates selecting a covariance matrix from a menu of pre-specified, mostly stationary covariance matrices including independence, CS, AR (1) and MA(1) models. The scatterplot matrix for the cattle data (not shown) demonstrates the strong positive correlations among the measurement taken at neighboring times and their decline for measurements taken far apart in time.

The ordinary scatterplot matrix (OSM) and its partial analogue PRISM (Partial Regression-On-Intervenors Scatterplot Matrix), introduced independently by Zimmerman (2000) and Davison and Sardy (2000), provide invaluable intuitions for graphical methods in longitudinal data. Their roles for stationary time series data was discussed in Section 2.4.4. The PRISM is the OSM of certain partial regression residuals. More specifically, its first diagonal plots are the scatterplots of consecutive (standardized) responses. The second diagonal plots are partial regression plots of standardized responses lagged two times apart, adjusted for the standardized response at the intervening time. The third diagonal plots are partial regression plots of standardized responses lagged three times apart, adjusted for the standardized responses at the two intervening times, and so on. Thus, the PRISM is the graphical equivalent of the matrix of partial correlations of the corresponding variables adjusted for the intervening variables.

The OSM and PRISM simplify considerably for stationary data. For equally spaced stationary data, the first row of OSM yields $n - 1$ scatterplots whose correlations coincide with the first $n - 1$ values of the autocorrelation function (ACF) of the data, so that the OSM is effective for identifying the general moving average (MA) structures since for this class the autocorrelations are zero after certain lags. Similarly, the first row of PRISM yields $n - 1$ scatterplots whose correlations coincide with the first $n - 1$ values of the partial autocorrelation function (PACF) and the PRISM is ideal for identifying general AR (Markov, antedependence) structures. Consequently, the PRISM and the OSM being effective for identifying two disjoint sets of dependence structures (Zimmerman, 2000) should be used both in tandem just as the correlogram and partial correlogram are used in Chapter 2. This complementary role of OSM and PRISM and connection with their antedated progenies in time series analysis, namely, correlogram and partial correlogram are of particular interest in the sequel.

It is of some historical significance to note that the correlogram (Yule, 1927) and partial correlogram (Quenouille, 1949) have been in use for quite a while and are an integral part of modern time series analysis. However, the analogues of OSM and PRISM, which logically precede correlogram and partial correlogram, have only briefly appeared in the textbooks (Box et al., 1994, p.25; Brockwell and Davis, 1991, pp. 99–101, Shumway and Stoffer, 2000, Example 1.25).

3.4 GENERALIZED CORRELOGRAMS

We introduce natural generalizations of the correlogram, partial and inverse correlograms to the nonstationary setup and indicate their roles in identifying parsimonious models for covariance and correlation matrices. When m is moderate (about 30) and the number of repeated measures per subject n is about 5 or more, our methodology can reduce considerably the number of covariance parameters. For notational simplicity, in the rest of this chapter we use $Y = (Y_1, \ldots, Y_n)$ to represent the vector of repeated measures on a typical subject.

3.4.1 Nonstationary Correlogram

Recall that the correlation coefficient between two random variables Y_s and Y_t, denoted by ρ_{st}, is a bivariate function of s and t. Plot of this function, in general, is a 3-D graph and hard to interpret. Using the factorization (3.14) of a covariance matrix, we construct two simpler 2-D graphs that are easy to interpret and could reveal a great deal about the dependence and heterogeneity of Y. These graphs turn out to be the natural generalizations of similar graphs for a stationary time series.

To make the connection clearer, note that the $\frac{n(n+1)}{2}$ possibly distinct parameters of an unstructured covariance matrix Σ can be reduced by introducing simplifying structures such as stationarity which forces all entries of the main diagonal and subdiagonals of Σ to be equal, that is,

$$\sigma_{tt} = \sigma_0, \rho_{t+k,t} = \rho_k, 1 \leq t \leq n, k = 0, \cdots, n-1, \tag{3.15}$$

so that Σ is a Toeplitz matrix and the ACF of Y is a univariate function of the lag variable k. The plot of ρ_k versus $k = 0, 1, \cdots, n-1$, called the *correlogram*, was shown in Chapter 2 to play crucial roles in time series model identification.

Our *generalized correlogram* replaces a 3-D ACF plot by a pair of 2-D plots of the entries of R and S in (3.14). The first plot corresponding to R is a *correlation cloud* constructed by plotting the multiple values $\{\hat{\rho}_{t+k,t}; t = 1, \cdots, n-k\}$ or the entries of the kth subdiagonal of R against the lag k, for $k = 0, \cdots, n-1$. As in the semivariogram cloud (Diggle et al., 1994, pp. 51–54) the degree and nature of nonstationarity of Y can be gauged graphically by scanning the spread of points plotted against various lags k. Finally, the graph corresponding to S constructed by plotting $\log s_t$ against $t = 1, 2, \cdots, n$, reveals the degree and nature of heterogeneity of Y. Note that for the stationary case, since $\rho_{t+k,t} = \rho_k$ and $\sigma_{tt} \equiv \sigma_0$, the first plot reduces to the familiar stationary correlogram and the latter is simply a horizontal line carrying no useful information about the structure of the data.

Cattle Data (cont.) Figure 3.3 provides the generalized correlograms for the cattle data. They clearly demonstrate the nonstationarity of the data and some delicate differences in the covariance structures for the treatments A and B are revealed graphically. Their recognizable patterns do suggest parametric models for the correlations, standard deviations and hence the covariances. For example, for the group A cattle

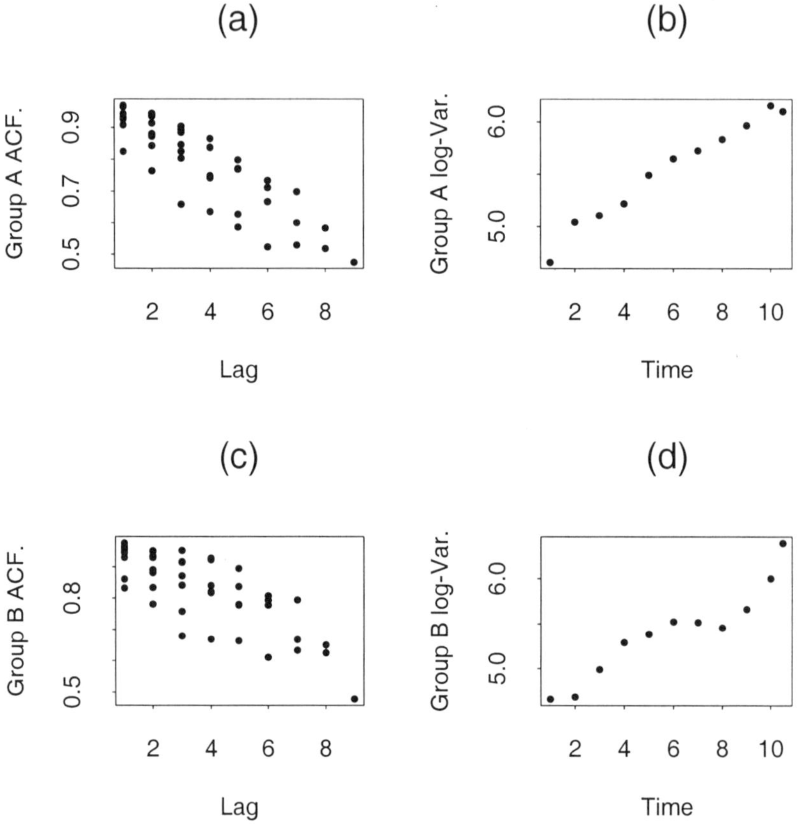

Fig. 3.3 Generalized correlograms for the two groups of cattle. (a) and (c) are the correlation clouds, (b) and (d) are the logarithm of the variances of the repeated measurements.

the model

$$\begin{cases} \hat{\rho}_{t+k,t} = \gamma_1 + \gamma_2 k + \varepsilon_{t,k,c}, \; k = 0, \cdots, n-1; t = 1, \cdots, n-k, \\ \log s_t = \lambda_1 + \lambda_2 t + \varepsilon_{t,v}, \; t = 1, 2, \cdots, n, \end{cases} \tag{3.16}$$

seems plausible; here $\varepsilon_{t,k,c}$ and $\varepsilon_{t,v}$ are the error terms with subscripts c and v signifying correlation and variance, respectively. This model with only four parameters $\theta = (\gamma_1, \gamma_2, \lambda_1, \lambda_2)$ is quite parsimonious compared to the 66-parameter unstructured 11×11 covariance matrix Σ. Moreover, from (3.16) it is evident that the postulated model for the correlation matrix of the data is Toeplitz and of the form LIN(2): $\boldsymbol{\rho} = \gamma_1 A_1 + \gamma_2 A_2$, where A_1 is the matrix of 1's and A_2 is a Toeplitz matrix with 0 on the main diagonal, 1 on the first sub(super)diagonal, and so on. Since the diagonal entries of A_2 are zero, for $\boldsymbol{\rho}$ to be a genuine correlation matrix (with 1's on the diagonal and positive-definite) we must have $\gamma_1 = 1$ and γ_2 restricted (Problem 3.3). However, the model for Σ is nonlinear in λ_1 and λ_2.

The OLS and maximum likelihood estimates (MLE) of the parameters of (3.16) are obtained, respectively, by minimizing

$$\sum_{k=0}^{n-1}\sum_{t=1}^{n-k}(\hat{\rho}_{t+k,t} - \gamma_1 - \gamma_2 k)^2 + \sum_{t=1}^{n}(\log s_t - \lambda_1 - \lambda_2 t)^2, \tag{3.17}$$

and maximizing the likelihood function over the range of the parameter values (Problem 3.4).

The OLS and MLE of (γ_1, γ_2) are, respectively, $(0.99, -0.05)$ and $(1.04, -0.07)$ and the OLS of (λ_1, λ_2) is (4.67, 0.15). The MLE of (γ_1, γ_2) are found using the standardized data and the SAS PROC MIXED. Since iteration is involved it is crucial to have good initial guesses for the relevant parameters. Otherwise, convergence may not occur because often during the iterations the parameter values may violate the required positive-definiteness of the correlation matrix, and hence making the likelihood function unbounded. For the cattle data, convergence did not occur for many apparently reasonable initial guesses, but it did converge after four iterations when the OLS of the parameters were used as initial guess. Perhaps because the fitted correlation matrix corresponding to the OLS of γ_1 and γ_2 were positive-definite with eigenvalues in the range (0.036, 8.89).

Unfortunately, a major drawback of the model (3.16) identified and estimated without heeding to the positive-definiteness constraint is that the estimated correlation matrix and hence Σ are not guaranteed to be positive-definite. Alternative methods of estimation of covariance parameters also suffering from the above drawback are discussed in Zhao and Prentice (1991).

3.4.2 Nonstationary Partial Correlogram

The partial regression residuals computed en route to the construction of the PRISM (Zimmerman, 2000) can be used to compute the partial correlation coefficients $\hat{\pi}_{st}$ between Y_s and Y_t adjusted for the intervening measurements. These are precisely the nonstationary analogues of the values of the PACF of a stationary time series introduced in Section 2.4 and discussed further in Chapter 7.

The *generalized partial correlogram* replaces the 3-D plot of $\hat{\pi}_{st}$ versus (s, t) by the 2-D plot of the multiple values $\{\hat{\pi}_{t+k,t}; t = 1, \ldots, n-k\}$ versus the lag k, $k = 1, \ldots, n-1$. The details of computing the partial correlations and constructing the generalized partial correlogram are similar to those in Sections 2.4 and 3.4.1 (Problem 3.5), and hence are omitted. Moreover, in Section 3.5.3 we introduce the regressogram, which is an improvement of the generalized partial correlogram.

3.4.3 Nonstationary Inverse Correlogram

While the elements of Σ provide information about the pairwise dependence of variables, those of the inverse covariance matrix $\Sigma^{-1} = (\sigma^{st})$ are capable of supplying

Table 3.3 Sample reciprocal of partial variances (along the main diagonal), sample inverse correlations (below the main diagonal), fitted inverse correlations using model (3.18) (above the main diagonal) and fitted reciprocal of partial variances (last row) using OLS for the group A cattle

106	0.94	0.89	0.84	0.79	0.74	0.68	0.63	0.58	0.53	0.48
0.82	155	0.94	0.89	0.84	0.79	0.74	0.68	0.63	0.58	0.53
0.76	0.91	165	0.94	0.89	0.84	0.79	0.74	0.68	0.63	0.58
0.66	0.84	0.93	185	0.94	0.89	0.84	0.79	0.74	0.68	0.63
0.64	0.80	0.88	0.94	243	0.94	0.89	0.84	0.79	0.74	0.68
0.59	0.74	0.85	0.91	0.94	284	0.94	0.89	0.84	0.79	0.74
0.52	0.63	0.75	0.82	0.87	0.93	307	0.94	0.89	0.84	0.79
0.53	0.67	0.77	0.84	0.89	0.94	0.93	341	0.94	0.89	0.84
0.52	0.60	0.71	0.77	0.84	0.90	0.93	0.97	389	0.94	0.89
0.48	0.58	0.70	0.73	0.80	0.86	0.88	0.94	0.96	470	0.94
0.48	0.55	0.68	0.71	0.77	0.83	0.86	0.92	0.96	0.98	445
124	143	165	191	222	256	297	343	397	460	495

information about conditional dependence structure. This motivates calculating the inverse covariance matrix, factorizing it and scrutinizing appropriate elements of its components for certain patterns of zeros. As a graphical tool the *generalized inverse correlogram* defined in this section facilitates the search for such patterns.

Unlike the residuals used in computing the PACF, here we need to rely on the residuals from the full regressions of Y_t on the rest $\{Y_u; u \neq t\}$. Let $\tilde{Y}_t$ stand for the linear least-square predictor of Y_t based on the *rest* $\{Y_u;\ u \neq t\}$. Then, $\mathrm{Var}(Y_t - \tilde{Y}_t)$ is referred to as the *partial variance* of Y_t and the $\mathrm{Corr}(Y_s - \tilde{Y}_s,\ Y_t - \tilde{Y}_t)$, denoted by π^{st}, as the *inverse correlation* between Y_s and Y_t adjusted for the *rest*. Note that π_{st} and π^{st} are similar notions, except that the former is obtained by adjusting for the *intervening* variables and the latter by adjusting for the *rest*. The π^{st} is the natural generalization of the notion of *inverse autocorrelation function* (IACF) for stationary processes; see Cleveland, 1972; Chatfield, 1979 and Chapter 8.

The following properties and interpretations of elements of the inverse covariance matrices (Whittaker, 1990, Chap. 5) are crucial for constructing the generalized inverse correlogram; see Problem 3.9.

Theorem 3.1 (a) Each diagonal element of Σ^{-1} is the reciprocal of a partial variance, that is, $\sigma^{tt} = 1/\mathrm{Var}(Y_t|Y_s, s \neq t)$.

(b) Each off diagonal element of Σ^{-1}, when scaled to have a unit diagonal, is the negative of the inverse correlation between the two corresponding variables adjusted for the *rest*.

(c) If an element in Σ^{-1} is zero, then the corresponding variables are conditionally independent given the *rest* (when Y is normally distributed).

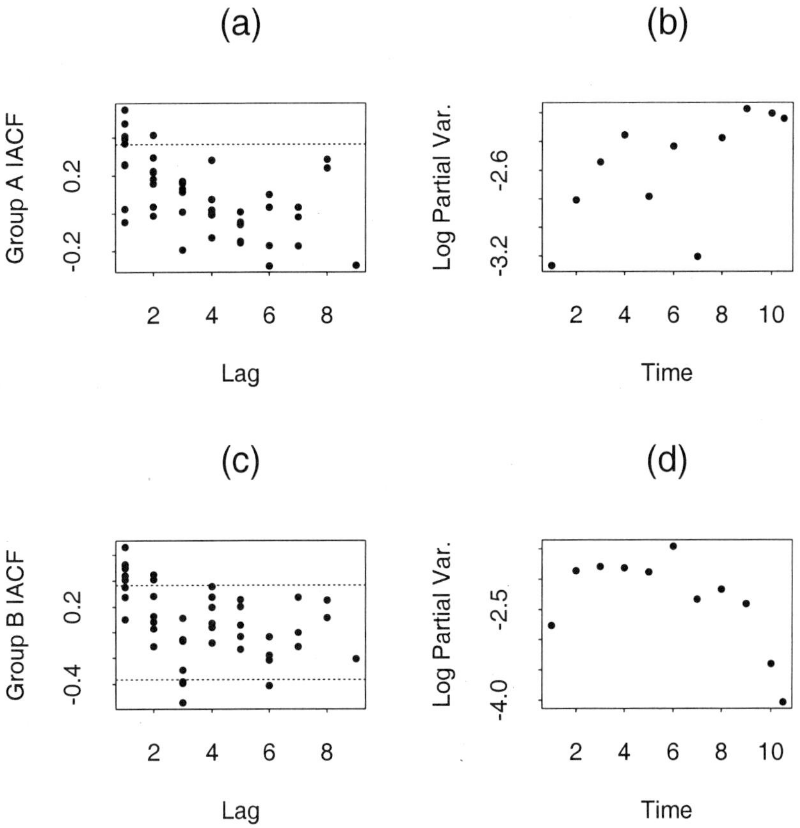

Fig. 3.4 Generalized inverse correlograms for the two groups of cattle. (a) and (c) are the inverse correlation clouds, (b) and (d) are the logarithm of the partial variances. The dotted lines are the approximate pointwise 95% confidence band for the inverse correlations.

Thus, π^{st} and σ^{tt} are estimated easily using the corresponding elements of $\hat{\Sigma}^{-1}$. Let $\hat{\pi}^{st}$ and $\hat{\sigma}^{tt}$, denote such estimates. Then, the *generalized inverse correlogram* is constructed in complete analogy with the generalized correlogram by plotting the multiple values $\{\hat{\pi}^{t+k,t}; t = 1, \ldots, n-k\}$ against $k = 1, \ldots, n-1$, and $0.5 \log \hat{\sigma}^{tt}$ against $t = 1, \ldots, n$.

Figure 3.4 provides the generalized inverse correlograms for the cattle data. While for the group A the inverse correlations for lags one and two seem to be significantly different from zero, this is not the case for the group B where up to lag six they seem to be nonzero. The patterns of $\log \hat{\sigma}^{tt}$ versus t are also quite different for the two treatments. Thus, comparing the generalized inverse correlograms reveals qualitative differences between the dependence structures of the data from the two groups. Based on Figure 3.4, one could postulate the following models for the inverse correlation coefficients and $\log \hat{\sigma}^{tt}$:

$$\begin{cases} \hat{\pi}_{t+k,t} = \gamma_1 + \gamma_2 k + \gamma_3 k^2 + \varepsilon_{t,k,c}, k = 0, \ldots, n-1; t = 1, \ldots, n-k, \\ \log \hat{\sigma}^{tt} = \lambda_1 + \lambda_2 \, t + \varepsilon_{t,v}, t = 1, \ldots, n, \end{cases} \tag{3.18}$$

with only five parameters $\theta = (\gamma_1, \gamma_2, \gamma_3, \lambda_1, \lambda_2)$, where $\varepsilon_{t,k,c}$ and $\varepsilon_{t,v}$ are the error terms. Note that the postulated model for the inverse correlation matrix of the group A cattle is Toeplitz and belongs to LIN(3). The OLS of the γ's and λ's are $(0.51, -0.19, 0.02)$ and $(-3.030.08)$, respectively (Problem 3.6).

As noted earlier, models (3.16) and (3.18) that are identified and estimated without heeding to the positive-definiteness requirement suffer from the drawback that the estimated (fitted) covariance matrix is not guaranteed to be positive-definite. We address this problem in the next section by relying on a special Cholesky factorization of the covariance matrix. We show that the entries of the first column of the lower triangular factor of the covariance matrix are scale multiples of partial autocorrelations.

3.5 UNCONSTRAINED REPARAMETERIZATION

In this section, we introduce an unconstrained and statistically meaningful reparameterization of a covariance matrix and a genuine enhancement of the generalized correlograms. The new graph relies on a special triangular factorization of Σ so that the estimated covariance matrix is guaranteed to be positive-definite. The reparameterization is in terms of the regression coefficients and prediction variances when a measurement is regressed on its predecessors; these are closely related to the PRISM and the relevant partial correlations.

3.5.1 Cholesky Decomposition: The Generalized AR Parameters

The idea of regression (Gram-Schmidt orthogonalization or the modified Cholesky decomposition) plays a central role in this section. For $1 \le t \le n$, let $\hat{Y}_t$ be the linear least-squares predictor of Y_t based on its predecessors $Y_{t-1}, \cdots, Y_1$, and let $\varepsilon_t = Y_t - \hat{Y}_t$ be its prediction error with variance $\sigma_t^2 = \text{var}(\varepsilon_t)$. For simplicity, assume that $E(Y) = 0$ and $\Sigma = \text{Cov}(Y)$ is positive-definite. Thus, for $t = 1, \hat{Y}_1 = E(Y_1) = 0$, and for $1 < t \le n$, there are unique scalars ϕ_{tj}'s (Problem 3.7) so that

$$Y_t = \sum_{j=1}^{t-1} \phi_{tj} Y_j + \varepsilon_t. \tag{3.19}$$

Let $\varepsilon = (\varepsilon_1, \ldots, \varepsilon_n)$, then (3.19) for $t = 1, \ldots, n$ can be written in matrix form as

$$\varepsilon = TY, \tag{3.20}$$

where T is a lower triangular matrix with 1's on the main diagonal and $-\phi_{tj}$ in the (t, j)th position for $2 \le t \le n$ and $j = 1, 2, \cdots, t-1$; see Pourahmadi (1999). Since

Table 3.4 A coherent display of responses, their GARP and IV

Order	$\boxed{Y_1}$	$\rightarrow$	$\boxed{Y_2}$	$\rightarrow$	$\boxed{Y_3}$	$\rightarrow$	$\cdot$	$\cdot$	$\cdot$	$\rightarrow$	$\boxed{Y_n}$
1	1										
2	ϕ_{21}		1								
3	ϕ_{31}		ϕ_{32}		1						
$\vdots$	$\vdots$		$\vdots$		$\vdots$	$\ddots$					
$\vdots$	$\vdots$		$\vdots$		$\vdots$		$\ddots$				
$\vdots$	$\vdots$		$\vdots$		$\vdots$			$\ddots$			
$\vdots$	$\vdots$		$\vdots$		$\vdots$				$\ddots$		
$\vdots$	$\vdots$		$\vdots$		$\vdots$					$\ddots$	
n	ϕ_{n1}		ϕ_{n2}		ϕ_{n3}	$\cdots$	$\cdots$	$\cdots$	$\cdots$	$\phi_{n,n-1}$	1
	σ_1^2		σ_2^2		σ_3^2	$\cdots$	$\cdots$	$\cdots$	$\cdots$	σ_{n-1}^2	σ_n^2

ε_t's are uncorrelated, it follows from (3.20) that the matrix T actually diagonalizes Σ, namely,

$$T\Sigma T' = D, \tag{3.21}$$

where $D = \text{diag}(\sigma_1^2, \ldots, \sigma_n^2)$. Note that $\sigma_t^2 = \text{var}(\varepsilon_t)$ is different from $\sigma_{tt} = \text{var}(Y_t)$, however, when the responses are independent, then $\phi_{tj} \equiv 0$ and $\sigma_t^2 = \sigma_{tt}$, so that the matrices T and D gauge the "dependence" and "heterogeneity" of Y or Σ.

The factorization of Σ in (3.21) is called the *modified Cholesky decomposition* of Σ. When Σ is unstructured, (3.21) makes it possible to swap the $\frac{1}{2}n(n+1)$ constrained parameters of Σ with the new unconstrained set of parameters ϕ_{tj}'s and $\log \sigma_t^2$'s of the same cardinality. Table 3.4 depicts in a coherent manner the regression coefficients ϕ_{tj}'s and the innovation variances σ_t^2's for the variables $Y_1, \cdots, Y_n$. We refer to the new parameters ϕ_{tj}'s and σ_t^2's as the *generalized autoregressive parameters* (GARP) and the *innovation variances* (IV) of Σ; see Pourahmadi (1999). A major advantage of using (3.21) is its ability to guarantee the positive-definiteness of the estimated covariance matrix. Since it gives rise to estimators of the form $\hat{T}^{-1}\hat{D}\hat{T}'^{-1}$, where the diagonal entries of $\hat{D}$ are positive. How does one compute and model T and D? The following examples shed some useful light on the question.

Example 3.1. (a) Consider the 3×3 matrix Σ with the factorization

$$
\begin{aligned}
\Sigma &= \begin{pmatrix} 1 & 1 & 1 \\ 1 & 5 & 5 \\ 1 & 5 & 14 \end{pmatrix} = \begin{pmatrix} 1 & 0 & 0 \\ 1 & 2 & 0 \\ 1 & 2 & 3 \end{pmatrix} \begin{pmatrix} 1 & 1 & 1 \\ 0 & 2 & 3 \\ 0 & 0 & 3 \end{pmatrix} \\
&= \begin{pmatrix} 1 & 0 & 0 \\ 2^{-1} & 1 & 0 \\ 3^{-1} & 2/3 & 1 \end{pmatrix} \begin{pmatrix} 1 & 0 & 0 \\ 0 & 4 & 0 \\ 0 & 0 & 9 \end{pmatrix} \begin{pmatrix} 1 & 2^{-1} & 3^{-1} \\ 0 & 1 & 2/3 \\ 0 & 0 & 1 \end{pmatrix} \\
&= LDL'.
\end{aligned}
$$

Comparing this to (3.21) and (3.19), we obtain

$$
T = L^{-1} = \begin{pmatrix} 1 & 0 & 0 \\ -1 & 1 & 0 \\ 0 & -1 & 1 \end{pmatrix}, \quad D = \text{diag}\,(1, 4, 9),
$$

and (3.19) reduces to

$$
Y_1 = \varepsilon_1, \; Y_t = Y_{t-1} + \varepsilon_t, \; t = 2, 3.
$$

(b) For an AR(1) covariance matrix with the correlation ρ, using direct computations leading to (3.19), we obtain for $2 \leq t \leq n$,

$$
\begin{aligned}
&\phi_{t,t-1} = \rho, \; \sigma_1^2 = \sigma^2(1 - \rho^2)^{-1}, \\
&\phi_{t,t-j} = 0, \; \sigma_t^2 = \sigma^2, \qquad\qquad 2 \leq j < t.
\end{aligned}
$$

Thus, only the first subdiagonal of T is non-zero for an AR(1) model and (3.19) reduces to $Y_1 = \varepsilon_1$ and $Y_t = \rho Y_{t-1} + \varepsilon_t$ for $2 \leq t \leq n$. More generally, it can be shown that for AR(p) covariance structures, the jth subdiagonal of T is zero and the prediction error variance σ_t^2 is constant as soon as t exceeds the order p of the model. These distinguishing features of the nonredundant entries of T and D are potentially useful for identifying parsimonious models for them and eventually for Σ via (3.21).

(c) For a CS covariance matrix with correlation ρ, from direct computations leading to (3.19), we obtain $\sigma_1^2 = \sigma^2$ and for $t \geq 2, \; 1 \leq j \leq t-1$

$$
\phi_{tj} = \rho\{1 + (t-1)\rho\}^{-1}, \; \sigma_t^2 = \sigma^2 \left\{1 - \frac{(t-1)\rho^2}{1 + (t-1)\rho}\right\},
$$

and $Y_1, \ldots, Y_n$ satisfy

$$
Y_t = \rho\{1 + (t-2)\rho\}^{-1} \sum_{j=1}^{t-1} Y_j + \varepsilon_t, \; 1 \leq t \leq n.
$$

Thus, in this case, unlike the AR(1) covariance structure, all the below-diagonal entries of T are non-zero and for a given t all the predecessors of Y_t receive identical regression coefficients.

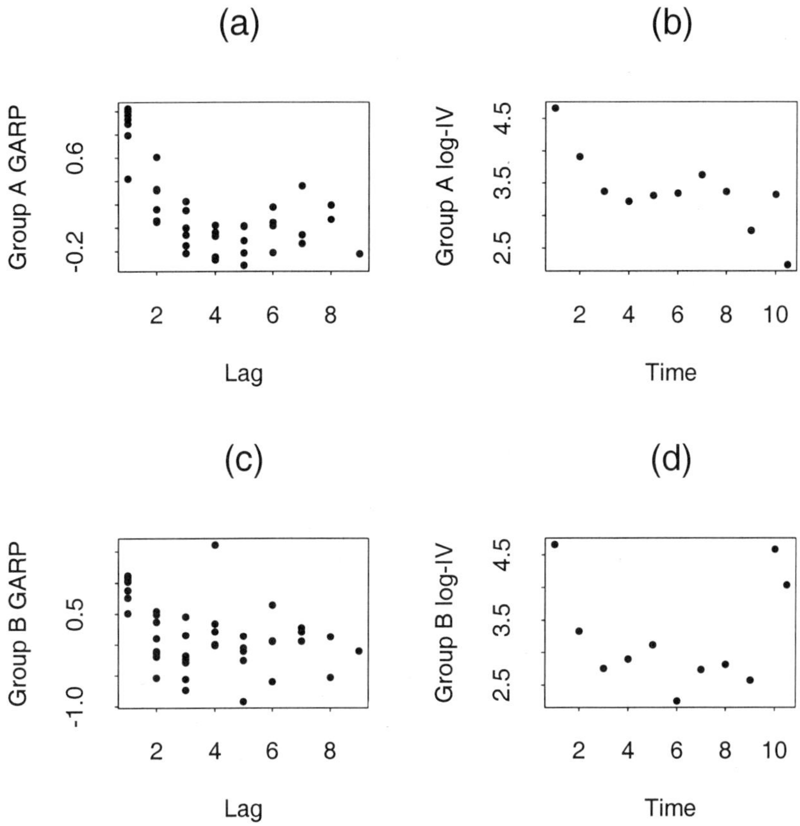

Fig. 3.5 Regressograms for the two groups of cattle. (a) and (c) are the GARP clouds, (b) and (d) are the logarithm of the innovation variances.

3.5.2 Regressograms

Now, we introduce plots similar to the correlograms for identifying parsimonious models for the GARP and IV and hence the covariance matrix Σ.

From (3.19), since ϕ_{tj} is the lag-$(t-j)$ regression coefficient, one expects it to be small for large $t-j$. A properly constructed graph of ϕ_{tj}'s and $\sigma_t^{2,s}$ are bound to reveal a great deal about the dependence structure of Y. We refer to the plot of $\{\phi_{t+k,t}; t = 1, \ldots, n-k\}$ versus $k = 1, 2, \ldots, n-1$, and the plot of $\log \sigma_t^2$ versus $t = 1, 2, \cdots, n$ as the *regressogram* of Σ (Pourahmadi, 1999). For AR(p) covariance structures, Example 3.1 (b) suggests that the first plot in the regressogram drops off to zero for lags $j > p$ and the second plot remains flat for times $t > p$. Note that this property of the first plot mimics that of the partial correlogram of a stationary AR(p) process. We show that this property holds and extends far beyond stationary data. Of course, in complete analogy with time series model identification, such distinguishing features of theoretical regressograms for special covariance structures are crucial at the model identification stage of a data-based covariance modeling strategy. The important role of the regressogram in identifying parsimonious models for (T, D) or Σ is illustrated using the cattle data.

Figure 3.5 provides the regressograms for the cattle data. From Figure 3.5 (a) and (b) the following cubic polynomial models for ϕ_{tj}'s and $\log \sigma_t^2$'s seem appropriate. For $t = 1, 2, \cdots, 11; j = 1, \cdots, t-1$,

$$\begin{cases} \hat{\phi}_{tj} & = & \gamma_1 + \gamma_2(t-j) + \gamma_3(t-j)^2 + \gamma_4(t-j)^3 + \epsilon_{t,j,c}, \\ \log \hat{\sigma}_t^2 & = & \lambda_1 + \lambda_2 t + \lambda_3 t^2 + \lambda_4 t^3 + \epsilon_{t,v}, \end{cases} \tag{3.22}$$

where $\hat{\sigma}_t^2$ and $\hat{\phi}_{tj}$ are the sample counterparts of σ_t^2 and ϕ_{tj} obtained from the sample covariance matrix $\hat{\Sigma}$ of group A cattle. This shows the important role of the regressogram in modeling a 66-parameter covariance matrix parsimoniously using merely 8 unconstrained parameters. The MLE of the parameters $\gamma = (\gamma_1, \gamma_2, \gamma_3, \gamma_4)$ and $\lambda = (\lambda_1, \lambda_2, \lambda_3, \lambda_4)$ are $(0.18, -1.71, 1.64, -1.11)$ and $(3.52, -1.14, 0.30, -0.85)$; see Pourahmadi (2000).

3.5.3 GARP, AR Coefficients and PACF

For an unstructured covariance matrix Σ, the GARP ϕ_{tj}'s capture the extent of dependence in Y in a manner similar to the PACF for stationary data. More precisely, in this section we show that the GARP values ϕ_{t1} for $t = 2, \ldots, n$ are multiples of π_{st}, the partial correlation coefficient between Y_1 and Y_t adjusted for the *intervening* variables $Y_2, \ldots, Y_{t-1}$.

For a fixed $t \geq 3$, let $S_i (i = 0, 1, 2)$ be, respectively, the sets of mean-zero random variables $\{Y_1, \ldots, Y_{t-1}\}$, $\{Y_2, \ldots, Y_{t-1}\}$ and $\{Y_1 - \hat{Y}_{1,S_1}\}$, where $\hat{Y}_{k,S_i}$ stands for the linear least squares predictor of Y_k based on S_i. Since S_1 and S_2 are uncorrelated sets of random variables and jointly generate the linear subspace generated by the elements of S_0, it follows that

$$\hat{Y}_{t,S_0} = \hat{Y}_{t,S_1} + \hat{Y}_{t,S_2} = \hat{Y}_{t,S_1} + \beta(Y_1 - \hat{Y}_{1,S_1}), \tag{3.23}$$

where β is the slope of regression of Y_t on $Y_1 - \hat{Y}_{1,S_1}$, namely,

$$\beta = \frac{\text{Cov}(Y_t, Y_1 - \hat{Y}_{1,S_1})}{\text{Var}(Y_1 - \hat{Y}_{1,S_1})}. \tag{3.24}$$

Let $v_1^2 = \text{Var}(Y_1 - \hat{Y}_{1,S_1})$ and $v_t^2 = \text{Var}(Y_t - \hat{Y}_{t,S_1})$, since the residual $Y_1 - \hat{Y}_{1,S_1}$ is uncorrelated with the predictor $\hat{Y}_{t,S_1}$ we can rewrite β as

$$\begin{aligned} \beta &= \frac{\text{Cov}(Y_t - \hat{Y}_{t,S_1}, Y_1 - \hat{Y}_{1,S_1})}{v_1 v_2} \cdot \frac{v_2}{v_1} \\ &= \text{Corr}(Y_t - \hat{Y}_{t,S_1}, Y_1 - \hat{Y}_{1,S_1}) \cdot \frac{v_2}{v_1}. \end{aligned} \tag{3.25}$$

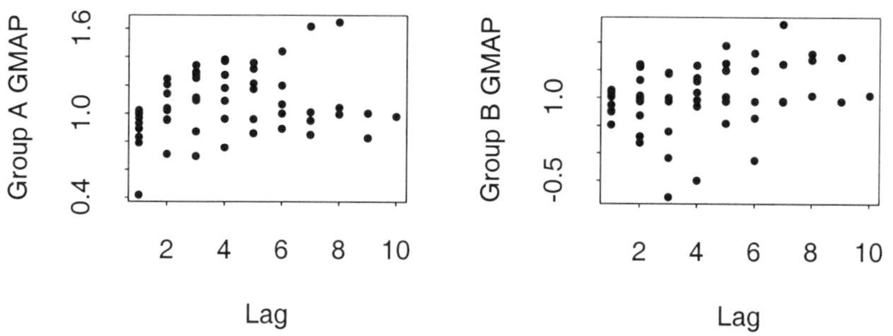

Fig. 3.6 GMAP regressograms for the cattle data. (a) treatment A, (b) treatment B.

Replacing $\hat{Y}_{1,S_1}$ and $\hat{Y}_{t,S_1}$ in (3.23) by their linear representations in term of $Y_2, \ldots, Y_{t-1}$ and collecting similar terms and comparing with (3.19), it follows that ϕ_{t1} the coefficient of Y_1 is precisely the β given in (3.25). When $Y_1, \ldots, Y_t$ are stationary, then $v_1 = v_t$, so that ϕ_{t1} is precisely the partial correlation coefficient between Y_1 and Y_t, adjusted for the intervening measurements in S_1. The idea of orthogonalization leading to (3.25) is also used in deriving the Kalman filter in Section 4.3.3 and the Durbin Levinson recursion in Chapter 7.

3.5.4 The Generalized MA Parameters and ARIMA Models

Having seen the success of the plot of GARP in identifying parsimonious models for covariances and in view of the traditional duality between the AR and MA models (parameters), we expect a similar role for the *generalized MA parameters* (GMAP) θ_{tj}. These are obtained, essentially, by "inverting" any of the identities in (3.19)–(3.21). Since $Y_1 = \varepsilon_1$, repeated substitution for Y_t's in (3.19) in terms of ε_t's gives

$$Y_t = \varepsilon_t + \sum_{j=1}^{t-1} \theta_{tj}\varepsilon_{t-j}, t = 1, \ldots, n. \tag{3.26}$$

Consequently, from (3.21) and (3.26) we have $\Sigma = T^{-1}DT'^{-1} = LDL'$, where $L = T^{-1}$ is a unit lower triangular matrix with θ_{tj} as its (t, j)th entry. The intimate relations among θ_{tj}'s and the GARP ϕ_{tj}'s are prescribed by the matrix identity $LT = I$, reminiscent of similar relations among AR and MA parameters in the stationary case discussed in Chapter 7.

Evidently, (3.26) suggests interpreting the GMAP θ_{tj}'s as the regression coefficients when Y_t is regressed on the past and present innovations $\varepsilon_t, \varepsilon_{t-1}, \ldots, \varepsilon_1$, so that for an unstructured Σ, these new parameters are unconstrained. The GMAP regressogram can be constructed in an obvious manner. Figure 3.6 provides the GMAP regressograms for the cattle data, these have similar patterns and indicate that most

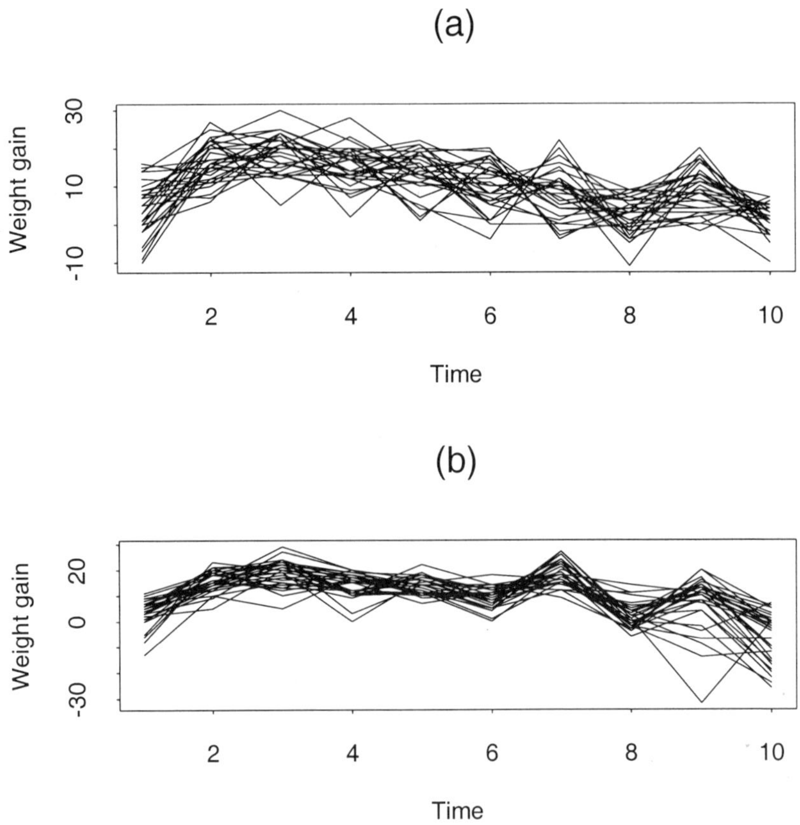

Fig. 3.7 Profile plots of the weight gains for the cattle, (a) treatment A, (b) treatment B.

of the θ_{tj}'s are hovering around one, so that differencing might be appropriate for the cattle data. Figure 3.7, the profile plots of the weight gains (difference of successive weight measurements) reveal some new features of the data and suggest random walk as a possible model for the data.

There are numerous instances of (3.26) appearing in the literature under different names. The most ubiquitous case of $\theta_{tj} \equiv 1$ and all σ_t^2's equal corresponds to a homogeneous random walk $Y_t = \sum_{j=0}^{t-1} \varepsilon_{t-j}$, $t = 1, 2, \ldots, n$; Zellner (1979). This is a special case of the ARIMA (p, d, q) models in Chapter 2 which is used frequently in times series analysis. Recall that a time series $\{Y_t\}$ follows an ARIMA (p, d, q) model, if its dth-difference $\{\Delta^d Y_t\}$ follows a stationary ARMA (p, q) model. One could also consider a continuous time analogue of (3.26), replacing the partial sums by integrals. For example, for the Wiener (WI) process with the covariance function cov $(Y_j, Y_k) = \sigma^2 \min(t_j, t_k)$, the within-subject covariance matrix Σ is of the form

$\sigma^2[\min(t_j, t_k)]_{t,j=1}^n$ with only one parameter and belongs to the LIN(1). A major advantage of a continuous-time analogue of (3.26) or WI process is that one can estimate σ^2 or the within-subject covariance matrix in severely unbalanced cases.

3.6 ANTEDEPENDENCE MODELS

In this section we study a nonstationary analogue of AR(p) models called *antedependence* models of order p, denoted by AD(p) for short. The order p serves as a memory gauge, where $p = 0$ corresponds to independence and $p = n - 1$ to the unstructured multivariate dependence.

3.6.1 Unstructured AD(p) Models

Following Gabriel (1962), we say the time-ordered random variables $Y_1, \ldots, Y_n$ are AD(p) if the conditional distribution of Y_t given $Y_{t-1}, \ldots, Y_1$ depends on $Y_{t-1}, \ldots, Y_{t-p}$ for all $t \geq p$. Evidently, this concept is equivalent to $Y_1, \ldots, Y_n$ having a Markovian dependence of order p, so that in view of (3.19) the requirement of AD(p) for normal data is equivalent to ϕ_{tj}'s of lag greater than p to be zero. More precisely, if the time-indexed, mean-zero multivariate normal random variables $Y_1, \ldots, Y_n$ are AD(p), then the following are equivalent:

(a) $$Y_t = \sum_{j=1}^{p_t} \phi_{tj} Y_j + \varepsilon_t, \tag{3.27}$$
where $p_t = \min(p, t-1),\ t = 1, \ldots, n,$

(b) the last $n - p - 1$ subdiagonals of T are zero,

(c) the last $n - p - 1$ subdiagonals of Σ^{-1} are zero (Gabriel, 1962).

The AD(p) models are nonstationary generalizations of AR(p) models in the sense that the innovation variances σ_t^2 and the same-lag correlations vary over time. It places no restrictions on how the variances and same-lag correlations up to lag p change over time, so that they are unstructured. It follows from (c) above that the the covariance matrix of an unstructured AD(p) model is specified by $(p+1)(2n-p)/2$ parameters, while the AR(p) and unstructured multivariate models are specified, respectively, by $p + 1$ and $\frac{n(n+1)}{2}$ parameters. Thus, the general AD(p) model requires more parameters than the former and less than the latter.

The AR(p) models are, indeed, special cases of the AD(p) models. Understanding this connection is helpful in relating and interpreting their respective parameters. More specifically, an AD(p) reduces to a stationary AR(p) model (Zimmerman and Núñez-Antón, 1997) when

(i) $\phi_{tj} = \phi_j,\ j = 1, \ldots, p,\ t = p + 1, \ldots, n;$

(ii) the roots of the polynomial equation

$$1 - \phi_1 z - \phi_2 z^2 - \cdots - \phi_p z^p = 0,$$

lie outside the unit circle in the complex plane;

(iii) $\sigma_{p+1}^2 = \cdots = \sigma_n^2 > 0$;

(iv) the "initial values" $\{\phi_{tj};\ j = 1, \ldots, t-1;\ t = 2, \ldots, p\}$ and $\sigma_1^2, \ldots, \sigma_p^2$ are chosen so that $(Y_1, \ldots, Y_n)$ is covariance stationary.

Let $\Sigma = (\sigma_{st})$ denote the covariance matrix of the AD(p) variables $Y_1, \ldots, Y_n$. We show that all entries of Σ are functions of ϕ_{tj}'s and σ_t^2's and can be computed recursively using analogues of the Yule-Walker equations (2.41) for the stationary AR(p) models (Problem 3.7). For example, multiplying both sides of (3.20) by Y_{t-k} and taking expectations, we obtain

$$EY_t Y_{t-k} = \sum_{j=1}^{p_t} \phi_{tj} EY_j Y_{t-k} + E\varepsilon_t Y_{t-k},$$

or

$$\sigma_{t,t-k} = \sum_{j=1}^{p_t} \phi_{tj}\sigma_{j,t-k} + \sigma_t^2 I_{\{k=0\}},\ k = 0, \cdots, t-1;\ t = p+1, \cdots, n, \quad (3.28)$$

where I_A stands for the indicator function of the set A. Thus, given ϕ_{tj}'s and σ_t^2's one can compute Σ and vice versa; see Example 3.1. Our results thus far show how closely the antedependence models and the representation (3.19) resemble the properties of stationary AR models.

3.6.2 Structured AD(p) Models

Even though the unstructured AD(p) model is considerably more parsimonious than an unstructured covariance matrix, still it may have far too many parameters to be useful in some applications. For example, in many growth studies variances increase over time and equidistant measurements become more highly correlated as the study progresses, due to the so-called "learning" effect. A model that incorporates such specific forms of nonstationarity is likely to be more useful and have greater applicability.

To allow for irregularly observed measurements, we denote the measurement times for an arbitrary subject by $0 \leq t_1 < t_2 < \ldots < t_n$. One may impose the following structures on the ϕ_{tj}'s and σ_t^2's of an AD(p) model and refer to it as *structured* AD(p) models, or SAD(p) for short, Núñez-Antón and Zimmerman (2000). For $i = p+1, \ldots, n;\ j = 1, \ldots, p$, set

$$\begin{cases} \phi_{ij} = \phi_j^{f(t_i;\lambda_j) - f(t_{i-j};\lambda_j)}, \\ \sigma_i^2 = \sigma^2 g(t_i; \psi), \end{cases} \quad (3.29)$$

Table 3.5 The # of parameters and BIC values for some models fitted to the group A cattle

Model	Number of Parameters	BIC
UN	66	2260.5
(3.9)	4	2198.7
(3.11)	5	2719.5
Poly (3,3)	8	2125.2
Poly (2,3)	7	2184.0
WI	1	2108.7
CS	2	2387.1
SAD(1)	4	2103.3
AD(1)	21	2241.6

where $\phi_1, \ldots, \phi_p$ are some scalars, $f(\cdot)$ and $g(\cdot)$ are known functions, λ and ψ are vectors of relatively few parameters. Common examples for $f(\cdot)$ are the Box-Cox power law:

$$f(t;\lambda) = \begin{cases} (t^\lambda - 1)/\lambda & \text{if } \lambda \neq 0, \\ \log t & \text{if } \lambda = 0, \end{cases}$$

and $g(t;\psi)$ is either a constant or quadratic function. Note that these models do not impose any structure on ϕ_{tj}'s and σ_t^2's corresponding to the first p measurement times. Indeed, it is prudent to leave these "initial" values unstructured rather than requiring them to adhere to structures supported by measurements at later times. We note that the Box-Cox form for $f(\cdot)$ prescribes the lag-j GARP ϕ_{tj} to be monotone increasing if $\lambda_j < 1$, monotone decreasing if $\lambda_j > 1$, or constant if $\lambda_j = 1$ for $j = 1, \ldots, p$. This method of introducing structures for ϕ_{tj}'s is not data-based as compared to analogues of (3.22) guided by the regressogram of the data.

3.6.3 Models for the Cattle Data

In this section we finally compare several models identified for and fitted to the covariance matrix of the group A cattle data. We note that most of our models are formulated using various sample generalized correlograms introduced thus far.

The results are summarized in Table 3.5, where Poly (3,3) stands for the model (3.22) and Poly (2,3) stands for the same model, but excluding the third power of $t - j$. We can test the null hypothesis that this submodel holds using the likelihood ratio test (Casella and Berger, 1990). Let L_1 and L_0 be the maximized values of the loglikelihood functions corresponding to these models. Comparing $2(L_1 - L_0)$ = 62.14 to the appropriate percentage point of the chi-squared distribution with one degrees of freedom, clearly reject the submodel Poly (2,3) so that $(t - j)^3$ is kept in (3.22).

The random walk model suggested by Figure 3.6 is, fortunately, compatible with the Núñez-Antón and Zimmerman's (2000) result showing that on the basis of BIC, the WI model fits best, and the Diggle et al. (1994, p. 119) suggestion of analyzing the difference or the successive weight gains of the animals. Table 3.5 also provides the BIC values for models (3.16), (3.18) and the unstructured (UN) covariance matrix. Its last four rows from Núñez-Antón and Zimmerman (2000) are included for ease of comparison. The BIC values for these models are converted from their's (or PROC MIXED's) larger-is-better form to the more commonly used smaller-is-better form. From the table it is evident that the WI, structured AD of order one (SAD(1)) and Poly(3,3) models have the smallest BIC values and therefore judged to be fitting the data better, and model (3.18) is among the worst.

3.7 GENERALIZATIONS

The idea of decomposing a general covariance matrix into two components capturing its "variance" and "dependence" and using analogues and generalizations of regression and time series graphic diagnostic to specify models for these components was shown to be effective in identifying a parsimonious model (with 1 to 8 parameters) for a 66-parameter unstructured 11×11 covariance matrix. However, the key ideas and graphical tools of this section may not carry over to the multivariate data setup where the variables are not necessarily (time-) ordered. This may require working with several tentative orders among the variables and searching for the most yielding order so far as substantive interpretation and emergence of discernible patterns in the graphs are concerned. For a particular order, the boxes and arrows in the first row of Table 3.4 suggest that these variables form a *univariate recursive regression graph* in the sense of Cox and Wermuth (1996, p.31 and Sec. 8.5) so that Y_t, for example, is a response to variables in boxes to its left. This might be useful for decomposing the joint distribution of a set of random variables into a running sequence of conditional distributions as in graphical models (Whittaker, 1990).

In the absence of a preferred order among the variables in a multivariate data, one may start with the default ordering of the variables according to their variances. This suggestion is motivated by the observation that in many growth studies the variability increases over time and in the principal component analysis (Anderson, 1984, Chap. 11) the components are ordered according to their variances. Alternatively, one could use the qualitative ordering proposed by Brown, Le and Zidek (1994, p.88) in the context of specifying priors for Σ. It is highly desirable to have a reparameterization of the covariance matrix not depending on the coordinate system. Some examples are provided in Pinheiro and Bates (1996), however, comparing these with (3.21) suggests that there might be a trade-off between coordinate-free parameterization and the statistical interpretability of the ensuing parameters. The exact nature of this trade-off is unknown and can benefit from further research.

The techniques of the last two sections are deeply rooted in time series analysis. In stark contrast to the multivariate analysis techniques which are based on the spectral decomposition of the covariance matrix, that is, finding an orthogonal matrix P so that $P\Sigma P'$ is diagonal, we use the modified Cholesky decomposition to find a unit lower triangular matrix T so that $T\Sigma T'$ is diagonal. Not only does this decomposition remove the notorious positive-definiteness constraint, but it turns out that the matrices T and its inverse are closely related to the celebrated Durbin-Levinson's algorithm, innovation algorithm and Kalman filter, ideas and tools forming the core of modern time series analysis; see Chapters 4 and 7. For more discussion on the role of time series techniques in the analysis of longitudinal data, see Diggle (1997).

Problems

3.1 Find the feasible values of θ_1 and θ_2 so that the 2×2 matrix

$$\Sigma = \theta_1 I_2 + \theta_2 \begin{pmatrix} 1 & 2 \\ 2 & 1 \end{pmatrix},$$

is a covariance matrix.

3.2 Show that the CS and AR(1) covariance matrices can be rewritten as (3.8). Identify the coefficients θ_i's and the matrices A_i's. Do the same for the covariance matrix V_i in (3.10) of a linear mixed model with known R_i and unknown D.

3.3 Show that the correlation matrices in (3.16) and (3.18) belong to LIN(2) and LIN(3), respectively. Identify the feasible regions for the parameters so that these matrices are positive-definite.

3.4 Derive the normal equations for the OLS estimates of the parameters $\gamma_1, \gamma_2, \lambda_1$ and λ_2 by minimizing (3.17). Use the data in Table 3.1 to compute these estimates. Verify whether the fitted correlation matrix is positive-definite.

3.5 For the cattle data construct the OSM, PRISM and generalized partial correlograms for the two groups. Use the techniques of Section 3.4.1 leading to (3.16)–(3.17) to formulate a similar model for the partial correlation of the group A data. Compute the OLS estimates of your identified model for the partial correlations.

3.6 Repeat Problem 3.4 for model (3.18).

3.7 For $t \geq 2$, set $\phi_t = (\phi_{t1}, \phi_{t2}, \ldots, \phi_{t,t-1})$ where ϕ_{tj}'s are as in (3.19).

(a) Show that $\Sigma_t \phi_t = \boldsymbol{\sigma}_t$ and $\sigma_t^2 = \sigma_{tt} - \boldsymbol{\sigma}_t' \Sigma_t^{-1} \boldsymbol{\sigma}_t = \sigma_{tt} - \boldsymbol{\sigma}_t' \phi_t$, where Σ_t is the $(t-1) \times (t-1)$ leading principal minor of Σ and $\boldsymbol{\sigma}_t$ is the column vector of the first $t-1$ entries of the tth column of Σ.

(b) Compare the two equations in (a) for nonstationary Y with the Yule-Walker equations (2.41) for fitting AR(p) models to stationary data.

(c) Use the formulas in (a) to compute ϕ_{tj}'s and σ_t^2 for $t \leq 4$ for the AR(1) and CS covariance structures.

3.8 Show that model (3.22) for components of the sample covariance matrix S of the group A cattle can be rewritten in matrix form as

$$g(S) = 2I - \hat{T} - \hat{T}' + \log \hat{D} = \sum_{k=1}^{8} \alpha_k A_k,$$

where g is a link function (McCullagh and Nelder, 1989, p. 27). Compare this with (3.8).

(a) Identify the design matrices A_k's and relate the α_k's to γ_i's and λ_i's in (3.22).

(b) Use the above model to compute the OLS estimates of the γ_i's and λ_i's as in (3.17).

3.9 For Y a zero-mean random vector with the positive-definite covariance matrix Σ, define a new random vector $Z = \Sigma^{-1}Y$. Show that

(a) $\text{Cov}(Z) = \Sigma^{-1}$ and $\text{Cov}(Z, Y) = I$.

(b) the components of Z are proportional to the *complete residuals* $Y_t - \tilde{Y}_t$, where $\tilde{Y}_t$ is the regression of Y_t on all other components of Y.

4

Stationary ARMA Models

Describing the second-order properties of a general stationary process requires an infinite number of parameters. In this chapter, we focus on the class of autoregressive and moving average (ARMA) processes whose second-order properties is described by a finite number of parameters. Such processes defined via difference equations enjoy transparent dynamics, direct interpretation of parameters and many other desirable features. In a sense, by increasing the number of parameters these concrete models could exhaust the whole class of second-order processes. Statistically these models are important and similar to the linear regression models; see (1.2) and Section 4.2.5. Their parameters can be estimated using this analogy (Durbin, 1960). Furthermore, the optimal predictors of ARMA processes involve a finite number of operations on the past data—a highly desirable feature for interpretation, computation, engineering implementation and instrumentation. In short, ARMA models are of central importance in the theory and practice of time series analysis, digital signal processing and other areas of applications dealing with correlated data (Pollock, 1999).

This chapter resumes with the study of solutions of stochastic difference equations of the AR-type, we find necessary and sufficient conditions on the coefficients so that the difference equation has a causal stationary solution (Doob, 1944). The procedure is then used in Section 4.2, without much change, to find necessary and sufficient conditions for causality and invertibility of ARMA models. These results are developed without relying on the traditional backward shift operator of a stationary time series. An unconstrained reparameterization of the causality and invertibility region of the ARMA models is given, and the AR, MA parameters and the covariance function of a stationary solution are expressed in terms of the ARMA coefficients. In Section 4.3, we focus on the prediction of ARMA models and obtain recursive and updating formulas for predictors, develop Akaike's Markovian representation and evaluate the Gaussian likelihood function for ARMA models via the Kalman filter. In Section 4.4, we show that entries of the inverse covariance matrix of AR models are quadratic functions of the AR coefficients and introduce a family of sufficient statistics and obtain the likelihood equations for the parameters of Gaussian AR models by explicitly differentiating the likelihood function.

In Section 4.5, the singularity of certain prediction error covariance matrices is used to characterize ARMA models. This is useful in understanding various model identification techniques based on recognizing patterns of zeros of certain determinants of matrices constructed from the sample correlations; see Woodward and Gray (1978); Tsay and Tiao (1985) and Choi (1992). The deeper problems of existence and construction of ARMA models with prescribed, but incomplete covariance information is also studied and presented in Sections A4.6 and A4.7. These topics are closely related to the problem of maximum entropy estimation and extension of autocovariance functions studied in Chapter 7.

4.1 CAUSAL STATIONARY SOLUTIONS OF AR MODELS

In this section, we find conditions on the coefficients of an AR(p) model so that it has a causal stationary solution. For the sake of clarity, we start with the simple and important case of $p = 1$. More general AR models are reduced to this simple case using the Schur decomposition of a matrix. Use of this result obviates the need to rely on formal manipulations involving the backward shift operator B of a time series and may prevent some of its potential pitfalls described in Wegman (1986) and Niemi (1984).

4.1.1 AR(1) Models

Consider the first-order stochastic difference equation

$$X_t = aX_{t-1} + e_t, \tag{4.1}$$

where a is arbitrary and $\{e_t\}$ is a white noise. We are interested in finding all second-order solutions of (4.1). To this end, substituting X_{t-1} by $aX_{t-2} + e_{t-1}$ in (4.1) and repeating for integers $n \geq 0$, we arrive at

$$X_t = e_t + a[aX_{t-2} + e_{t-1}] = \cdots = \sum_{k=0}^{n} a^k e_{t-k} + a^{n+1} X_{t-n-1}. \tag{4.2}$$

Let $S_n = \sum_{k=0}^{n} a^k e_{t-k}$, then $X_t - S_n = a^{n+1} X_{t-n-1}$ and the existence of limit of S_n is equivalent to the existence of limit of $a^{n+1} X_{t-n-1}$. Since $\{e_t\}$ is a white noise, $\lim_{n\to\infty} \sum_{k=0}^{n} a^k e_{t-k}$ exists, if and only if $\sum_{k=0}^{\infty} |a|^{2k} < \infty$ or

$$|a| < 1, \tag{4.3}$$

(Problem 4.3). Assume (4.3) holds and define $V_t = \lim_{n\to\infty} a^{n+1} X_{t-n-1}$. The process $\{V_t\}$ satisfies the equation $V_t = aV_{t-1}$, and hence $V_t = Xa^t$, where X is an arbitrary random variable with $EX^2 < \infty$. Thus, from (4.2) and under (4.3) all solutions of (4.1) are of the form

$$X_t = \sum_{k=0}^{\infty} a^k e_{t-k} + a^t X. \tag{4.4}$$

These are nonstationary except when $X \equiv 0$. In this case, $X_t = \sum_{k=0}^{\infty} a^k e_{t-k}$ is the unique stationary solution of (4.1) with the additional properties:

$$\begin{cases} \text{Cov}(e_{t+k}, X_t) = 0, \text{ for } \textit{all } k \geq 1 \text{ and } t, \\ E\, X_t = 0. \end{cases} \tag{4.5}$$

Next, we show that the condition (4.3) is necessary for (4.1) to have a stationary solution satisfying (4.5). Taking the variance of both sides of (4.1) and using the first equation in (4.5) we arrive at

$$\text{Var}(X_t) = a^2 \text{Var}(X_{t-1}) + Var(e_t) > a^2 \text{Var}(X_{t-1}), \tag{4.6}$$

since $\gamma_0 = \text{Var}(X_t) = \text{Var}(X_{t-1})$, it follows that $|a| < 1$.

We note that the first property in (4.5) does play a crucial role in the proof of necessity of (4.3). It simply says that any solution of (4.1) with such property does not depend on the *future inputs*. This is certainly a property of systems and devices that do not *anticipate the future inputs* (changes)—unlike human beings physical systems do not learn from the past to anticipate the future! Such systems are referred to as *nonanticipative*, or *causal*. Hereafter we shall use the term *causal* for solutions of ARMA models having this property.

For ease of reference a more general version of the previous result is stated in the next theorem.

Theorem 4.1. Consider the difference equation (4.1) with $\{e_t\}$ a zero-mean stationary process. Then, $|a| < 1$ is necessary and sufficient for (4.1) to have a *causal stationary solution*.

For an example of an AR(1) model with $|a| > 1$ having a noncausal stationary solution, see Problem 4.4.

4.1.2 AR(p) Models

For the general case $p > 1$, one could use repeated substitutions as in (4.2), but it is lengthy and does not lead to neat results. Instead, it is convenient and instructive to write a pth order stochastic difference equation (2.40) in the vector form

$$\begin{bmatrix} X_t \\ X_{t-1} \\ \vdots \\ X_{t-p+1} \end{bmatrix} = \begin{bmatrix} a_1 & a_2 & a_3 & \cdot & \cdot & a_p \\ 1 & 0 & 0 & \cdot & \cdot & 0 \\ 0 & 1 & 0 & \cdot & \cdot & 0 \\ \vdots & \vdots & \vdots & \vdots & \vdots & \vdots \\ 0 & 0 & \cdot & 0 & 1 & 0 \end{bmatrix} \begin{bmatrix} X_{t-1} \\ X_{t-2} \\ \vdots \\ X_{t-p} \end{bmatrix} + \begin{bmatrix} e_t \\ 0 \\ \vdots \\ 0 \end{bmatrix},$$

or

$$\boldsymbol{X}_t = A\boldsymbol{X}_{t-1} + \boldsymbol{e}_t, \tag{4.7}$$

where $\boldsymbol{X}_t = (X_t, X_{t-1}, \cdots, X_{t-p+1}), \boldsymbol{e}_t = (e_t, 0, \cdots, 0)$, and A is the above $p \times p$ matrix called the *companion* matrix of the AR(p) model. Note that the first component equation of (4.7) is precisely the AR(p) model, and the rest is merely a restatement of the obvious identities $X_{t-i} = X_{t-i}$, $i = 2, \cdots, p$. The idea of writing a scalar equation as a first-order vector equation is useful in the maximum likelihood estimation of its coefficients and the Akaike's state-space representation; see Akaike (1975); Jones (1980); Pollock (1999) and Section 4.3.2.

By repeated substitutions in (4.7), we arrive at an analogue of (4.2),

$$\boldsymbol{X}_t = \sum_{k=0}^{n} A^k \boldsymbol{e}_{t-k} + A^{n+1}\boldsymbol{X}_{t-n-1}, n \geq 0.$$

It is evident that $\lim_{n\to\infty} \sum_{k=0}^{n} A^k \boldsymbol{e}_{t-k}$ exists, if and only if $\lim_{n\to\infty} A^{n+1}\boldsymbol{X}_{t-n-1}$ exists. Assuming the convergence of the latter and defining $\boldsymbol{V}_t = \lim_{n\to\infty} A^{n+1}\boldsymbol{X}_{t-n-1}$, it can be shown that

$$\boldsymbol{V}_t = A\boldsymbol{V}_{t-1}, \text{ for } all\ t, \text{ or } \boldsymbol{V}_t = A^t\boldsymbol{X},$$

where $\boldsymbol{X}$ is an arbitrary random p-vector whose components have finite second moments. In this case any causal solution of (4.7) can be represented as

$$\boldsymbol{X}_t = \sum_{k=0}^{\infty} A^k \boldsymbol{e}_{t-k} + A^t\boldsymbol{X},$$

which is nonstationary, in general. However, when $\boldsymbol{X} \equiv 0$, then a causal stationary solution of (4.7) and its covariance function are given by

$$\begin{cases} \boldsymbol{X}_t = \sum_{k=0}^{\infty} A^k \boldsymbol{e}_{t-k}, \\ \text{Cov}(\boldsymbol{X}_{t+k}, \boldsymbol{X}_t) = A^k \sum_{j=0}^{\infty} A^j \boldsymbol{C} A'^j, \end{cases} \tag{4.8}$$

where $\boldsymbol{C} = \text{Cov}(\boldsymbol{e}_t)$, provided that the infinite series converge. Their convergence, however, depends on the eigenvalues of A.

For any $p \times p$ matrix A, let $S(A)$ denote the set of eigenvalues of $A, \lambda(A) = \max_{\lambda \varepsilon S(A)} |\lambda|$ be the spectral radius of A, and $a_{i,j}$ or $(A)_{i,j}$ be the (i, j)th element of A. The condition for the convergence of the series in (4.8) involves the eigenvalues of the asymmetric matrix A. To highlight this, first we introduce the Schur decomposition of matrices (Golub and Van Loan, 1989), established in 1909.

Theorem 4.2. (Schur decomposition). Any square matrix A can be written as $A = PTP^*$, where P is a unitary (orthogonal) matrix with P^* its conjugate transpose and T is a lower triangular matrix.

Proof. We use induction. Note that the statement is true for $n = 1$. Suppose it holds for all matrices of order $n - 1$ or less. Let λ be an eigenvalue of A and q be the corresponding normalized eigenvector, that is, $Aq = \lambda q$ and $q^*q = 1$. Consider any unitary (orthogonal) matrix Q with q as its last column and partitioned as $Q = (Q_1 : q)$, then $Q_1^* q = 0$ and we have

$$Q^*AQ = \begin{pmatrix} Q_1^* \\ \cdots \\ q^* \end{pmatrix} (AQ_1 : Aq) = \begin{pmatrix} Q_1^*AQ_1 & 0 \\ q^*AQ_1 & \lambda \end{pmatrix},$$

where $A_1 = Q_1^*AQ_1$ is an $(n-1) \times (n-1)$ matrix. By induction there is a unitary (orthogonal) matrix Q_2 such that $Q_2^*A_1Q_2$ is lower triangular. Then, with $P = Qdiag(1, Q_2), P^*AP$ is lower triangular. Further, P can be chosen so that diagonal entries of this lower triangular matrix are ordered according to their modulus from the largest to the smallest.

Q.E.D.

We use the Schur decomposition to prove the following theorem. For alternative proofs, see Anderson (2000); Pollock (1999) and Brockwell and Davis (1991).

Theorem 4.3. The vector difference equation (4.7) with A an arbitrary $p \times p$ matrix and $\{e_t\}$ a white noise process, has a zero-mean causal stationary solution, if and only if $\lambda(A) < 1$.

Proof. Using Theorem 4.2, A can be decomposed as $P^*AP = T$, where T is a lower triangular matrix and $S(A) = \{t_{i,i}; i = 1, 2, \cdots, p\}$ is the set of diagonal elements of T with $|t_{1,1}| \geq \cdots \geq |t_{p,p}|$. Then (4.7) can be written as $\boldsymbol{X}_t = PTP^*\boldsymbol{X}_{t-1} + \boldsymbol{e}_t$, or

$$P^*\boldsymbol{X}_t = TP^*\boldsymbol{X}_{t-1} + P^*\boldsymbol{e}_t.$$

Let $\boldsymbol{Y}_t = P^*\boldsymbol{X}_t, \boldsymbol{\eta}_t = P^*\boldsymbol{e}_t$, then

$$\boldsymbol{Y}_t = T\boldsymbol{Y}_{t-1} + \boldsymbol{\eta}_t. \tag{4.9}$$

Suppose (4.7) has a causal stationary solution in terms of $\boldsymbol{e}_t$, then (4.9) has a causal stationary solution in terms of $\boldsymbol{\eta}_t$. Since T is (lower) triangular, the first equation in (4.9), that is $Y_{t,1} = t_{1,1}Y_{t-1,1} + \eta_{t,1}$ must have a causal stationary solution. Thus, by Theorem 4.1, $|t_{1,1}| < 1$, and $|t_{i,i}| < 1$, for all $2 \leq i \leq p$, that is, $\lambda(A) < 1$. Proof of the converse is not hard and is left to the reader.

Q.E.D.

It follows from Theorem 4.3 that all solutions of an AR(p) model with $\{e_t\}$ a white noise and A its companion matrix with $\lambda(A) < 1$, are of the form $X_t = \sum_{k=0}^{\infty}(A^k)_{1,1}e_{t-k} + V_t$, where $\{V_t\}$ satisfies the homogeneous difference equation

$$V_t = \sum_{k=1}^{p} a_k V_{t-k},\ t \in \mathbb{Z}.$$

Furthermore, it is evident from (4.8) that the covariance function $\{\gamma_k\}$ of a causal stationary solution decays to zero exponentially fast. This property of the covariance function is helpful in distinguishing AR and MA models.

For a given AR(p) model we need to examine the roots of the *characteristic polynomial* of its companion matrix A given by $|zI - A| = 0,\ z \in \mathbb{C}$, to see if it has a causal stationary solution. Since A has a special form, the coefficients of the characteristic polynomial is related to the a_k's in the following simple manner: Let $D_p(z) = |zI - A|$, note that $D_1(z) = -a_1 + z$. By expanding $|zI - A|$ on the last column we get $D_p(z) = -a_p + zD_{p-1}(z)$, using this recursion repeatedly we arrive at

$$\begin{aligned} D_p(z) &= -a_p - a_{p-1}z - a_{p-2}z^2 - \cdots - a_1 z^{p-1} + z^p \\ &= z^p(1 - a_1 z^{-1} - a_2 z^{-2} - \cdots - a_p z^{-p}) = z^p \alpha(z^{-1}), \end{aligned}$$

where

$$\alpha(z) = 1 - a_1 z - a_2 z^2 - \cdots - a_p z^p. \tag{4.10}$$

In fact, if $z_1, \cdots, z_p$ are the roots of $\alpha(z)$ and $\lambda_1, \cdots, \lambda_p$ are the eigenvalues of A, we have

$$z_i = \lambda_i^{-1},\ i = 1, 2, \cdots, p,$$

so that $\lambda(A) < 1$, if and only if all roots of $\alpha(z) = 0$ are greater than one in absolute value.

In the sequel we refer to $\alpha(z)$ as the *characteristic polynomial* and the equation $\alpha(z) = 0,\ z \in \mathbb{C}$, as the *characteristic equation* of an AR(p) process. Now, Theorem 4.3 restated in terms of the roots of the characteristic equation of an AR(p) model: a necessary and sufficient condition for an AR(p) model to have a causal stationary solution is that all roots of its characteristic polynomial to lie outside the unit circle, that is,

$$|z_i| > 1, i = 1, \cdots, p. \tag{4.11}$$

We illustrate this result with few examples.

Example 4.1. In the following $\{e_t\}$ is a white noise process.

(a) For $X_t = 2X_{t-1} + e_t$, the only root of $\alpha(z) = 1 - 2z = 0$, is $z_1 = \dfrac{1}{2}$ and lies inside the unit circle, hence it does not have a causal stationary solution; see Problem 4.4.

(b) For $X_t = 2.25\, X_{t-1} - .5X_{t-2} + e_t$, the two roots of

$$\alpha(z) = 1 - 2.25z + .5z^2 = (1-2z)(1-.25z) = 0$$

are $z_1 = \frac{1}{2}$, $z_2 = 4$, since $z_1 = \frac{1}{2}$ lies inside the unit circle, it cannot have a causal stationary solution.

(c) For $X_t = X_{t-1} - \frac{3}{16}X_{t-2} + e_t$, the roots of $\alpha(z) = 1 - z + \frac{3}{16}z^2 = \left(1 - \frac{3}{4}z\right)\left(1 - \frac{1}{4}z\right) = 0$ are $z_1 = \frac{4}{3}$ and $z_2 = 4$. Since both lie outside the unit circle, it does have a unique causal stationary solution.

(d) For $X_t = aX_{t-1} - X_{t-2} + e_t$, a real, the roots of $\alpha(z) = 1 - az + z^2 = 0$, are

$$z = \frac{a \pm \sqrt{a^2 - 4}}{2} = \frac{a}{2} \pm \sqrt{\frac{a^2}{4} - 1}.$$

When $|a| < 2$, these roots are complex and lie on the unit circle, so that the model does not have any stationary solution.

For more examples of the above type, see Problems 4.7, 4.8 and 4.10.

4.2 ARMA MODELS

In this section, we give conditions on the ARMA coefficients for the model to have a causal and invertible solution. An unconstrained reparameterization of the causality and invertibility regions are presented. The AR, MA parameters, the covariance function and matrix of n consecutive values of a stationary solution are expressed in terms of the ARMA coefficients.

4.2.1 Invertible ARMA Models

A stochastic process $\{X_t\}$ satisfying the difference equation

$$X_t = \sum_{k=1}^{p} \alpha_k X_{t-k} + \varepsilon_t + \sum_{j=1}^{q} \beta_j \varepsilon_{t-j}, \tag{4.12}$$

is said to be an *autoregressive and moving average* process of orders p and q, or an ARMA(p,q) model for short. First, we find conditions on $\alpha_1, \cdots, \alpha_p$ so that (4.12) has a *causal* stationary solution in terms of $\{\varepsilon_t\}$. Then, conditions on $\alpha_1, \cdots, \alpha_p, \beta_1, \cdots, \beta_q$ are given so that $\{\varepsilon_t\}$ is actually the *innovation* of a solution of (4.12).

By writing (4.12) in vector form as in (4.7) with $e_t = \varepsilon_t + \sum_{j=1}^{q} \beta_i \varepsilon_{t-j}$, and repeating the arguments in the proof of Theorem 4.3, we find that (4.11) is a necessary and sufficient condition for (4.12) to have a causal stationary solution, in terms of $\{\varepsilon_t\}$. For $\{\varepsilon_t\}$ to be the innovation process of $\{X_t\}$, we need to express ε_t in terms of the past $X_t, X_{t-1}, \ldots$. More precisely, an ARMA(p, q) model is said to be *invertible*, if there exists a sequence $\{d_k\}_{k=0}^{\infty}$ of scalars such that

$$\varepsilon_t = \sum_{k=0}^{\infty} d_k X_{t-k}, \tag{4.13}$$

and the series is mean-convergent.

A simple way to find conditions for invertibility is to write (4.12) first as

$$\varepsilon_t = \sum_{j=1}^{q}(-\beta_j)\varepsilon_{t-j} + \left(X_t - \sum_{k=1}^{p} \alpha_k X_{t-k}\right),$$

and view it as a difference equation for $\{\varepsilon_t\}$ with $\left\{X_t - \sum_{k=1}^{p} \alpha_k X_{t-k}\right\}$ as the input. Then, in vector form we have

$$\begin{bmatrix} \varepsilon_t \\ \varepsilon_{t-1} \\ \vdots \\ \varepsilon_{t-q+1} \end{bmatrix} = \begin{pmatrix} -\beta_1 & -\beta_2 & \cdots & & -\beta_q \\ 1 & 0 & \cdots & & 0 \\ \vdots & \vdots & & & \vdots \\ 0 & 0 & & 1 & 0 \end{pmatrix} \begin{bmatrix} \varepsilon_{t-1} \\ \varepsilon_{t-2} \\ \vdots \\ \varepsilon_{t-q} \end{bmatrix} + \begin{bmatrix} X_t - \sum_{k=1}^{p} \alpha_k X_{t-k} \\ 0 \\ \vdots \\ 0 \end{bmatrix}.$$

Using repeated substitutions as in (4.2), for any $n \geq 0$ and with obvious choice of notation, we obtain

$$\varepsilon_t = \sum_{k=0}^{n} \boldsymbol{\beta}^k \boldsymbol{X}_{t-k} + \boldsymbol{\beta}^{n+1} \varepsilon_{t-n-1}. \tag{4.14}$$

Applying Theorem 4.3 to (4.14), it follows that a necessary and sufficient condition for invertibility of (4.12) is that all roots of $\beta(z) = 1 + \sum_{j=1}^{q} \beta_j z^j = 0$ must lie outside the unit circle. In this case, since $\lim_{n\to\infty} \boldsymbol{\beta}^{n+1} \varepsilon_{t-n-1} = 0$, all solutions are of the form

$$\varepsilon_t = \sum_{k=0}^{\infty} \boldsymbol{\beta}^k \boldsymbol{X}_{t-k}.$$

We summarize these results in the following.

Theorem 4.4. The ARMA model (4.12) is invertible and has a causal stationary solution, if and only if all roots of the polynomial equations

$$\begin{cases} \alpha(z) = 1 - \sum_{k=1}^{p} \alpha_k z^k = 0, \\ \beta(z) = 1 + \sum_{j=1}^{q} \beta_j z^j = 0, \end{cases} \tag{4.15}$$

lie outside the unit circle.

From (4.15) and an analogue of (4.8), it is evident that the covariance function $\{\gamma_k\}$ of a causal stationary solution of an ARMA model decays to zero exponentially fast. For more results of this nature, see Sections 4.2.3 and 4.2.4.

4.2.2 Reparameterization of the Causality Region

In this section, the shape of the causality region of ARMA(p, q) models is studied. Being an irregular subset of the Euclidean space R^p, we transform it to the whole R^p-space. This new configuration of the region is ideal for selecting admissible parameter values for computer simulation, maximum likelihood calculation and Bayesian analysis of ARMA models; see Osborn (1976), Jones (1987) and Marriott and Smith (1992) for more details on the roles and applications of the transformed region.

The causality and invertibility conditions in Theorem 4.4 constrain the parameter vectors $\boldsymbol{\alpha} = (\alpha_1, \cdots, \alpha_p)$ and $\boldsymbol{\beta} = (\beta_1, \cdots, \beta_q)$ to be in regions $C_p \in R^p$ and $C'_q \in R^q$. Thus, the admissible values of $(\boldsymbol{\alpha}, \boldsymbol{\beta})$ lie in the region $C_p \times C'_q$, the shape of which for low-order ARMA (p, q) models, with $p+q \leq 2$, are easy to visualize and work with. What is the shape of $C_p \times C'_q$ for $p+q > 2$? Since C'_q is obtainable from C_p upon replacing $\boldsymbol{\beta}$ by $-\boldsymbol{\beta}$, it suffices to focus on C_p. Note that for an AR(1) model, the admissible region is $C_1 = (-1, 1)$ and for an AR(2) model C_2 is the triangular region in R^2 determined by the requirement on the roots of $1 - \alpha_1 z - \alpha_2 z^2 = 0$, namely,

$$-1 < \alpha_2 < 1, -1 + \alpha_2 < \alpha_1 < 1 - \alpha_2. \tag{4.16}$$

We show that C_p is a certain Cartesian product of the simpler C_1 and C_2 regions. For this we rely on the quadratic factorization of the polynomial $\alpha(z)$. Since the roots of this polynomial can be divided into pairs that are either complex conjugate or real, each pair generates a quadratic polynomial of the form $1 - \alpha_{1j} z - \alpha_{2j} z^2$ with real coefficients. Thus, $\alpha(z)$ can be factored as

$$\alpha(z) = \begin{cases} \prod_{j=1}^{p/2} (1 - \alpha_{1j} z - \alpha_{2j} z^2) & \text{for } p \text{ even}, \\ (1 - \alpha_0 z) \prod_{j=1}^{(p-1)/2} (1 - \alpha_{1j} z - \alpha_{2j} z^2) & \text{for } p \text{ odd}. \end{cases} \tag{4.17}$$

Consequently, $C_p = C_2 \times \cdots \times C_2(p/2$ times) when p is even and $C_p = C_2 \times \cdots \times C_2 \times C_1$ when p is odd. Expanding (4.17) and matching the like powers of z expresses the components of $\boldsymbol{\alpha} = (\alpha_1, \cdots, \alpha_p)$ in terms of the new parameters $\boldsymbol{\alpha}' = (\alpha_0, \alpha_{11}, \alpha_{12}, \alpha_{21}, \alpha_{22}, \cdots)$. This reparameterization of AR(p) models is evidently not one-to-one, but any given $\boldsymbol{\alpha}'$ determines $\boldsymbol{\alpha}$ uniquely.

Next, we introduce a transformation between C_p and R^p. To motivate the idea, we note that for $\alpha_0 \in (-1, 1) = C_1$, the logarithmic or "Fisher type" transformation sends α_0 to R, that is,

$$\alpha_0^* = \log \frac{1 + \alpha_0}{1 - \alpha_0} \in R. \tag{4.18}$$

For AR(2) models, the region C_2 in (4.16) is first transformed to the square $C_1 \times C_1$ using

$$\alpha_1' = \frac{\alpha_1}{1 - \alpha_2}, \alpha_2' = \alpha_2,$$

then applying (4.18) to α_1' and α_2' separately, that is,

$$\alpha_1^* = \log \frac{1 + \alpha_1'}{1 - \alpha_1'} = \log \frac{1 - \alpha_2 + \alpha_1}{1 - \alpha_2 - \alpha_1}, \alpha_2^* = \log \frac{1 + \alpha_2'}{1 - \alpha_2'}, \tag{4.19}$$

gives the desired result. Finally, for the general AR(p) model (4.17), the transformation (4.19) is applied to each pair $(\alpha_{1j}, \alpha_{2j})$ to obtain the unconstrained reparameterization $(\alpha_{1j}^*, \alpha_{2j}^*) \in R^2$.

As a potential application of the above unconstrained reparameterization of ARMA models, we consider the problem of choosing appropriate parameter configurations to simulate data from such models. Often, the results of simulation studies in time series analysis depend critically on the particular choices of the parameter configurations. When little or no information about important segments of $C_p \times C_q'$ is available, it seems prudent to simulate series corresponding to a number of choices of parameter values that are representative of the admissible region. This is simple to do for C_1 and C_2. However, for $p + q > 2$ it is much more difficult to set up appropriate grid of parameter values. Nevertheless, using the above reparameterization one may start with a grid of parameter values in R^{p+q} and back-transform them via (4.18)–(4.19) to obtain the corresponding parameter configurations in the $C_p \times C_q'$ setup.

An alternative reparameterization of ARMA coefficients in terms of the partial correlations of stationary processes and their roles in applications are discussed in Chapter 7.

4.2.3 AR and MA Parameters of ARMA Processes

In this section, the autoregressive parameters $\{a_k\}$ and moving average parameters $\{b_k\}$ of the causal, invertible model (4.12) are expressed in terms of its AR coefficients $\alpha_1, \ldots, \alpha_p$ and MA coefficients $\beta_1, \ldots, \beta_q$. As an illustration of proper use of the shift operator (symbol) B, we exploit it effectively for the first time in this section.

When (4.12) is invertible, then $\{\varepsilon_t\}$ and $\{X_t\}$ are stationarily correlated, that is, $\text{Cov}(X_{t+k}, \varepsilon_t)$ depends only on k, and hence by Corollary 5.3 (b) they have the same shift operator B defined by $BX_t = X_{t-1}$ and $B\varepsilon_t = \varepsilon_{t-1}$. Using this symbol, (4.12) can be written compactly as

$$\alpha(B)X_t = \beta(B)\varepsilon_t, \tag{4.20}$$

where

$$\alpha(B) = I - \sum_{k=1}^{p} \alpha_k B^k, \ \beta(B) = I + \sum_{j=1}^{q} \beta_j B^j,$$

are the operators corresponding to the AR and MA parts of the model, also compare this with (4.32) . Causality and invertibility imply that both $\alpha(B)$ and $\beta(B)$ are invertible operators (Chapter 9). Consequently, from (4.20) we have

$$\begin{cases} X_t &= \alpha^{-1}(B)\beta(B)\varepsilon_t, \\ \varepsilon_t &= \beta^{-1}(B)\alpha(B)X_t, \end{cases} \tag{4.21}$$

where the first identity is the $MA(\infty)$ and the second is related to the AR representation of $\{X_t\}$; see Appendix A, Section 2.1. From the first, we may identify the MA parameters $\{b_k\}$ as the Taylor coefficients of $\alpha^{-1}(z)\beta(z)$:

$$\sum_{k=0}^{\infty} b_k z^k = \left(1 - \sum_{k=1}^{p} \alpha_k z^k\right)^{-1} \left(1 + \sum_{j=1}^{q} \beta_j z^j\right), \ |z| < 1.$$

Let $m = \max\{p, q+1\}$, then the MA parameters are computed recursively using

$$b_\ell = \begin{cases} \sum_{k=1}^{\ell} \alpha_k b_{\ell-k} + \beta_\ell & , \quad 0 \le \ell < m, \\ \sum_{k=1}^{p} \alpha_k b_{\ell-k} & , \quad \ell \ge m. \end{cases} \tag{4.22}$$

The Taylor coefficients of $\beta^{-1}(z)\alpha(z)$ are the negatives of the AR parameters of the process, so that a slight modification of (4.22) can be used to compute them.

Example 4.2. For the ARMA(1,1) model

$$X_t = \alpha X_{t-1} + \varepsilon_t + \beta\varepsilon_{t-1}, |\alpha| < 1, |\beta| < 1, \alpha \neq -\beta,$$

its MA and AR parameters calculated using (4.22) are

$$b_0 = 1, b_j = (\alpha + \beta)\alpha^{j-1}, j \ge 1,$$

$$a_0 = 1, a_j = (\alpha + \beta)(-1)^j \beta^{j-1}.$$

Setting $\beta = 0$ in the first formula, one obtains the MA parameters of an AR(1) process. Similarly, setting $\alpha = 0$ in the second we obtain the AR parameters of an MA(1) process.

4.2.4 Covariance Function of ARMA Processes

In this section, for a causal and invertible ARMA model its covariance function is expressed in terms of its parameters $\theta = (\alpha_1, \cdots, \alpha_p, \beta_1, \cdots, \beta_q, \sigma^2)$. There are several ways to do this, the choice of a method is usually dictated by the application at hand.

The first and the most general method relies on the MA representation of the process. From the moving average representation of a stationary solution of (4.12), see (4.21), it follows that

$$\gamma_k = \sigma^2 \sum_{j=0}^{\infty} b_j b_{j+k}, k = 0, 1, 2, \cdots$$

where $b_0 = 1, b_1, b_2, \cdots$ are computed using (4.22). For the ARMA(1,1) model of Example 4.2, using this formula we obtain

$$\gamma_k = \sigma^2 \left(b_k + \sum_{j=1}^{\infty} b_j b_{j+k} \right), \quad k \geq 0.$$

Thus, for $k = 0$,

$$\gamma_0 = \sigma^2 (1 - \alpha^2)^{-1} (1 + \beta^2 + 2\alpha\beta),$$

and for $k \geq 1$,

$$\begin{aligned} \gamma_k &= \sigma^2 \left[(\alpha + \beta)\alpha^{k-1} + (\alpha + \beta)^2 \sum_{j=1}^{\infty} \alpha^{2j+k-2} \right] \\ &= \sigma^2 (\alpha + \beta)\alpha^{k-1} \left[1 + (\alpha + \beta) \sum_{j=1}^{\infty} \alpha^{2j-1} \right] \\ &= \sigma^2 (1 - \alpha^2)^{-1} (\alpha + \beta)(1 + \alpha\beta)\alpha^{k-1}. \end{aligned}$$

For some applications the following recursive method of computing the covariance function is preferable. It is closely related to the recursion (4.22) for computing the MA parameters.

On multiplying both sides of (4.12) by $X_{t-k}, k = 0, 1, 2, \cdots$, and taking expectation, we obtain

$$EX_t X_{t-k} = \sum_{i=1}^{p} \alpha_i E X_{t-i} X_{t-k} + E\varepsilon_t X_{t-k} + \sum_{j=1}^{q} \beta_j E\varepsilon_{t-j} X_{t-k}.$$

Since $\{X_t\}$ is a zero-mean causal stationary process, using the first representation in (4.21) it follows that $E\varepsilon_{t-j} X_{t-k} = b_{j-k}$ and hence

$$\gamma_k = \begin{cases} \sum_{i=1}^{p} \alpha_i \gamma_{k-i} + \sigma^2 \sum_{k \leq j \leq q} \beta_j b_{j-k}, & 0 \leq k < m, \\ \sum_{i=1}^{p} \alpha_i \gamma_{k-i} & k \geq m. \end{cases} \tag{4.23}$$

In using this we need to compute first $b_1, b_2, \cdots, b_q$ from (4.22) and then find $\gamma_0, \cdots, \gamma_p$ from the first $p+1$ equations in (4.23). The remaining values of the covariance function, namely, $\gamma(p+1), \gamma(p+2), \cdots$, are obtained recursively using the second equation in (4.23).

Example 4.3. For the ARMA(1,1) process of Example 4.2, the set of equations (4.23) reduces to

$$\begin{cases} \gamma_0 = \alpha\gamma_1 + \sigma^2[1+\beta(\alpha+\beta)], \\ \gamma_1 = \alpha\gamma_0 + \sigma^2\beta, \\ \gamma_k = \alpha^{k-1}\gamma_1, k \geq 2. \end{cases}$$

The first two equations give

$$\begin{cases} \gamma_0 = & \sigma^2(1-\alpha^2)^{-1}(1+\beta^2+2\alpha\beta), \\ \gamma_1 = & \sigma^2(1-\alpha^2)^{-1}(\alpha+\beta)(1+\alpha\beta). \end{cases}$$

Other γ_k's are computed using the recursion $\gamma_k = \alpha^{k-1}\gamma_1, k \geq 2$.

Remark 4.1 Equations (4.23) reveal an important property of stationary ARMA(p,q) models, namely given only $\gamma_0, \cdots, \gamma_p, b_1, \cdots, b_q$ and σ^2, it is possible to reconstruct the whole covariance function $\{\gamma_k\}$. Is the converse true? That is, given the above information, does there exist an ARMA(p,q) model with $\gamma_0, \cdots, \gamma_p$, as its first $p+1$ autocovariances, $b_1, \cdots, b_q$ as its first q MA parameters and σ^2 as its innovation variance? An affirmative answer to this problem is given in Section 4.5.

4.2.5 Covariance Matrix of ARMA Processes

The next two methods provide useful representations for the covariance matrix Γ_n of n consecutive observations from (4.12) in terms of simpler matrices involving its AR and MA coefficients.

To develop the third alternative for expressing γ_k in terms of the AR and MA coefficients, note that from (4.12) for $t = 1, 2, \cdots, n$, we have

$$\begin{array}{lcl} X_1 = \alpha_1 X_0 + \alpha_2 X_{-1} + & \cdots & +\alpha_p X_{1-p} + \varepsilon_1 + \beta_1\varepsilon_0 + \cdots + \beta_q\varepsilon_{1-q}, \\ \quad\vdots & \vdots & \\ X_t = \alpha_1 X_{t-1} + \alpha_2 X_{t-2} + & \cdots & +\alpha_p X_{t-p} + \varepsilon_t + \beta_1\varepsilon_{t-1} + \cdots + \beta_q\varepsilon_{t-q}, \\ \quad\vdots & \vdots & \\ X_n = \alpha_1 X_{n-1} + \alpha_2 X_{n-2} + & \cdots & +\alpha_p X_{n-p} + \varepsilon_n + \beta_1\varepsilon_{n-1} + \cdots + \beta_q\varepsilon_{n-q}. \end{array}$$

Taking the observables $X = (X_1, \cdots, X_n)$ to the left, these n equations relating the values and innovations $\varepsilon = (\varepsilon_1, \ldots, \varepsilon_n)$ of an ARMA(p,q) model, can be written in matrix form as

$$L_{\alpha,n}X = L_{-\beta,n}\varepsilon + L_{\alpha,\beta}e,$$

where $L_{\alpha,n}$ is an $n \times n$ lower triangular matrix with 1's on the leading diagonal, $-\alpha_1$ on the first subdiagonal, $-\alpha_2$ on the second subdiagonal and so on with the provision that $\alpha_i = 0$ for $i > p$ and

$$L_{\alpha,\beta} = \begin{bmatrix} A_\alpha & A_\beta \\ 0 & 0 \end{bmatrix}, A_\alpha = \begin{bmatrix} \alpha_p & \alpha_{p-1} & \cdots & \alpha_1 \\ 0 & \alpha_p & \cdots & \alpha_2 \\ & \vdots & & \vdots \\ 0 & 0 & \cdots & \alpha_p \end{bmatrix}. \tag{4.24}$$

Moreover, $e = (X_{1-p}, \cdots, X_0, \varepsilon_{1-q}, \cdots, \varepsilon_0)$ is the $(p+q)$-vector of "initial" values of the model (4.12) with the covariance matrix

$$\text{Cov}(e) = \sigma^2 \begin{bmatrix} \Gamma_p & T_{p,q} \\ T'_{p,q} & I_q \end{bmatrix} = \sigma^2 \Omega, \tag{4.25}$$

where $T_{p,q} = (b_{i-j})_{i=1,p}^{j=1,q}$. In the sequel when there is no danger of confusion we use L_α in place of $L_{\alpha,n}$. Since ε and e are independent random vectors, it follows from above linear relationships among X, ε and e that

$$\Gamma_n = \sigma^2 L_\alpha^{-1} (L_{-\beta} L'_{-\beta} + L_{\alpha,\beta} \Omega L'_{\alpha,\beta}) L_\alpha^{'^{-1}}. \tag{4.26}$$

This representation of Γ_n, due to Ljung and Box (1979), see also Galbraith and Galbraith (1974), and the next one are useful in the numerical computation of the exact likelihood function of a Gaussian stationary time series.

The fourth method, due to Ansley (1979), is obtained using a simple transformation of the observables $X_1, \ldots, X_n$ which unlike the previous one, does not rely on "initial" values. For $m = \max(p, q)$, set

$$\begin{cases} Z_t = X_t, \quad t = 1, \cdots, m, \\ Z_t = X_t - \alpha_1 X_{t-1} \cdots \alpha_p X_{t-p}, t > m. \end{cases} \tag{4.27}$$

Then, with $Z = (Z_1, \cdots, Z_n)$ and $X = (X_1, \cdots, X_n)$, (4.27) can be written in matrix form as

$$Z = \Phi X = \begin{pmatrix} I_m & 0 \\ \Phi_{21} & L_{\alpha,\, n-m} \end{pmatrix} X, \tag{4.28}$$

where I_m is the $m \times m$ identity matrix and $\Phi_{21} = \begin{bmatrix} -A_\alpha \\ 0 \end{bmatrix}$. From (4.28) it follows that

$$\Gamma_n = \text{Cov}(X) = \Phi^{-1} \text{Cov}(Z) \Phi^{-1'} = \Phi^{-1} K \Phi^{-1'}, \tag{4.29}$$

where $K = (k_{ij}) = \text{Cov}(Z)$ is a symmetric band matrix with the bandwidth m. Its entries computed from (4.27) and (4.12) are:

$$k_{ij} = \begin{cases} \gamma_{i-j} & 1 \le i, j \le m, \\ \gamma_{i-j} - \sum_{\ell=1}^{p} \alpha_\ell \gamma_{\ell-|i-j|} & \min(i,j) \le m < \max(i,j) \le 2m, \\ \sigma^2 \sum_{\ell=0}^{q} \beta_\ell \beta_{\ell+|i-j|} & \min(i,j) > m, \\ 0 & \text{otherwise,} \end{cases} \tag{4.30}$$

where $\beta_0 = 1, \beta_j = 0$ for $j > q$.

As concrete examples, the reader may obtain the representations (4.26) and (4.30) for the ARMA(1,1) model in Example 4.3.

What happens to the covariance decompositions (4.26) and (4.29) when $n \to \infty$? This question is of particular interest when computing the likelihood function of ARMA models for large samples. It is known, from Wise (1955), that the semi-infinite covariance matrix Γ of $(X_1, X_2, \ldots)$ has the simpler representation

$$\Gamma = \sigma^2 L_\alpha^{-1} L_{-\beta} L'_{-\beta} L_\alpha^{-1\prime}, \tag{4.31}$$

where, now, L_α is a semi-infinite lower triangular matrix. Comparing (4.31) with (4.26) and (4.30) reveals that the second term in (4.26) and Γ_m at the upper-left corner of K are merely ‘ corrections ’ for a finite segment of the process. Their presence creates additional computational burdens when computing Γ_n^{-1}, $|\Gamma_n|$ in the Gaussian likelihood; see Section 4.4.

Wise's (1955) derivation of (4.31) is based on using the semi-infinite shift matrix $U = (\delta_{i+1,j})$ where $\delta_{i,j}$ is the Kronecker's delta. Writing (4.12) as

$$\left(I - \sum_{i=1}^{p} \alpha_i U^i\right) X = \left(I + \sum_{i=1}^{q} \beta_i U^i\right) \varepsilon, \tag{4.32}$$

with $X = (X_1, X_2, \ldots)$ and $\varepsilon = (\varepsilon_1, \varepsilon_2, \ldots)$, it follows that

$$L_\alpha \Gamma L'_\alpha = \sigma^2 L_{-\beta} L'_{-\beta}, \tag{4.33}$$

where $L_\alpha = I - \sum_{i=1}^{p} \alpha_i U^i$ and $L_{-\beta} = I + \sum_{i=1}^{q} \beta_i U^i$ are the semi-infinite analogues of those appearing in (4.26). It is instructive to compare (4.32) and its consequences with (4.20) using the shift operator B.

For pure AR models, $L_{-\beta}$ is the identity matrix, hence from (4.33) the formal inverse of $\sigma^{-2}\Gamma$ has entries that are quadratic in the AR coefficients. This result, proved more rigorously in Section 4.4, plays a central role in inference for the parameters of Gaussian AR models (Siddiqui , 1958).

4.3 PREDICTION AND MARKOVIAN REPRESENTATION

In this section, we focus on the linear prediction of ARMA models and obtain recursive and updating formulas for their predictors. These formulas are then used to motivate and develop the Akaike's (1975) Markovian or state-space representation of ARMA processes. The Kalman filter is introduced and used in the numerical evaluation of the likelihood function of a Gaussian ARMA model. The latter is useful in the estimation of parameters using nonlinear optimization routines (Jones, 1980).

4.3.1 Prediction of ARMA Models

For a causal, invertible ARMA model we compute the linear least square predictor of the future values of $\{X_t\}$. Owing to the simple structure of (4.12), we are able to obtain recursive formulas for the predictors and their updates when new observations become available.

For $\ell = 0, \pm 1, \pm 2, \cdots$, and t arbitrary let $\hat{X}_{t+\ell|t}$ denote the linear least square predictor of $X_{t+\ell}$ based on the past $X_t, X_{t-1}, \cdots$. Note that $X_{t+\ell|t} = X_{t+\ell},$ for $\ell \leq 0$, but for $\ell \geq 1$, since

$$X_{t+\ell} = \sum_{k=1}^{p} \alpha_k X_{t+\ell-k} + \varepsilon_{t+\ell} + \sum_{k=1}^{q} \beta_k \varepsilon_{t+\ell-k}, \tag{4.34}$$

from the linearity of the orthogonal projection, the ℓ-step ahead predictor of the process with $m = \max\{p, q+1\}$ satisfies the recursion:

$$\hat{X}_{t+\ell|t} = \begin{cases} \sum_{k=1}^{\ell-1} \alpha_k \hat{X}_{t+\ell-k|t} + \sum_{k=\ell}^{p} \alpha_k X_{t+\ell-k} + \sum_{k=\ell}^{q} \beta_k \varepsilon_{t+\ell-k}, & 0 \leq \ell < m, \\ \sum_{k=1}^{p} \alpha_k \hat{X}_{t+\ell-k|t}, & \ell \geq m. \end{cases} \tag{4.35}$$

Note that this is obtained from (4.34) by replacing the unknown values by their best linear predictions. The nature of the recursion (4.35) for the predictors is quite similar to those in (4.22) and (4.23) for the MA parameters and the covariance function of ARMA models.

Example 4.4. For an ARMA(1,1) model, (4.35) reduces to

$$\begin{cases} \hat{X}_{t+1|t} &= \alpha X_t + \beta \varepsilon_t, \\ \hat{X}_{t+\ell|t} &= \alpha^\ell \hat{X}_{t+1|t}, \ell \geq 2. \end{cases}$$

Thus, for an AR(1) model, $\hat{X}_{t+\ell|t} = \alpha^\ell X_t,\ \ell = 1, 2, \cdots$, while for an MA (1) model $\hat{X}_{t+1|t} = \beta \varepsilon_t, \hat{X}_{t+\ell|t} = 0,$ for $\ell \geq 2$. Note that in both cases, $\hat{X}_{t+\ell|t}$ converges to $0 = E(X_{t+\ell})$ as $\ell \to \infty$. In fact, it can be shown that this phenomenon holds true for all zero-mean, causal and invertible stationary ARMA models.

When new observations arrive one needs to update the forecasts, that is, upon observing X_{t+1}, new predictors of future values based on $X_{t+1}, X_t, X_{t-1}, \cdots$ must be calculated. From (4.12) it is evident that

$$\hat{X}_{t+\ell|t+1} = \sum_{k=1}^{\ell-2} \alpha_k \hat{X}_{t+\ell-k|t+1} + \sum_{k=\ell-1}^{p} \alpha_k X_{t+\ell-k} + \sum_{k=\ell-1}^{q} \beta_k \varepsilon_{t+\ell-k}. \tag{4.36}$$

Subtracting (4.36) from (4.35), we arrive at

$$\hat{X}_{t+\ell|t+1} - \hat{X}_{t+\ell|t} = \sum_{k=1}^{\ell-1} \alpha_k (\hat{X}_{t+\ell-k|t+1} - \hat{X}_{t+\ell-k|t}) + \beta_{\ell-1} \varepsilon_{t+1}. \tag{4.37}$$

Substituting the predictors in the right-hand side by their MA representations, from (2.73), one obtains a recursion involving only the innovation at time $t+1$; namely,

$$\hat{X}_{t+\ell|t+1} = \hat{X}_{t+\ell|t} + b_{\ell-1}\varepsilon_{t+1}, \tag{4.38}$$

where the b_ℓ's are the MA parameters of the ARMA model (4.12); see (4.22).

From (4.38) and (4.35) with $\ell = m$ we obtain

$$\hat{X}_{t+m|t+1} = \sum_{k=1}^{p} \alpha_k \hat{X}_{t+m-k|t} + b_{m-1}\varepsilon_{t+1},$$

this combined with other equations of (4.38) for $\ell = 1, \cdots, m-1$, give rise to the following simple (vectorial) updating equation (Jones, 1980):

$$\begin{bmatrix} X_{t+1|t+1} \\ X_{t+2|t+1} \\ \vdots \\ X_{t+m|t+1} \end{bmatrix} = \begin{bmatrix} 0 & 1 & 0 & \cdot & \cdot & \cdot & 0 \\ 0 & 0 & 1 & \cdot & \cdot & \cdot & 0 \\ \cdot & \cdot & \cdot & \cdot & \cdot & \cdot & \cdot \\ \alpha_m & \cdot & \cdot & \cdot & \cdot & \alpha_2 & \alpha_1 \end{bmatrix} \begin{bmatrix} X_{t|t} \\ X_{t+1|t} \\ \vdots \\ X_{t+m-1|t} \end{bmatrix} + \begin{bmatrix} 1 \\ b_1 \\ \vdots \\ b_{m-1} \end{bmatrix} \varepsilon_{t+1}, \tag{4.39}$$

where $\alpha_i = 0$, for $i > p$.

Note that (4.35) and (4.39) show that the predictors of an ARMA process and their updates are found by performing a finite number of operations on the past of the process. This is in sharp contrast with the prediction of a general stationary process which requires infinite number of operations on the past; see (2.79), (2.73), Chapters 5 and 6.

For the ARMA(1, 1) model in Example 4.4, with $m = 2$, the updating equation (4.39) reduces to

$$\begin{bmatrix} X_{t+1|t+1} \\ X_{t+2|t+1} \end{bmatrix} = \begin{bmatrix} 0 & 1 \\ 0 & \alpha_1 \end{bmatrix} \begin{bmatrix} X_{t|t} \\ X_{t+1|t} \end{bmatrix} + \begin{bmatrix} 1 \\ \alpha+\beta \end{bmatrix} \varepsilon_{t+1}.$$

Such simple dynamic representations of the predictors are used in the next three subsections to provide a fast method of evaluating the Gaussian likelihood functions.

4.3.2 Markovian Representation of ARMA Processes

In this section, we present the Markovian representation of ARMA processes using the updating equations (4.39). Among many other applications, the representation is useful for fast calculation of the likelihood function of a Gaussian ARMA model (Jones, 1980).

In the Akaike's (1975) Markovian representation for the ARMA process (4.12), the state of the process is the column vector of length m appearing on the right-hand side of (4.39):

$$Z_t = (X_{t|t}, X_{t+1|t}, \cdots, X_{t+m-1|t})'.$$

Thus, with obvious notation, the updating equations (4.39) becomes the *state equation*:

$$Z_{t+1} = AZ_t + B\varepsilon_{t+1}, \tag{4.40}$$

where $\{Z_t\}$ is evidently a vector-valued Markov process and A is an $m \times m$ state transition matrix involving the AR coefficients in (4.12). The entries of B are, indeed, the MA parameters of a stationary solution of (4.12).

The *observation equation* is

$$Y_t = G'Z_t + W_t, \tag{4.41}$$

where $G = (1, 0, \cdots, 0)$. The observation Y_t is the sum of the specific linear combination of the state vector, which picks off its first element, and the random variable W_t which stands for the observational error. Note that when there is no observational error, that is, $W_t = 0$, then (4.40) and (4.41) reduce to (4.39).

4.3.3 The Kalman Filter

This section develops the much revered Kalman filter (Kalman, 1960; Kalman and Bucy, 1961) for the state-space model (2.57). The interest and novelty of the Kalman filter lies in the recursive nature of the predictor of the unobserved state $X_{t+\ell}$ based on the observed data $Y_0, Y_1, \cdots, Y_t$. Its application to the fast calculation of the likelihood function of a Gaussian ARMA models via their state-space representations (4.40)–(4.41) is discussed in the next section.

The object of the Kalman filter is to compute $\hat{X}_{t+\ell|t}$, the linear least-squares estimator of $X_{t+\ell}$ based on the past and present data $Y_t, Y_{t-1}, \cdots, Y_0$. For $\ell \geq 1$ this is called the *prediction problem* and for $\ell = 0$ it is referred to as the *filtering problem*. The *smoothing problem* is that of computing the projection of X_t onto H_n, the span of all the data $Y_0, Y_1, \cdots, Y_n$. Recall that $\hat{X}_{t+\ell|t}$ is simply the orthogonal projection of $X_{t+\ell}$, onto

$$H_t = H_t(Y) = sp\{Y_t, Y_{t-1}, \cdots, Y_0\},$$

the linear space of all linear combinations of the form

$$\sum_{j=0}^{t} C_j Y_j,$$

where C_j's are $d \times d$ matrix coefficients.

The key idea behind the Kalman filter computation is to write H_t as the sum of two orthogonal subspaces:

$$H_t = H_{t-1} \oplus sp\{Y_t - \hat{Y}_{t|t-1}\}, H_{t-1} \perp sp\{Y_t - \hat{Y}_{t|t-1}\}, \tag{4.42}$$

where $\hat{Y}_{t|t-1}$ is the linear least-squares predictor of Y_t based on its own past $Y_{t-1}, \cdots, Y_0$. Thus, the orthogonality of the *innovation* $\varepsilon_t = Y_t - \hat{Y}_{t|t-1}$ to H_{t-1} and to each other is self-evident and central to the conceptual development of the Kalman filter. The close relationships among these concepts and the Cholesky decomposition of a covariance matrix are also explored in Section 3.5.

Using the additivity of the projection on the sum of two orthogonal subspaces from (4.42), we obtain

$$\hat{X}_{t+1|t} = P_{H_{t-1}}^{X_t+1} + P_{\varepsilon_t} X_{t+1} = \hat{X}_{t+1|t-1} + K_t \varepsilon_t, \tag{4.43}$$

where the $d \times d$ matrix K_t given by

$$K_t = \text{Cov}\,(X_{t+1}, \varepsilon_t)[\,\text{Cov}\,(\varepsilon_t)]^{-1},$$

is called the *Kalman gain* matrix.

Next, we find the two covariance matrices needed for computing K_t. To this end, let $P_t = \text{Cov}\,(X_t - \hat{X}_{t|t-1})$ be the covariance matrix of the state prediction error. From the state-space equation (2.57) and properties of the orthogonal projection we have

$$\begin{cases} \hat{X}_{t+1|t-1} &= A\hat{X}_{t|t-1}, \\ \hat{Y}_{t|t-1} &= G\hat{X}_{t|t-1}. \end{cases} \tag{4.44}$$

Consequently,

$$\varepsilon_t = Y_t - \hat{Y}_{t|t-1} = GX_t + W_t - G\hat{X}_{t|t-1} = G(X_t - \hat{X}_{t|t-1}) + W_t,$$

from which

$$\begin{aligned} \text{Cov}\,(X_{t+1}, \varepsilon_t) &= \text{Cov}[AX_t + V_{t+1}, G(X_t - \hat{X}_{t|t-1}) + W_t] \\ &= A\,\text{Cov}(X_t - \hat{X}_{t|t-1})G' = AP_tG', \end{aligned}$$

and

$$\text{Cov}\,(\varepsilon_t) = \text{Cov}\,[G(X_t - \hat{X}_{t|t-1}) + W_t] = GP_tG' + R.$$

Thus, K_t is expressed in terms of the coefficients A, G, covariances R and P_t by

$$K_t = AP_tG'[GP_tG' + R]^{-1}, \tag{4.45}$$

and from (4.43)–(4.45) the state predictor is computed recursively using

$$\hat{X}_{t+1|t} = A\hat{X}_{t|t-1} + K_t\varepsilon_t, \ \hat{X}_{1|0} = 0. \tag{4.46}$$

Finally, using the state equation (2.57) and (4.42) it follows that

$$\hat{X}_{t|t} = \hat{X}_{t|t-1} + P_tG'(GP_tG' + R)^{-1}\varepsilon_t, \tag{4.47}$$

which connects the solutions of the filtering and prediction problems.

To use (4.46)–(4.47) for recursive computation, one needs the sequence $\{P_t\}$ of state prediction error covariances. Fortunately, this can be computed recursively as well. Note that

$$\begin{aligned} X_{t+1} - \hat{X}_{t+1|t} &= AX_t + V_{t+1} - A\hat{X}_{t|t-1} - K_t\varepsilon_t \\ &= (A - K_tG)(X_t - \hat{X}_{t|t-1}) + V_t - K_tW_t, \end{aligned}$$

since $X_t - \hat{X}_{t|t-1}, V_t$ and W_t are uncorrelated, we have

$$P_{t+1} = (A - K_tG)P_t(A - K_tG)' + \Sigma + K_tRK_t', \; P_1 = \text{Cov}(X_1). \tag{4.48}$$

The major computational burden of the Kalman filter occurs in computing $\{P_t\}$ via (4.48). This is a highly nonlinear recursive equation known as the *discrete matrix Riccati equation*. Barring the computational concerns, once P_1 is given, using (4.48), (4.46)–(4.47) one can proceed to handle efficiently several optimization problems that are otherwise computationally intensive and expensive. This is illustrated in the next section. For more details on implementational issues and recent developments in Kalman Filtering; see Harvey (1989) and Durbin (2000).

4.3.4 Evaluation of the Gaussian Likelihood Functions

In this section, we show the role of the Kalman Filter in evaluating the Gaussian likelihood function (2.59) for an ARMA(p, q) model. For the maximum likelihood estimation of the parameters using the grid method, after choosing a fine grid in the parameter space one needs to evaluate the likelihood function at all the grid points. Often, computing the determinant and inverse of the $n \times n$ matrices $\Gamma_n(\theta)$ for each grid point is the most costly part. This computational burden can be reduced by orthogonalizing the data or diagonalizing $\Gamma_n(\theta)$. Fortunately, the Kalman filter is ideal for this purpose.

To be precise, let $X_1, X_2, \ldots, X_n$ be n consecutive values of the causal and invertible ARMA process (4.12) and that X_t's are measured with observational errors W_t so that the Markovian (state space) representation (4.40)–(4.41) is in effect. Note that this is a special case of (2.57) with $d = \max\{p, q+1\}$ and $G = (1, 0, \ldots, 0)$ so that R is merely a scalar and no matrix inversion is required to run the Kalman filter.

We assume that $\{\varepsilon_t\}, \{V_t\}and\{W_t\}$ are Gaussian. Consequently, the observations $Y_1, \ldots, Y_n$ are zero-mean Gaussian and so are the innovations $\varepsilon_1, \ldots, \varepsilon_n$ appearing in (4.42). Furthermore, the latter are independent random variables with

$$\text{Var}(\varepsilon_t) = GP_tG' + R = P_{t,11} + R,$$

where $P_{t,11}$ is the $(1,1)$ entry of the matrix P_t. Thus, dropping the constant 2π, $-2\,ln$ likelihood function of $Y_1, \ldots, Y_n$ is

$$L(\theta) = \sum_{t=1}^{n} \left[ln(P_{t,11} + R) + \frac{\varepsilon_t^2}{P_{t,11} + R} \right].$$

This can be minimized with respect to the parameters $\beta_1, \ldots, \beta_p$, $\alpha_1, \ldots, \alpha_q$, σ^2 and R using a grid method, however, the Kalman filter is essential for speeding up the calculations at each grid point.

To start up the Kalman filter, P_1 the initial state covariance matrix in (4.48) is needed. To this end, recall that

$$\begin{aligned} Z_1 &= (X_1, X_{2|1}, \ldots, X_{m|1}) \\ &= (X_1, X_2, \ldots, X_m) - (0, X_2 - X_{2|1}, \ldots, X_m - X_{m|1}). \end{aligned}$$

Since for $j > 1$, $X_j - X_{j|1} \perp X_1$, and $\text{Cov}(X_j - X_{j|1}, X_j) = \text{Var}(X_j - X_{j|1})$, it follows from the definition of Z_1 above and (2.73) that

$$P_1 = \text{Cov}(Z_1) = \Gamma_m - \sigma^2 \begin{bmatrix} 0 & \mathbf{0} \\ \mathbf{0} & T'_{m-1}T_{m-1} \end{bmatrix}, \tag{4.49}$$

where $\Gamma_m = (\gamma_{i-j})_{i,j=1}^m$ with $\{\gamma_k\}$ the covariance function of the ARMA(p, q) process (4.12) and $T_{m-1} = (b_{i-j})_{i,j=1}^{m-1}$ with $\{b_i\}$ its MA parameters. Note that both γ_k's and b_k's are functions of the ARMA parameters; see Sections 4.2.3 and 4.2.4.

4.4 INFERENCE FOR GAUSSIAN ARMA MODELS

In this section, for a Gaussian ARMA process we focus on the functional form of its likelihood function (2.59) and discuss some inferential consequences such as finding sufficient statistics for the parameters and the likelihood equations (Casella and Berger, 1990, Chap. 7). As stated before the likelihood function is hard to work with mostly because of the difficulty in computing or differentiating $|\Gamma_n|$ and Γ_n^{-1}. We pay particular attention to finding workable expressions for these quantities and making explicit their dependence on the ARMA coefficients.

4.4.1 Inverse Covariance Matrix

Many inferential issues concerning ARMA models invariably relate to the inverse and determinant of the covariance matrix Γ_n. We provide explicit formulas for computing Γ_n^{-1} and $|\Gamma_n|$ when $\{X_t\}$ is an ARMA model and discuss their roles in computing the exact likelihood function and deriving the likelihood equations.

The fact that Γ_n is *persymmetric*, that is, is symmetric about both its diagonals plays a key role in computing Γ_n^{-1} and establishing the celebrated Durbin-Levinson algorithm in Chapter 7. More precisely, an $n \times n$ matrix matrix A is *persymmetric* if

$$a_{ij} = a_{ji} = a_{n-i+1, n-j+1}, i, j = 1, 2, \ldots, n.$$

It seems plausible that the inverse of a persymmetric matrix should be persymmetric, this turns out to be true. To show it and for later use we introduce the $n \times n$ *exchange* matrix $E_n = (e_n, e_{n-1}, \ldots, e_1)$ where $e_i = (0, \ldots, 0, 1, 0, \ldots, 0)'$ is a column vector with 1 in the ith position. Clearly, $E_n = E_n'$, $E_n^2 = I_n$ and $E_n A$ is obtained from A by simply exchanging the rows of A, the last row becoming the first and so on. Now, it is evident that a matrix A is persymmetric, if and only if $E_n A E_n = A$. Consequently, when A is invertible, then $E_n A^{-1} E_n = A^{-1}$. So that A^{-1} is also persymmetric.

First, we compute Γ_n^{-1} and $|\Gamma_n|$ when $\{X_t\}$ is an AR(p) process and show that the number of arithmetic operations required to compute them is independent of the sample size n. Therefore, increasing n would not increase the computational complexity of problems involving these quantities.

Theorem 4.5. Let $\{X_t\}$ be a causal stationary AR(p) process with the covariance function $\{\gamma_k\}$ and $\Gamma_n = (\gamma_{k-\ell})_{k,\ell=1}^n$. Then, for $n > p$,
(a) $|\Gamma_n| = |\Gamma_p|$,
(b) $\sigma^2 \Gamma_p^{-1} = E_p L_\alpha' L_\alpha E_p - A_\alpha' A_\alpha$,

where L_α and A_α are as in (4.24)–(4.26). More precisely, the (i, j)-th entry of $\sigma^2 \Gamma_p^{-1}$ is given by

$$\gamma_{(p)}^{ij} = \sum_{k=0}^{(i-1)\wedge(j-1)} \alpha_k \alpha_{k+|i-j|} - \sum_{k=1}^{i\wedge j} \alpha_{p-i+k} \alpha_{p-j+k},$$

where $i \wedge j$ stands for the minimum of i and j.
(c)

$$\sigma^2 X' \Gamma_n^{-1} X = \sum_{i=0}^{p} \sum_{j=0}^{p} \alpha_i \alpha_j s_{ij},$$

where $\alpha_0 = -1$ and

$$s_{00} = (X_1, \ldots, X_p)' \, \Gamma_p^{-1} \, (X_1, \ldots, X_p) + n\hat{\gamma}_0 - \sum_{t=1}^{p} X_t^2$$

depends on the AR parameters $\alpha = (\alpha_1, \ldots, \alpha_p)$. The rest of the s_{ij}'s, that is, those for $i \neq 0$ or $j \neq 0$, depend only on the data. In fact,

$$s_{ij} = n\hat{\gamma}_{i-j} - \sum_{k=1+i\vee j}^{p} X_{k-i} X_{k-j} - \sum_{k=n+1}^{n+i\wedge j} X_{k-i} X_{k-j},$$

where $i \vee j = \max(i, j)$.

Proof. (a) For simplicity take $\sigma^2 = 1$ from (4.29) and partitioning K conformally with Φ the result follows, since for AR(p) models we have $K_{11} = \Gamma_p$, $K_{12} = 0, K_{22} = I_{n-p}$.

(b) Using the knowledge of K for AR(p) models as described above and (4.29) we obtain

$$\Gamma_n^{-1} = \Phi' K^{-1} \Phi = \begin{pmatrix} \Gamma_p^{-1} + A_\alpha' A_\alpha & \Phi_{21}' L_{\alpha,n-p} \\ L_{\alpha,n-p}' \Phi_{21} & L_{\alpha,n-p}' L_{\alpha,n-p} \end{pmatrix}, \tag{4.50}$$

so that Γ_n^{-1} is known up to Γ_p^{-1}. To find the latter we use the fact that Γ_n^{-1} is persymmetric and hence its upper-left $p \times p$ submatrix, that is, $\Gamma_p^{-1} + A'_\alpha A_\alpha$ is the image of its lower-right $p \times p$ submatrix $L'_{\alpha,p} L_{\alpha,p}$, that is,

$$\Gamma_p^{-1} + A'_\alpha A_\alpha = E_p L'_{\alpha,p} L_{\alpha,p} E_p, \tag{4.51}$$

from which the result follows. Part (c) is proved in the next section; see (4.57)–(4.59). Q.E.D.

An important and useful consequence of (4.50)–(4.51) is that all entries of Γ_p^{-1} and hence those of Γ_n^{-1} are quadratic functions of the AR parameters (Siddiqui, 1958). This plays crucial roles in finding sufficient statistics and the likelihood equations for the AR parameters. Part (b) essentially establishes the *Gohberg-Semencel formula* (Landau, 1987a) for the inverse of the Toeplitz matrix Γ_p. In fact, this holds for any $n \geq 1$, that is,

$$\Gamma_n^{-1} = A_n A'_n - B_n B'_n = A'_n A_n - B'_n B_n, \tag{4.52}$$

where A_n and B_n are lower triangular matrices with $-\alpha_{i-j}$, $-\alpha_{n-i+j}$ as their (i, j)-th entry with $\alpha_k = 0$ if $k > p$ (Pollock, 1999).

Example 4.5. For an AR(2) model we have

$$L_{\alpha,2} = \begin{pmatrix} 1 & 0 \\ -\alpha_1 & 1 \end{pmatrix}, A_2 = \begin{pmatrix} \alpha_2 & \alpha_1 \\ 0 & \alpha_2 \end{pmatrix},$$

and from (4.51)

$$\Gamma_2^{-1} = \begin{pmatrix} 1 - \alpha_2^2 & -\alpha_1 + \alpha_1\alpha_2 \\ -\alpha_1 + \alpha_1\alpha_2 & 1 - \alpha_2^2 \end{pmatrix}.$$

For $n \geq 3, \Gamma_n^{-1}$ is found using (4.50).

Most of the above arguments are applicable to ARMA(p, q) models. In the rest of this section we use (4.26) which represents Γ_n as a rank-$(p+q)$ perturbation of $L_{-\beta} L'_{-\beta}$. Thus, applying the inversion formula in Problem 4.14 (a) to (4.26) with $L = L_{-\beta} L'_{-\beta}$ we obtain that

$$\sigma^2 \Gamma_n^{-1} = L'_\alpha \left[L^{-1} - L^{-1} L_{\alpha,\beta} (L'_{\alpha,\beta} L^{-1} L_{\alpha,\beta} + \Omega^{-1})^{-1} L'_{\alpha,\beta} L^{-1} \right] L_\alpha. \tag{4.53}$$

First, we compute L^{-1} explicitly. To this end, note that $L^{-1} = L'^{-1}_{-\beta} L^{-1}_{-\beta}$ and $L^{-1}_{-\beta} = (a_{k-\ell}) = L_{\mathrm{a}}$ where the a_i's are obtained from $\sum_{i=0}^{\infty} a_i z^i = \left(\sum_{j=0}^{q} \beta_j z^j \right)^{-1}$, $\beta_0 = 1, a_0 = 1, a_i = 0$ for $i < 0$; see Section 4.2.3. Now, substituting for L^{-1} in (4.53) yields

$$\sigma^2 \Gamma_n^{-1} = L'_\alpha L'_{\mathrm{a}} \left[I_n - Z(Z'Z + \Omega^{-1})^{-1} Z' \right] L_{\mathrm{a}} L_\alpha, \tag{4.54}$$

where

$$\begin{aligned} Z = L_{\mathrm{a}} L_{\alpha,\beta} &= \begin{pmatrix} L_m & 0 \\ L_{n-m,m} & L_{n-m} \end{pmatrix} \begin{pmatrix} A_{\alpha,\beta} \\ 0 \end{pmatrix} \\ &= \begin{pmatrix} L_m A_{\alpha,\beta} \\ L_{n-m,m} A_{\alpha,\beta} \end{pmatrix}, \end{aligned} \tag{4.55}$$

and $A_{\alpha,\beta} = (-A_\alpha, A_\beta)$. Thus, (4.54) requires only inversion of $p \times p$ matrices, since the matrix Ω can be inverted by applying the inversion formula in Problem 4.14(b) to the partitioned Ω in (4.25).

4.4.2 Likelihood Function of ARMA Models

We give explicit expressions for the likelihood function of Gaussian ARMA models using Ansley's (1979) transformation or the factorization (4.29) of the $n \times n$ covariance matrix Γ_n, and the explicit expressions for $|\Gamma_n|$ and the quadratic form $X'\Gamma_n^{-1}X$. Such representations of the likelihood function are useful in computing the gradient and Hessian matrix with respect to the parameters with considerably fewer operations than using directly the likelihood function (2.59). Since these expressions, in a sense, "separate" the AR and MA coefficients, they are also useful in a profile likelihood approach where the likelihood is first maximized with respect to the AR parameters conditioned on the current values of the MA parameters and then maximized with respect to the former.

A natural method for evaluating the quadratic form $X'\Gamma_n^{-1}X$, suggested by Ansley (1979) and Luceño (1994), is based on writing (4.28) as a linear combination of the observables X_t's with α_i's as their coefficients:

$$Z = -\sum_{i=0}^{p} \alpha_i Y_i, \tag{4.56}$$

where $\alpha_0 = -1$, $Y_0 = X$ and the entries of the vectors $Y_i = (Y_{i1}, \ldots, Y_{in})$, for $i = 1, \ldots, p$, are given by $Y_{it} = 0$, for $1 \le t \le p$, and by $Y_{it} = X_{t-i}$, for $p+1 \le t \le n$. Partitioning $K = (K_{ij})$ conformally with Φ in (4.28), it is evident that K_{22} depends only on the moving average parameters $\beta = (\beta_1, \cdots, \beta_q)$.

Consequently, for ARMA(p, q) models using (4.29) we have

$$X'\Gamma_n^{-1}X = Z'K^{-1}Z = \sum_{i=0}^{p}\sum_{j=0}^{p} \alpha_i \alpha_j Y_i' K^{-1} Y_j. \tag{4.57}$$

The importance of this expression in dealing with analytical and computational aspects of maximum likelihood estimation of ARMA parameters is discussed next.

For AR(p) models, the matrix K takes a particularly simple form because $K_{11} = \Gamma_p$, $K_{12} = 0$ and $K_{22} = I_{n-p}$. Then, the bilinear forms $Y_i'K^{-1}Y_j$ can be written in terms of certain familiar functions of the data. To this end, let us partition $Y_i = (Y_i^{(1)}, Y_i^{(2)})$, where $Y_i^{(1)}$ is the p-vector of zeros when $i \neq 0$. Therefore, with

$X^{(1)} = (X_1, \ldots, X_p)$ we have,

$$\begin{aligned} Y_0' K^{-1} Y_0 &= (Y_0^{(1)'}, Y_0^{(2)'}) \begin{pmatrix} K_{11}^{-1} & 0 \\ 0 & I_{n-p} \end{pmatrix} \begin{pmatrix} Y_0^{(1)} \\ Y_0^{(2)} \end{pmatrix} \\ &= X^{(1)'} K_{11}^{-1} X^{(1)} + n\hat{\gamma}_0 - \sum_{t=1}^{p} X_t^2. \end{aligned} \tag{4.58}$$

Similarly, for $i, j \neq 0$ taking into account that $Y_i^{(1)'} K_{11}^{-1} Y_j^{(1)} = 0$, we obtain

$$Y_i' K^{-1} Y_j = n\hat{\gamma}_{i-j} - \sum_{t=1+i\vee j}^{p} X_{t-i} X_{t-j} - \sum_{t=n+1}^{n+i\wedge j} X_{t-i} X_{t-j}, \tag{4.59}$$

where $i \vee j$ and $i \wedge j$ stand for the maximum and minimum of i and j, respectively, and $\hat{\gamma}_k$ is the sample covariance (2.2) given by $n\hat{\gamma}_k = \sum_{t=1}^{n-k} X_t X_{t+k}$.

Thus, for AR(p) models, the above calculations show that $Y_i' K^{-1} Y_j$ depends only on the data and not the parameters except when $i = j = 0$. In the latter, it depends on the parameters only through $X^{(1)'} K_{11}^{-1} X^{(1)}$ involving the $p \times p$ matrix $K_{11} = \Gamma_p$, which is also needed in computing $|\Gamma_n| = |\Gamma_p|$. Thus, the number of arithmetic operations required to evaluate the likelihood function of an AR model is independent of the sample size n.

For the general ARMA models since the matrix K has a more complicated structure, we exploit its Cholesky decomposition and the fact that K_{22} depends only on the MA parameters β (Luceño, 1994; Kohn and Ansley, 1985). To preserve the last property, we factorize K starting from its last element so that

$$K = L'L = \begin{pmatrix} L_{11}' & L_{21}' \\ 0 & L_{22}' \end{pmatrix} \begin{pmatrix} L_{11} & 0 \\ L_{21} & L_{22} \end{pmatrix}, \tag{4.60}$$

where only L_{22} depends on the MA parameters. Then, the bilinear forms appearing in (4.57) can be written as

$$Y_i' K^{-1} Y_j = Y_i' L^{-1} L'^{-1} Y_j = U_i' U_j = \sum_{t=1}^{n} u_{it} u_{jt}, \tag{4.61}$$

where $U_i = L'^{-1} Y_i = (u_{it})_{t=1}^{n}, i = 0, \cdots, p$. Since L_{22} does not depend on the AR parameters, the last $n - p$ components of U_i, that is, u_{it} for $t = p + 1, \cdots, n$, do not depend on these parameters. Therefore, $Y_i' K^{-1} Y_j$ can be split into the sum of two components:

$$Y_i' K^{-1} Y_j = \sum_{t=1}^{p} u_{it} u_{jt} + \sum_{t=p+1}^{n} u_{it} u_{jt} = Q_{ij}^{(1)}(\alpha, \beta) + Q_{ij}^{(2)}(\beta). \tag{4.62}$$

Finally, from (4.57) and (4.62) we obtain

$$\begin{cases} X'\Gamma_n^{-1}X = \sum_{i=0}^{p}\sum_{j=0}^{p} Q_{ij}^{(1)}(\alpha,\beta)\alpha_i\alpha_j + \sum_{i=0}^{p}\sum_{j=0}^{q} Q_{ij}^{(2)}(\beta)\alpha_i\alpha_j, \\ |\Gamma_n| = |K| = |L_{11}|^2|L_{22}|^2. \end{cases} \tag{4.63}$$

In view of the "almost" quadratic nature of $X'\Gamma_n^{-1}X$ in α in (4.63), it seems advantageous to minimize this or maximize the likelihood function with $Q_{ij}^{(1)}(\alpha,\beta)$ replaced by $Q_{ij}^{(1)}(\tilde{\alpha}_s,\beta)$ where $\tilde{\alpha}_s$ is a proxy for α obtained, for example, from an earlier stage of a Newton-Raphson iterative method. We note that with such substitution for $Q_{ij}^{(1)}$, the $X'\Gamma_n^{-1}X$ is quadratic in α, so that its minimum is found using standard methods; for more details see Luceño (1994).

4.4.3 Sufficient Statistics for AR Models

A family of *sufficient statistics* for a probability model reveals those functionals of data that are useful in making inferences about the model. If we write a probability density $f(x|\theta)$ as a multiple of $\exp\left\{\sum_{k=1}^{n_0} g_k(\theta)T_k(x)\right\}$ where n_0 does not depend on the sample size n, then using the factorization theorem (Casella and Berger, 1990, p. 250) the family $\{T_1,\ldots,T_{n_0}\}$ is *sufficient* for θ. We use this principle to find sufficient statistics for the parameters $\alpha = (\alpha_1,\ldots,\alpha_p)$ and σ^2 of a Gaussian stationary AR model.

Theorem 4.6. Let $\{X_t\}$ be a real-valued, zero-mean Gaussian stationary AR(p) process of known order p. Then, the family of statistics

$$\sum_{t=p+1}^{n-p+j} x_t x_{t-j}, j = 0,1,2,\ldots,p,$$

$$x_i x_j + x_{n+1-i}x_{n+1-j}, i,j = 1,\ldots,p, i \geq j,$$

is sufficient for the parameters $(\alpha_1,\ldots,\alpha_p,\sigma^2)$ when $n > 2p$.

Remark 4.2. Before proving the theorem, a few remarks are in order. For an AR(1) process a family of sufficient statistics for (α,σ^2) is $\left\{\sum_{t=2}^{n-1} x_t^2, \sum_{t=2}^{n} x_t x_{t-1}, x_1^2 + x_n^2\right\}$ containing 3 statistics. For the general AR(p) process, the family consists of

$$p + 1 + \frac{p(p+1)}{2} = \frac{(p+1)(p+2)}{2}$$

statistics, which is independent of the number of observations n. This is considerably more than $p+1$, the number of parameters in the model and differs from the known results for the i.i.d. data from the exponential family where the number of sufficient statistics equals the number of parameters, Casella and Berger (1990, p.254). When n is large compared to p, the sums $\sum_{t=1}^{n} x_t x_{t-j}, j = 0, 1, \ldots, p$, could dominate the terms $x_i x_j + x_{n+1-i} x_{n+1-j}$, so that ignoring these one would end up with a family of $p+1$ sufficient statistics $\sum_{t=1}^{n} x_t x_{t-j}, j = 0, 1, \ldots, p$, which are needed to estimate the first $p+1$ sample correlations as in (2.2). (Here, $x_t = 0$ when $t \leq 0$ or $t > n$). The increase in the number of sufficient statistics for AR models is possibly due to the time-order and the inherent difficulty in inference for correlated data. Study of sufficient statistics for Gaussian AR processes were initiated by Arato (1961); see also Forchini (2000). He also showed that an MA(1) process does not admit a sufficient statistics of dimension less than n. For ARMA processes, Dickinson (1982) finds sufficient statistics for the AR coefficients with MA coefficients kept fixed using ideas leading to (4.63).

Proof. A quick proof with a slightly different family of sufficient statistics can be obtained using Theorem 4.5 (c). The proof presented here is more elementary and revealing (Anderson, 1971).

Let $X = (X_1, \ldots, X_p, X_{p+1}, \ldots, X_n) = (X^{(1)\prime}, X^{(2)\prime})$ be n consecutive observations from an AR(p) model and let $f_X, f_{X^{(1)}}, f_{X^{(2)}}$ stand for the probability density functions of X and its subvectors, respectively. Note that the density of $X^{(1)}$ is determined by the $p \times p$ covariance matrix $\Gamma_p = \sigma^2 R_p$ with the inverse $\Gamma_p^{-1} = \sigma^{-2}(\rho_{(p)}^{ij})_{i,j=1}^{p}$, and

$$f_{X^{(1)}}(x_1, \ldots, x_p) = (2\pi\sigma^2)^{-\frac{p}{2}} |R_p|^{-\frac{1}{2}} \exp\left\{ -\frac{1}{2\sigma^2} \sum_{i,j=1}^{p} \rho_{(p)}^{ij} x_i x_j \right\}.$$

Since $\varepsilon_{p+1}, \ldots, \varepsilon_n$ are independent $N(0, \sigma^2)$ random variables, their joint probability density function is given by

$$(2\pi\sigma^2)^{-\frac{n-p}{2}} \exp\left\{ -\frac{1}{2\sigma^2} \sum_{t=p+1}^{n} \varepsilon_t^2 \right\}.$$

Moreover, since the Jacobian of the transformation from $(X_1, \ldots, X_p, \varepsilon_{p+1}, \ldots, \varepsilon_n)'$ to $(X_1, \ldots, X_n)'$ is one, we have

$$f_X(x) = f_{X^{(2)}|X^{(1)}} f_{X^{(1)}} = (2\pi\sigma^2)^{-n/2} |R_p|^{-1/2} \exp\left(-\frac{Q}{2\sigma^2} \right), \qquad (4.64)$$

where

$$Q = \sum_{i,j=1}^{p} \rho_{(p)}^{ij} x_i x_j + \sum_{t=p+1}^{n} \left(x_t - \sum_{k=1}^{p} \alpha_k x_{t-k} \right)^2,$$

is quadratic both in x_t's and α_k's. Our goal is to write Q as $\sum_{k=1}^{n_0} g_k(\alpha)T_k(x)$, where n_0 is fixed and does not depend on n:

$$\begin{aligned} Q = \sum_{i,j=1}^{p} \rho_{(p)}^{i,j} x_i x_j \quad &+ \quad \sum_{t=p+1}^{n} (x_t^2 + \alpha_1^2 x_{t-1}^2 + \ldots + \alpha_p x_{t-p}^2 \\ &- \quad 2\alpha_1 x_t x_{t-1} - \ldots - 2\alpha_p x_t x_{t-p} \\ &+ \quad 2\alpha_1\alpha_2 x_{t-2} x_{t-3} + \ldots + 2\alpha_{p-1}\alpha_p x_{t-p+1} x_{t-p}) \\ &= \quad \left[(1 + \alpha_1^2 + \ldots + \alpha_p^2) \sum_{t=p+1}^{n-p} x_t^2 \right. \\ &+ \quad 2(-\alpha_1 + \alpha_1\alpha_2 + \ldots + \alpha_{p-1}\alpha_p) \\ &\left. \sum_{t=p+1}^{n-p+1} x_t x_{t-1} + \ldots - 2\alpha_p \sum_{t=p+1}^{n} x_t x_{t-p} \right] \\ &+ \quad \sum_{i,j=1}^{p} a_{i,j} x_i x_j + \sum_{i,j=1}^{p} b_{i,j} x_{n+1-i} x_{n+1-j}, \end{aligned} \tag{4.65}$$

where $A = (a_{i,j})$ and $B = (b_{i,j})$ are some symmetric $p \times p$ matrices. Next, we show that $A = B$. Since Γ_n^{-1} is persymmetric and Q is also of the form $X'\Gamma_n^{-1}X$, it follows that the quadratic form in $x_1, \ldots, x_p$ is the same as the quadratic form in $x_n, x_{n-1}, \ldots, x_{n+1-p}$, namely, $A = B$. Thus the last two sums in (4.65) can be replaced by $\sum_{i,j=1}^{p} a_{i,j}(x_i x_j + x_{n-i+1}x_{n-j+1})$ and from the factorization theorem, Casella and Berger (1990, p. 250), the result follows.

Q.E.D.

Although the entries of A can be found from (4.65), they are not needed in the proof; only it is important to realize that they depend on the AR parameters. We note that the quadratic form Q contains no term $x_i x_j$ with $|i - j| > p$.

4.4.4 Likelihood Equations for AR Parameters

In this section, we explicitly differentiate the exact Gaussian likelihood function of an AR(p) model with respect to its model parameters. The likelihood equations obtained are closely related to the familiar Yule-Walker equations. Such differentiation of the likelihood function is usually avoided in the time series literature mostly because the derivative of $|\Gamma_n(\theta)|$ is too complicated to obtain and is unyielding for computational purposes. Following Miller (1995) our presentation relies on (4.65) and takes advantage of the fact that Q is quadratic in the AR parameters; see also Theorem 4.5. We show that differentiation of $\log|\Gamma_n(\theta)|$ can be reduced to that of computing the expectation of derivative of the quadratic form Q.

Using (4.65), and viewing Q as a quadratic form in the AR parameters $\alpha_1, \cdots, \alpha_p$ we have

$$f_X(x|\alpha, \sigma^2) = (2\pi\sigma^2)^{-\frac{n}{2}} |R_p|^{\frac{-1}{2}} \exp\left\{-\frac{Q}{2\sigma^2}\right\}, \tag{4.66}$$

with

$$Q = Q(\alpha) = \boldsymbol{\alpha}' D \boldsymbol{\alpha}, \tag{4.67}$$

where $\boldsymbol{\alpha} = (-1, \alpha_1, \ldots, \alpha_p)$ and the $(p+1) \times (p+1)$ symmetric matrix $D = (d_{ij})$ is given by

$$d_{ij} = \sum_{t=1}^{n-i-j} x_{t+i} x_{t+j}, 0 \le i \le j \le p. \tag{4.68}$$

Note that $\frac{d_{k0}}{n}$ is the estimate of the covariance function γ_k given in (2.2).

Theorem 4.7. Let $\{X_t\}$ be a mean-zero Gaussian AR(p) stationary process with the covariance function $\{\gamma_k\}$ and parameters $\alpha_1, \ldots, \alpha_p$ and σ^2. Then, the likelihood equations for the parameters are

$$\begin{cases} \sum_{j=1}^{p} (d_{0j} + j\gamma_j)\alpha_j + n\sigma^2 &= d_{00}, \\ \sum_{j=1}^{p} (d_{ij} + j\gamma_{j-i})\alpha_j &= d_{i0}, \quad i = 1, \ldots, p. \end{cases} \tag{4.69}$$

Before proving the theorem the following remarks regarding the above equations are noteworthy:

(a) The likelihood equations (4.69) are strikingly similar to the Yule-Walker equations (2.41), where d_{i0} takes the place of γ_i and $d_{ij} + j\gamma_{j-i}$ that of γ_{j-i}. The addition of $j\gamma_{i-j}$ to each d_{ij} could be considered as a 'correction' for the differences in the number of terms in d_{ij} which happens to be $n - i - j$. In fact, dividing both sides of (4.69) by n and letting $n \to \infty$, it reduces to the Yule-Walker equations (2.52). This shows, in a straightforward manner, that the Yule-Walker estimates of the AR parameters are asymptotically equivalent to the MLEs. Consequently, Theorem V of Chapter 2 implies the asymptotic normality of the Yule-Walker estimates.

(b) Since $\{\gamma_k\}$ is a function of α and σ^2, the likelihood equations (4.69) are nonlinear in the AR parameters. The following iterative algorithm in Miller (1995) could be used to compute $\alpha_1, \ldots, \alpha_p$ and σ^2. The algorithm uses the data only through the matrix D, computed only once before going into the iterative loop.

(i) Compute the matrix D and the initial estimates of $\gamma_0, \ldots, \gamma_p$ using the sample covariances (2.2).

(ii) Use D and the current estimates of γ_k's to solve (4.69) as a linear system to obtain estimates of α and σ^2.

(iii) Use (2.52) to obtain updated estimates of γ_k's given current estimates of α and σ^2. [This requires rewriting (2.52) with γ_k's treated as unknown, and α and σ^2 as known.]

(iv) Iterate (ii)–(iii) until convergence.

Proof of Theorem 4.7 relies on the following useful lemma. The problem of finding likelihood equations for the covariance parameters of a multivariate normal is encountered in many areas of statistics. The chief difficulty, as mentioned before, is that of computing the derivatives of $\log|\Sigma|$. The next lemma shows a simple way around this problem.

Lemma 4.8. Let X be an n-vector with the density function

$$f_X(x|\theta) = \frac{1}{(2\pi)^{n/2}|\Sigma|^{\frac{1}{2}}}\exp\left\{-\frac{1}{2}(x-\mu)'\Sigma^{-1}(x-\mu)\right\},$$

where $\mu = \mu(\theta)$ and $\Sigma = \Sigma(\theta)$. Then, with $Q(\theta) = (x-\mu)'\Sigma^{-1}(x-\mu)$, we have

$$\left\{\begin{array}{lcl} \dfrac{\partial \log|\Sigma|}{\partial\theta} & = & -E\left(\dfrac{\partial Q}{\partial\theta}\right), \\ 2\dfrac{\partial \log f_X}{\partial\theta} & = & E\left(\dfrac{\partial Q}{\partial\theta}\right) - \dfrac{\partial Q}{\partial\theta}. \end{array}\right. \tag{4.70}$$

Proof. Since integral of a probability density function over its support is unity, it follows that

$$(2\pi)^{\frac{n}{2}}|\Sigma|^{\frac{1}{2}} = \int \exp\left(-\frac{Q}{2}\right)dx.$$

Taking logarithms of both sides and differentiating each side with respect to θ gives the first desired identify:

$$\begin{aligned} \frac{1}{2}\frac{\partial}{\partial\theta}\log|\Sigma| &= \frac{-\frac{1}{2}\int \frac{\partial Q}{\partial\theta}\exp\left(-\frac{Q}{2}\right)dx}{\int\exp\left(-\frac{Q}{2}\right)dx} \\ &= -\frac{1}{2}E\left(\frac{\partial Q}{\partial\theta}\right). \end{aligned} \tag{4.71}$$

Differentiating $\log f_X$ with respect to θ gives

$$\frac{\partial}{\partial \theta} \log f_X = -\frac{1}{2}\frac{\partial}{\partial \theta} \log |\Sigma| - \frac{1}{2}\frac{\partial Q}{\partial \theta}. \tag{4.72}$$

Substituting from (4.71) into (4.72) we obtain

$$\frac{\partial}{\partial \theta} \log f_X = \frac{1}{2} E\left(\frac{\partial}{\partial \theta} Q\right) - \frac{1}{2}\frac{\partial Q}{\partial \theta},$$

as desired.

Q.E.D.

Proof of Theorem 4.7. Note that $Q(\alpha)$ in (4.67) can be expanded as

$$Q(\alpha) = d_{00} - 2\sum_{j=1}^{p} \alpha_j d_{0j} + \sum_{j=1}^{p}\sum_{k=1}^{p} \alpha_j \alpha_k d_{jk},$$

so that

$$\frac{\partial Q}{\partial \alpha_i} = -2\left(d_{0i} - \sum_{j=1}^{p} \alpha_{ji} d_{ij}\right), i = 1, \ldots, p, \tag{4.73}$$

and

$$\begin{aligned} E\left(\frac{\partial Q}{\partial \alpha_i}\right) &= -2\left\{(n-i)\gamma_i - \sum_{j=1}^{p} \alpha_j (n-i-j)\gamma_{j-i}\right\} \qquad (4.74) \\ &= -2\left\{(n-i)\gamma_i - (n-i)\sum_{j=1}^{p} \alpha_j \gamma_{j-i} + \sum_{j=1}^{p} j\alpha_j \gamma_{j-i}\right\} \\ &= -2\sum_{j=1}^{p} j\alpha_j \gamma_{j-i}, i = 1, \ldots, p. \end{aligned}$$

Now, substituting from (4.73) and (4.74) into (4.70), the latter reduces to

$$\frac{\partial}{\partial \alpha_i} \log f_X = \frac{1}{\sigma^2}\left\{d_{0i} - \sum_{j=1}^{p} \alpha_j (d_{ij} + j\gamma_{j-i})\right\}, \ i = 1, \ldots, p,$$

from which the last equations in (4.69) are immediate.

To obtain the first equation in (4.69), from (4.66) we note that

$$\frac{\partial}{\partial \sigma^2} \log f_X = \frac{1}{2\sigma^4}\{Q - n\sigma^2\}. \tag{4.75}$$

Rewriting Q as

$$Q = d_{00} - \sum_{j=1}^{p} \alpha_j d_{0j} - \sum_{i=1}^{p} \alpha_i \left(d_{i0} - \sum_{j=1}^{p} \alpha_j d_{ij} \right)$$

and substituting into (4.75) we obtain

$$\sum_{j=1}^{p} \alpha_j d_{0j} + \sum_{i=1}^{p} \alpha_i \left(d_{i0} - \sum_{j=1}^{p} \alpha_j d_{ij} \right) + n\sigma^2 = d_{00}. \tag{4.76}$$

Using the last equations from (4.69) we obtain

$$d_{i0} - \sum_{j=1}^{p} \alpha_j d_{ij} = \sum_{j=1}^{p} j\alpha_j \gamma_{j-i}, i = 1, 2, \ldots, p,$$

substituting these in (4.76), it reduces to the desired equation after some simplifications.

Q.E.D.

4.5 CHARACTERIZATION OF ARMA MODELS

The rest of this chapter is concerned with the characterization, and question of existence of ARMA processes with prescribed autocovariances up to lag p and MA parameters up to lag q; see Remark 4.1. More specifically, given $\gamma_0, \cdots, \gamma_p$ such that Γ_p is positive definite, and arbitrary $b_1, \cdots, b_q$, we are interested in the existence of an ARMA(p, q) process such that, for an appropriate $\sigma^2 > 0, \gamma_0, \cdots, \gamma_p, b_1, \cdots, b_q, \sigma^2$ are, respectively, its first $p+1$ autocovariances, first q MA parameters and innovation variance. We show that the answer to this question is positive. The proof, however, is long, revealing and employs most ideas and techniques discussed so far and develops new ones that are useful in time series analysis. It highlights the prominent position of ARMA models within the broad class of stationary processes. Our presentation follows closely the methods and ideas of Huang (1990) and Franke (1985).

In this section we characterize ARMA models in terms of singularity of certain prediction error covariance matrices. The key idea is the simple observation that in the ARMA(p, q) model (4.12), the collection of random variables $X_{-m}, \cdots, X_0, \varepsilon_0, \cdots, \varepsilon_{-n}, m, n \geq 0$, become linearly dependent as soon as $m \geq p$ and $n \geq q$. The ideas and techniques developed here are useful in motivating several time series model identification techniques based on determinants of appropriately formed matrices of the sample correlations; for more details see Woodward and Gray (1978); Piccolo and Wilson (1984); and Choi (1992).

We begin with a nondeterministic process $\{X_t\}$ with the covariance function $\{\gamma_k\}$, MA parameters $\{b_k\}_{-\infty}^{\infty}, b_0 = 1, b_k = 0, k < 0$, and innovation variance σ^2. Let us introduce the following notation:

$$\begin{aligned}
\boldsymbol{X} &= (X_0, \cdots, X_{-m}),\ \boldsymbol{\varepsilon} = (\varepsilon_0, \cdots, \varepsilon_{-n}), \\
\boldsymbol{T}_{n,m} &= (b_{i-j})_{\substack{i=0,\cdots,n \\ j=0,\cdots,m}} = \sigma^{-2}\mathrm{Cov}(\boldsymbol{\varepsilon}, \boldsymbol{X}), \Gamma_m = \mathrm{Cov}(\boldsymbol{X}), \quad (4.77) \\
H_{n,m} &= T_{n,m}\Gamma_m^{-1}T'_{n,m}, \Gamma_{m,n} = \Gamma_m - \sigma^2 T'_{n,m}T_{n,m}.
\end{aligned}$$

Define $\tilde{\boldsymbol{X}} = (\tilde{X}_0, \cdots, \tilde{X}_{-m}), \tilde{\varepsilon} = (\tilde{\varepsilon}_0, \cdots, \tilde{\varepsilon}_{-n})$ where $\tilde{X}_{-k}$ is the linear predictor of X_{-k} based on ε and $\tilde{\varepsilon}_{-k}$ is the linear predictor of ε_{-k} based on $\boldsymbol{X}$. In the following, subscripts will be used to indicate size of matrices only when necessary.

Lemma 4.9. With notation as above we have,

(a) $\mathrm{Cov}\begin{pmatrix} \boldsymbol{X} \\ \boldsymbol{\varepsilon} \end{pmatrix} = \begin{pmatrix} \Gamma_m & \sigma^2 T' \\ \sigma^2 T & \sigma^2 I_{n+1} \end{pmatrix} \geq 0.$

(b) $\tilde{\boldsymbol{X}} = T'\varepsilon, \mathrm{Cov}(\boldsymbol{X} - \tilde{\boldsymbol{X}}) = \Gamma_m - \sigma^2 T'T = \Gamma_{m,n}.$

(c) $0 \leq \Gamma_{m,n} \leq \Gamma_{m,j}, m \geq 0, 0 \leq j \leq n.$

(d)

$$\begin{aligned}
\begin{pmatrix} \boldsymbol{X} - \tilde{\boldsymbol{X}} \\ \boldsymbol{\varepsilon} \end{pmatrix} &= \begin{pmatrix} I_{m+1} & -T' \\ 0 & I_{n+1} \end{pmatrix} \begin{pmatrix} \boldsymbol{X} \\ \boldsymbol{\varepsilon} \end{pmatrix}, \\
\mathrm{Cov}\begin{pmatrix} \boldsymbol{X} - \tilde{\boldsymbol{X}} \\ \boldsymbol{\varepsilon} \end{pmatrix} &= \begin{pmatrix} \Gamma - \sigma^2 TT & 0 \\ 0 & \sigma^2 I_{n+1} \end{pmatrix} \geq 0.
\end{aligned}$$

(e) $\tilde{\varepsilon} = \sigma^2 T\Gamma^{-1}\boldsymbol{X}$ and $\mathrm{Cov}(\varepsilon - \tilde{\varepsilon}) = \sigma^2(I_{n+1} - \sigma^2 T\Gamma^{-1}T')$.

Proof. (a) is immediate from (2.73) and the rest follows from the basic least squares theory.

Q.E.D.

The following result gives a simple characterization of ARMA processes.

Theorem 4.10. A nondeterministic stationary process $\{X_t\}$ is an ARMA(p, q) process, if and only if the matrix $I_q - \sigma^2 T_{q,p}\Gamma_p^{-1}T'_{q,p}$ is singular.

Proof. Suppose $\{X_t\}$ is a stationary ARMA(p, q) process. Then, there are coefficients $\alpha_0 = 1, \alpha_1, \cdots, \alpha_p, \beta_0 = 1, \beta_1, \cdots, \beta_q$ such that

$$\sum_{k=0}^{p} \alpha_k X_{-k} = \sum_{j=0}^{q} \beta_j \varepsilon_{-j}.$$

From the definition of $\tilde{\varepsilon}_{-k}$ and projecting all components of the above equation onto $\boldsymbol{X}$ we get,

$$\sum_{k=0}^{p} \alpha_k X_{-k} = \sum_{j=0}^{q} \beta_j \tilde{\varepsilon}_{-j}.$$

Now, subtracting the two equations we obtain

$$\sum_{j=0}^{q} \beta_j(\varepsilon_{-j} - \tilde{\varepsilon}_{-j}) = 0,$$

with $\beta_0 = 1$. This and Lemma 4.9 (e) imply that $\text{Cov}(\varepsilon - \tilde{\varepsilon})$ is singular.

Conversely, $\text{Cov}(\varepsilon - \tilde{\varepsilon})$ is singular if and only if there exists a $(q+1)$ - vector $\boldsymbol{u} \neq 0$ such that

$$\boldsymbol{u}'(\varepsilon - \tilde{\varepsilon}) = 0 \text{ or } \boldsymbol{u}'\varepsilon = \boldsymbol{u}'\tilde{\varepsilon} \in sp\{X_{-p}, \cdots, X_0\}.$$

Thus, there must exist coefficients $v_0, v_1, \cdots, v_p$, not all vanishing, such that

$$\sum_{k=0}^{p} v_k X_{-k} = \sum_{k=0}^{q} u_k \varepsilon_{-k}. \tag{4.78}$$

Let $m = \min\{k; v_k \neq 0\}$. Since ε_n and X_j are uncorrelated for $n > j$, we conclude that $u_k = 0$ for all $k < m$ and $u_m, v_m \neq 0$. Thus,

$$\sum_{k=m}^{p} v_k X_{-k} = \sum_{k=m}^{q} u_k \varepsilon_{-k}, \ u_m, v_m \neq 0,$$

that is, $\{X_t\}$ is an ARMA(p, q) process.

Q.E.D.

An analogue of Theorem 4.10 can be proved in terms of singularity of $\text{Cov}(\boldsymbol{X} - \hat{\boldsymbol{X}})$; see Problem 4.12. How does one construct such ARMA processes? Or, express the u_k's and v_k's above in terms of the covariance function. In the rest of this chapter the details of construction and existence of ARMA models with prescribed covariances and MA parameters are presented following the work of Huang (1990). Since the proofs rely on more advanced and core topics from the theory of stationary processes, Sections A4.6 and A4.7 are labeled as appendix and the reader may skip them in the first reading.

A4.6 CONSTRUCTION OF ARMA PROCESSES

For given $\gamma_j, j = 0, 1, \cdots, p$ so that $\Gamma_p = (\gamma_{i-j})_{i,j=0}^{p} > 0$ and arbitrary $b_k, k = 1, 2, \cdots, q$, for the moment assume that there exists a stationary ARMA process $\{X_t\}$ as in (4.12) satisfying

$$\begin{cases} EX_jX_0 = \gamma_j, j = 0, 1, \cdots, p \\ \dfrac{EX_0\varepsilon_{-k}}{E\varepsilon_0^2} = b_k, k = 1, 2, \cdots, q, \quad E\varepsilon_t V_s = 0, \end{cases} \tag{4.79}$$

where $\{V_t\}$ is the deterministic component in its Wold decomposition (2.69). In this section, first we give an algorithm to express coefficients of (4.12) and $\sigma^2 = E\varepsilon_0^2$ in terms of the γ_j's and b_k's.

For $0 \leq m \leq p, 0 \leq n \leq q$, let the matrices $T_{nm}, \Gamma_{mn}, \cdots$, be as in (4.77) with σ^2 replaced by

$$\sigma_0^2 = 1/\lambda_{max}(\Gamma_p^{-\frac{1}{2}} T'_{qp} T_{qp} \Gamma_p^{-\frac{1}{2}}), \tag{4.80}$$

where $\lambda_{\max}(A)$ is the largest eigenvalue of a symmetric matrix A. Let $B_{k,m}$ be the kth column of T_{nm}, then $B_{k,m} = (b_k, b_{k-1}, \ldots, b_{k-m})' = (B_{k,m-1}, b_{k-m})' = (b_k, B_{k-1,m})'$ and since Γ_m is a Toeplitz matrix, we have

$$\begin{aligned} \Gamma_{m,n} &= \Gamma_m - \sigma_0^2 \sum_{k=0}^{n} B_{k,m} B'_{k,m} = \begin{bmatrix} \Gamma_{m-1,n} & \vdots \\ \cdots & \cdots \end{bmatrix} \\ &= \begin{bmatrix} \vdots & \cdots \\ \vdots & \Gamma_{m-1,n-1} \end{bmatrix}, \end{aligned} \tag{4.81}$$

where $\cdot$ denotes some scalar, not always the same. Since $\lambda_{\max}(\sigma_0^2 \Gamma_p^{-\frac{1}{2}} T'_{qp} T_{qp} \Gamma_p^{-\frac{1}{2}}) = 1$, we have

$$|\Gamma_{p,q}| = 0, \tag{4.82}$$

and from Lemma 4.9 (c), it follows that there exist a unique nonnegative integer $p_0 \leq p$ such that

$$|\Gamma_{p_0,q}| = 0, \; \Gamma_{p_0-1,q-1} \geq \Gamma_{p_0-1,q} > 0. \tag{4.83}$$

Consequently, there exists a unique vector $\boldsymbol{\alpha}_{p_0} = (1, \alpha_1, \cdots, \alpha_{p_0})$ such that

$$\Gamma_{p_0,q} \boldsymbol{\alpha}_{p_0} = (\Gamma_{p_0} - \sigma_0^2 T'_{qp_0} T_{qp_0}) \boldsymbol{\alpha}_{p_0} = 0. \tag{4.84}$$

Let $\alpha_k = 0, \; k > p_0$, and define

$$\begin{cases} \boldsymbol{\alpha} &= (1, -\alpha_1, \cdots, -\alpha_p), \\ \boldsymbol{\beta} &= (\beta_0, \beta_1, \cdots, \beta_q) = T_{qp} \boldsymbol{\alpha}, \end{cases} \tag{4.85}$$

we show in Lemma 4.11 below that these $\boldsymbol{\alpha}, \boldsymbol{\beta}$ are the desired coefficients of (4.12). But, first an example is helpful in illustrating the above computations.

Example 4.6. For $p = q = 1$, with $\gamma_0 = 1, \gamma_1 = \gamma, |\gamma| < 1, b_0 = 1$ and $b_1 = b$ we have

$$\Gamma_1 = \begin{pmatrix} 1 & \gamma \\ \gamma & 1 \end{pmatrix}, \; T_{11} = \begin{pmatrix} 1 & 0 \\ b & 1 \end{pmatrix},$$

$$|\Gamma_1 - \lambda T'_{11} T_{11}| = \lambda^2 + (2b\gamma - b^2 - 2)\lambda + 1 - \gamma^2 = 0,$$

with roots $\lambda = \dfrac{b^2 + 2 - 2b\gamma \pm \sqrt{(b^2 + 2 - 2b\gamma)^2 - 4(1 - \gamma^2)}}{2}$.

For $\gamma = 0$ and $b = 1$, the largest root is $\lambda_{\max} = \dfrac{3+\sqrt{5}}{2}$, so that $\sigma_0^2 = \dfrac{2}{3+\sqrt{5}}$. It is easy to check that $|\Gamma_{11}| = 0$ and

$$\Gamma_{00} > \Gamma_{01} = 1 - \frac{4}{3+\sqrt{5}} > 0,$$

thus $p_0 = 1 = q$. From (4.84)–(4.85), we find

$$\boldsymbol{\alpha} = \left(1, \frac{1-\sqrt{5}}{2}\right), \ \boldsymbol{\beta} = \left(1, \frac{1+\sqrt{5}}{2}\right),$$

and

$$X_t = \frac{1-\sqrt{5}}{2} X_{t-1} + \varepsilon_t + \frac{1+\sqrt{5}}{2}\varepsilon_{t-1},$$

with $E\varepsilon_t^2 = \dfrac{2}{3+\sqrt{5}}$, is the promised ARMA(1,1) model.

Lemma 4.11. Suppose there is a stationary ARMA(p, q) process $\{X_t\}$ satisfying (4.79) with $\Gamma_p > 0$. Then, its coefficients are given by the solutions of (4.84) and (4.85) and $E\varepsilon_t^2 = \sigma_0^2$. Furthermore, for any stationary ARMA(p', q') process satisfying (4.79) with $q' \le q$, we must have $p' \ge p_0$.

Proof. Let $E\varepsilon_0^2 = \sigma^2$. Since $\{X_t\}$ is an ARMA(p, q) process, there are vectors $\boldsymbol{a}$ and $\boldsymbol{b}$, with first components 1, such that $(X_0, X_{-1}, \cdots, X_{-p})\boldsymbol{a} = (\varepsilon_0, \varepsilon_{-1}, \cdots, \varepsilon_{-q})$ $\boldsymbol{b}$, and therefore

$$E\boldsymbol{X}(\boldsymbol{X}'\boldsymbol{a} - \boldsymbol{\varepsilon}'\boldsymbol{b}) = \Gamma_p \boldsymbol{a} - \sigma^2 T'_{qp}\boldsymbol{b} = 0, \tag{4.86}$$

$$E\boldsymbol{\varepsilon}(\boldsymbol{X}'\boldsymbol{a} - \boldsymbol{\varepsilon}'\boldsymbol{b}) = \sigma^2 T_{qp}\boldsymbol{a} - \sigma^2\boldsymbol{b} = 0. \tag{4.87}$$

From (4.86), it follows that $\sigma^2 > 0$, otherwise $|\Gamma_p| = 0$ contradicting $\Gamma_p > 0$. Substituting for $\boldsymbol{b}$ in (4.86) from (4.87), we get

$$(\Gamma_p - \sigma^2 T'_{qp} T_{qp})\boldsymbol{a} = 0,$$

for a non-zero $\boldsymbol{a}$, thus the smallest eigenvalue of $(I_{p+1} - \sigma^2 \Gamma_p^{-1/2} T'_{qp} T_{qp} \Gamma_p^{-1/2})$ must be zero, hence from (4.80), $\sigma^2 = \sigma_0^2$. Since (4.84) implies that $\boldsymbol{\alpha}'\Gamma_{pq}\boldsymbol{\alpha} = 0$, then using (4.85) we obtain

$$(\boldsymbol{\alpha}', -\boldsymbol{\beta}') \begin{bmatrix} \Gamma_p & \sigma_0^2 T'_{qp} \\ \sigma_0^2 T_{qp} & \sigma_0^2 I_{q+1} \end{bmatrix} \begin{bmatrix} \boldsymbol{\alpha} \\ -\boldsymbol{\beta} \end{bmatrix} = \boldsymbol{\alpha}'\Gamma_{pq}\boldsymbol{\alpha} = 0. \tag{4.88}$$

From this, Lemma 4.9 (a) and $\sigma_0^2 = \sigma^2$, it follows that

$$E|(X_0, X_{-1}, \cdots, X_{-p})\boldsymbol{\alpha} - (\varepsilon_0, \varepsilon_{-1}, \cdots, \varepsilon_{-q})\boldsymbol{\beta}|^2 = 0,$$

that is, $\{X_t\}$ satisfies (4.12) with coefficients $\boldsymbol{\alpha}$ and $\boldsymbol{\beta}$.

Finally, suppose there is an ARMA(p',q') process $\{X_t\}$ satisfying (4.79) with $\Gamma_p > 0$ such that

$$\sum_{k=0}^{p'} a_k X_{t-k} = \sum_{j=0}^{q'} b_j \varepsilon_{t-j},\ a_0 = b_0 = 1,\ p' < p_0, q' \leq q.$$

Repeating the above arguments, we get that

$$(\Gamma_{p'} - \sigma^2 T'_{q'p'} T_{q'p'})\boldsymbol{a} = 0,$$

where $\boldsymbol{a} = (1, a_1, \cdots, a_{p'})$ and $\sigma^2 = \sigma_0^2$. This contradicts (4.83) and the proof is complete.

Q.E.D.

For an ARMA process with coefficients as in (4.84) and (4.85), the following lemma gives information about the roots of its characteristic polynomials.

Lemma 4.12. If γ_j's and b_k's satisfy (4.79) with $b_0 = 1$ and $\Gamma_p > 0$, then

$$\sum_{k=0}^{p_0} \alpha_k z^k \neq 0,\ |z| < 1,$$

where α_k's are as in (4.84). Furthermore, a root of $\sum_{k=0}^{p_0} \alpha_k z^k$ with modulus 1 and multiplicity m is also a root of $\sum_{k=0}^{q} \beta_k z^k$ with multiplicity not smaller than m.

Proof. For any root λ and with $\boldsymbol{a} = (1, a_1, \cdots, a_{p_0-1})'$, we have

$$\begin{aligned}\sum_{k=0}^{p_0} \alpha_k z^k &= (1, z, \cdots, z^{p_0})' \boldsymbol{\alpha}_{p_0} = (1 - \lambda^{-1} z)(1, z, \cdots, z^{p_0-1})' \boldsymbol{a} \\ &= (1, z, \cdots, z^{p_0-1})' \boldsymbol{a} - \lambda^{-1}(z, z^2, \cdots, z^{p_0})\boldsymbol{a}.\end{aligned}$$

Thus,

$$\boldsymbol{\alpha}_{p_0} = \begin{pmatrix} \boldsymbol{a} \\ 0 \end{pmatrix} - \lambda^{-1} \begin{pmatrix} 0 \\ \boldsymbol{a} \end{pmatrix},$$

and from (4.81) and (4.84) we get

$$\begin{cases} \boldsymbol{a}'\Gamma_{p_0-1,q}\boldsymbol{a} - \lambda^{-1}(\boldsymbol{a}', 0)\Gamma_{p_0 q}\begin{pmatrix} 0 \\ \boldsymbol{a} \end{pmatrix} = 0 , \\ (0, \boldsymbol{a}')\Gamma_{p_0 q}\begin{pmatrix} \boldsymbol{a} \\ 0 \end{pmatrix} - \lambda^{-1}\boldsymbol{a}'\Gamma_{p_0-1,q-1}\boldsymbol{a} = 0. \end{cases}$$

Consequently,

$$\boldsymbol{a}'\Gamma_{p_0-1,q}\boldsymbol{a} = \frac{1}{|\lambda|^2}\boldsymbol{a}'\Gamma_{p_0-1,q-1}\boldsymbol{a}, \tag{4.89}$$

and from (4.83) we obtain $|\lambda| \geq 1$.

If $|\lambda| = 1$, from Lemma 4.9 (c) and (4.89) we get that $B'_{q,p_0-1}a = 0$, where B'_{q,p_0-1} is the last row of T_{qp_0}. Since

$$T_{qp_0} = \begin{bmatrix} T_{q-1,p_0-1} & \vdots \\ B'_{q,p_0-1} & \vdots \end{bmatrix} = \begin{bmatrix} \vdots & 0\cdots 0 \\ \vdots & T_{q-1,p_0-1} \end{bmatrix},$$

with $\boldsymbol{b} = T_{q-1,p_0-1}\boldsymbol{a}$, we have

$$\boldsymbol{\beta} = T_{qp_0}\{\begin{pmatrix} \boldsymbol{a} \\ 0 \end{pmatrix} - \lambda^{-1}\begin{pmatrix} 0 \\ \boldsymbol{a} \end{pmatrix}\} = \begin{pmatrix} \boldsymbol{b} \\ 0 \end{pmatrix} - \lambda^{-1}\begin{pmatrix} 0 \\ \boldsymbol{b} \end{pmatrix},$$

showing that λ is also a root of $\sum_{k=0}^{q} \beta_k z^k$.

If λ is a root of $\sum_{k=0}^{p_0} \alpha_k z^k$ with multiplicity $m > 1$, then there is a vector $\boldsymbol{c} = (1, c_1, \cdots, c_{p_0-m})'$ such that $\boldsymbol{a} = \begin{pmatrix} \boldsymbol{c} \\ 0 \end{pmatrix} - \lambda^{-1}\begin{pmatrix} 0 \\ \boldsymbol{c} \end{pmatrix}$. Define the $n \times (n-1)$ matrix

$$\Lambda_n = \begin{bmatrix} 1 & 0 & \cdots & 0 \\ -\lambda^{-1} & 1 & \cdots & 0 \\ 0 & -\lambda^{-1} & \cdots & 0 \\ \vdots & \vdots & & \vdots \\ 0 & 0 & \cdots & 1 \\ 0 & 0 & \cdots & -\lambda^{-1} \end{bmatrix}.$$

Then,

$$\Lambda'_{p_0}\Gamma_{p_0q}\Lambda_{p_0}\boldsymbol{a} = \Lambda'_{p_0}\Gamma_{p_0q}\boldsymbol{\alpha} = 0,$$

$$\Lambda'_{p_0}\Gamma_{p_0q}\Lambda_{p_0} = \begin{bmatrix} \Lambda'_{p_0-1}\Gamma_{p_0-1,q}\Lambda_{p_0-1} & \vdots \\ \cdots & \end{bmatrix} = \begin{bmatrix} \vdots & \cdots \\ \vdots & \Lambda'_{p_0-1}\Gamma_{p_0-1,q-1}\Lambda_{p_0-1} \end{bmatrix},$$

$$0 < \Lambda'_{p_0-1}\Gamma_{p_0-1,q}\Lambda_{p_0-1} \leq \Lambda'_{p_0-1}\Gamma_{p_0-1,q-1}\Lambda_{p_0-1}.$$

Now, the argument leading to (4.89) can be repeated to show that λ is a root of $\sum_{k=0}^{q} \beta_k z^k$. This proves that if λ is a root of $\sum_{k=0}^{p_0} \alpha_k z^k$ with modulus 1 and multiplicity m, then it is also a root of $\sum_{k=0}^{q} \beta_k z^k$ with multiplicity not smaller than m.

Q.E.D.

A4.7 EXISTENCE OF ARMA PROCESSES

In this section, we show that given $\gamma_j, j = 0, \cdots, p$ with $\Gamma_p > 0$ and arbitrary constants $b_k, k = 0, 1, \cdots, q$ with $b_0 = 1$, there exists a general ARMA(p, q) process satisfying (4.79) with parameters given by (4.80), (4.84) and (4.85). A necessary and sufficient condition is given for the existence of a purely nondeterministic ARMA solution. The proof is based on more advanced results from Chapters 5 and 9 on the positive-definite functions, spectral representation and the Wold decomposition of stationary processes.

Theorem 4.13. (a) There always exists a general ARMA(p, q) process $\{X_t\}$ satisfying (4.79) with σ^2 and coefficients α_k's and β_j's given by (4.80), (4.84) and (4.85).
(b) Let

$$\sum_{k=0}^{p_0} \alpha_k z^k = \sum_{k=0}^{p_0-m} a_k z^k \sum_{k=0}^{m} u_k z^k,$$

where $\sum_{k=0}^{p_0-m} a_k z^k \neq 0, |z| \leq 1$, and $\sum_{k=0}^{m} u_k z^k = 0, |z| = 1$.
Then, the deterministic component $\{V_t\}$ and the purely nondeterministic component $\{U_t\}$ of the Wold decomposition (2.69) of $\{X_t\}$ satisfy

$$\sum_{k=0}^{m} u_k V_{t-k} = 0, \quad \sum_{k=0}^{p_0-m} a_k U_{t-k} = \sum_{j=0}^{q-m} d_j \varepsilon_{t-j},$$

where the d_k's satisfy

$$\sum_{k=0}^{q-m} d_k z^k = \left(\sum_{k=0}^{q} \beta_k z^k\right) \left(\sum_{k=0}^{m} u_k z^k\right)^{-1}.$$

(c) $\{X_t\}$ is purely nondeterministic, if and only if

$$\sum_{k=0}^{p_0} \alpha_k z^k \neq 0, \quad |z| = 1.$$

Proof. (a) Using the α_k's and β_j's given by (4.84)–(4.85), let us extend γ_j's and b_k's via

$$\gamma_m = \sum_{k=1}^{p_0} \alpha_k \gamma_{m-k} + \sigma_0^2 \sum_{j=0}^{q} \beta_j b_{j-m}, \quad m = p+1, p+2, \cdots \tag{4.90}$$

$$b_n = \sum_{k=1}^{p_0} \alpha_k b_{n-k}, \quad n = q+1, q+2, \cdots; \tag{4.91}$$

see (4.22)–(4.23). Then, Γ_m, T_{mn} and Γ_{mn} can be defined for all integers m, n. First, we show that for all $m, n \geq 0$,

$$C_{mn} = \begin{bmatrix} \Gamma_m & \sigma_0^2 T'_{nm} \\ \sigma_0^2 T_{nm} & \sigma_0^2 I_{n+1} \end{bmatrix} \geq 0. \tag{4.92}$$

Since for all $m' \le m$ and $n' \le n, C_{m'n'}$ is a sub-block of C_{mn}, we only need to show that $C_{p+k,q+k} \ge 0$, for all integers $k \ge 0$. Furthermore, since

$$\begin{bmatrix} I_{m+1} & -T'_{nm} \\ 0 & I_{n+1} \end{bmatrix} C_{mn} \begin{bmatrix} I_{m+2} & 0 \\ -T_{nm} & I_{n+1} \end{bmatrix} = \begin{bmatrix} \Gamma_{mn} & 0 \\ 0 & I_{n+1} \end{bmatrix},$$

it is sufficient to show that

$$\Gamma_{p+k,q+k} \ge 0, \; k = 0, 1, 2, \cdots. \tag{4.93}$$

For $k = 0$, (4.93) follows from Lemma 4.9. For all $k \ge 1$, let $\alpha_{p_0+k} = \beta_{q+k} = 0, \boldsymbol{\alpha}_{p+k} = (1, \alpha_1, \cdots, \alpha_{p+k})$ and $\boldsymbol{\beta}_{q+k} = (1, \beta_1, \cdots, \beta_{q+k})$. Then, from (4.91),

$$\boldsymbol{\beta}_{q+k} = T_{q+k,p+k}\boldsymbol{\alpha}_{p+k},$$

and from (4.92) we get that

$$\Gamma_{p+k,q+k}\boldsymbol{\alpha}_{p+k} = \Gamma_{p+k}\boldsymbol{\alpha}_{p+k} - \sigma_0^2 T'_{q+k,p+k}\boldsymbol{\beta}_{q+k} = 0.$$

For the moment suppose that $\Gamma_{p+k-1,q+k-1} \ge 0$, then for any $\boldsymbol{a} = (a_1, a_2, \cdots, a_{p+k+1})$, we have from Lemma 4.9 (c) that

$$\begin{aligned} \boldsymbol{a}'\Gamma_{p+k,q+k}\boldsymbol{a} &= \left(a_1\alpha_{p+k} + \begin{bmatrix} 0 \\ b \end{bmatrix}\right)' \Gamma_{p+k,q+k} \left(a_1\boldsymbol{\alpha}_{p+k} + \begin{bmatrix} 0 \\ \boldsymbol{b} \end{bmatrix}\right) \\ &= \boldsymbol{b}'\Gamma_{p+k-1,q+k-1}\boldsymbol{b} \ge 0, \end{aligned}$$

where $\boldsymbol{b} = (a_2 - a_1\alpha_1, \cdots, a_{p+k+1} - a_1\alpha_{p+k})$. Thus, (4.93) holds by induction.

From the positive-definiteness of C_{mn} in (4.92) and Theorem 5.2, there exists a bivariate stationary series $\left\{\begin{bmatrix} X_t \\ \varepsilon_t \end{bmatrix}\right\}$ with

$$\mathrm{Cov}(X_t, X_{t-1}, \cdots, X_{t-m}, \varepsilon_t, \cdots, \varepsilon_{t-n}) = C_{mn}. \tag{4.94}$$

Consequently,

$$EX_tX_{t-j} = \gamma_j, EX_t\varepsilon_{t-k} = b_k E\varepsilon_t^2, \text{ for all integers } j, k,$$

and

$$E\varepsilon_t X_{t-k} = 0, k > 0, E\varepsilon_t X_t = E\varepsilon_t^2 = \sigma_0^2. \tag{4.95}$$

Furthermore, (4.94) and (4.87) imply that

$$E\left|\sum_{k=0}^{p} \alpha_k X_{t-k} - \sum_{j=0}^{q} \beta_j \varepsilon_{t-j}\right|^2 = (\boldsymbol{\alpha}', -\boldsymbol{\beta}')\Gamma_{pq}\begin{pmatrix} \boldsymbol{\alpha} \\ -\boldsymbol{\beta} \end{pmatrix} = 0,$$

that is $\{X_t\}$ satisfies (4.12) with coefficients given by (4.84) and (4.85).

(b) Let $X_t = U_t + V_t$ be the Wold decomposition of any stationary ARMA process where $\{V_t\}$ is its deterministic component. Since V_s, for any s, belongs to the remote past of $\{X_t\}$, from (4.95) it follows that

$$E\varepsilon_t V_s = 0, \quad \text{for all } s, t. \tag{4.96}$$

Now, using (4.96) and since $\{U_t\}$, $\{V_t\}$ are uncorrelated, we obtain

$$0 = E\left(\sum_{k=0}^{p_0} \alpha_k X_{t-k} - \sum_{j=0}^{q} \beta_j \varepsilon_{t-j}\right)\left(\sum_{k=0}^{p_0} \alpha_k V_{t-k}\right) = E\left(\sum_{k=0}^{p_0} \alpha_k V_{t-k}\right)^2,$$

that is,

$$\sum_{k=0}^{p_0} \alpha_k V_{t-k} = 0. \tag{4.97}$$

Thus, from the Wold decomposition of X_t, (4.12) and (4.97), it follows that

$$\sum_{k=0}^{p_0} \alpha_k U_{t-k} = \sum_{j=0}^{q} \beta_j \varepsilon_{t-j}. \tag{4.98}$$

Let $W_t = \sum_{k=0}^{m} u_k V_{t-k}$ and $\boldsymbol{W}_t = (W_t, W_{t-1}, \cdots, W_{t-p_0+m+1})'$. Since $\alpha_k = \sum_{\ell=0}^{k} a_{k-\ell} u_\ell$, from (4.97) we obtain

$$\sum_{j=0}^{p_0-m} a_j W_{t-j} = \sum_{k=0}^{p_0}\left(\sum_{\ell=0}^{k} a_{\ell-k} u_\ell\right) V_{t-k} = 0. \tag{4.99}$$

Defining A appropriately as in (4.7) and rewriting (4.99) as $\boldsymbol{W}_{t+1} = A\boldsymbol{W}_t$, it follows from the stationarity of $\{\boldsymbol{W}_t\}$ that

$$\begin{aligned}
\Gamma &= E\boldsymbol{W}_{t+1}\boldsymbol{W}'_{t+1} = AE(\boldsymbol{W}_t\boldsymbol{W}'_t)A' \\
&= A^k E(\boldsymbol{W}_{t-k}\boldsymbol{W}'_{t-k})A'^k = A^k \Gamma A'^k \to 0,
\end{aligned}$$

as $k \to \infty$, since the eigenvalues of A, the inverses of the roots of $\sum_{k=0}^{p_0-m} a_k z^k$, are less than one in absolute value. Thus, $\boldsymbol{W}_t = 0$ and the assertion about $\{V_t\}$ in (b) follows.

Proof of the assertions regarding $\{U_t\}$ and $\{\varepsilon_t\}$ exploits their spectral representations (2.35). Let $U_t = \int_{-\pi}^{\pi} e^{it\lambda} dU(\lambda)$ and $\varepsilon_t = \int_{-\pi}^{\pi} e^{it\lambda} d\varepsilon(\lambda)$, where $\{U(\lambda)\}$ and $\{\varepsilon(\lambda)\}$ are orthogonal increment processes with $E|dU(\lambda)|^2 = f(\lambda)d\lambda$, $E|d\varepsilon(\lambda)|^2 =$

$\sigma^2 d\lambda$, and $f(\lambda)$ is the spectral density of the purely nondeterministic stationary process $\{U_t\}$. Substituting these in (4.98), it follows that

$$\sum_{k=0}^{p_0} \alpha_k e^{-ik\lambda} dU(\lambda) = \sum_{j=0}^{q} \beta_j e^{-ij\lambda} d\varepsilon(\lambda).$$

Since

$$\sum_{k=0}^{p_0} \alpha_k z^k = \sum_{k=0}^{m} u_k z^k \sum_{j=0}^{p_0-m} a_j z^j,$$

it follows from Lemma 4.9 that

$$\sum_{j=0}^{q} \beta_j z^j = \sum_{k=0}^{m} u_k z^k \sum_{j=0}^{q-m} d_j z^j,$$

then

$$\sum_{k=0}^{m} u_k e^{-ik\lambda} \sum_{j=0}^{p_0-m} a_j e^{-ij\lambda} dU(\lambda) = \sum_{k=0}^{m} u_k e^{-ik\lambda} \sum_{j=0}^{q-m} d_j e^{-ij\lambda} d\varepsilon(\lambda).$$

However, since $\sum_{k=0}^{m} u_k e^{-ik\lambda} = 0$ at most on finite number of points of λ, and on any finite set $dU(\lambda) = 0$, we arrive at

$$\sum_{j=0}^{p_0-m} a_j e^{-ij\lambda} dU(\lambda) = \sum_{j=0}^{q-m} d_j e^{-ij\lambda} d\varepsilon(\lambda), a.e.(Leb.).$$

Multiplying both sides by $e^{it\lambda}$ and integrating, we obtain

$$\sum_{j=0}^{p_0-m} a_j \int e^{i(t-j)\lambda}\, dU(\lambda) = \sum_{j=0}^{q-m} d_j \int e^{i(t-j)\lambda}\, d\varepsilon(\lambda),$$

or

$$\sum_{j=0}^{p_0-m} a_j U_{t-j} = \sum_{j=0}^{q-m} d_j \varepsilon_{t-j},$$

as desired. Proof of (c) is immediate from (b).

Q.E.D.

This theorem and other results of this section are of independent interest in theory of stationary processes. However, their immediate applications to maximum entropy problems and Padé approximations are discussed in detail in Huang (1990) and Franke (1985).

Problems

4.1 Show that $\{V_t\}$ defined in Section 4.1 satisfies $V_t = aV_{t-1}$, for *all* t, and find all its solutions.

4.2 Consider the difference equation $X_t = aX_{t-1} + e_t$, where $\{e_t\}$ is a sequence of uncorrelated random variables with $Ee_t = \mu \neq 0$ and $Var(e_t) = \sigma^2 > 0$, for all t. Find condition(s) on a, μ and σ^2 so that it has a causal stationary solution.

4.3 Let $\{e_t\}$ be a zero-mean stationary process.

(a) Show that $\sum_{t=0}^{\infty} a^t e_t$ is mean-convergent, if and only if $|a| < 1$.

(b) Prove Theorem 4.1.

4.4 Show that $\{X_t\}$ defined by $X_t = \sum_{k=0}^{\infty} 2^{-k} e_{t+k}$ is a mean-zero, noncausal stationary solution of the difference equation $X_t = 2X_{t-1} + e_t$.

4.5 Find the companion matrix and its eigenvalues for each model in Example 4.1.

4.6 Use (4.21)–(4.22) to obtain recursive formulas for computing $\{a_k\}$ similar to (4.22). Compare these recursions with (4.23).

4.7 For the difference equation $X_t - aX_{t-1} + X_{t-2} = e_t$, discuss the role of a on the nature of solutions of this equation as far as stationarity is concerned. (Note that for $a = 1.62$, this equation reduces to model (2) in Problem 4.10). Express the moving average parameters of a stationary solution of this difference equation in terms of a.

4.8 For the difference equation $X_t - aX_{t-1} + a^2 X_{t-2} = e_t$, find the roots of its characteristic equation in terms of a. Find the values of a for which it has a stationary solution. Express its moving average parameters and covariance function in terms of a.

4.9 Find the values a for which the following MA models are invertible.

(a) $X_t = e_t - ae_{t-1} + e_{t-2}$,

(b) $X_t = e_t - ae_{t-1} + a^2 e_{t-1}$.

Compare your results with those in Problems 4.7 and 4.8.

4.10 (PROJECT) The sunspot data (yearly averages) has been studied and modeled by many authors over many years starting with G. Yule (1927). The following is a partial list of fitted models, cf. Woodward and Gray (1978, p.100):

Model No.	Model	Source in Literature
1.	$X_t = 1.34X_{t-1} - .65X_{t-2} + \varepsilon_t$	(Yule,Moran,Bailey)
2.	$X_t = 1.62X_{t-1} - X_{t-2} + \varepsilon_t$	(Yule)
3.	$X_t = 1.30X_{t-1} - .54X_{t-2} - .15X_{t-3}$ $+.19X_{t-4} - .24X_{t-5} + .14X_{t-6} + \varepsilon_t$	(Bailey)
4.	$X_t = 1.57X_{t-1} - 1.02X_{t-2} + .21X_{t-3} + \varepsilon_t$	(Box and Jenkins)
5.	$X_t = 1.42X_{t-1} - .72X_{t-2} + \varepsilon_t - .15\varepsilon_{t-1}$	(Phadke and Wu)
6.	$X_t = 1.25X_{t-1} - .54X_{t-2} - .19X_{t-3} + \varepsilon_t$	(Morris, Schaerf)

(a) Which of these ARMA models have stationary solutions?

(b) Which of these ARMA models are causal? Invertible?

(c) Find the AR and MA parameters of those ARMA models having a unique stationary solution.

(d) Compute the covariance function of those ARMA models having a unique stationary solution.

(e) Write the forecast, updating equations (4.35) and (4.39) and the Akaike's state-space representation for each model.

(f) Plot and compare the functions obtained in (c) and (d) for each appropriate model.

4.11 Discuss (4.27)–(4.30) for a pure moving average model of order q.

4.12 Show that a nondeterministic stationary process $\{X_t\}$ is an ARMA(p, q) process if and only if Cov$(\boldsymbol{X} - \tilde{\boldsymbol{X}})$ is singular for $m \geq p$ and $n \geq p$.

4.13 Let $\hat{X}_0$ be the predictor of X_0 based on $X_{-m}, \cdots, X_{-1}, \varepsilon_{-n}, \cdots, \varepsilon_{-1}$.

(a) Show that

$$\hat{X}_0 = \gamma'\Gamma^{-1}\boldsymbol{X} + \boldsymbol{d}'(\varepsilon - \tilde{\varepsilon}),$$

$$\text{Var}(X_0 - \hat{X}_0) = \gamma_0\gamma'\Gamma^{-1}\gamma - \sigma^2\boldsymbol{d}'(I_n - \sigma^2 H)\boldsymbol{d},$$

where $\boldsymbol{d} \in R^n$ satisfies

$$(I_n - \sigma^2 H)d = \boldsymbol{b} - T\Gamma^{-1}\gamma$$

with $\boldsymbol{b} = (b_1, \cdots, b_n)$ and $\boldsymbol{\gamma} = (\gamma_1, \cdots, \gamma_m)$.

(b) Show that

$$\begin{aligned} \hat{X}_0 &= \mathbf{c}'(X - \tilde{X}) + b'\varepsilon, \\ \text{Var}\,(X_0 - \tilde{X}_0) &= \gamma_0 - \sigma^2 \mathbf{b}'\mathbf{b} - \mathbf{c}'(\Gamma - \sigma^2 T'T)\mathbf{c}, \end{aligned}$$

where $\mathbf{c} \in R^m$ satisfies

$$(\Gamma - \sigma^2 T'T)\mathbf{c} = \gamma - \sigma^2 T'\mathbf{b}.$$

4.14 (a) If A and D are invertible $n \times n$ and $m \times m$ matrices, show that

$$(A + BDB')^{-1} = A^{-1} - A^{-1}B(B'A^{-1}B + D^{-1})^{-1}B'A^{-1}.$$

(b) If A and D are as above and the inverses which occur in the expression exits, then

$$\begin{pmatrix} A & B \\ C & D \end{pmatrix}^{-1} = \begin{pmatrix} A^{-1} + FE^{-1}F & -FE^{-1} \\ -E'^{-1}F' & E^{-1} \end{pmatrix},$$

where $E = D - CA^{-1}B$ and $F = A^{-1}B$.

(c) If A and D are as above and A is invertible, then

$$\begin{vmatrix} A & B \\ C & D \end{vmatrix} = |A| \cdot |D - CA^{-1}B|.$$

5

Stationary Processes

The notion of *stationarity* has its roots in physical sciences and phenomena characterized by an invariance with respect to shifts of origin of time. A stochastic process whose probabilistic laws enjoy the same property is called a *strictly stationary process*. It turns out that for many applications stochastic processes whose first two moments remain unchanged through shifts in time provide appropriate and flexible models. Such processes are called *covariance stationary* or simply *stationary processes*. They play a central role in prediction theory and time series analysis. Their importance in applications is mostly due but not limited to the reduction in the number of parameters needed to describe a system, their theoretical importance can be traced to readily defined notion of spectrum, ergodic property, spectral representation and other elegant results discussed in the sequel.

We shall focus on discrete-time covariance stationary processes. This is defined in Section 5.1, where properties of their covariance functions are reviewed. Section 5.2 contains an exhaustive list of examples of stationary processes, showing that stationary processes are built from linear combinations of white noise or uncorrelated random variables. The characterization of a stationary process in terms of its shift operator, the spectral representation of the process and its covariance function are given in Section 5.3. The notions of predictor, its error, innovation process, deterministic and nondeterministic processes are studied in Sections 5.4 and 5.5 where also the Wold decomposition theorem is proved and used to compute the multi-step ahead predictors and their prediction error variances. Section 5.6 is devoted to the estimation of mean, covariance and spectrum of ergodic stationary processes.

This chapter covers most of the core and standard results in the theory of stationary processes. More complete coverage with specific theoretical or applied goals in mind can be found in Kolmogorov (1941); Doob (1953); Wiener and Masani (1957, 1958); Robinson (1954); Yaglom (1962, 1986); Whittle (1963); Rozanov (1967); Dym and McKean (1976) and Rosenblatt (1985).

5.1 COVARIANCE OF STATIONARY PROCESSES

The natural framework for studying the second-order properties of a collection of random variables is the space of all complex-valued random variables X (defined on a probability space) with

$$EX = 0 \quad \text{and} \quad E|X|^2 < \infty,$$

where E is the expectation operator. This space denoted by H and endowed with the inner product and norm

$$(X, Y) = EX\bar{Y} \ , \ \|X\| = (E|X|^2)^{\frac{1}{2}} \ , \ X, Y \in H,$$

is a Hilbert space. For more on Hilbert spaces, see Chapter 9. Note that for any two random variables $X, Y \in H$, the inner product of X and Y is the same as their covariance, that is,

$$(X, Y) = \text{Cov}(X, Y).$$

Thus, the geometric notion of orthogonality in H is equivalent to the statistical notion of uncorrelatedness. Since (X, Y) is a more economical notation than $\text{Cov}(X, Y)$, we shall use the former often, particularly when emphasizing the geometric aspect of a problem or its solution. The terms covariance and inner product will be used interchangeably.

Let $\mathbb{Z}$ denote the set of all integers. A collection of random variables $\{X_t;\ t \in \mathbb{Z}\}$ with $X_t \in H$ is called a discrete-time *second-order* stochastic process. Its *covariance function* is defined as the inner product of X_s and X_t:

$$\gamma_{s,t} = (X_s,\ X_t)\ s, t \in \mathbb{Z}.$$

So that for each $s, t, \gamma_{s,t}$ is a measure of dependence (angle) between X_s and X_t. The infinite matrix $\Gamma = (\gamma_{s,t})_{s,t \in \mathbb{Z}}$ is called the *covariance matrix* of $\{X_t\}$. When there is no danger of confusion we may use X or $\{X_t\}$ in place of $\{X_t; t \in \mathbb{Z}\}$; similarly γ or $(\gamma_{s,t})$ will be used for $(\gamma_{s,t})_{s,t \in \mathbb{Z}}$.

The function γ or the matrix Γ is too general to be of any practical use. Next, we introduce the class of stationary processes for which γ or Γ has a simpler structure. It is useful to view stationary processes as a curve or the path of a moving object with certain time-invariance property.

A second-order stochastic process $\{X_t\}$ is said to be *covariance stationary* or simply *stationary* if the covariance of X_s and X_t depends only on the time difference or the *lag* $s - t$, for any s, t. In this case, $\{\gamma_k\}$ defined by

$$\gamma_k = (X_{t+k}, X_t) \ , \ k \in \mathbb{Z},$$

is called the *(auto) covariance function* (ACF) of the stationary process $\{X_t\}$.

The sequence $\{\gamma_k\}$, viewed as a function on the set of all integers $\mathbb{Z}$, has a special *positivity* property, namely, for all integers $n \geq 1$ and complex scalars $c_1, \cdots, c_n$,

$$\sum_{i=1}^{n} \sum_{j=1}^{n} c_i \bar{c}_j \gamma_{i-j} \geq 0, \tag{5.1}$$

which follows rather easily from the definitions of covariance and stationarity:

$$\begin{aligned}\sum_{i=1}^{n} \sum_{j=1}^{n} c_i \bar{c}_j \gamma_{i-j} &= \sum_{i=1}^{n} \sum_{j=1}^{n} c_i \bar{c}_j (X_i, X_j) = \left(\sum_{i=1}^{n} c_i X_i, \sum_{j=1}^{n} c_j X_j \right) \\ &= \left\| \sum_{i=1}^{n} c_i X_i \right\|^2 \geq 0.\end{aligned}$$

This property of the function $\{\gamma_k\}$ is evidently equivalent to the *positive - definiteness* of all $n \times n$ principal submatrices of Γ,

$$\Gamma_n = (\gamma_{i-j})_{i,j=1}^{n} , \; n \geq 1. \tag{5.2}$$

For this reason, a $\{\gamma_k\}$ satisfying (5.1) is said to be a *positive-definite* function. This is a notion of central importance in probability, statistics and analysis. For example, recall that the characteristic function of any random variable is a positive-definite function on the real line R.

It is rather surprising that any function satisfying (5.1) must be the covariance function of a stationary process. Thus, such *positive-definite functions on $\mathbb{Z}$ can arise only as the covariance functions of stationary processes.* This is established by the following theorem of Kolmogorov.

Theorem 5.1. *(Kolmogorov, 1941).* A function $\{\gamma_k\}$ on the set of integers is positive-definite if and only if, there exists a Hilbert space H and a stationary process $\{X_t\}$ with $X_t \in H$ such that $\gamma_k = (X_{t+k}, X_t)$, for all integers t, k.

The "if" part is proved above; for a proof of the "only if" part of this theorem and more information on positive-definite functions see Section 9.7. Often it is desirable, but difficult to check the positive-definiteness of a given function. The next two properties of covariance functions that follow from (5.1) are useful for ruling out certain functions as candidates: If $\{\gamma_k\}$ is a covariance function, then for any k,

$$\begin{cases} \gamma_k = \bar{\gamma}_{-k} & , \quad \textit{(conjugate symmetry)}, \\ |\gamma_k| \leq \gamma_0 & , \quad \textit{(boundedness)}. \end{cases} \tag{5.3}$$

Example 5.1. We shall frequently use the following examples to check various statements and results on covariance functions and stationary processes.

(a) $\gamma_k = a^{|k|}$, $k = 0, \pm 1, \cdots$, where a is a real number.

(b) $\gamma_0 = 1 + b^2$, $\gamma_{\pm 1} = b$, $\gamma_k = 0, |k| \geq 2$, where b is a real number.

(c) $\gamma_k = \rho + \delta_{k0}$, where ρ is a real number with $|\rho| \leq 1$ and δ_{k0} is the *Kronecker's delta function.*

Note that when $|a| > 1$, the $\{\gamma_k\}$ in (a) is not a covariance function because the boundedness property in (5.3) is violated. For $|a| \leq 1$, although it is a covariance function, the task of establishing it directly using (5.1) is rather complicated. In view of Theorem 5.1, a useful alternative is to construct a stationary process $\{X_t\}$ with $\{\gamma_k\}$ as its covariance function. The next section provides various examples of stationary processes useful for this purpose; see (5.4)–(5.7).

5.2 EXAMPLES OF STATIONARY PROCESSES

The simplest stationary process is a sequence of uncorrelated (orthogonal) random variables. In this section, we show that essentially all stationary processes are appropriate linear combinations of elements of this simple process. As such it is the building blocks of the family of stationary processes and deserves a special name.

Example 5.2. *White Noise Process:* A sequence $\{e_t; t \in \mathbb{Z}\}$ of uncorrelated random variables with $E(e_t) = 0$ and $\text{Var}(e_t) = E|e_t|^2 = \sigma^2$, $t \in \mathbb{Z}$, $0 < \sigma^2 < \infty$, is called a *white noise process.* It is stationary, because

$$(e_{t+k}, e_t) = \sigma^2 \delta_{k0}, \ k, \ t \in \mathbb{Z},$$

depends only on k.

Note that the covariance function $\{\gamma_k\}$ of a white noise process takes only the two distinct values 0 and σ^2, and as such it is the simplest of all nontrivial stationary process; see also Example 5.1 (c).

In the next few examples, more general stationary processes with general covariance functions are constructed using the white noise $\{e_t\}$.

Example 5.3. *Linear Process:* Let $\{e_t\}$ be a white noise process and $\{c_j\}_{j=-\infty}^{\infty}$ be a sequence of scalars with

$$\sum_{j=-\infty}^{\infty} |c_j|^2 < \infty.$$

Then, the stochastic process $\{X_t\}$ given by

$$X_t = \sum_{j=-\infty}^{\infty} c_j e_{t-j}, \ t \in \mathbb{Z}, \tag{5.4}$$

is well defined since the infinite series converges in norm by the Riesz-Fischer Theorem 9.7. It is stationary, because using continuity of inner products (Theorem 9.2),

$$
\begin{aligned}
(X_{t+k}, X_t) &= \lim_{m,n\to\infty} \left(\sum_{i=-m}^{m} c_i e_{t+k-i}, \sum_{j=-n}^{n} c_j e_{t-j} \right) \\
&= \sum_{i=-\infty}^{\infty} \sum_{j=-\infty}^{\infty} c_i \bar{c}_j (e_{t+k-i}, e_{t-j}) \qquad (5.5) \\
&= \sigma^2 \sum_{i=-\infty}^{\infty} \sum_{j=-\infty}^{\infty} c_i \bar{c}_j \delta_{t+k-i,t-j} \\
&= \sigma^2 \sum_{i=-\infty}^{\infty} c_i \bar{c}_{i+k},
\end{aligned}
$$

depends only on k.

The class of linear processes includes many important and interesting subclasses. For example, when $c_j = 0, \ j < 0$, one obtains the important class of *one-sided moving average* or MA (∞) processes:

$$X_t = \sum_{j=0}^{\infty} c_j e_{t-j}, \ t \in \mathbb{Z}. \qquad (5.6)$$

Furthermore, when $c_j = a^j, \ j \geq 0$, with $|a| < 1$, it can be shown that X_t in (5.4) satisfies the difference equation

$$X_t = aX_{t-1} + e_t.$$

From (5.5) its covariance function is given by $\gamma_k = \frac{\sigma^2}{1-a^2} a^k$ which is a multiple of the function in Example 5.1 (a), so the latter is a positive-definite function. Similarly, when $c_0 = 1, c_1 = \beta, c_k = 0, k \geq 2$ and $\sigma^2 = 1$, then (5.6) reduces to

$$X_t = e_t + \beta e_{t-1},$$

its covariance function is given in Example 5.1 (b). Thus, the first two functions in Example 5.1 are positive-definite. An important class of stochastic processes that subsumes these two simple examples is the class of ARMA models studied in Chapter 4.

The class of linear processes is rather broad in the sense that their covariance functions can assume infinitely many distinct values tending to zero for k large, that is,

$$\gamma_k \to 0 \text{ as } k \to \infty. \qquad (5.7)$$

We prove this for the class of $MA(\infty)$ processes. The proof for the linear processes is similar; see Problem 5.4. The covariance function $\{\gamma_k\}$ of the process in (5.4) is given by

$$\gamma_k = \sigma^2 \sum_{j=0}^{\infty} c_j \bar{c}_{j+k}, \tag{5.8}$$

applying the Cauchy-Schwarz inequality to (5.8) gives

$$\begin{aligned} |\gamma_k| &\leq \sigma^2 \left(\sum_{j=0}^{\infty} |c_j|^2 \right)^{\frac{1}{2}} \left(\sum_{j=0}^{\infty} |c_{j+k}|^2 \right)^{\frac{1}{2}} \\ &= \sigma^2 \left(\sum_{j=0}^{\infty} |c_j|^2 \right)^{\frac{1}{2}} \left(\sum_{j=k}^{\infty} |c_j|^2 \right)^{\frac{1}{2}}. \end{aligned} \tag{5.9}$$

Now, since $\{c_j\}_{j=0}^{\infty}$ is square-summable, its tail $\sum_{j=k}^{\infty} |c_j|^2$ converges to zero and hence γ_k tend to zero as $k \to \infty$.

Next, we introduce another class of stationary processes which unlike the previous class, their covariance functions do not converge to zero as $k \to \infty$.

Example 5.4. *Harmonic or Oscillatory Process:* Let $\{e_t\}$ be a white noise and $\lambda \varepsilon (-\pi, \pi]$ be a fixed number. The process $\{X_t\}$ defined by

$$X_t = e_1 cos\lambda t + e_2 sin\lambda t, t \in \mathbb{Z},$$

is stationary, because

$$\begin{aligned} (X_{t+k}, X_t) &= \sigma^2 [cos\lambda(t+k)cos\lambda t + sin\lambda(t+k)sin\lambda t] \\ &= \sigma^2 cos\lambda(t+k-t) = \sigma^2 cos\lambda k, \end{aligned}$$

depends on k alone. This is a special case of more general stationary processes called *harmonic* or *oscillatory process* defined by

$$X_t = \sum_{j=-\infty}^{\infty} c_j e^{i\lambda_j t} e_j, t \in \mathbb{Z}, \tag{5.10}$$

where $\{c_j\}_{j=-\infty}^{\infty}$ is a sequence of scalars with $\sum_{j=-\infty}^{\infty} |c_j|^2 < \infty$, and $\{\lambda_j\}_{j=-\infty}^{\infty}$ a sequence of real numbers in $(-\pi, \pi]$. The process $\{X_t\}$ is stationary with covariance function; see Problem 2.3,

$$\gamma_k = \sigma^2 \sum_{j=-\infty}^{\infty} |c_j|^2 e^{i\lambda_j k}, k \in \mathbb{Z}. \tag{5.11}$$

In sharp contrast with the general linear processes, this covariance function does not tend to zero as $k \to \infty$. Other differences between harmonic and linear processes will become apparent later.

Example 5.5. *General Linear Process:* Since the classes of processes in Examples 5.3 and 5.4 are disjoint, their orthogonal sum provides a larger class of stationary processes. More precisely, let $\{e_t\}, \{c_j\}_{j\geq 0}$ be as in Example 5.3 and $\{e'_t\}$ be a white noise process orthogonal to $\{e_t\}$, that is, $(e_s, e'_t) = 0$, for all s, t, let $\{c'_j\}$ and $\{\lambda j\}$ be as in the Example 5.4, then the process $\{X_t\}$ defined by

$$X_t = \sum_{j=0}^{\infty} c_j e_{t-j} + \sum_{j=-\infty}^{\infty} c'_j e^{i\lambda jt} e'_{t-j}, \tag{5.12}$$

is stationary; see Problem 5.4. The simplest nontrivial member of this class is the process

$$X_t = e_t + e, \tag{5.13}$$

where e is a random variable uncorrelated with $\{e_t\}$. Note that with $\sigma_e^2 = \text{Var}(e)$,

$$\gamma_k = \begin{cases} \sigma^2 + \sigma_e^2, & k = 0, \\ \sigma_e^2, & k \neq 0, \end{cases} \tag{5.14}$$

does not converge to zero as $k \to \infty$. Compare (5.14) with the Example 5.1 (c) and Example 1.2.

The celebrated Wold decomposition theorem essentially says that all stationary processes are of the form (5.12); see Section 5.5.

The stationary processes constructed so far are obtained as limits of linear combinations of a white noise $\{e_t\}$. There might be processes which do not admit this form. Next, we construct the most general example of a stationary process using a unitary operator U on a Hilbert space H. It is known that certain members of this class have representations more general than (5.12).

Example 5.6. *General Stationary Process:* Let $X \in H$ and

$$U : H \to H,$$

be a *unitary* operator, that is, $U^* = U^{-1}$ where U^* stands for the adjoint of U and U^{-1} for its inverse, see Section 9.5.3. The stochastic process $\{X_t\}$ defined by

$$X_t = U^t X, \ \ t = 0, \pm 1, \cdots, \tag{5.15}$$

is stationary, because for all t, k,

$$\begin{aligned} (X_{t+k}, X_t) &= (U^{t+k}X, U^t X) = (U^{*t}U^{t+k}X, X) \\ &= (U^{-t+t+k}X, X) = (U^k X, X), \end{aligned}$$

depends only on k.

In the next section we show that *all* stationary processes are of the form (5.15).

5.3 THE SHIFT OPERATOR AND SPECTRAL REPRESENTATION

In this section, we establish a correspondence between stationary processes and unitary operators and use the simple structure of the latter to obtain the spectral representation of stationary processes and their covariance functions.

5.3.1 The Shift Operator

In the time series literature the *(backward) shift symbol* B introduced by

$$BX_t = X_{t-1},\ t \in \mathbb{Z}, \tag{5.16}$$

is used usually as a formal mathematical tool to "prove" results regarding ARMA models. There is a similar situation for the (lag) operators L, Z and D in the econometrics and engineering literature. Of course, it is desirable to define and use B as a bona fide mathematical object. This requires extending the transformation B from the set $\{X_t; t \in \mathbb{Z}\}$ to its span or space of linear combinations, denoted by $sp\{X_t; t \in \mathbb{Z}\}$, and then to its closure, denoted by $\overline{sp}\{X_t; t \in \mathbb{Z}\} = H(X)$, in the Hilbert space H. The space $H(X)$ is called the *time-domain* of $\{X_t\}$.

One may require B to be linear, namely,

$$B\left(\sum_{k=1}^{n} c_k X_{t_k}\right) = \sum_{k=1}^{n} c_k B X_{t_k} = \sum_{k=1}^{n} c_k X_{t_k - 1}, \tag{5.17}$$

for all integers $n \geq 1, t_1, \cdots, t_n$ and scalars $c_1, \cdots, c_n$. Can B be extended uniquely to $H(X)$ as a continuous linear operator? If the answer is yes, then this extension denoted by U is called the *shift operator* of the stochastic process $\{X_t\}$. Kolmogorov (1941) was the first to recognize the importance of the shift operator in the study of stationary processes.

For $\{X_t\}$ stationary, the transformation B in (5.17) is an *isometry*, namely, it preserves inner products and norms and hence for a sequence $\{Y_n\}$ in $sp\{X_t; t \in \mathbb{Z}\}$ we have $||BY_n - BY_m|| = ||Y_n - Y_m||$, for all m, n. For any $Y \in H(X)$, define UY as follows: Take a sequence $\{Y_n\}$ with $Y_n \in sp\{X_t\}$ so that $||Y_n - Y|| \to 0$, consequently $||BY_n - BY_m|| \to 0$ as $m, n \to \infty$ and $\{BY_n\}$ is a Cauchy sequence and convergent, let us set $UY = \lim_{n\to\infty} UY_n$. It can be shown that such U is well-defined, preserves inner products and is onto, namely $B : H(X) \to H(X)$ defined by (5.17) can be extended to a *unitary* operator U. This is *the shift operator of the stationary process* $\{X_t\}$. For more details, see Section 9.5.3.

A characterization of stationary processes in terms of unitary operators is given in the following fundamental theorem due to Kolmogorov (1941). Its proof is immediate from the above argument and the Example 5.6.

Theorem 5.2. A stochastic process $\{X_t\}$ in H is stationary, if and only if there exist a unique unitary operator U on its time-domain and an $X \in H(X)$ such that

$$X_t = U^t X,\ t \in \mathbb{Z}. \tag{5.18}$$

The shift operator is useful in showing that the parameters of a stationary process are not time-dependent; see Problem 5.7. As another application, we show that any linear transformation of a stationary process with constant coefficients is stationary.

Corollary 5.3. Let $\{X_t\}$ be a stationary process in H with the shift operator U. Then the stochastic process $\{Y_t; t \in \mathbb{Z}\}$ defined by

$$Y_t = \lim_{n\to\infty} \sum_{k=-n}^{n} a_{k,n} X_{t-k}, t \in \mathbb{Z}$$

is stationary with the same shift operator.

Proof. Since for each n and t

$$\sum_{k=-n}^{n} a_{k,n} X_{t-k} = U^t \sum_{k=-n}^{n} a_{k,n} X_{-k},$$

it follows from the continuity of the operator U^t that

$$Y_t = \lim_{n\to\infty} \sum_{k=-n}^{n} a_{k,n}\, X_{t-k} = U^t \left(\lim_{n\to\infty} \sum_{k=-n}^{n} a_{k,n} X_{-k} \right) = U^t Y_0.$$

Thus, by Theorem 5.2, $\{Y_t; t \in \mathbb{Z}\}$ is stationary with the same shift operator U.

Q.E.D.

It is a simple exercise to compute (X_{t+k}, Y_t) and note that it also depends only on k. It turns out that this condition is necessary and sufficient for two stationary processes to have the same shift operator. This is discussed in the next section.

5.3.2 The Shift Operator of Several Processes and ARMA Models

In this section for two or more processes the notion of shift operator is studied. The results are applied to ARMA models whose definition involves two processes.

Suppose $\{X_t\}$ and $\{Y_t\}$ are stationary, and *stationarily correlated* in the sense that (X_{t+k}, Y_t) depends only on k. The symbol B defined by

$$BX_t = X_{t-1}, \quad BY_t = Y_{t-1}, \tag{5.19}$$

can be extended, using linearity and continuity as in Section 5.3.1, to a unitary operator on $H(X, Y) = \overline{sp}\{X_t, Y_t; t \in \mathbb{Z}\}$. This extension denoted by U is called the *joint shift operator* of $\{X_t\}$ and $\{Y_t\}$. Conversely, if B in (5.19) is extendible to a unitary operator on $H(X, Y)$, then $\{X_t\}$ and $\{Y_t\}$ must be stationary and stationarily correlated. The latter is equivalent to $\{(X_t, Y_t); t \in \mathbb{Z}\}$ being *bivariate stationary* as defined below.

We say that a process $\{\boldsymbol{X}_t\}$ where $\boldsymbol{X}_t = (X_{t,1}, \cdots, X_{t,d})$ with $X_{t,i} \in H$, is a *d-variate (multivariate) stationary process* if for all t, the $d \times d$ covariance matrices

$$(\boldsymbol{X}_{t+k}, \boldsymbol{X}_t) = [(X_{t+k,i}, X_{t,j})]_{i,j=1}^{d}, \tag{5.20}$$

depend only on k. In this case, the $d \times d$ matrix-valued function $\{\boldsymbol{\Gamma}_k\}$ defined by $\boldsymbol{\Gamma}_k = (\boldsymbol{X}_{t+k}, \boldsymbol{X}_t)$ is called the covariance function of the d-variate stationary process $\{\boldsymbol{X}_t\}$ and a single unitary operator U defined on $H(\boldsymbol{X}) = \overline{sp}\{X_{t,i}; 1 \leq i \leq d, t \in \mathbb{Z}\}$ by $UX_{t,i} = X_{t-1,i}, 1 \leq i \leq d$, serves as the shift operator of $\{\boldsymbol{X}_t\}$.

Study of ARMA models (2.55) requires dealing with the bivariate time series $\{(X_t, \varepsilon_t); t \in \mathbb{Z}\}$. In view of the preceding discussion, bivariate stationarity of $\{(X_t, \varepsilon_t); t \in \mathbb{Z}\}$ is necessary and sufficient for extending the shift symbol B defined by $BX_t = X_{t-1}$ and $B\varepsilon_t = \varepsilon_{t-1}, t \in \mathbb{Z}$ to a unitary operator U on $\overline{sp}\{X_t, \varepsilon_t; t \in \mathbb{Z}\}$. While $\{\varepsilon_t\}$ is usually assumed a priori to be white noise and hence stationary, a condition on the roots of $\alpha(z)$ guarantees the existence of a stationary solution $\{X_t\}$ and hence *bivariate stationarity* of $\{(X_t, e_t); t \in \mathbb{Z}\}$. Thus, only under such a condition a *joint shift operator* U exists and may be used to write (2.55) more compactly as

$$\alpha(U)X_t = \beta(U)\varepsilon_t, \tag{5.21}$$

where $\alpha(z)$ and $\beta(z)$ are the associated polynomials of the model. This compact representation and the subsequent symbolical simplification in computing the ARMA parameters, the covariance function and predictors, see Chapter 4, are the main reasons for using B, L, Z and D in the literature related to time series analysis.

5.3.3 The Spectral Representation

In this section, the spectral theorem of unitary operators is used formally to obtain spectral representations for stationary processes and their covariances. Before embarking on this important task, we develop some insight into the structure of unitary operators through their "eigenvalues" and "eigenvectors" in the finite-dimensional cases. The following examples are useful in this direction.

Example 5.7. (a) Consider a unitary operator U satisfying $U^p = I$ for a fixed integer $p \geq 1$. The corresponding process $\{X_t\}$ satisfies

$$X_t = U^t X = U^p(U^{t-p}X) = U^{t-p}X = X_{t-p}.$$

Its time domain given by

$$H(X) = sp\{X, UX, \cdots, U^{p-1}X\} = sp\{X_0, X_1, \cdots, X_{p-1}\},$$

is at most a p-dimensional space. It is p-dimensional precisely when $X_0, X_1, \cdots, X_{p-1}$ are linearly independent, in this case choosing $X_0, \cdots, X_{p-1}$ as a basis for $H(X)$, U is identified with the $p \times p$ (orthogonal) matrix

$$U = \begin{bmatrix} 0 & 1 & 0 & 0 & \cdot & \cdot & \cdot & 0 & 0 \\ 0 & 0 & 1 & 0 & \cdot & \cdot & \cdot & 0 & 0 \\ 0 & 0 & 0 & 1 & \cdot & \cdot & \cdot & \cdot & \cdot \\ \cdot & \cdot & \cdot & \cdot & \cdot & \cdot & \cdot & \cdot & \cdot \\ \cdot & \cdot & \cdot & \cdot & \cdot & \cdot & \cdot & 0 & 0 \\ \cdot & \cdot & \cdot & \cdot & \cdot & \cdot & \cdot & 0 & 1 \\ 1 & 0 & 0 & 0 & \cdot & \cdot & \cdot & 0 & 0 \end{bmatrix}. \tag{5.22}$$

The eigenvalues of U satisfy the equation $z^p - 1 = 0$, and are the p roots of the identity:

$$\lambda_k = e^{2\pi i k/p}, k = 0, 1, \cdots, p-1. \tag{5.23}$$

The corresponding normalized eigenvectors are

$$e_k = \frac{1}{\sqrt{p}}(1, \bar{\lambda}_k, \bar{\lambda}_k^2, \cdots, \bar{\lambda}_k^{p-1}), k = 0, \cdots, p-1. \tag{5.24}$$

Furthermore, it can be verified directly, see Problem 5.11, that

$$U = \sum_{k=0}^{p-1} \lambda_k e_k e_k^* = \sum_{k=0}^{p-1} \lambda_k P_k, \tag{5.25}$$

where $P_k = e_k e_k^*, k = 0, \cdots, p-1$, are the orthogonal projection operators onto the eigenspaces $sp\{e_k\}$, respectively, with the following properties:

$$\begin{cases} P_k P_\ell = 0, & k \neq \ell, \\ \sum_{k=0}^{p-1} P_k = I. \end{cases} \tag{5.26}$$

Let us use P_k's to introduce a function $E(\cdot)$ on the subsets of the unit circle in the complex plane or the interval $(-\pi, \pi]$ by

$$E(A) = \sum_{\lambda_k \in A} P_k. \tag{5.27}$$

For example, if $A = \{\lambda_0, \lambda_5\}$, then $E(A) = P_0 + P_5$. Now, using any basic definition of an integral, the sum in (5.25) can be written as

$$U = \int_{-\pi}^{\pi} e^{i\lambda} dE(\lambda). \tag{5.28}$$

The next example shows the complicated nature of the projection-valued function $E(\cdot)$ when U is not a matrix.

(b) Suppose X and Y are two real-valued random variables with $EY = 0, EY^2 = 1$. The shift operator of the process $X_t = Ye^{itX}$ is the operator of multiplication by e^{-iX}, that is

$$UZ = e^{-iX}Z, \text{ for } all\ Z \in H(X).$$

Note that

$$UX_t = X_{t-1} = Ye^{i(t-1)X} = e^{-iX}Ye^{itX} = e^{-iX}X_t,$$

thus, for all $n \geq 1, c_1, \cdots, c_n, t_1, \cdots, t_n$,

$$U\left(\sum_1^n c_k X_{t_k}\right) = e^{-iX}\sum_1^n c_k X_{t_k},$$

from which the result follows. What are the "eigenvalues" of U? How many are there? See Problem 5.10 (b).

In Chapter 9 we give a rigorous proof the integral representation (5.28) for any unitary operator. Here, we use this representation formally to obtain the spectral representation of a stationary process and its covariance function.

The projection-valued set function E in (5.28) is called the *spectral measure* of the unitary operator U and the integral is called its *spectral integral* with respect to E. To understand these, let $\Omega = (-\pi, \pi]$ and $\mathcal{F}$ be the σ-algebra of Borel subsets of Ω. A function $E(\cdot)$ on $\mathcal{F}$ whose values are orthogonal projection operators onto subspaces of H is called a *spectral measure* if

(a) $E(\Omega) = I$, where I is the identity operator on H, and

(b) $E(\bigcup_{n=1}^{\infty} A_n) = \sum_{n=1}^{\infty} E(A_n)$, whenever $\{A_n\}$ is a sequence of disjoint sets in $\mathcal{F}$.

A spectral measure enjoys many properties of the familiar probability measures, see (5.27) and Section 9.8. Of particular importance is the countable additivity property (b).

The meaning of the spectral integral is made more clear by the following operational property of the integral: for functions $f, g \in L^{\infty}$ (the class of complex-valued, bounded and measurable functions on Ω),

$$\left(\int f dE\right)\left(\int g dE\right) = \int fg dE. \tag{5.29}$$

Assuming (5.28) for the time being and using (5.29) repeatedly with $f = g = e^{i\lambda}$, it follows that for any integer t,

$$U^t = \int_{-\pi}^{\pi} e^{it\lambda} dE(\lambda). \tag{5.30}$$

From (5.18) and (5.30) the *spectral representation* of $\{X_t\}$ is immediate:

$$X_t = U^t X = \int_{-\pi}^{\pi} e^{it\lambda} dE(\lambda) X = \int_{-\pi}^{\pi} e^{it\lambda} dZ_X(\lambda), \tag{5.31}$$

where the H-valued measure $Z_X(A) = E(A)X \in H, A \in \mathcal{F}$, is called the *random spectral measure* of $\{X_t\}$. This measure inherits the additivity and orthogonality of $E(\cdot)$ and is used to obtain the *spectral representation* of the covariance function of $\{X_t\}$. These are summarized in the next theorem and proved in Chapter 9.

Theorem 5.4. Let $\{X_t\}$ be stationary process with the spectral representation (5.31) and random spectral measure Z_X. Then,

(a) Z_X is *countably additive*; $Z_X(\bigcup_{n=1}^{\infty} A_n) = \sum_{n=1}^{\infty} Z_X(A_n)$, whenever $\{A_n\}$ is a sequence of disjoint sets in $\mathcal{F}$. The infinite series on the right converges in the norm of H.

(b) Z_X has orthogonal increments: $(Z_X(A), Z_X(B)) = 0$, whenever $A, B \in \mathcal{F}$ with $A \bigcap B = \phi$.

(c) The function F defined by

$$F_X(\lambda) = ||Z_X(-\pi, \lambda)||^2, \lambda \in \Omega$$

is nondecreasing and right-continuous with $F_X(-\pi) = 0$ and $F_X(\pi) = ||X||^2 < \infty$. It is called the *spectral distribution* of the process $\{X_t\}$.

(d) The covariance function of $\{X_t\}$ is the Fourier transform of its spectral distribution F_X:

$$\gamma_k = (X_{t+k}, X_t) = \int_{-\pi}^{\pi} e^{ik\lambda} dF_X(\lambda), k = 0, \pm 1, \cdots. \tag{5.32}$$

The derivative $F'_X = f_X$ of F_X is called the *spectral density* of $\{X_t\}$.

The representation (5.32) is the bridge between the *time-domain analysis* of a time series based on the covariance function $\{\gamma_k\}$ and its *spectral-domain analysis* based on the spectral function $F_X(\lambda)$. It shows the mathematical equivalence of these two apparently distinct parameterizations of a time series. The first is more popular in economics, social sciences where the main interest is in forecasting. The spectral-domain approach is more popular in engineering, physical and biological sciences, where the main focus is in detecting cycles and understanding the mechanism generating the data.

Through the spectral representations of the process or its covariance function it is evident that a linear combination $Y_t = \sum_{j \in J} a_j X_{t-j}$ in $H(X)$ corresponds to the trigonometric polynomial $\sum_{j \in J} a_j e^{i(t-j)\lambda}$ in $L^2(F_X)$. Furthermore, since

$$\left(\sum_J a_j X_j, \sum_K b_k X_k \right) = \int \left(\sum_J a_j e^{ij\lambda} \right) \left(\overline{\sum_K b_k e^{ik\lambda}} \right) dF_X(\lambda), \tag{5.33}$$

for all finite sets J and K, and scalars a_j and b_k, it follows that the mapping

$$V : X_t \to e^{it\lambda}, t = 0, \pm 1, \cdots,$$

can be extended to an isomorphism between $H(X)$ and $L^2(F_X)$, the so called *Kolmogorov isomorphism*. This enables one to replace linear approximation problems in $H(X)$ by trigonometric polynomial approximation problems in $L^2(F_X)$ which can be solved using techniques of Fourier analysis. For instance, starting from

$$\left\| X_0 - \sum_{k=1}^{n} c_k X_{-k} \right\|^2 = \int_{-\pi}^{\pi} \left| 1 - \sum_{k=1}^{n} c_k e^{ik\lambda} \right|^2 dF_X(\lambda), \tag{5.34}$$

we show in Chapter 10 that

$$\begin{aligned} \inf \left\| X_0 - \sum_{k=1}^{n} c_k X_{-k} \right\|^2 &= \inf \int \left| 1 - \sum_{k=1}^{n} c_k e^{ik\lambda} \right|^2 dF_X(\lambda) \\ &= \exp\left(\int logF'_X(\lambda) d\lambda \right), \end{aligned} \tag{5.35}$$

where the infimum is over all $n \geq 1, c_1, \cdots, c_n$. This is the celebrated Szegö-Kolmogorov theorem of prediction theory, see Theorem VII.

Another important application of the spectral representation is computing the spectral distribution of a transformed (filtered) series. Let $\{X_t\}$ be stationary with the spectral distribution F_X and $\{Y_t\}$ be defined as in (5.33), then with obvious notation, the covariance and spectral distribution of $\{Y_t\}$ are related to those of $\{X_t\}$ via

$$\gamma_k^Y = (Y_{t+k}, Y_t) = \int e^{ik\lambda} \left| \sum_J a_j e^{ij\lambda} \right|^2 dF_X(\lambda) = \int e^{ik\lambda} dF_Y(\lambda). \tag{5.36}$$

We use this and the uniqueness of the Fourier transform to compute the spectral density of several useful classes of stationary processes introduced in Section 5.2.

Example 5.8. (a) For a *white noise process* using (5.32) we obtain

$$\int e^{ik\lambda} dF(\lambda) = \sigma^2 \gamma_{k0}, k \in \mathbb{Z},$$

so that $dF(\lambda) = \sigma^2 d\lambda$, where $d\lambda$ is the normalized Lebesgue measure on $(-\pi, \pi]$. This implies that $F'(\lambda) = \frac{\sigma^2}{2\pi}, -\pi < \lambda \leq \pi$, i.e. the spectral density of a white noise is *flat* or has the same intensity at all frequencies just like the white light in physics; and hence the name white noise.

(b) For the linear process in Example 5.3, using the spectral density of the white noise obtained in (a) and (5.36), we find

$$f_X(\lambda) = \frac{\sigma^2}{2\pi} \left| \sum_{j=-\infty}^{\infty} c_j e^{ij\lambda} \right|^2. \tag{5.37}$$

Thus, for the AR (1) process its spectral density is

$$f_X(\lambda) = \frac{\sigma^2}{2\pi} \frac{1}{|1 - ae^{i\lambda}|^2},$$

and for the MA (1) process we obtain

$$f_X(\lambda) = \frac{\sigma^2}{2\pi} |1 + \beta e^{i\lambda}|^2.$$

More generally, applying (5.36) twice to the left- and right-hand side of (2.55), one obtains

$$\left|1-\sum_{k=1}^{p}\alpha_k e^{ik\lambda}\right|^2 f_X(\lambda)=\frac{\sigma^2}{2\pi}\left|1+\sum_{k=1}^{q}\beta_k e^{ik\lambda}\right|^2$$

or

$$f_X(\lambda)=\frac{\sigma^2}{2\pi}\left|\frac{1+\sum_{k=1}^{q}\beta_k e^{ik\lambda}}{1-\sum_{k=1}^{p}\alpha_k e^{ik\lambda}}\right|^2, \tag{5.38}$$

which is a rational function of its argument, see (2.65).

5.4 PREDICTION OF STOCHASTIC PROCESSES

In this section we introduce some basic concepts, theoretical and computational challenges and successes pertinent to linear and nonlinear prediction problems.

Stochastic processes are often used to model the behavior of physical systems or time-dependent phenomena such as the water-level of a river, the price of a stock, and so on. Usually, given the past history of a system it is desirable to predict or control its future. Several important concepts are needed before a precise mathematical statement of the prediction problem and its solution can be given. For example, with no criteria and/or model governing a system it is hard to judge the goodness of a given prediction. In this section we introduce the needed concepts. The objective and/or subjective information about the past of a system is usually represented by a subset M of a Hilbert space H, any element of M is called a *predictor*. It is useful to view M as the space of allowable predictors for the future among which we are to find the "best" predictor.

If $\hat{X}$ is our predictor for a future random variable X, the prediction error is $X-\hat{X}$, and certainly it is desirable to make this error as small as possible in some reasonable sense. But, since $X-\hat{X}$ is a random quantity and not observable in general, all we can hope for is to choose $\hat{X}$ so that $X-\hat{X}$ is small on the *average*. This is done, for example, by choosing $\hat{X}$ so that either $P(|X-\hat{X}|\geq\epsilon)$ or $E|X-\hat{X}|^p$ is small, for appropriate values of ϵ or p. For a detailed discussion on assessing a prediction system, see Section 2.5.5.

In what follows our criterion for the goodness of a predictor is its *mean square error*

$$E|X-\hat{X}|^2=||X-\hat{X}||^2.$$

This criterion is quite satisfactory as far as mathematical tractability of problems related to finding explicit formula for $\hat{X}$ are concerned. On the other hand, it reflects the requirement that large errors are more serious than small ones as it should be.

For $X \in H$ and M a subset of H, *the best predictor of X based on M* is an element $\hat{X} \in M$ which is closest to X, that is

$$\|X - \hat{X}\|^2 \leq \|X - Y\|^2, \text{ for all } Y \in M. \tag{5.39}$$

The distance from X to M, i.e.

$$\|X - \hat{X}\|^2 = \inf_{Y \in M} \|X - Y\|^2, \tag{5.40}$$

is called the *minimum mean square error* (MMSE) of prediction of X based on M.

The existence and uniqueness of $\hat{X}$ is not guaranteed. However, when M is a (closed linear) subspace of H, then there exists a unique $\hat{X}$ satisfying (5.39). A useful characterization of $\hat{X}$ is given in the following theorem, see Section 9.3.

Theorem 5.5 *(Projection Theorem).* Let M be a subspace of H and $X \notin M$. Then,

(a) there is a unique element $\hat{X} \in M$ satisfying (5.40).

(b) an $\hat{X} \in M$ satisfies (5.40), if and only if $X - \hat{X} \perp M$ or

$$(X - \hat{X}, Y) = 0, \text{ for all } Y \in M. \tag{5.41}$$

The unique element $\hat{X}$ of M is called the *(orthogonal) projection* of X onto M and is denoted by P_M^X, $P(X|M)$ or simply $(X|M)$ depending on the typographical convenience. The set of equations (5.41) is called the *normal equations*. Computing $\hat{X}$ using (5.41) simplify considerably when an appropriate orthogonal basis for M is given. Prediction theory is devoted to just finding such bases for M. Indeed, the Wold decomposition and spectral representation of stationary processes are the most powerful tools for this task.

Next, we focus on two particular choices of M corresponding to nonlinear and linear prediction of $\{X_t\}$.

5.4.1 The Nonlinear Prediction

Let $\{X_t\}$ be a stochastic process with known infinite past $X_{-1}, X_{-2}, \cdots$, consider the problem of predicting the next future value X_0. We may take the space of allowable predictors M to be the set of all random variables in $H = L^2(\Omega, \mathcal{F}, P)$ that are measurable with respect to the σ-algebra generated by the random variables $X_{-1}, X_{-2}, \cdots$. The best predictor of X_0 based on such M is called its *nonlinear predictor* and is given by the conditional expectation

$$E(X_0|X_{-1}, X_{-2}, \cdots);$$

see Problem 5.18. Its computation, however, requires the full knowledge of the (joint) distribution of the random variables $X_0, X_{-1}, X_{-2}, \cdots$. Mathematically, the nonlinear prediction problem can be stated as that of computing $E(X_0|X_{-1}, X_{-2}, \cdots)$ given the past and the joint distribution of $X_0, X_{-1}, \ldots$. Even when this is available, the resulting formulas are often too complicated to interpret or implement in practice and in engineering designs.

Of course, it is desirable to develop a prediction theory that leads to simple prediction formulas and requires less statistical information than the full distribution of $X_0, X_{-1}, \cdots$. The linear prediction or the Kolmogorov-Wiener prediction theory provides such a setting in that only the knowledge of the past and the first two moments of the distribution of $X_0, X_{-1}, \cdots$ are required.

A spark that started the linear prediction theory is the following striking property of normal (Gaussian) random variables. It is one of the most important consequences of equivalence of uncorrelatedness and independence for these random variables and shows that solution of the nonlinear prediction problem for normal processes is actually a linear function of the observations. We state this more precisely in the following theorem.

Theorem 5.6. If $X, X_1, X_2, \cdots, X_n, n \geq 1$, have a multivariate normal distribution with mean zero, then

$$E(X|X_1, \cdots, X_n) = a_1 X_1 + \cdots + a_n X_n,$$

where $a_1, \cdots, a_n$ are some constants depending on the covariance matrix of $X, X_1, \cdots, X_n$.

Proof. First, we choose the constants $a_1, \cdots, a_n$ so that $X - \sum_{j=1}^{n} a_j X_j$ is orthogonal to $X_1, \cdots, X_n$, cf. Theorem 5.5 (b). That is,

$$\left(X - \sum_{j=1}^{n} a_j X_j, X_k \right) = 0, \ k = 1, 2, \cdots, n,$$

or

$$\sum_{j=1}^{n} a_j \gamma_{j,k} = EXX_k, k = 1, 2, \cdots, n, \tag{5.42}$$

where $\gamma_{j,k} = (X_j, X_k)$. Since $X, X_1, \cdots, X_n$ are jointly normal, $X - \sum_{j=1}^{n} a_j X_j$ being uncorrelated with each of $X_1, \cdots, X_n$ implies independence of $X - \sum_{j=1}^{n} a_j X_j$ and $X_1, \cdots, X_n$. Therefore,

$$E \left(X - \sum_{j=1}^{n} a_j X_j | X_1, \cdots, X_n \right) = E \left(X - \sum_{j=1}^{n} a_j X_j \right) = 0,$$

or

$$E(X|X_1, \cdots, X_n) = E\left(\sum_{j=1}^{n} a_j X_j | X_1, \cdots, X_n\right) = \sum_{j=1}^{n} a_j X_j.$$

Q.E.D.

Various forms of (5.42) with finite and infinite n are used frequently in the sequel in solving the linear prediction problems. There are other situations where the nonlinear predictors are easy to compute or are linear, mostly because of a particular dependence structure of the process. Important examples are processes with independent increments, Markov processes, martingales; see Problem 5.15.

5.4.2 The Linear Prediction Problem

Let $\{X_t\}$ be a stochastic process with known infinite past $X_{-1}, X_{-2}, \cdots$. Let us take M to be the set of all linear combinations of the form $\sum_{k=1}^{n} c_k X_{-k}$ and their limits in the norm of H. The best predictor of X_0 based on this particular M is called the *linear least squares predictor* or simply the *predictor* of X_0 based on the past and is denoted by $\hat{X}_0$. While the Projection Theorem identifies $\hat{X}_0$ as the orthogonal projection of X_0 onto M, it does not provide a workable recipe for computing it. Computation of $\hat{X}_0$ in terms of the observable past is at the core of the linear prediction problem.

Since $\hat{X}_0 \in M$, from the definition of M it follows that for each $n \geq 1$, there exists (not necessarily unique) scalars $a_{k,n}$, $1 \leq k \leq n$, such that

$$\hat{X}_0 = \lim_{n\to\infty} \sum_{k=1}^{n} a_{k,n} X_{-k}. \tag{5.43}$$

But, this is not a satisfactory formula for computing $\hat{X}_0$ because of its dependence on $a_{k,n}$ and n. In spite of this, (5.42) and Theorem 5.5 (b) suggest the need and importance of the knowledge of $\{\gamma_k\}$ in computing the linear predictor. Now, we may state the linear prediction problem more precisely: Given the infinite past and the covariance function $\{\gamma_k\}$ of a mean-zero stationary process, find the (unique) sequence of scalars $a_1, a_2, \cdots$ and condition(s) on $\{\gamma_k\}$ such that

$$\hat{X}_0 = \lim_{n\to\infty} \sum_{k=1}^{n} a_k X_{-k} = \sum_{k=1}^{\infty} a_k X_{-k}, \tag{5.44}$$

that is, the predictor has a mean-convergent series representation.

The following example shows that the representation (5.44) is not always possible.

Example 5.9. For $\{X_t\}$ defined in (5.13), we show that $\hat{X}_0 = e$. Since $\frac{1}{n}\sum_{k=1}^{n} X_{-k} = e + \frac{1}{n}\sum_{k=1}^{n} e_{-k}$, and $\left\|\frac{1}{n}\sum_{k=1}^{n} e_{-k}\right\| = \frac{\sigma}{\sqrt{n}} \to 0$ as $n \to \infty$, it follows that

$e = \lim_{n\to\infty} \frac{1}{n} \sum_{k=1}^{n} X_{-k} \in M$. The predictor of X_0 based on M is e, because its prediction error $X_0 - e = e_0$ is orthogonal to M, see Theorem 5.5 (b). But,

$$\hat{X}_0 = \lim_{n\to\infty} \sum_{k=1}^{n} \frac{1}{n} X_{-k},$$

is not of the form (5.44), since $a_{k,n} = \frac{1}{n}$ depends on n.

To solve the linear prediction problem effectively, we introduce the concepts of history and innovation of stochastic processes. The latter plays a crucial role in computing the predictor since its appears in the Wold decomposition and forms a base for the history of the process.

The subspace H_t defined by

$$H_t = H_t(X) = \bar{sp}\{X_s; s \leq t\}$$

is called the *history or past and present subspace* of the process up to time t. The *prediction error and its variance* for $X_t,\ t \in \mathbb{Z}$, are defined by

$$\varepsilon_t = X_t - \hat{X}_t,\ \sigma^2 = \text{Var}(\varepsilon_t), \tag{5.45}$$

where $\hat{X}_t$ is the orthogonal projection of X_t onto H_{t-1}.

The ε_t is also called the *innovation* at time t and the process $\{\varepsilon_t; t \in Z\}$ is called the *innovation process* of $\{X_t\}$. Consequently, σ^2 is called its *innovation variance* or *intrinsic variance*. Both of these concepts play fundamental roles in prediction theory and time series analysis. The name innovation (something new) for ε_t is quite suggestive, since (5.45) can be rewritten

$$X_t = \hat{X}_t + \varepsilon_t, \tag{5.46}$$

as the orthogonal sum of two components: the predictor $\hat{X}_t$ coming from the past and the innovation ε_t coming from the future. The $\{\varepsilon_t\}$ or its variance σ^2 is a measure of dependence among the infinite number of random variables in the set $\{X_t; t \in \mathbb{Z}\}$, as we show in the next section.

There is a price to be paid, however, for simplicity and elegance of linear prediction theory. This is usually assessed in terms of its innovation variance σ^2. Theorem 5.6 shows that for Gaussian processes the linear and nonlinear predictors are the same. But the farther a process is from Gaussianity the larger is the ratios of the linear and nonlinear prediction error variances, as the next simple example shows.

Example 5.10. Let X and Z be independent standard normal random variables, $k \geq 1$ an integer and $Y = X^{2k} + Z$. The nonlinear predictor of Y based on X is $E(Y|X) = X^{2k}$, and its prediction error is $Y - E(Y|X) = Z$ with

$$\text{Var}[Y - E(Y|X)] = \text{Var}(Z) = 1.$$

However, the linear predictor of Y based on X is the constant $\hat{Y} = EX^{2k}$, and its prediction error is $Y - \hat{Y} = X^{2k} - EX^{2k} + Z$ with

$$\mathrm{Var}(Y - \hat{Y}) = 1 + \frac{(2k)!}{2^k k!}, \tag{5.47}$$

which goes to infinity as $k \to \infty$. Note that larger k represents larger departure from normality of Y.

For more concrete examples comparing performance of linear and nonlinear predictors, see Shepp, Slepian and Wyner (1980).

5.5 LINEAR INDEPENDENCE AND PREDICTION

The infinite number of random variables in a stationary process $\{X_t; t \in \mathbb{Z}\}$ are not necessarily linearly independent. For instance, the process in Example 5.7 (a) has at most p linearly independent random variables. In this section, we define notions of linear independence for stochastic processes that are useful in the linear prediction theory and time series analysis.

5.5.1 Nondeterministic Processes and Their Innovations

The standard definition of linear independence for a finite number of vectors $X_1, \cdots, X_n$ can be seen to be equivalent to

$$X_{t+1} \notin sp\{X_s;\ s \leq t\}, t = 1, 2, \cdots, n-1.$$

Motivated by this and the definition of H_t, a stationary process $\{X_t\}$ is said to be *nondeterministic* if

$$X_0 \notin H_{-1},$$

otherwise, $\{X_t\}$ is said to be *deterministic*. In the latter case, since $X_0 \in H_{-1}$, it follows that $\hat{X}_0 = X_0, \varepsilon_0 = 0$ and $\sigma^2 = 0$. Thus, the predictor of future values of a deterministic process is always correct (with zero error), while for a nondeterministic process the predictor is never correct and $\sigma^2 > 0$.

Deterministic processes having too much redundancy are of little interest in prediction theory. Nevertheless, they are important as models for cyclic and seasonal phenomena in engineering, physical and social sciences. The harmonic processes introduced in Example 5.4 are deterministic, though this is not easy to establish at this time. The problem of finding conditions on the covariance function $\{\gamma_k\}$ so that $\{X_t\}$ is (non)deterministic is rather challenging. However, finding such conditions in terms of the spectrum seems to be easier and more natural, see Theorem VII. In the rest of this section we prove a few basic properties of $\{\varepsilon_t\}$.

First, we show that the innovation of a nondeterministic process $\{X_t\}$ is a white noise.

Lemma 5.7. Let $\{X_t\}$ be a nondeterministic stationary process with innovation $\{\varepsilon_t\}$ and variance σ^2. Then,

(a) $\{\varepsilon_t\}$ *is a white noise*. That is,

$$(\varepsilon_s, \varepsilon_t) = \sigma^2 \delta_{s,t},\ s, t \in \mathbb{Z},\ 0 < \sigma^2 < \infty.$$

Furthermore, $\{X_t\}, \{\varepsilon_t\}$ and $\{\hat{X}_t\}$ have the same shift operator.

(b) Future innovations are orthogonal to the past of the process. That is, for all $t, k \geq 1$,

$$(\varepsilon_t, X_{t-k}) = 0, (\varepsilon_t, X_t) = \sigma^2.$$

Proof. (a) Note that for $s < t$, $\varepsilon_s \in H_s$ and $\varepsilon_t = X_t - \hat{X}_t$ is orthogonal to H_s, therefore $(\varepsilon_s, \varepsilon_t) = 0$. Let U denote the shift operator of the stationary process $\{X_t\}$. Since $\hat{X}_t \in H_{t-1}$ and $\hat{X}_t = \lim_{n\to\infty} \sum_{k=1}^{n} a_{k,n} X_{t-k}$, for some $a_{k,n}$'s, it follows from Corollary 5.3 that $\{\hat{X}_t; t \in Z\}$ is stationary with the shift operator U. Also, $\varepsilon_t = X_t - \hat{X}_t = U^t X_0 - U^t \hat{X}_0 = U^t(X_0 - \hat{X}_0) = U^t \varepsilon_0$, i.e. $\{\varepsilon_t\}$ is stationary with the shift operator U.

(b) For any $k \geq 1$, since $X_{t-k} \in H_{t-1}$ and ε_t is orthogonal to H_{t-1}, we have $(\varepsilon_t,\ X_{t-k}) = 0$. From the definitions of σ^2 and ε_t we have

$$\sigma^2 = (\varepsilon_t, \varepsilon_t) = (\varepsilon_t,\ X_t - \hat{X}_t) = (\varepsilon_t\ X_t).$$

Q.E.D.

The values of $(\varepsilon_t,\ X_{t+k})$, $k \geq 1$, are given in (2.70) and Theorem 5.11 (b), they depend on the moving average parameters of $\{X_t\}$ yet to be defined. Nevertheless, at this early stage it follows from (a) that $\{(X_t, \varepsilon_t); t \in \mathbb{Z}\}$ is a bivariate stationary process, since they share the same shift operator.

5.5.2 Wold Decomposition

In this section we define purely nondeterministic (PND) processes and characterize them in terms of MA (∞) representations. The Wold decomposition proved here shows that any stationary process is the orthogonal sum of PND and deterministic processes, this confirms the statement made in Example 5.5. The proofs are in the spirit of Wiener and Masani (1957).

The nondecreasing chain of subspaces H_t holds the key to many geometric and analytic properties of $\{X_t\}$. This is made more explicit using the following (recursive) property of these subspaces, i.e.,

$$H_t = \bar{sp}\{X_t, H_{t-1}\} = \bar{sp}\{\varepsilon_t, H_{t-1}\}. \tag{5.48}$$

The next lemma proves and provides a more general version of (5.46)

Lemma 5.8. Let $\{X_t\}$ be a nondeterministic stationary process with innovation $\{\varepsilon_t\}$. Then,

(a) for $s < t,\ H_t(X) = H_s(X) \oplus sp\{\varepsilon_{s+1}, \cdots, \varepsilon_t\}$,

(b) $H_t(X) = H_t(\varepsilon) \oplus H_{-\infty}(X)$, where $H_{-\infty}(X) = \bigcap_t H_t(X)$.

Proof. (a) Since $H_{t-1}(X) \subseteq H_t(X)$ and $\varepsilon_t \in H_t(X)$, we have

$$H_{t-1}(X) \oplus sp\{\varepsilon_t\} \subseteq H_t(X). \tag{5.49}$$

From (5.46), $X_t = \varepsilon_t + \hat{X}_t \in sp\{\varepsilon_t\} \oplus H_{t-1}(X)$. Also, for $s < t, X_s \in sp\{\varepsilon_t\} \oplus H_{t-1}(X)$, thus it follows that (5.49) holds with $\supseteq$ replacing $\subseteq$, and therefore with equality as in (5.48). By iteration of this equality we get

$$\begin{aligned} H_t(X) &= H_{t-1}(X) \oplus sp\{\varepsilon_t\} = H_{t-2}(X) \oplus sp\{\varepsilon_t, \varepsilon_{t-1}\} \\ &= \cdots = H_s(X) \oplus sp\{\varepsilon_{s+1}, \cdots, \varepsilon_t\}. \end{aligned}$$

(b) If $Y \in H_{-\infty}(X)$, then for each $t, Y \in H_{t-1}(X) \perp \varepsilon_t$, hence $H_{-\infty}(X) \perp H(\varepsilon)$. Next, since $\varepsilon_t \in H_t(X)$, we obtain $H_t(\varepsilon) \subseteq H_t(X)$. Also, $H_{-\infty}(X) \subseteq H_t(X)$. Hence,

$$H_{-\infty}(X) \oplus H_t(\varepsilon) \subseteq H_t(X). \tag{5.50}$$

To prove equality in (5.50), take $Y \in H_t(X)$. Since $H_{-\infty}(X) \perp H_t(\varepsilon)$ and by (a) $H_s(X) \perp sp\{\varepsilon_{s+1}, \cdots, \varepsilon_t\}$, we have

$$\begin{aligned} P^Y_{H_{-\infty}(X) \oplus H_t(\varepsilon)} &= P^Y_{H_{-\infty}(X)} + P^Y_{H_t(\varepsilon)} \\ &= \lim_{s \to -\infty} \left[P^Y_{H_s(X)} + P^Y_{sp\{\varepsilon_k; s+1 \le k \le t\}} \right] \\ &= \lim_{s \to -\infty} P^Y_{H_t(X)} = Y. \end{aligned}$$

Hence $Y \in H_{-\infty}(X) \oplus H_t(\varepsilon)$, and (5.50) holds with $\supseteq$ replacing $\subseteq$, and therefore with equality. Q.E.D.

The subspace $H_{-\infty}(X)$ appearing in Lemma 5.8 (b) is called the *remote past (history)* of the process $\{X_t; t \in \mathbb{Z}\}$. We say $\{X_t\}$ is *purely nondeterministic* (PND) if its remote past is the trivial subspace, i.e, $H_{-\infty}(X) = \{0\}$. In this case,

$$H_t(X) = H_t(\varepsilon) = \sum_{k=0}^{\infty} sp\{\varepsilon_{t-k}\}, \text{ for all } t. \tag{5.51}$$

While every purely nondeterministic process is nondeterministic, the converse is not (necessarily) true; see Example 5.5.

Given a stationary process $\{X_t\}$ with covariance function $\{\gamma_k\}$, it is desirable to find conditions on $\{\gamma_k\}$ so that $\{X_t\}$ is PND. This is a challenging problem in prediction theory. The key to its solution is showing that white noise, the simplest stationary process, is PND.

Lemma 5.9. Every white noise $\{e_t\}$ is PND.

Proof. It suffices to show that every $Y \in H_{-\infty}(e)$ is identically zero.

Suppose $0 \neq Y \in H_{-\infty}(e)$. Then,

$$Y \in H_{t-1}(e) \subset H_t(e), \text{ for } all\ t. \tag{5.52}$$

Since e_t is orthogonal to $H_{t-1}(e)$, it follows from (5.52) that

$$(Y, e_t) = 0, \text{ for } all\ t. \tag{5.53}$$

On the other hand, since $\{e_t\}$ is an orthogonal basis of $H(e)$ and $Y \in H(e)$, it has a Fourier expansion in terms of $\{e_t\}$:

$$Y = \sum_{s=-\infty}^{\infty} c_s e_s, \sum_{s=-\infty}^{\infty} |c_s|^2 < \infty, \tag{5.54}$$

with

$$c_s = (Y, e_s), \text{ for } all\ s. \tag{5.55}$$

Thus, from (5.53)–(5.55) it follows that $Y \equiv 0$. Q.E.D.

Using this lemma we show that every MA(∞) process is PND.

Theorem 5.10. Let $\{e_t\}$ be a white noise. Then, a non-zero stationary process $\{X_t\}$ defined by

$$X_t = \sum_{j=0}^{\infty} c_j e_{t-j}, \sum_{j=0}^{\infty} |c_j|^2 < \infty,$$

is PND.

Proof. From the representation of $\{X_t\}$ it is evident that

$$H_s(X) \subseteq H_s(e), \text{ for } all\ s,$$

hence

$$\bigcap_{s \in \mathbb{Z}} H_s(X) \subseteq \bigcap_{s \in \mathbb{Z}} H_s(e).$$

From Lemma 5.9, $H_{-\infty}(e) = \{0\}$ and thus $\{X_t\}$ is PND. Q.E.D.

One of the most fundamental result in the theory of stationary processes is the Wold decomposition theorem expressing the process as the sum of two orthogonal components, one of which is a PND process and the other is deterministic. It is the cornerstone of the time domain approach to time series analysis, digital signal processing, and so on. The rest of this section is devoted to its proof and some of its applications.

Theorem 5.11. *(Wold Decomposition Theorem).* Let $\{X_t\}$ be a nondeterministic stationary process with innovation $\{\varepsilon_t\}$. Then, for each t,

(a) $X_t = U_t + V_t$, where $U_t = P^{X_t}_{H_t(\varepsilon)}$, $V_t = P^{X_t}_{H_{-\infty}(X)}$ and $V_t \perp U_s$, for all s, t.

(b) The process $\{U_t\}$ has a unique one-sided moving average representation:

$$U_t = \sum_{k=0}^{\infty} b_k \varepsilon_{t-k},\ b_0 = 1,\ \sum_{k=0}^{\infty} |b_k|^2 < \infty.$$

and is PND. The unique scalars $b_1, b_2, \cdots$ are called the *moving average parameters* of $\{X_t\}$ and satisfy $(\varepsilon_t, X_{t+k}) = \sigma^2 b_k,\ k \geq 0$.

(c) The process $\{V_t\}$ is deterministic with $H_t(V) = H_{-\infty}(X)$.

Proof. (a) From Lemma 5.8 (b), since $X_t \in H_t(X)$, it is evident that

$$X_t = P^{X_t}_{H_t(X)} = P^{X_t}_{H_t(\varepsilon)} + P^{X_t}_{H_{-\infty}(X)} = U_t + V_t.$$

(b) Since $U_t \in H_t(\varepsilon)$ and $\{\varepsilon_s; s \leq t\}$ is an orthogonal basis for $H_t(\varepsilon)$, it follows that there exists a unique sequence of scalars $\{b_k\}_{k=0}^{\infty}$ with $\sum_{k=0}^{\infty} |b_k|^2 < \infty$, and $\sum_{k=0}^{\infty} b_k \varepsilon_{t-k}$ is mean-convergent. Moreover, the process $\{U_t\}$ defined by $U_t = \sum_{k=0}^{\infty} b_k \varepsilon_{t-k}$ is stationary and PND (Theorem 5.10).

(c) Since for any $s < t, V_s \in H_{-\infty}(X) \subseteq H_t(X)$, and $H_t(\varepsilon), \subseteq H_t(X)$, it follows that

$$H_t(\varepsilon) \oplus H_t(V) \subseteq H_t(X). \tag{5.56}$$

Also from (a) for any $s < t,\ X_s = U_s + V_s \in H_t(\varepsilon) \oplus H_t(V)$, hence $H_t(X) \subseteq H_t(\varepsilon) \oplus H_t(V)$ and from (5.56) we obtain

$$H_t(X) = H_t(\varepsilon) \oplus H_t(V).$$

Comparing this with Lemma 5.8 (b) we conclude that $H_t(V) = H_{-\infty}(X)$ and thus $\{V_t\}$ is deterministic.

Q.E.D.

5.5.3 Linear Prediction and PND Processes

In this section the Wold decomposition is used to provide the first solution of the prediction problem in terms of the moving average parameters and the innovation process $\{\varepsilon_t\}$. Also, a characterization of PND processes is given in terms of $MA(\infty)$ representations.

Corollary 5.12. Let $\{X_t\}$ be a nondeterministic stationary process with the Wold decomposition

$$X_t = \sum_{k=0}^{\infty} b_k \varepsilon_{t-k} + V_t,\ b_0 = 1,\ \sum_{k=1}^{\infty} |b_k|^2 < \infty.$$

Then the ℓ-step ahead predictor of $X_\ell, \ell \geq 1$, based on the past $X_0, X_{-1}, \cdots$, its prediction error and variance of prediction error are given by

$$\begin{cases} \hat{X}_\ell = \sum_{k=\ell}^{\infty} b_k \varepsilon_{\ell-k} + V_\ell, \\ X_\ell - \hat{X}_\ell = \sum_{k=0}^{\ell-1} b_k \varepsilon_{\ell-k}, \\ \mathrm{Var}(X_\ell - \hat{X}_\ell) = \sigma^2 \sum_{k=0}^{\ell-1} |b_k|^2. \end{cases} \tag{5.57}$$

Proof. From the Wold decomposition of X_t we have

$$X_\ell = \sum_{k=0}^{\ell-1} b_k \varepsilon_{\ell-k} + \sum_{k=\ell}^{\infty} b_k \varepsilon_{\ell-k} + V_\ell.$$

Since $\sum_{k=0}^{\ell-1} b_k \varepsilon_{\ell-k}$ is orthogonal to $H_0(X)$, $\sum_{k=\ell}^{\infty} b_k \varepsilon_{\ell-k} \in H_0(X)$ and $V_\ell \in H_{-\infty}(X) \subset H_0(X)$ we get

$$\hat{X}_\ell = \sum_{k=\ell}^{\infty} b_k \varepsilon_{\ell-k} + V_\ell,$$

as desired. The remaining two assertions are easy to establish.

Q.E.D.

The next corollary provides two characterizations of $\{X_t\}$ being purely nondeterministic.

Corollary 5.13. Let $\{X_t\}$ be a nondeterministic stationary process with the innovation $\{\varepsilon_t\}$. Then, the following are equivalent:

(a) $\{X_t\}$ is purely nondeterministic.

(b) $\{X_t\}$ has a one-sided moving average representation as

$$X_t = \sum_{k=0}^{\infty} b_k \varepsilon_{t-k},\ b_0 = 1,\ \sum_{k=0}^{\infty} |b_k|^2 < \infty.$$

(c) $\lim_{n\to\infty} P_{H_{t-n}}^{X_t} = 0$.

Proof. (a) $\Longrightarrow$ (b). Suppose $\{X_t\}$ is purely nondeterministic. Then, $H_{-\infty}(X) = \{0\}$, and from Theorem 5.11 (c) we get that $V_t = 0$, and hence (b).
(b) $\Longrightarrow$ (c). Suppose that (b) holds. It follows from Corollary 5.12 that

$$P^{X_t}_{H_{t-n}} = \sum_{k=n+1}^{\infty} b_k \varepsilon_{t-k}, ||P^{X_t}_{H_{t-n}}||^2 = \sigma^2 \sum_{k=n+1}^{\infty} |b_k|^2 \to 0,$$

as $n \to \infty$, and hence (c).
(c) $\Longrightarrow$ (a). Suppose that $\lim_{n\to\infty} P^{X_t}_{H_{t-n}} = 0$. Since $H_{t-n}(\varepsilon) \subseteq H_{t-n}(X)$, therefore $\lim_{n\to\infty} P^{X_t}_{H_{t-n}(\varepsilon)} = 0$. Now, from Lemma 5.8 (b)

$$P^{X_t}_{H_{t-n}(X)} = P^{X_t}_{H_{t-n}(\varepsilon)} + P^{X_t}_{H_{-\infty}(V)},$$

and taking limits of both sides as $n \to \infty$, we get

$$V_t = P^{X_t}_{H_{-\infty}(X)} = 0, \text{ for all } t,$$

and from Theorem 5.11 (c), $H_{-\infty}(X) = \{0\}$.

Q.E.D.

To finish the task of prediction of $\{X_t\}$, at least in theory, it is important to relate the moving average parameters $b_1, b_2, \cdots$ to the covariance function $\{\gamma_k\}$. In this direction we have the following result.

Corollary 5.14 . Let $\{X_t\}$ be a purely nondeterministic, real stationary process. Then $\sigma^2, b_1, b_2, \cdots$ satisfy the following (quadratic) equations:

$$\gamma_k = \sigma^2 \sum_{j=0}^{\infty} b_j b_{j+k}, \; b_0 = 1, \; k \in \mathbb{Z}. \tag{5.58}$$

In matrix form (5.58) can be written as

$$\begin{bmatrix} \gamma_0 & \gamma_1 & \gamma_2 & \cdot & \cdot & \cdot \\ \gamma_1 & \gamma_0 & \gamma_2 & \cdot & \cdot & \cdot \\ \cdot & \cdot & \cdot & \cdot & \cdot & \cdot \\ \cdot & \cdot & \cdot & \cdot & \cdot & \cdot \end{bmatrix} = \sigma^2 \begin{bmatrix} 1 & b_1 & b_2 & \cdot & \cdot & \cdot \\ 0 & 1 & b_2 & \cdot & \cdot & \cdot \\ \cdot & \cdot & \cdot & \cdot & \cdot & \cdot \\ \cdot & \cdot & \cdot & \cdot & \cdot & \cdot \end{bmatrix} \begin{bmatrix} 1 & 0 & \cdot & \cdot \\ b_1 & 1 & \cdot & \cdot \\ b_2 & b_1 & \cdot & \cdot \\ \cdot & \cdot & \cdot & \cdot \end{bmatrix},$$

or more compactly

$$\Gamma = \sigma^2 T' \, T, \tag{5.59}$$

where $\Gamma = (\gamma_{i-j})_{i,j=0}^{\infty}$, $T = (b_{i-j})_{i,j=0}^{\infty}$ with $b_0 = 1$ and $b_j = 0$ for $j < 0$.

This shows that for a PND process its moving average parameters $b_1, b_2, \cdots$ can be obtained from the triangular or Cholesky factorization of its covariance matrix Γ. Next, we seek the characterizing properties of $\{b_k\}$ and $\{\varepsilon_t\}$. To begin, consider a stationary process $\{X_t\}$ given by

$$X_t = \sum_{j=0}^{\infty} b_j e_{t-j}, \; \sum_{j=0}^{\infty} |b_j|^2 < \infty,$$

where $\{e_t\}$ is an arbitrary white noise. Is this the unique moving average representation of $\{X_t\}$ promised by the Wold decomposition? Is $\{e_t\}$ the innovation of $\{X_t\}$? The following theorem sheds light on these questions. The proof is from Doob (1953).

Theorem 5.15. Let $\{e_t\}$ be a white noise with $Ee_t^2 = \sigma^2$, $\{b_j\}_{j=0}^{\infty}, b_0 \neq 0$, $\sum_{j=0}^{\infty} |b_j|^2 < \infty$, be any a sequence of scalars. Define a process $\{X_t\}$ by

$$X_t = \sum_{j=0}^{\infty} b_j e_{t-j},\ t \in \mathbb{Z}. \tag{5.60}$$

Then,

(a) $\{X_t\}$ is a PND process with covariance function

$$\gamma_k = \sigma^2 \sum_{j=0}^{\infty} b_j b_{j+k},\ k \in \mathbb{Z},$$

and

$$E|X_{t-1+\ell} - \hat{X}_{\ell|t-1}|^2 \geq \sigma^2 \sum_{j=0}^{\ell-1} |b_j|^2, \ell = 1, 2, \cdots,$$

(b) There cannot be equality in (a) for any $\ell \geq 1$, if the function $\varphi(\cdot)$ defined by

$$\varphi(z) = \sum_{j=0}^{\infty} b_j z^j,\ |z| \leq 1,$$

has any zeros in the open disc $|z| < 1$.

(c) For (5.60) to be the Wold decomposition of $\{X_t\}$ or $\{e_t\}$ to be its innovation process, it is necessary that the function $\varphi(z)$ does not have any zeros inside the unit disc $|z| < 1$.

Proof. (a) It follows from Corollary 5.13 that $\{X_t\}$ is purely nondeterministic. Since $H_0(X) \subseteq H_0(e)$, it follows that

$$E|X_\ell - \hat{X}_\ell|^2 \geq E|X_\ell - P_{H_0(e)} X_\ell|^2 = \sigma^2 \sum_{j=0}^{\ell-1} |b_j|^2,$$

using Corollary 5.12, which gives the result.

(b) We prove (b) for $\ell = 1$, the proof for general case is similar. Assume that $\sum_{j=0}^{\infty} b_j z^j$ has a zero z_0 in $|z| < 1$. Then,

$$\sum_{j=0}^{\infty} b_j z^j = (z - z_0) \sum_{j=0}^{\infty} b'_j z^j,\ b'_0 = -\frac{b_0}{z_o},$$

and on the unit circle $|z| = 1$, i.e. replacing z by $e^{-i\lambda}$, we get (recalling that $\dfrac{1 - e^{i\lambda} z_0}{e^{i\lambda} - \bar{z}_0}$ has absolute value one)

$$\begin{aligned}\left|\sum_{j=0}^{\infty} b_j e^{-ij\lambda}\right| &= \left|\sum_{j=0}^{\infty} b_j e^{ij\lambda}\right| = \left|\frac{1 - e^{i\lambda} z_0}{e^{i\lambda} - z_0}\right| \left|\sum_{j=0}^{\infty} b_j e^{-ij\lambda}\right| \\ &= \left|1 - e^{i\lambda} z_0\right| \left|\sum_{j=0}^{\infty} b'_j e^{ij\lambda}\right| = \left|\sum_{j=0}^{\infty} b''_j e^{ij\lambda}\right| = \left|\sum_{j=0}^{\infty} b''_j e^{-ij\lambda}\right|,\end{aligned}$$

where $b''_0 = -\frac{b_0}{z_0}$. Since $\sum_{j=0}^{\infty} |b_j|^2 < \infty$, it follows that $\sum_{j=0}^{\infty} b_j e^{-ij\lambda}$ is convergent in the norm of L^2 (of the Lebesgue measure on the unit circle), and therefore from the above equality we get that $\sum_{j=0}^{\infty} b''_j e^{-ij\lambda}$ is convergent in the norm of L^2 as well, so that $\sum_{j=0}^{\infty} |b''_j|^2 < \infty$. Also from

$$\left|\sum_{j=0}^{\infty} b_j e^{ij\lambda}\right|^2 = \left|\sum_{j=0}^{\infty} b''_j e^{-ij\lambda}\right|^2,$$

as function in L^1 and uniqueness of Fourier coefficients it follows that

$$\sum_{j=0}^{\infty} b_j b_{j+k} = \sum_{j=0}^{\infty} b''_j b''_{j+k}, \ k \in \mathbb{Z}.$$

This shows that the purely nondeterministic process $\{X'_t\}$ defined by

$$X'_t = \sum_{j=0}^{\infty} b''_j e_{t-j}, \ t \in \mathbb{Z},$$

has the same covariance function as the $\{X_t\}$ process. Therefore they have the same prediction error, but from (a) we have (for $\ell = 1$),

$$E|X_1 - \hat{X}_1|^2 = E|X'_1 - \hat{X}'_1|^2 \geq |b''_0|^2 = \left|\frac{b_0}{z_0}\right|^2 > |b_0|^2,$$

since $|z_0| < 1$. This shows that if $\varphi(z)$ has a zero in $|z| < 1$, then the inequalities in (a) are strict.

(c) If (5.60) is the Wold decomposition of $\{X_t\}$, it follows from Corollary 5.12 that (a) holds with equality for all $\ell \geq 1$. Thus, by (b) the function $\varphi(z)$ cannot have any zeros in $|z| < 1$.

Q.E.D.

For a nondeterministic stationary process $\{X_t\}$ with the moving average parameters $b_1, b_2, \cdots$ the function

$$\varphi(z) = \sum_{j=0}^{\infty} b_j z^j, z \in C,$$

is called the *transfer function* of $\{X_t\}$. Since it does not have any zeros in $|z| < 1$, the function $\varphi^{-1}(z) = \frac{1}{\varphi(z)}$ has a Taylor expansion

$$\varphi^{-1}(z) = 1 - \sum_{j=1}^{\infty} a_j z^j, \ |z| < 1.$$

From $\varphi^{-1}(z)\varphi(z) \equiv 1$, it follows that

$$a_0 = 1, \ a_\ell = b_\ell - \sum_{k=1}^{\ell-1} a_{\ell-k} b_k = 0, \ \ell = 1, 2, \cdots, \tag{5.61}$$

the coefficients a_k's the autoregressive parameters of $\{X_t\}$ introduced in Chapters 1 and 6.

Example 5.11. (a) For $\{X_t\}$ defined by $X_t = e_t - 3e_{t-1} + e_{t-2}$, since $\varphi(z) = 1-3z+z^2 = 0$ has the root $z = \dfrac{3-\sqrt{5}}{2}$ inside the unit disc, $\{e_t\}$ is not its innovation process.
(b) For the process $\{X_t\}$ defined by $X_t = e_t - 2e_{t-1} + e_{t-2}$, $\{e_t\}$ is its innovation process since $\varphi(z) = 1 - 2z + z^2 = (1-z)^2 = 0$, has no roots inside the unit disc.

5.6 STATISTICS OF STATIONARY PROCESSES

This section is concerned with certain asymptotic properties of estimators of the mean, covariance and the spectral density of a stationary time series.

Let $\{X_t\}$ be a stationary process with mean μ and covariance function $\{\gamma_k\}$. Up to now, we have assumed that μ and $\{\gamma_k\}$ are known, however, this is seldom the case in practice. In almost all practical situations the data $X_1, \cdots, X_n$ are used first to estimate μ and $\{\gamma_k\}$, then wherever needed these are replaced by their estimated counterparts and one proceeds as if the estimated values were correct. How does this recipe work in practice? It all depends on how "good" these estimators are. In this section we show that the natural estimators of μ and γ_k are *consistent*, that is, their variances are small when n is large, provided that $\gamma_k \to 0$ for k large. This requirement of asymptotic (independence) of X_s and X_t, for $|s-t|$ large is the trademark of statistical time series analysis and inference for stochastic processes. Asymptotic normality of these estimators and various mixing conditions are reviewed later in the section.

Of course, there are several measures assessing the degree of dependence between past and future of a stationary process. We review some of these, starting with the *ergodic property* which enables us to state and prove analogues of law of large numbers for stationary processes. In estimating parameters of ergodic processes, it is not necessary to collect data on many replications of the process. Instead, a single realization of long duration - $X_1, \cdots, X_n$ with n large - will suffice. We pay particular attention to *mean-ergodicity* and *mean-ergodic theorem* due to their reliance on the second-order statistics of stationary processes and Fourier analysis. Proofs in this setup are much easier, more transparent and reveal a great deal about the depth and beauty of the concept of stationarity and its spectral representation. Asymptotic normality of the estimators of μ, γ_k and the periodogram are useful in finding confidence intervals for these parameters. Proving that the sample mean $\bar{X}$ is asymptotically normal requires more stringent conditions on the process than ergodicity. In this section, we review the *mixing conditions* needed for proving central limit theorem (CLT) for estimators of parameters of a stationary process.

Ideally, as in the i.i.d. case, it is desirable to have a universal condition on a stationary process under which all the required asymptotic distributions for statistical inference can be established with the appropriate rates of convergence. The strict stationarity, pointwise ergodicity and existence of the first four moments are all that we need for most of the results stated and proved in this section.

5.6.1 Mean Ergodic Theorem: Consistency of Estimators

Natural estimators of the mean, covariance and spectral density of a stationary process are appropriate time averages of the data, that is,

$$\begin{aligned} \bar{X} = \frac{1}{n}\sum_{t=1}^{n} X_t,\ \hat{\gamma}_k &= \frac{1}{n}\sum_{t=1}^{n-k}(X_{t+k}-\bar{X})(X_t-\bar{X}), \\ I(\lambda) &= \frac{1}{n}\left|\sum_{t=1}^{n} X_t e^{it\lambda}\right|^2. \end{aligned} \tag{5.62}$$

We show that $\bar{X}$ and $\hat{\gamma}_k$ are *consistent* for the parameters μ and γ_k, when $\{X_t\}$ is *mean-ergodic*. But, the *periodogram* $I(\lambda)$ is not a consistent estimator of the spectral density. We shall begin with the *mean-ergodic theorem* due to J. Von Neumann, showing that the sample average of any stationary process is convergent in norm.

Surprisingly, the simple fact that the averages

$$\frac{1}{n}\sum_{t=1}^{n} e^{it\lambda} \tag{5.63}$$

form a convergent sequence is crucial in proving the *mean-ergodic theorem*. Consider separately the cases $\lambda = 0$, and $\lambda \neq 0$. In the first case, each average is one and hence the limit is 1. If $\lambda \neq 0$, then

$$\left|\frac{1}{n}\sum_{t=1}^{n} e^{it\lambda}\right| = \left|\frac{1-e^{in\lambda}}{n(1-e^{i\lambda})}\right| \leq \frac{2}{n|1-e^{i\lambda}|}, \tag{5.64}$$

and the limit is zero.

Theorem 5.16. *(Mean Ergodic Theorem).* Let $\{X_t\}$ be stationary with mean μ and the spectral representation

$$X_t - \mu = \int e^{it\lambda} dZ(\lambda), ||dZ(\lambda)||^2 = dF(\lambda).$$

Then,

(a) $\lim_{n\to\infty} \frac{1}{n}\sum_{t=1}^{n} X_t = \mu + Z(0) - Z(0^-)$.

(b) $\lim_{n\to\infty} \frac{1}{n}\sum_{k=1}^{n} \gamma_k = ||Z(0) - Z(0^-)||^2 = F(0) - F(0^-)$.

Proof. Note that

$$\frac{1}{n}\sum_{t=1}^{n}(X_t - \mu) = \int \frac{1}{n}\sum_{t=1}^{n} e^{it\lambda} dZ(\lambda) = \int g_n(\lambda) dZ(\lambda), \tag{5.65}$$

where $g_n(\lambda)$ is the average in (5.62). Similarly,

$$\frac{1}{n}\sum_{k=1}^{n}\gamma_k = \int g_n(\lambda) dF(\lambda). \tag{5.66}$$

Since $|g_n(\lambda)| \leq 1$ and $g_n(\lambda)$ converges, as $n \to \infty$, to

$$g(\lambda) = \begin{cases} 1 & \lambda = 0, \\ 0 & \lambda \neq 0, \end{cases}$$

(a) and (b) follow by applying the bounded convergence theorem to (5.65) and (5.66). Q.E.D.

Now, conditions for the mean-consistency of $\bar{X}$ for μ is immediate from part (b) of the theorem.

Corollary 5.17. Let $\{X_t\}$ be stationary with mean μ and covariance $\{\gamma_k\}$. Then, $\bar{X}$ *converges to* μ *in the mean*, that is, $\lim_n E(\bar{X} - \mu)^2 = 0$, if and only if

$$\lim_n \frac{1}{n}\sum_{k=0}^{n}\gamma_k = 0.$$

The next corollary of the *mean-ergodic theorem* reveals the importance of the *discrete Fourier transform* of the data $X_1, \cdots, X_n$ defined in Section 2.4.

Corollary 5.18. Let $\{X_t\}$ be as in Theorem 5.16. Then for any $\lambda \in (-\pi, \pi]$,

$$\lim_{n\to\infty} \frac{1}{n}\sum_{t=1}^{n}(X_t - \mu)e^{i\lambda t} = Z(\lambda) - Z(\lambda^-),$$

moreover,

$$\|Z(\lambda) - Z(\lambda^-)\|^2 = F(\lambda) - F(\lambda^-) = \lim_{n\to\infty} \frac{1}{n}\sum_{k=0}^{n}\gamma_k e^{i\lambda k}.$$

Proof. Apply the Theorem 5.16 to the stationary process $\{(X_t - \mu)e^{i\lambda_0 t}\}$ with the covariance function $\{\gamma_k e^{ik\lambda_0}\}$ and the spectral distribution function $F(\lambda_0 + \lambda)$. Q.E.D.

For the time being we take $\mu = 0$ and allow X_t to be complex-valued. In view of the Corollary 5.17, the sample mean $\bar{X}$ is nearly zero and would not appear in the formula for $\hat{\gamma}_k$ so that the average $\frac{1}{n}\sum_{t=1}^{n-k} X_{t+k}\bar{X}_t$ provides an estimator for γ_k. Note that in this formula and the rest of this section $\bar{X}_t$ stands for the complex-conjugate of X_t. To establish the mean-consistency of $\hat{\gamma}_k$ as an estimator of γ_k the following condition on the fourth-order statistics of $\{X_t\}$ are required:

$$\begin{aligned} &E|X_t|^4 < \infty, t \in \mathbb{Z}, \text{ and} \\ &\{X_{t+k}\bar{X}_t; t \in \mathbb{Z}\} \text{ is stationary for } k = 0, \pm 1, \cdots. \end{aligned} \tag{5.67}$$

Note that for $\{X_t\}$ a Gaussian stationary process, (5.67) holds. A simple application of the mean-ergodic theorem to the stationary process $\{X_{t+k}\bar{X}_t; t \in \mathbb{Z}\}$ shows that $\hat{\gamma}_k$ is consistent for γ_k.

Theorem 5.19. Let $\{X_t\}$ be a stationary process satisfying (5.67). Then,

(a) $\lim_{n\to\infty} \frac{1}{n}\sum_{t=1}^{n-k} X_{t+k}\bar{X}_t$ exists for $k = 0, 1, 2, \cdots$

(b) $\lim_{n\to\infty} \frac{1}{n}\sum_{t=1}^{n-k} X_{t+k}\bar{X}_t = \gamma_k$, if and only if

$$\lim_{n\to\infty} \frac{1}{n}\sum_{t=1}^{n} EX_{t+k}\bar{X}_t\overline{X_k\bar{X}_0} = |\gamma_k|^2. \tag{5.68}$$

(c) If $\{X_t\}$ is a real and Gaussian stationary process with $EX_t = 0$, then (b) holds, if and only if

$$\lim_{n\to\infty} \frac{1}{n}\sum_{k=0}^{n} |\gamma_k|^2 = 0. \tag{5.69}$$

Proof. (a) and (b) follow from Theorem 5.16 upon noting that for a fixed k with $Y_t = X_{t+k}\bar{X}_t$, the process $\{Y_t - \gamma_k\}$ is a mean-zero stationary process and

$$\lim_{n\to\infty} \frac{1}{n}\sum_{t=1}^{n-k}(Y_t - \gamma_k) = \lim_{n\to\infty} \frac{1}{n}\sum_{t=1}^{n-k} X_{t+k}\bar{X}_t - \gamma_k$$

exists. Moreover, this limit is zero if and only if the limit of the average of the covariance function of $\{Y_t - \gamma_k\}$ is zero, this gives (5.68).
(c) When $\{X_t\}$ is real and Gaussian, (5.68) reduces to (Problem 5.13 (c))

$$\lim_{n\to\infty} \frac{1}{n}\sum_{j=0}^{n}(\gamma_j^2 + \gamma_{j-k}\gamma_{j+k}) = 0, k = 0, 1, 2, \cdots. \tag{5.70}$$

For $k = 0$, this is precisely (5.69). On the other hand, (5.69) implies (5.70), since using the inequality $2ab \leq a^2 + b^2$, we get

$$\begin{aligned} \frac{1}{n}\sum_{j=0}^{n}(\gamma_j^2 + \gamma_{j-k}\gamma_{j+k}) &\leq \frac{1}{n}\sum_{j=0}^{n}\gamma_j^2 + \frac{1}{2n}\left(\sum_{j=0}^{n}\gamma_{j-k}^2 + \sum_{j=0}^{n}\gamma_{j+k}^2\right) \\ &= \frac{2}{n}\sum_{j=0}^{n}\gamma_j^2 + r_n, \end{aligned}$$

where $r_n \to 0$ as $n \to \infty$. The remaining part of (c) follows from the next theorem which is of independent interest. Q.E.D.

The condition (5.69) is equivalent to the spectral distribution function of $\{X_t\}$ having no discontinuities. This is shown using the following theorem due to N. Wiener. Recall that a bounded monotone function F can have at most a countable number of jumps of size $F(\lambda^+) - F(\lambda^-)$ at points λ.

Theorem 5.20. For any covariance function $\{\gamma_k\}$ with the spectral representation

$$\gamma_k = \int e^{ik\lambda} dF(\lambda),\ k \in \mathbb{Z},$$

we have,

$$\lim_{n\to\infty} \frac{1}{n}\sum_{k=0}^{n}|\gamma_k|^2 = \sum_{\lambda}[F(\lambda^+) - F(\lambda^-)]^2.$$

Proof. Note that

$$\begin{aligned} \frac{1}{n}\sum_{k=0}^{n}|\gamma_k|^2 &= \frac{1}{n}\int\int\sum_{k=0}^{n} e^{ik(\lambda-\lambda')} dF(\lambda)dF(\lambda') \\ &= \int\int g_n(\lambda,\lambda')dF(\lambda)dF(\lambda'), \end{aligned}$$

where $g_n(\lambda, \lambda') = \dfrac{1 - e^{i(n+1)(\lambda-\lambda')}}{n(1 - e^{i(\lambda-\lambda')})}$. Since $|g_n(\lambda, \lambda')| \le 1$ and it converges to the function $g(\lambda, \lambda')$ which is zero except that $g(\lambda, \lambda) \equiv 1$. Thus, from the bounded convergence theorem

$$\begin{aligned} \lim_{n\to\infty} \frac{1}{n} \sum_{k=0}^{n} |\gamma_k|^2 &= \int\int g(\lambda, \lambda') dF(\lambda) dF(\lambda') \\ &= \int [F(\lambda^+) - F(\lambda^-)] dF(\lambda) = \sum_{\lambda} [F(\lambda^+) - F(\lambda^-)]^2. \end{aligned}$$

Q.E.D.

5.6.2 Periodogram and the Spectral Density

Due to the close relationship between the covariance and the spectral density, one would expect that accurate estimation of covariance should lead to accurate estimation of the spectral density function and vice versa. Surprisingly, this is not the case and the *periodogram* of the data $X_1, \cdots, X_n$,

$$I_n(\lambda) = \sum_{k=-(n-1)}^{n-1} \hat{\gamma}_k e^{ik\lambda}, \tag{5.71}$$

is a notoriously bad estimator of the spectral density, see Section 2.4. However, appropriate *averaging (smoothing)* of values of $I_n(\lambda)$ would give a consistent estimator of the spectral density. Since integration is a smoothing operation, the situation is much better for the spectral distribution $F(\lambda)$.

Theorem 5.21. Let $\{X_t\}$ be a mean-zero stationary process with covariance $\{\gamma_k\}$ and spectral distribution F such that (5.67) holds and for every $k = 0, 1, 2, \cdots$

$$\lim_{n\to\infty} \frac{1}{n} \sum_{t=1}^{n-k} X_{t+k} \bar{X}_t = \gamma_k,$$

in the mean. Then, for λ a continuity point of F, the integrated periodogram is a consistent estimator, that is,

$$\lim_{n\to\infty} \int_{-\pi}^{\lambda} I_n(\lambda') d\lambda' = F(\lambda), \tag{5.72}$$

in the mean.

Proof. Let us define F_n by

$$F_n(\lambda) = \int_{-\pi}^{\lambda} I_n(\lambda') d\lambda'.$$

For $|k| < n$, note that

$$\int_{-\pi}^{\pi} e^{ik\lambda} dF_n(\lambda) = \frac{1}{n} \sum_{t=1}^{n-|k|} X_{t+k} \bar{X}_t.$$

From the assumption of the theorem it follows that

$$\lim_n \int_{-\pi}^{\pi} e^{ik\lambda} dF_n(\lambda) = \lim_n \frac{1}{n} \sum_{t=1}^{n-|k|} X_{t+k} \bar{X}_t = \gamma_k = \int_{-\pi}^{\pi} e^{ik\lambda} dF(\lambda),$$

in the mean, that is the Fourier-Stieltjes coefficients of F_n converge to those of F. By Levy's continuity theorem, $F_n \to F+$ const., at all the continuity points of F as desired. Q.E.D.

It is important to note that (5.72) *does not necessarily* imply that $I_n(\lambda)$ converges to $F'(\lambda)$, as the following simple example shows.

Example 5.12. Let $\{X_t\}$ be a real Gaussian white noise with $EX_t = 0, EX_t^2 = 1$, for all t. Then,

$$\gamma_k = \delta_{k0} \text{ and } F'(\lambda) \equiv 1.$$

Note that $F_n'(0) = I_n(0) = \frac{1}{n} \left| \sum_{t=1}^{n} X_t \right|^2 \sim Z^2$ for all n, does not converge to $F'(0) = 1$. (Here, Z is the standard normal random variable and Z^2 has a chi-squared distribution with one degree of freedom.)

5.6.3 Pointwise Ergodic Theorem

In this section some applications of the pointwise ergodic theorem of strictly stationary processes to the previous estimation problems are reviewed briefly. Here, the mean-convergence of estimators are replaced by the stronger almost sure convergence.

A real-valued stochastic process $\{X_t\}$ defined on a probability space $(\Omega, \mathcal{F}, P)$ is said to be *strictly stationary* if for all integers $n \geq 1, t_1, \cdots, t_n, h$, the distribution of $(X_{t_1}, \cdots, X_{t_n})$ is the same as that of $(X_{t_1+h}, \cdots, X_{t_n+h})$. Evidently any sequence of i.i.d. random variables $\{e_t\}$ is strictly stationary and can be used to generate more general strictly stationary processes via linear combinations as in Example 5.3. In view of the Wold decomposition, we consider only the one-sided moving averages

$$X_t = \mu + \sum_{k=0}^{\infty} c_k e_{t-k}, \tag{5.73}$$

where $\{c_k\}$ is so that the infinite series converges with probability one. For example, when e_0 is zero-mean with $Ee_0^2 < \infty$, then $\sum_{k=0}^{\infty} |c_k|^2 < \infty$ is sufficient for this purpose.

The ergodicity property of strictly stationary processes guarantees an analogue of the strong law of large numbers. The definition given here is more suitable for applications in time series analysis; for more information see Rosenblatt (1985; Chapter 2).

A strictly stationary process $\{X_t\}$ with $E|X_t| < \infty$ is said to be *ergodic*, if for all (Borel measurable) functions $f : R^{k+1} \rightarrow R, k \geq 0$ with $E|f(X_1, \cdots, X_{1+k})| < \infty$,

$$\frac{1}{n} \sum_{t=1}^{n-k} f(X_t, \cdots, X_{t+k}) \rightarrow E\, f(X_1, \cdots, X_{1+k}) \tag{5.74}$$

with probability one as $n \rightarrow \infty$.

As before, it is not necessary to observe many independent replications of the process. Instead, a single realization of a long duration will suffice. To see the relevance of (5.74) to our estimation problems, consider an ergodic process $\{X_t\}$ with $EX_t^2 < \infty$, $\mu = EX_t$, and $\gamma_k = E(X_{t+k} - \mu)(X_t - \mu)$. Then, using (5.74) with $k = 0, f(x) = x$, and $k = 1, f(x_1, x_2) = (x_1 - \mu)(x_2 - \mu)$, it follows that $\bar{X}$ and $\hat{\gamma}_k$ are strongly consistent estimators of μ and γ_k.

This, in particular, shows that for most MA(∞) processes $\bar{X}$ and $\hat{\gamma}_k$ are good estimators of μ and γ_k. In fact, the matrix $\hat{\Gamma}_n$ the estimator of Γ_n in (5.2) is a positive-definite matrix, see Problem 5.19, and plays crucial roles in almost every aspect of time series data analysis. For instance, least squares and maximum likelihood estimators of parameters and orders of ARMA models are implicit nonlinear functions of $\hat{\Gamma}_n$, see Chapter 4. Establishing consistency and asymptotic normality of such estimators would require fast and uniform rate of convergence of $\hat{\gamma}_k, k = 0, 1, 2, \cdots$, to their theoretical counterparts. For this, one must impose more stringent conditions on $\{\gamma_k\}$ than those in (5.67)–(5.68), and allow the number of γ_k's estimated for each sample size n to grow gradually with n. To be clear, let $p(n)$ be a nonnegative integer denoting the number of γ_k's to be estimated. Then, with $\dfrac{p(n)}{n} \rightarrow 0$ it can be shown that

$$\max_{0 \leq k \leq p(n)} |\hat{\gamma}_k - \gamma_k| \rightarrow 0, \tag{5.75}$$

with probability one under some mixing conditions similar to those needed to establish the Central Limit Theorem (CLT) for the sample mean $\bar{X}$; see Section 2.6.1.

5.6.4 Regularity

The Kolmogorov's 0-1 law shows the tail σ-algebra of a sequence of i.i.d. random variables is trivial, that is, for any event A in the tail σ-algebra, $P(A) = 0$ or 1. The notion of regularity for strictly stationary processes imposes similar restriction on its tail σ-algebra. This is somewhat stronger than requiring a covariance stationary process to $\{X_t\}$ to be purely nondeterministic (Corollary 5.13).

For a process $\{X_t\}$, define $\mathcal{F}_n$ as the σ-algebra generated by the past values up to n, that is, $\mathcal{F}_n = \sigma\{X_t; t \leq n\}$. The tail (remote past) σ-algebra of $\{X_t\}$ is defined by $\mathcal{F}_{-\infty} = \bigcap_{n=-\infty}^{0} \mathcal{F}_n$. The process $\{X_t\}$ is said to be *regular* if $\mathcal{F}_{-\infty}$ is a trivial

σ-algebra, that is, for any event $A \in \mathcal{F}_{-\infty}, P(A) = 0$ or 1. The next result reveals the relation between ergodicity and regularity of MA(∞) processes.

Theorem 5.22. The strictly stationary process (5.73) is regular. Furthermore, if $E|X_t| < \infty$, then it is ergodic.

Proof. Let $\mathcal{F}_n^X (\mathcal{F}_n^e)$ stand for the σ-algebra generated by $\{X_t; t \leq n\}(\{e_t; t \leq n\})$, respectively. In view of (5.73), $\mathcal{F}_n^X \subseteq \mathcal{F}_n^e$, hence $\mathcal{F}_{-\infty}^X \subseteq \mathcal{F}_{-\infty}^e$ and the result follows from the Kolmogorov's $0-1$ law for i.i.d. random variables. That regularity plus moment conditions imply ergodicity and other related results see Bradley (1986). Q.E.D.

This theorem underscores the importance of MA(∞) processes in prediction theory and time series analysis. Indeed, for Gaussian process the notions of regularity and $\{X_t\}$ being purely nondeterministic are equivalent. Spectral characterizations of regularity and other similar asymptotic independence properties are given in Chapter 8.

5.6.5 Asymptotic Normality

Although $\bar{X}$ and $\hat{\gamma}_k$ are consistent estimators of μ and γ_k under broad conditions, such estimators are deemed more valuable when they are normally distributed, since then confidence intervals for the unknowns can be constructed. We review briefly conditions on $\{X_t\}$ that lead to the asymptotic normality of $\bar{X}_n$.

For simplicity of exposition and to reveal the key ideas we restrict attention to the $MA(q)$ process:

$$X_t = \mu + e_t + \beta_1 e_{t-1} + \cdots + \beta_q e_{t-q}.$$

The distribution of $\bar{X}_n - \mu$ depends only on $\{e_t\}$ and the parameters $\beta_1, \cdots, \beta_q$, since

$$\bar{X}_n - \mu = \frac{1}{n} \sum_{t=1-q}^{n} \left(\sum_{j=1-t}^{n-t} \beta_j \right) e_t. \tag{5.76}$$

In particular, for $q = 1$,

$$\sqrt{n}(\bar{X}_n - \mu) = \frac{\beta_1 e_0 + e_n}{\sqrt{n}} + (1 + \beta_1)\sqrt{n}\bar{e}_{n-1}, \tag{5.77}$$

where $\bar{e}_{n-1}$ is the average of $e_1, \cdots, e_{n-1}$. Note that $\left\| \frac{\beta_1 e_0 + e_n}{\sqrt{n}} \right\|^2 = \sigma^2(1 + \beta_1^2)/n \to 0$ as $n \to \infty$. The second term and hence $\sqrt{n}(\bar{X} - \mu)$ converge to a normal random variable provided that $\{e_t\}$ satisfies a Central Limit Theorem (CLT) and $1 + \beta_1 \neq 0$.

Thus, for the CLT to hold for $\bar{X}$ in (5.76), the sequence $\{e_t\}$ must satisfy a CLT and the coefficients β_j's must satisfy $\sum_{j=0}^{q} \beta_j \neq 0$. How can one state these conditions for a general stationary process? Fortunately, the second condition on the coefficients is there to ensure that the asymptotic variance of $\bar{X}$ is not zero. In general, this can be replaced by

$$\lim_{n\to\infty} n\mathrm{Var}(\bar{X}) = \gamma_0 + 2\sum_{k=1}^{\infty}\gamma_k = F'(0) > 0. \tag{5.78}$$

There are three distinct approaches to imposing conditions on $\{e_t\}$ or $\{X_t\}$ so that CLT holds for $\bar{X}$. The first and the most common approach in the time series literature is to assume that $\{e_t\}$ is a sequence of i.i.d. random variables with finite variance and then rely on the CLT for such sequences. Since the MA (∞) process (5.73) can be approximated closely by MA(q) processes with q large, ideas leading to (5.77) can be used to establish CLT for $\bar{X}$ for a large subclass of MA(∞) processes, the details are somewhat technical and can be found in Brockwell and Davis (1991, Chapter 7) and Rosenblatt (1985, Chap. III).

The second approach, that has been gaining popularity in recent years, requires that $\{e_t\}$ be a *martingale difference* with respect to its past, that is,

$$E(e_t|e_{t-1}, e_{t-2}, \cdots) = 0, \text{ for all } t,$$

and then relies on the following *martingale* CLT.

Theorem 5.23. *(Billingsley, 1961).* If $\{e_t\}$ is an ergodic strictly stationary martingale difference with $Ee_t^2 = \sigma^2$, then $\dfrac{1}{\sigma\sqrt{n}}\sum_{t=1}^{n} e_t$ converges to a normal random variable with mean zero and variance one.

For instance, if $\{e_t\}$ is a strictly stationary ergodic martingale difference with $\sigma^2 = Ee_t^2 < \infty$, then for an MA (q) model $\sqrt{n}(\bar{X} - \mu)$ converges to a normal random variable provided that $\sum_{j=0}^{q}\beta_j \neq 0$.

The third approach is based on the idea of *little-block, big-block* due to Bernstein for establishing CLT for partial sums of a strictly stationary process under one of the many mixing conditions introduced briefly in the next section; see Rosenblatt (1985).

5.6.6 Mixing Conditions

In this section we introduce some asymptotic independence and mixing conditions under which a CLT can be proved.

To introduce such mixing conditions, for an integer n let $\mathcal{F}_0 = \sigma\{X_0, X_{-1}, \cdots\}$ and $\mathcal{F}^n = \sigma\{X_n, X_{n+1}, \cdots\}$ be the σ-algebras generated by the past and future of the process $\{X_t\}$ separated by n time units. Recall that $\mathcal{F}_0$ and $\mathcal{F}^n$ are independent if $P(A \cap B) - P(A)P(B) = 0$ for all $A \in \mathcal{F}_0, B \in \mathcal{F}^n$. Therefore, various measures of dependence between the past and future of $\{X_t\}$ can be defined as follows:

$$\alpha(n) \quad = \quad \sup|P(A \cap B) - P(A)P(B)|,$$

$$\begin{aligned}\phi(n) &= \sup|P(B|A) - P(B)|, \quad P(A) > 0 \\ \psi(n) &= \sup \frac{|P(A \cap B) - P(A)P(B)|}{P(A)P(B)}, \quad P(A) > 0, P(B) > 0 \\ \rho(n) &= \sup|\mathrm{Cov}(X, Y)|, \quad X \in L^2(\mathcal{F}_0), Y \in L^2(\mathcal{F}^n) \\ \beta(n) &= \sup \frac{1}{2} \sum_{i=1}^{I} \sum_{j=1}^{J} |P(A_i \cap B_j) - P(A_i)P(B_j)|,\end{aligned}$$

where the first three sups are taken over all $A \in \mathcal{F}_0, B \in \mathcal{F}^n$, the fifth over all pairs of partitions $\{A_1, \cdots, A_I\}$ and $\{B_1, \cdots, B_J\}$ of Ω such that $A_i \in \mathcal{F}_0$ and $B_j \in \mathcal{F}^n$ for all i, j. The maximal correlation $\rho(n)$ between past and future is obtained by taking supremum over all $\mathcal{F}_0$ - measurable X and $\mathcal{F}^n$-measurable Y with norm one. This is quite appealing from geometrical point of view and is related to the notion of angle between past and future studied in Chapter 8.

Corresponding to the above measures of dependence we say a strictly stationary process $\{X_t\}$ is

(a) *strongly mixing (α-mixing)*	if	$\lim_{n\to\infty} \alpha(n) = 0$,
(b) *ϕ-mixing*	if	$\lim_n \phi(n) = 0$,
(c) *ψ-mixing*	if	$\lim_n \psi(n) = 0$,
(d) *ρ-mixing*	if	$\lim_n \rho(n) = 0$,
(e) *absolutely regular (β-mixing)*	if	$\lim_n \beta(n) = 0$.

The following inequalities reveal some relations among these conditions.

$$2\alpha(n) \leq \beta(n) \leq \phi(n) \leq \frac{1}{2}\psi(n), 4\alpha(n) \leq \rho(n) \leq \psi(n). \tag{5.79}$$

Note that α-mixing or "strong mixing" is the weakest of the five. It is implied by the other four and many other mixing conditions found in the literature, it has the distinction of being the first such condition introduced by M. Rosenblatt (1956).

The first four of these measures of dependence are subsumed by the following family of measures of dependence:

$$\alpha_{r,s}(n) = \sup \frac{|P(A \cap B) - P(A)P(B)|}{P(A)^r P(B)^s}, \quad 0 \leq r, s \leq 1. \tag{5.80}$$

Note that $\alpha_{0,0}(n) = \alpha(n)$ and $\alpha_{1,1}(n) = \psi(n), \alpha_{1,0}(n) = \phi(n)$. These measures of dependence also fit into a more general and geometrically appealing framework of "maximal correlation". For a σ-algebra $\mathcal{A}$, let $S(\mathcal{A})$ denote the set of all simple $\mathcal{A}$-measurable random variables. For $1 \leq p,\ q \leq \infty$, define

$$\rho_{p,q}(n) = \sup \left\{ \frac{|\mathrm{Cov}(X, Y)|}{\|X\|_p \, \|Y\|_q};\ X \in S(\mathcal{F}_0), Y \in S(\mathcal{F}^n) \right\}, \tag{5.81}$$

where $\|X\|_p = (E|X|^p)^{1/p}$. It can be shown that $\alpha_{r,s}(n)$ is a variant of $\rho_{1/r,1/s}(n)$ when only indicator functions are used. In fact, $\rho(n) = \alpha_{1/2,1/2}(n) = \rho_{2,2}(n)$ and in general it can be shown that

$$\rho_{p,q}(n) \leq C\alpha_{1/p,1/q}(n)[1 - \log \alpha_{1/p,1/q}(n)], \tag{5.82}$$

where C is a positive constant depending only on $1 < p, q < \infty, \dfrac{1}{p} + \dfrac{1}{q} = 1$. For more detailed information on the relationships among these mixing conditions, their history and relevant literature, see Bradley (1986); the papers in Eberlein and Taqqu (1986) and Doukhan (1994).

Problems

5.1 Express the covariance function $\{\gamma_k\}$ of the stationary process in Example 5.7 (b) in terms of the characteristic function of the random variable X.

- **(a)** Does $|\gamma_k| \to 0$ as $k \to \infty$?
- **(b)** When X is a standard normal random variable, calculate $\{\gamma_k\}$ and check whether $|\gamma_k| \to 0$ as $k \to \infty$. Do the same when X takes only two distinct values with $P(X = 1) = P(X = -1) = \frac{1}{2}$.
- **(c)** Which of the following functions defined on $\mathbb{Z}$ is positive-definite? (i) $\gamma_k = (-1)^{|k|}, (ii)\gamma_k = e^{-k^2}, (iii)\gamma_k = (1 + k^2)^{-1}$.

5.2 Let Y be a real-valued random variable with $EY = 0$ and $EY^2 = 1$ and f an arbitrary bounded function defined on the set of integers. Consider the stochastic process $\{X_t\}$ defined by

$$X_t = f(t)Y.$$

- **(a)** Show that $\{X_t\}$ is stationary, if and only if

$$f(t) = ce^{i\lambda t},$$

where c and λ are real constants.
- **(b)** Find $\hat{X}_{0,2}$, the best linear predictor of X_0 based on the two past values X_{-1}, X_{-2}. Is there a unique (linear) representation of $\hat{X}_{0,2}$ in terms of these past values of the process?
- **(c)** Is the process $\{X_t\}$ nondeterministic? Why?
- **(d)** For $\lambda = 0$ and $\lambda = \pi$ calculate and plot the covariance functions $\{\gamma_k\}$ of the stationary process in (a) and compare their graphs.

5.3 Show that the harmonic process defined in Example 5.4 is stationary. Does $|\gamma_k|$ converge to zero as $k \to \infty$? How does this process relate to the stationary process in Problem 5.2?

5.4 (a) For the linear process of Example 5.3, show that

$$|\gamma_k| \to 0 \; as \; k \to \infty.$$

(b) Show that the process $\{X_t\}$ in (5.12) is stationary and compute its covariance function.

5.5 For fixed $m \geq 0$, consider the (finite order) harmonic process $\{X_t\}$ defined by

$$X_t = \sum_{j=0}^{m} c_j e^{i\lambda_j t} e_j.$$

(a) Find $\hat{X}_t$, the best linear predictor of X_0 based on the infinite past X_{t-1}, $X_{t-2}, \cdots$, and σ^2 the variance of the prediction error.

(b) Is the process $\{X_t\}$ nondeterministic? Why?

(c) Repeat (a) and (b) when $m = \infty$. Is the general harmonic process of Example 5.4 deterministic?

(d) Is there any relation between the (finite order) harmonic processes and the circular processes of Example 5.7 (a)?

5.6 Let $\{e_t\}$ be a white noise process and $\{c_j\}$ be a sequence of scalars.

(a) Prove that the infinite series $\sum_{j=-\infty}^{\infty} c_j e_{t-j}$ is mean-convergent, if and only if

$$\sum_{j=-\infty}^{\infty} |c_j|^2 < \infty.$$

(b) If $\sum_{j=-\infty}^{\infty} c_j e_{t-j}$ is mean-convergent, then

$$\left(\sum_{i=-\infty}^{\infty} c_i e_{s-i}, \sum_{j=-\infty}^{\infty} c_j e_{t-j} \right) = \sum_{i=-\infty}^{\infty} \sum_{j=-\infty}^{\infty} c_i \, \bar{c}_j \, (e_{s-i}, e_{t-j}).$$

5.7 Let $\{X_t\}$ be stationary and $\{\varepsilon_t\}$ its innovation process, suppose that X_0, X_{-1}, X_{-2} and ε_0 satisfy

$$X_0 = a_1 X_{-1} + a_2 X_{-2} + \varepsilon_0,$$

for some constants a_1 and a_2. Show that, for all $t \in \mathbb{Z}$

$$X_t = a_1 X_{t-1} + a_2 X_{t-2} + \varepsilon_t.$$

5.8 Let $\hat{X}_{t+\ell}$ for $\ell \geq 0$, be the best linear predictor of $X_{t+\ell}$ based on the past $X_{t-1}, X_{t-2}, \cdots$. For ℓ fixed and $\{X_t\}$ stationary, show that the process $\{\hat{X}_{t+\ell}\}$ is also stationary. How does the shift operator of $\{\hat{X}_{t+\ell}\}$ relate to that of $\{X_t\}$? How does the covariance of $\{\hat{X}_{t+\ell}\}$ relate to that of $\{X_t\}$?

5.9 Show that a second-order stochastic process $\{X_t\}$ is stationary if and only if, for all $n \geq 1$, integers $t_1, \cdots, t_n \in \mathbb{Z}$ and scalars $c_1, \cdots, c_n, h$

$$\left\| \sum_{k=1}^{n} c_k X_{t_k+h} \right\| = \left\| \sum_{k=1}^{n} c_k X_{t_k} \right\|.$$

5.10 (a) Complete the proof of the statement and answer all questions, in Example 5.7 (b).

(b) Show that the set of "eigenvalues" of U is precisely the set $\{e^{ix};\ x$ a possible value of the random variable $X\}$.

5.11 For the matrix U in (5.22), find its eigenvalues, eigenvectors and write its spectral decomposition.

5.12 Express the covariance function of $\{Y_t\}$ to that of $\{X_t\}$ in Corollary 5.3. Re-examine the calculation of covariance functions in Examples 5.3-5.5 and state a general rule for calculation of covariance functions of linear transformation of stationary processes.

5.13 Suppose $X = (X_1, X_2, X_3, X_4) \sim N_4(\mu, \Sigma)$, that is, X is a four-dimensional normally distributed random vector with mean $\mu = (\mu_1, \mu_2, \mu_3, \mu_4)$ and invertible covariance matrix $\Sigma = (\sigma_{ij})_{i,j=1}^{4}$.

(a) Use the normal equations obtained in the proof of Theorem 5.6, to show that

$$E(X_1|X_2) = \mu_1 + \frac{\sigma_{12}}{\sigma_{22}}(X_2 - \mu_2).$$

Express Var$[X_1 - E(X_1|X_2)]$ in terms of elements of $\sum$.

(b) Compute $E(X_1|X_2, X_3)$ and $E(X_1|X_2, X_3, X_4)$ and variances of errors of approximating X_1 by $E(X_1|X_2, X_3)$ and $E(X_1|X_2, X_3, X_4)$.

(c) Use the results in (a) and (b) to express $E(X_1X_2), E(X_1X_2X_3)$ and $E(X_1X_2 X_3X_4)$ in terms of μ and Σ. In particular, show that $E(X_1 - \mu_1)(X_2 - \mu_2)(X_3 - \mu_3)(X_4 - \mu_4) = \sigma_{12}\sigma_{34} + \sigma_{13}\sigma_{24} + \sigma_{14}\sigma_{23}$.

(d) Let $\hat{X}_i = E(X_i|X_j, 1 \leq j \leq 4, j \neq i), e_i = X_i - \hat{X}_i$ and $Y = (e_1, \ldots, e_4)$. Find the distribution of Y and show that

$$\text{Cov}(Y) = \Sigma^{-1}.$$

Compare this result with those in Problem 3.9.

5.14 Let $\{X_t\}$ be a stochastic process in $H = L^2(\Omega, \mathcal{F}, P)$. Its *nonlinear innovation* process is defined by $N_t = X_t - E(X_t|X_{t-1}, X_{t-1} \ldots), t \in \mathbb{Z}$.

(a) Show that $\{N_t\}$ is a *martingale difference* sequence with respect to the past of $\{X_t\}$, that is

$$E(N_t|X_{t-1}, X_{t-2}, \ldots) = 0, \text{ for all } t.$$

(b) Show that $\{N_t\}$ is an orthogonal (uncorrelated) process, that is,

$$\text{Cov}(N_s, N_t) = 0, \text{ for } s \neq t.$$

Is $\{N_t\}$ a white noise process? Is $\{N_t\}$ an i.i.d. sequence?

5.15 Consider a second order stochastic process $\{X_t\}$ satisfying the stochastic difference equation

$$X_t = a_1 X_{t-1} + \ldots + a_p X_{t-p} + e_t,$$

where $p \geq 1, a_1, \ldots a_p$ are known, $\{e_t\}$ is a stochastic process with $E(e_t) = 0, \text{Var}(e_t) = \sigma^2$.

(a) Show that $\hat{X}_t$, the linear predictor of X_t based on the past $X_{t-1}, X_{t-2}, \cdots$, is given by

$$\hat{X}_t = \sum_{k=1}^{p} a_k X_{t-k},$$

provided that e_t is uncorrelated with the $X_{t-1}, X_{t-2}, \cdots$.

(b) Show that the best predictor of X_t based on the past, is given

$$E(X_t|X_{t-1}, X_{t-2}, \cdots) = \sum_{k=1}^{p} a_k X_{t-k},$$

if $E(e_t|X_{t-1}, X_{t-2}, \cdots) = 0$, that is, when $\{e_t\}$ is a *martingale difference* with respect to the past of $\{X_t\}$

(c) Compute the best predictor of X_t based on the past when e_t is independent of $X_{t-1}, X_{t-2}, \cdots$.

(d) Compute the best (linear and nonlinear) predictor of $X_{t+\ell}, \ell \geq 0$, based on the past in (a)–(c). Find the variance of prediction errors for $\ell = 0, 1$.

5.16 Consider the stationary process $\{X_t\}$ with covariance function $\{\gamma_k\}$ and let $\hat{X}_{t,1}$ be the linear predictor of X_t based on only the single observation X_{t-1} from the past.

(a) Show that

$$\hat{X}_{t,1} = \frac{\gamma_1}{\gamma_0} X_{t-1} \text{ and } \text{Var}(X_t - \hat{X}_{t,1}) = (\gamma_0^2 - |\gamma_1|^2)/\gamma_0.$$

(b) Evidently

$$\text{Var}[X_t - E(X_t|X_{t-1})] \geq \frac{\gamma_0^2 - |\gamma_1|^2}{\gamma_0},$$

find a better lower bound for this variance by showing that

$$\text{Var}[X_t - E(X_t|X_{t-1})] \geq \frac{1}{E\left[\frac{\partial}{\partial x} log\ f(x, y)\right]^2},$$

where $f(x, y)$ is the joint density of (X_t, X_{t-1}). Note that this is a Cramer-Rao type inequality for the variance of the best predictor.

5.17 Consider the process $\{X_t\}$ in Example 5.7 (b) with $\phi(\cdot)$ denoting the characteristic function of X.

(a) Show that $E(X_t|X_{t-1}, X_{t-2}) = X_t$ and $\text{Var}[X_t - E(X_t|X_{t-1}, X_{t-2})] = 0$. *Hint:* Solve the system $X_{t-j} = Ye^{i(t-j)X}, j = 1, 2$, for X and Y.

(b) Show that the linear predictor of X_t based on X_{t-1}, X_{t-2} is given by

$$\hat{X}_{t,2} = a_{1,2}X_{t-1} + a_{2,2}X_{t-2},$$

where $(a_{1,2}, a_{2,2})$ is the solution of

$$\begin{bmatrix} 1 & \phi(1) \\ \bar{\phi}(1) & 1 \end{bmatrix} \begin{bmatrix} a_{1,2} \\ a_{2,2} \end{bmatrix} = \begin{bmatrix} \phi(1) \\ \phi(2) \end{bmatrix}.$$

(c) Compute $\text{Var}(X_t - \hat{X}_{t,2})$ and compare it with the variance of the best nonlinear predictor in (a).

(d) Repeat (a)–(c) for n observations $X_{t-1}, \cdots, X_{t-n}$ from the past.

5.18 Show that $E(X_0|X_{-1}, X_{-2}, \cdots,)$ is the minimum mean square error predictor of X_0 based on the past. *Hint:* Use Problem 5.14 (a).

5.19 With $\hat{\gamma}_k$ as in (5.62), show that the matrix $\hat{\Gamma}_n = (\hat{\gamma}_{i-j})_{i,j=1}^n$ is a positive definite matrix. *Hint:* Write $\hat{\Gamma}_n$ as XX' for an appropriate matrix X involving $X_1, \cdots, X_n$.

5.20 If $X_t = \sum_{i=0}^{\infty} b_i \varepsilon_{t-i}$, where $\{\varepsilon_t\}$ is i.i.d. with $E(\varepsilon_t^4) < \infty$ and fourth cumulant k_4. Then, for $k = 0, 1, 2, \cdots$.

$$(n-k)E\left[\frac{1}{n-k}\sum_{t=1}^{n-k} X_t X_{t+k} - \gamma_k\right]^2 = \sum_{i=-(n-k)+1}^{n-k-1}\left(1 - \frac{|i|}{n-k}\right)$$

$$\left(r_i^2 + r_{i-k}r_{i+k} + k_4 \sum_{j=0}^{\infty} b_j\, b_{j+k}\, b_{j+i}\, b_{j+k+i}\right).$$

This converges, as $n \to \infty$, provided that

$$\sum_{i=0}^{\infty} |r_i|^2 < \infty.$$

(Hannan 1960, p. 39.).

6

Parameterization and Prediction

The covariance and spectrum of a stationary process provide information about its *global* dependence. However, for the purpose of prediction, control and other related problems it is desirable to work with alternative parameters which reveal more about its *local* dependence and structure. The main goal of this chapter is to present a simple and generic method of introducing parameters for and decomposing a stationary process in the time-domain. The proposed method works even when the traditional notions of covariance and spectrum are not defined as is the case for infinite variance processes.

Historically, Fourier transforms of the covariance and the process itself have been the starting point of decomposition, parameterization and prediction of stationary processes. This *spectral-domain approach* is a powerful and fully developed method of solving problems in prediction theory and time series analysis, aspects of this theory is covered in Chapters 5, 9 and 10. On the other hand, the Wold decomposition is the starting point of the *time-domain approach* to prediction theory and time series analysis. The real force behind this approach is the simple and familiar idea of regression, exploited successfully by Wold in 1938 for the first time, to decompose and unify various known time series models and to understand the structure of general stationary processes.

Our starting point in this chapter is also the familiar idea of linear regression. A rigorous statement of it, in Section 6.1, leads to several decompositions of the process. Corresponding to each decomposition of $\{X_t\}$ a set of parameters is generated, the most important of these are autoregressive, and moving average parameters. This approach suggests a natural duality between the AR and MA representations as well as their parameters. To emphasize this point we depart from the traditional approach and first derive the AR representation. Then, using precisely the same steps, we obtain the MA representation showing in effect the similarity between AR and MA representations. This becomes even more evident for infinite variance processes where there is no apparent preference for MA representation over the AR representation since the innovation process is no longer a white noise.

As mentioned earlier, the traditional approach to prediction theory developed by Kolmogorov (1941), Doob (1953), Wiener and Masani (1957, 1958) and Rozanov (1967) is based on the spectral representation and the Wold decomposition of the process. The important practical problem of AR representation of the process could not be discussed and resolved in the early stages of the development. Wiener and Masani (1957, 1958) and Akutowicz (1957) seem to be the first to address this problem in the context of a general stationary process, showing that its solution is related to the problem of convergence of a certain Fourier series in the spectral domain $L^2(F)$. Our approach here puts the AR and MA representations on equal footing.

In Section 6.2, the process and its multi-step predictors are expressed in terms of its observed past values. In Section 6.3, we shall decompose the process in terms of the past and present values of its innovation process. Section 6.4 is devoted to ARMA-type decompositions of stationary processes, which indicates why the popular ARMA(p, q) models may provide appropriate approximants for nondeterministic stationary processes. In Section 6.5, we show the idea of regression remains effective in introducing AR and MA representations and parameters and solving prediction problems for infinite variance processes. The key idea of this section, namely, the Regression Lemma and its implications were developed first in trying to parameterize infinite variance processes (Miamee and Pourahmadi, 1988b).

6.1 A REGRESSION LEMMA

In almost every area of physical and engineering sciences the characteristics or variables of interest are not observed directly, instead a noisy version of them are recorded, analyzed and inferences are made about the unobserved. This section gives a rigorous setup for linking components of such a problem.

More precisely, consider a quantity (message) X which we like to measure, but instead the (signal) Y a contaminated version of X with error (noise) coming from a source E is observed. We want to connect X, Y and elements of E in a practically meaningful manner. This is a prototype of many problems in probability and statistics, filtering and signal processing, etc. To state it mathematically, consider a Hilbert space H with E a subspace, $X \notin E$ and $Y \in \bar{sp}\{X, E\}$. We want to express Y in terms of X and elements of E . The following simple and important lemma shows a way to solve this generic problem. Figure 6.1 provides a schematic view of the problem and its solution suggested by the Regression Lemma.

Lemma 6.1. (Regression Lemma.) Let H be a Hilbert space, X an element of H and E a closed subspace of H with $X \notin E$. Then, for any $Y \in \bar{sp}\{X, E\}$ there

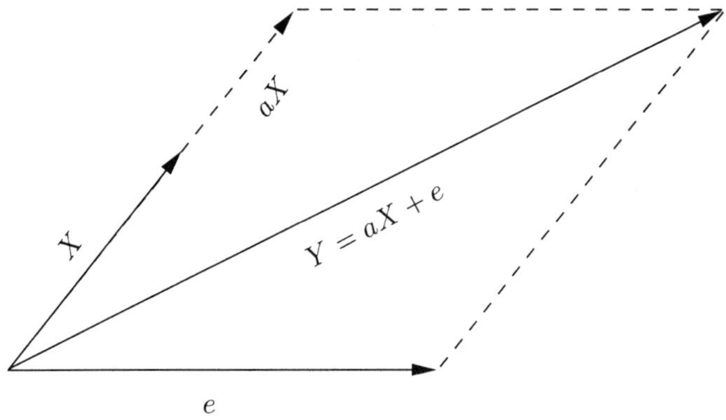

Fig. 6.1. A schematic view of the Regression Lemma.

are unique scalar a and vector $e \in E$ such that

$$Y = aX + e.$$

Proof. (Uniqueness) If Y has two different representations

$$Y = a_1X + e_1 = a_2X + e_2,$$

by subtraction we get

$$(a_1 - a_2)X = e_2 - e_1 \in E.$$

Since $X \notin E$, this is impossible unless $a_1 = a_2$, and therefore $e_1 = e_2$.

Existence. Since $Y \in \bar{sp}\{X, E\}$ there exists a sequence $\{Y_n\}$ such that

$$Y_n = a_nX + e_n \in sp\{X, E\} \text{ and } Y = \lim_{n\to\infty} Y_n. \tag{6.1}$$

We show that the sequence $\{a_n\}$ is bounded. If $\{a_n\}$ is not bounded, then there exists a subsequence $\{a_{n_k}\}$ such that $0 \neq a_{n_k} \to \infty$ as $k \to \infty$. From (6.1) we have

$$Y_{n_k} = a_{n_k}X + e_{n_k}. \tag{6.2}$$

Since $\{Y_{n_k}\}$ is bounded (being a subsequence of a convergent sequence), it follows from (6.2) that

$$X = \lim_{k\to\infty} \frac{Y_{n_k}}{a_{n_k}} - \lim_{k\to\infty} \frac{e_{n_k}}{a_{n_k}} = 0 - \lim_{k\to\infty} \frac{e_{n_k}}{a_{n_k}} \in E,$$

which contradicts $X \notin E$. Thus, $\{a_n\}$ is a bounded sequence. And it has a convergent subsequence (Bolzano-Weierstrass Theorem) with $\lim_{k\to\infty} a_{n_k} = a$. Now, from (6.2) the result follows. Q.E.D.

How does the Regression Lemma relate to linear models and least square theory? Let $X_1, X_2, \cdots, X_n$ be n linearly independent random variables in H and $Y \in sp\{X_1, X_2, \cdots, X_n\}$. For each $1 \leq p \leq n$, define $E_p = sp\{X_p, \cdots, X_n\}$ as the space of errors or variables not included in approximating Y. Since $Y \in sp\{X_1, E_2\}$ with $X_1 \notin E_2$, by the Regression Lemma there are unique scalar a_1 and $e_2 \in E_2$ such that

$$Y = a_1 X_1 + e_2.$$

Note that a_1 is the minimizer of $Q(c) = E(Y - cX_1)^2$, if and only if X_1 is orthogonal to E_2; see Problem 6.1. Similarly, since $e_2 \in sp\{X_2, E_3\}$ with $X_2 \notin E_3$, it follows that there are unique scalar a_2 and $e_3 \in E_3$ such that

$$e_2 = a_2 X_2 + e_3.$$

Combining the previous two equations we have

$$Y = a_1 X_1 + a_2 X_2 + e_3, \; e_3 \in E_3.$$

Continuing in this fashion we get the following theorem.

Theorem 6.2. Let $X_1, X_2, \cdots, X_n$ be n linearly independent random variables in H with $Y \in sp\{X_1, \cdots, X_n\}$. Then,

(a) For any $1 \leq p < n$, there are unique scalars $a_1, \cdots, a_p$ and $e_{p+1} \in E_{p+1}$ such that

$$Y = a_1 X_1 + \cdots + a_p X_p + e_{p+1}.$$

(b) There exist unique scalars $a_1, \cdots, a_n$ such that

$$Y = a_1 X_1 + \cdots + a_n X_n.$$

Such a theorem can be proved even when $n = \infty$, provided that $\sum_{k=1}^{\infty} a_k X_k$ converges in some sense. In the next four sections analogs of Theorem 6.2 are proved for nondeterministic processes.

6.2 AUTOREGRESSIVE PARAMETERS AND REPRESENTATIONS

Heuristically, for prediction, control and other related problems information from the immediate past is more relevant. Hence, it is desirable to express future value of a process in terms of its immediate past and an "error" component related to its remote past. The following theorem, an important consequence of the Regression Lemma, provides such a representation for stationary processes.

Theorem 6.3. Let $\{X_t\}$ be a nondeterministic stationary process, $\hat{X}_t$ the predictor of X_t based on $X_{t-1}, X_{t-2}, \cdots$, and $\{\varepsilon_t\}$ its innovation. Then, for any integer $n \geq 1$, there exist unique scalars $a_1, \cdots, a_n$ and $e_{t,n} \in H_{t-n-1}$ such that

$$\hat{X}_t = \sum_{k=1}^{n} a_k X_{t-k} + e_{t,n}, \; t \in \mathbb{Z}, \tag{6.3}$$

$$X_t = \sum_{k=1}^{n} a_k X_{t-k} + \varepsilon_t + e_{t,n}, \;\; t \in \mathbb{Z}. \tag{6.4}$$

Furthermore, $\{e_{t,n}\}$ is stationary.

Proof. Since $\hat{X}_0 \in H_{-1} = \bar{sp}\{X_{-1}, H_{-2}\}$ with $X_{-1} \notin H_{-2}$, by applying the Regression Lemma with

$$Y = \hat{X}_0, \; X = X_{-1}, E = H_{-2},$$

it follows that there exist a unique scalar a_1 and $e_{0,1} \in H_{-2}$ such that

$$\hat{X}_0 = a_1 X_{-1} + e_{0,1}.$$

But, since $e_{0,1} \in H_{-2} = \bar{sp}\{X_{-2}, H_{-3}\}$ with $X_{-2} \notin H_{-3}$, by applying the Regression Lemma again to $e_{0,1}$ and repeating we get $\hat{X}_0 = a_1 X_{-1} + \cdots + a_n X_{-n} + e_{0,n}$, $e_{0,n} \in H_{-n-1}$, as desired (6.4) follows from (6.3).

Q. E. D.

The scalars $a_1, a_2, \cdots, a_n$ in Theorem 6.3 do not depend on n and appear as regression coefficients when X_t is regressed on its own past $X_{t-1} \cdots, X_{t-n}$. Hence, to each nondeterministic process is associated a unique sequence of parameters with the following suggestive names:

The unique sequence of scalars $\{a_k\}_{k=1}^{\infty}$ in (6.3) is called the *autoregressive (AR) parameters* of the nondeterministic stationary process $\{X_t\}$. For each fixed $n \geq 1$, the decomposition (6.4) is called the finite or nth order *predictive decomposition* of $\{X_t\}$.

The AR parameters are of fundamental importance in prediction theory and time series analysis. They relate to the covariance function in a simple manner, particularly for the class of AR(p) processes defined next.

A stationary process $\{X_t\}$ is said to be an *autoregressive process of order* p, $1 \leq p < \infty$, in symbols AR(p), if $e_{t,p} = 0$, for all $t \in \mathbb{Z}$, in its predictive decomposition (6.4). In this case, we have

$$X_t = \sum_{k=1}^{p} a_k X_{t-k} + \varepsilon_t, \; t \in \mathbb{Z}. \tag{6.5}$$

Since autoregressive processes have a finite number of parameters, they are extremely useful in analyzing time series data and approximating more general stationary processes.

To express $a_1, a_2, \cdots, a_p$ and σ^2 in terms of $\{\gamma_k\}$, multiply both sides of (6.5) by $X_{t-j}, 0 \le j \le p$, to get

$$(X_t, X_{t-j}) = \sum_{k=1}^{p} a_k (X_{t-k}, X_{t-j}) + (\varepsilon_t, X_{t-j}),\ 0 \le j \le p. \tag{6.6}$$

For $j = 0$, from the definition of γ_k and Lemma 5.7 (b) one obtains

$$\sigma^2 = ||\varepsilon_t||^2 = \gamma_0 - \sum_{k=1}^{p} a_k \gamma_k, \tag{6.7}$$

and for $1 \le j \le p$, we have

$$\gamma_j = \sum_{k=1}^{p} a_k \gamma_{j-k}. \tag{6.8}$$

Writing (6.8) in matrix form gives

$$\begin{bmatrix} \gamma_0 & \gamma_1 & \cdots & \gamma_{p-1} \\ \gamma_1 & \gamma_0 & \cdots & \gamma_{p-2} \\ \vdots & \vdots & \vdots & \vdots \\ \gamma_{p-1} & \gamma_{p-2} & \cdots & \gamma_0 \end{bmatrix} \begin{bmatrix} a_1 \\ a_2 \\ \vdots \\ a_p \end{bmatrix} = \begin{bmatrix} \gamma_1 \\ \gamma_2 \\ \vdots \\ \gamma_p \end{bmatrix}. \tag{6.9}$$

The system of linear equations in (6.7) and (6.9) is referred to as the *Yule-Walker equations* for an AR(p) process. They are used to compute $a_1, \cdots, a_p$ and σ^2, once the covariances $\gamma_0, \cdots, \gamma_p$ are given; see Problem 6.2.

6.2.1 Autoregressive Representation

When $\{X_t\}$ is not an AR(p) process, expressing $\{a_k\}_{k=1}^{\infty}$ and σ^2 in terms of $\{\gamma_k\}$ is more challenging and hinges on the mode of convergence of the sequence $\left\{\sum_{k=1}^{n} a_k X_{t-k}\right\}_{n=1}^{\infty}$.

A stationary process $\{X_t\}$ with the predictive decomposition (6.4)) is said to have *autoregressive representation* (ARR), if the infinite series $\sum_{k=1}^{\infty} a_k X_{t-k}$ is convergent in the norm of H.

Note that the ARR is a truly second-order property and does not depend on the distribution of the stochastic process beyond the second moments. The following result shows the roles of $\{\gamma_k\}$ and $\{a_k\}$ in characterizing the ARR property.

Lemma 6.4. Let $\{X_t\}$ be a nondeterministic stationary process with covariance $\{\gamma_k\}$ and AR parameters $\{a_k\}$ Then,

(a) $\{X_t\}$ has the ARR property, if and only if

$$\sum_{k=1}^{\infty} \sum_{l=1}^{\infty} a_k a_l \gamma_{k-l} < \infty.$$

(b) $\{X_t\}$ has the ARR property if

$$\sum_{k=l}^{\infty} |a_k| < \infty.$$

Proof. (a) Recall that $\sum_{k=1}^{\infty} a_k X_{t-k}$ is mean-convergent, if and only if its partial sums sequence is Cauchy in H, that is for all $m < n$,

$$\left\| \sum_{k=m}^{n} a_k X_{t-k} \right\|^2 = \sum_{k=m}^{n} \sum_{l=m}^{n} a_k a_l \gamma_{k-l}$$

converges to zero as m and n go to infinity, or equivalently $\sum_{k=1}^{\infty} \sum_{l=1}^{\infty} a_k a_l \gamma_{k-l} < \infty$.

(b) Since $\sum_{k=1}^{\infty} \sum_{l=1}^{\infty} a_k a_l \gamma_{k-\ell}$ is nonnegative, it follows from the boundedness of $\{\gamma_k\}$, see (5.3), that

$$\sum_{k=1}^{\infty} \sum_{l=1}^{\infty} a_k a_l \gamma_{k-l} \leq \gamma_0 \sum_{k=1}^{\infty} \sum_{l=1}^{\infty} |a_k||a_l| = \gamma_0 \left(\sum_{k=1}^{\infty} |a_k| \right)^2,$$

and the result follows from (a).

Q.E.D.

For stationary processes with the ARR property, the predictive decomposition (6.4) takes the following form as $n \to \infty$.

Theorem 6.5. Let $\{X_t\}$ be a nondeterministic stationary process with ARR. Then, there exists a unique stationary process $\{e_t\}$ with $e_t \in H_{-\infty}(X)$ such that

$$X_t = \sum_{k=1}^{\infty} a_k X_{t-k} + \varepsilon_t + e_t, t \in \mathbb{Z}.$$

Proof. Since $\{X_t\}$ has ARR and $e_{t,n} \in H_{t-n-1}$, it follows from (6.4) that for each $t \in \mathbb{Z}$, as $n \to \infty$, $e_{t,n}$ converges in the mean to an element e_t belonging to $\bigcap_{n=1}^{\infty} H_{t-n-1} = H_{-\infty}$. Uniqueness and stationarity of $\{e_t\}$ are easy to establish.

Q.E.D.

When $\{X_t\}$ is a purely nondeterministic process with ARR, then $e_t = 0$ and

$$X_t = \sum_{k=1}^{\infty} a_k X_{t-k} + \varepsilon_t, \ t \in \mathbb{Z}. \tag{6.10}$$

In this case, we say that $\{X_t\}$ *has an infinite order autoregressive representation in the time-domain.* The class of such processes, denoted by AR(∞), can be viewed as Markov processes with the state-space $R^\infty = R \times R \times R \cdots$. For this class the parameters σ^2, $a_1, a_2, \cdots$ can be expressed in terms of the covariance function just like (6.7) and (6.8).

Theorem 6.6. Let $\{X_t\}$ be a purely nondeterministic stationary process with ARR. Then, $a_1, a_2, \cdots$ and σ^2 satisfy the following equations:

$$\begin{bmatrix} \gamma_0 & \gamma_1 & \gamma_2 & \cdot & \cdot & \cdot & \cdot & \cdot \\ \gamma_1 & \gamma_0 & \gamma_1 & \cdot & \cdot & \cdot & \cdot & \cdot \\ \gamma_2 & \gamma_1 & \gamma_0 & \cdot & \cdot & \cdot & \cdot & \cdot \\ \vdots & \vdots & \vdots & & & & & \end{bmatrix} \begin{bmatrix} a_1 \\ a_2 \\ \vdots \end{bmatrix} = \begin{bmatrix} \gamma_1 \\ \gamma_2 \\ \vdots \end{bmatrix}, \tag{6.11}$$

and

$$\sigma^2 = ||\varepsilon_t||^2 = \gamma_0 - \sum_{k=1}^{\infty} a_k \gamma_k. \tag{6.12}$$

The system of equations (6.11)–(6.12) are the discrete version of the famous Wiener-Hopf integral equation, and the "limit" of the Yule-Walker equations (6.7)–(6.9), when $p \to \infty$. For three concrete examples these equations are solved at the end of this section.

6.2.2 Computing Multi-Step Predictors

The ARR of a stationary process is ideal for expressing its predictors and innovations in terms of the past. We give recursive and updating formulas for the multi-step predictors.

Theorem 6.7. Let $\{X_t\}$ be a purely nondeterministic stationary process with the ARR and innovation $\{\varepsilon_t\}$. Then,

(a) The predictor $\hat{X}_t$ and innovation ε_t are given by

$$\begin{aligned} \hat{X}_t &= \sum_{k=1}^{\infty} a_k X_{t-k}, \\ \varepsilon_t &= X_t - \hat{X}_t = X_t - \sum_{k=1}^{\infty} a_k X_{t-k}. \end{aligned}$$

(b) The ℓ-step ahead predictor of X_ℓ based on $X_0, X_{-1}, \cdots$, denoted by $\hat{X}_{\ell|0}, \ell \geq 1$, is given by the *recursive* formula

$$\hat{X}_{\ell|0} = \sum_{k=1}^{\ell-1} a_k \hat{X}_{\ell-k|0} + \sum_{k=\ell}^{\infty} a_k X_{\ell-k}.$$

(c) (*Updating Formula*) Given the additional observations $X_1, \cdots, X_m$, $1 \leq m < \ell$, the predictor of X_ℓ based on $X_m, X_{m-1}, \cdots$ is given by

$$\hat{X}_{\ell|m} = \hat{X}_{\ell|0} + \sum_{k=1}^{m} b_{\ell-k}(X_k - \hat{X}_k).$$

Proof. (a) follows from (6.3) by letting $n \to \infty$. (b) follows from $X_\ell = \sum_{k=1}^{\ell-1} a_k X_{\ell-k} + \sum_{k=\ell}^{\infty} a_k X_{\ell-k}$, the definition of predictors as projections and its linearity. To prove (c), note that from Lemma 5.8 (a),

$$H_m(X) = H_0(X) \oplus sp\{\varepsilon_1, \cdots, \varepsilon_m\}$$

and therefore

$$\hat{X}_{\ell|m} = \hat{X}_{\ell|0} + \sum_{k=1}^{m} \beta_k \varepsilon_k,$$

with $\beta_k = \dfrac{(X_\ell, \varepsilon_k)}{||\varepsilon_k||^2} = b_{\ell-k}, 1 \leq k \leq m$; see Theorem 5.11 (b). Q.E.D.

In the following we compute predictors for three concrete examples corresponding to some of the covariances given in Example 5.1.

Example 6.1. (a) Let $\{X_t\}$ be a stationary process with the covariance

$$\gamma_k = a^k, |a| < 1, \ k \geq 0.$$

Then, it can be checked directly that $a_1 = a, \ a_k = 0, \ k \geq 2$, satisfies the equations (6.11). From Theorem 6.7 (a) and (b) we have

$$\hat{X}_t = aX_{t-1}, \hat{X}_{\ell|t-1} = a^\ell X_{t-1}, \ell \geq 1,$$

and $\sigma^2 = 1 - a^2$. Note that such a process has an ARR as $X_t = aX_{t-1} + \varepsilon_t, \ t \in \mathbb{Z}$.
(b) Let $\{X_t\}$ be stationary with the covariance

$$\gamma_0 = 1 + b^2, \gamma_k = b, k = \pm 1, \gamma_k = 0, \ |k| \geq 2 \text{ and } |b| < 1.$$

It can be shown directly that

$$a_k = -(-b)^k, \ k = 1, 2, \cdots,$$

satisfies equations (6.11). From Theorem 6.7 (a) we have

$$\hat{X}_t = -\sum_{k=1}^{\infty} (-b)^k X_{t-k}, \text{ and } \sigma^2 = 1 + b^2 - b^2 = 1.$$

Note that such a process has an ARR as

$$X_t = -\sum_{k=1}^{\infty}(-b)^k X_{t-k} + \varepsilon_t.$$

But, when $|b| = 1$ the $\{X_t\}$ does not have ARR since $\sum_{k=1}^{\infty}\sum_{l=1}^{\infty} a_k a_l \gamma_{k-l} = +\infty$.

(c) Let $\{X_t\}$ be the AR(p) process in (6.5) with known $a_1, \cdots, a_p$. Then, from Theorem 6.7 (b) we have

$$\hat{X}_{\ell|t-1} = \begin{cases} \sum_{j=1}^{\ell-1} a_j \hat{X}_{\ell-j|t-1} + \sum_{k=\ell}^{p} a_k X_{t-1+\ell-k}, & \text{if } \ell \le p, \\ \sum_{j=1}^{p} a_j \hat{X}_{\ell-j|t-1} & \text{if } \ell > p. \end{cases}$$

The (recursive) formulas for the multi-step predictors of an AR(p) process involve only finite number of operations on the past, this is in sharp contrast with the formulas for the MA (6.1) process in Example 6.1 (b).

In general, solving equations (6.11) is not easy and is related to triangular (Cholesky) factorization of the infinite Toeplitz matrix $\Gamma = (\gamma_{k-l})_{k,l=1}^{\infty}$, or the canonical factorization of the spectral density (the Fourier transform of the sequence $\{\gamma_k\}$); see Chapters 2 and 10.

6.3 MOVING AVERAGE PARAMETERS AND REPRESENTATIONS

In this section we represent a stationary process in terms of the past and present values of its innovation process $\{\varepsilon_t\}$, that is, as a sum of uncorrelated (orthogonal) random variables. Orthogonal decompositions are quite common in many areas of mathematics, statistics and physical sciences where it is desired to identify or decompose a quantity of interest as a sum of orthogonal components where each component can be measured without any interference from the others. Fourier series and spectral analysis of a stationary time series are important examples of such orthogonal decompositions.

The following theorem, the second important application of the Regression Lemma, gives the desired decomposition .

Theorem 6.8. Let $\{X_t\}$ be a nondeterministic stationary process, $\hat{X}_t$ the predictor of X_t based on the infinite past and $\{\varepsilon_t\}$ its innovation. Then, for any integer $n \ge 1$,

(a) there exist unique scalars $b_1, \cdots, b_n$ and a unique element $V_{t,n} \in H_{t-n-1}$ such that

$$\hat{X}_t = \sum_{k=1}^{n} b_k \varepsilon_{t-k} + V_{t,n},\ t \in \mathbb{Z}, \tag{6.13}$$

(b)

$$X_t = \sum_{k=0}^{n} b_k \varepsilon_{t-k} + V_{t,n},\ t \in \mathbb{Z},\ b_0 = 1, \tag{6.14}$$

where $V_{t,n} = P^{X_t}_{H_{t-n-1}}$ and is orthogonal to $U_{t,n} = \sum_{k=0}^{n} b_k \varepsilon_{t-k}$.

(c) For $k \geq 0,\ (\varepsilon_{t-k}, X_t) = \sigma^2\, b_k$.

Proof. (a) Since $\hat{X}_t \in H_{t-1} = \bar{sp}\{\varepsilon_{t-1}, H_{t-2}\}$ with $\varepsilon_{t-1} \notin H_{t-2}$, cf. (5.48)), by applying the Regression Lemma with $Y = \hat{X}_t,\ X = \varepsilon_{t-1},\ E = H_{t-2}$ and proceeding as in the proof of Theorem 6.3 the desired result follows. (6.14) follows from (6.13). (c) follows from (6.14) and noting that for $n \geq k,\ \varepsilon_{t-k}$ is orthogonal to H_{t-n-1}.

Q.E.D.

We refer to (6.14) as the *finite Wold decomposition* of the process $\{X_t\}$. Since $V_{t,n} \in H_{t-n-1}$ is orthogonal to $\varepsilon_t,\ \varepsilon_{t-1}, \cdots, \varepsilon_{t-n}$, it expresses X_t as a sum of $n+2$ orthogonal random variables. The finite Wold decomposition associates to each nondeterministic stationary process a unique set of parameters $b_1, b_2, \cdots$ called *the moving average (MA) parameters of* $\{X_t\}$. Like the autoregressive parameters, they are of fundamental importance in prediction theory and time series analysis. Next, we relate the new parameters $\{b_k\}$ to the covariance function $\{\gamma_k\}$ and the AR parameters of $\{X_t\}$. This is particularly simple for the following class of stationary processes.

A nondeterministic stationary process $\{X_t\}$ is said to be a *moving average process of order* q, $1 \leq q < \infty$, if $V_{t,q} = 0,\ t \in \mathbb{Z}$, in its finite Wold decomposition. In this case, from (6.14) we have

$$X_t = \sum_{k=0}^{q} b_k \varepsilon_{t-k},\ b_0 = 1,\ t\ \in\ \mathbb{Z}. \tag{6.15}$$

We use MA(q) to denote a moving average process of order q.

For an MA(q) process it is easy to relate $\sigma^2,\ b_1, \cdots, b_q$ to $\{\gamma_k\}$. In fact, from (6.15) it is immediate that

$$\gamma_k = \begin{cases} \sigma^2 \sum_{j=0}^{q} b_j b_{j+k}, & |k| \leq q, \\ 0, & |k| > q, \end{cases} \tag{6.16}$$

where $b_0 = 1, b_j = 0$ for $j < 0$ and $j > q$. With $T = (b_{i-j})_{i,j=0}^{\infty}$ a lower triangular matrix, (6.16) is equivalent to

$$\Gamma = (\gamma_{i-j}) = \sigma^2\, T'\, T. \tag{6.17}$$

While for an AR(p) process the Yule-Walker equations are linear in the parameters $a_1, \cdots, a_p$, the equations (6.16) for an MA(q) process are nonlinear (quadratic) in $b_1, \cdots, b_q$. This is the source of considerable difficulties in computing these parameters when covariances are given.

The two sets of parameters for a nondeterministic process introduced so far are related to each other via recursive formulas, so that knowing one, it is possible to compute the other.

Corollary 6.9. Let $\{X_t\}$ be a nondeterministic stationary process with the AR parameters $\{a_k\}_{k=1}^{\infty}$ and MA parameters $\{b_k\}_{k=1}^{\infty}$. Then,

$$b_l = \sum_{k=0}^{l-1} b_k a_{l-k}, \quad l = 1, 2, \cdots, \quad b_0 = 1. \tag{6.18}$$

Proof. For any integer $n \geq 1$, from (6.3) $a_1, \cdots, a_n$ satisfy

$$\hat{X}_t - \sum_{k=1}^{n} a_k X_{t-k} \in H_{t-n-1}.$$

Substituting for $\hat{X}_t$ and X_{t-k} from (6.14) we obtain

$$\hat{X}_t - \sum_{k=1}^{n} a_k X_{t-k} = \sum_{l=1}^{n} \left(b_l - \sum_{k=0}^{l-1} b_k a_{t-k} \right) \varepsilon_{t-l} + R_{n,t}, \tag{6.19}$$

where $R_{n,t} \in H_{t-n-1}$ and the first term is orthogonal to H_{t-n-1}. Thus, it follows that $\hat{X}_t - \sum_{k=1}^{n} a_k X_{t-k} \in H_{t-n-1}$, if and only if the first term in (6.19) is zero. Since ε_t's are orthogonal, (6.18) must then be satisfied.

Q.E.D.

For given $\{b_k\}$, one may write (6.18) as a (Toeplitz) system of equations with $\{a_k\}$ as the unknowns:

$$\begin{bmatrix} 1 & 0 & 0 & \cdot & \cdot & \cdot & 0 \\ b_1 & 1 & 0 & \cdot & \cdot & \cdot & 0 \\ b_2 & b_1 & 1 & 0 & \cdot & \cdot & 0 \\ \vdots & \vdots & \vdots & \vdots & \vdots & \vdots & \vdots \\ \cdot & \cdot & \cdot & \cdot & \cdot & \cdot & 1 \end{bmatrix} \begin{bmatrix} a_1 \\ a_2 \\ \vdots \\ \vdots \\ \cdot \end{bmatrix} = \begin{bmatrix} b_1 \\ b_2 \\ \vdots \\ \vdots \\ \cdot \end{bmatrix}. \tag{6.20}$$

Similarly, when $\{a_k\}$ is given, one can rewrite (6.18) as a system of equations with $\{b_k\}$ as the unknowns:

$$\begin{bmatrix} 1 & 0 & 0 & \cdot & \cdot & \cdot & 0 \\ -a_1 & 1 & 0 & \cdot & \cdot & \cdot & 0 \\ -a_2 & -a_1 & 1 & 0 & \cdot & \cdot & 0 \\ \vdots & \vdots & \vdots & \vdots & \vdots & \vdots & \vdots \\ \cdot & \cdot & \cdot & \cdot & \cdot & \cdot & 1 \end{bmatrix} \begin{bmatrix} b_1 \\ b_2 \\ b_3 \\ \vdots \\ \cdot \end{bmatrix} = \begin{bmatrix} a_1 \\ a_2 \\ a_3 \\ \vdots \\ \cdot \end{bmatrix}. \tag{6.21}$$

For the covariances in Example 6.1 (a), (b), using the $\{a_k\}$ found earlier and solving (6.18) we get, respectively,

$$\begin{cases} b_k = a^k & , & k \geq 1, \\ b_1 = b, & b_k = 0, & k \geq 2. \end{cases} \tag{6.22}$$

Finally, using (6.14) we give an alternative proof of the Wold decomposition theorem that relies solely on the orthogonality of the innovation $\{\varepsilon_t\}$.

Theorem 6.10. (Wold Decomposition Theorem). Let $\{X_t\}$ be a nondeterministic process with the innovation $\{\varepsilon_t\}$. Then, for each t,

(a) $X_t = U_t + V_t$, where $U_t = P^{X_t}_{H_t(\varepsilon)}$, $V_t = P^{X_t}_{H_{-\infty}(X)}$ and $\{U_t\}$ is orthogonal to $\{V_t\}$.

(b) The process $\{U_t\}$ is purely nondeterministic with the MA (∞) representation

$$U_t = \sum_{k=0}^{\infty} b_k \varepsilon_{t-k}, \quad \sum_{k=0}^{\infty} b_k^2 < \infty, \quad b_0 = 1,$$

where $b_1, b_2, \cdots$ are the MA parameters of $\{X_t\}$.

(c) The process $\{V_t\}$ is deterministic with $H_t(V) = H_{-\infty}(X)$.

Proof. Proofs of (a) and (c) are the same as in Theorem 5.11. (b) From (6.14), since $U_{t,q}$ is orthogonal to $V_{q,t}$, we have for $q \geq 1$,

$$||X_t||^2 = \left\| \sum_{k=0}^{q} b_k \varepsilon_{t-k} \right\|^2 + ||V_{q,t}||^2 \geq \sigma^2 \sum_{k=0}^{q} b_k^2.$$

Therefore,

$$\sum_{k=0}^{\infty} b_k^2 \leq \sigma^{-2} ||X_t||^2 < \infty,$$

and $\sum_{k=0}^{\infty} b_k \varepsilon_{t-k}$ is mean-convergent so that the process $\{U_t\}$ is purely nondeterministic by Theorem 5.10.

Q.E.D.

6.4 ARMA REPRESENTATIONS

Thus far, the value X_t of a process has been expressed either in terms of its past values or the past values of its innovation. In this section, our goal is to find representations expressing X_t in terms of the past values of the process as well as its innovations. For arbitrary integers $0 \leq p,\ q < \infty$, such representations and consequently the parameterizations of the process depend on the partition $\{P, Q\}$ of the set of integers $\{1, 2, \cdots, p+q\}$ into two disjoint sets P and Q with p and q elements, respectively. Whenever it is necessary to indicate the number of elements of P or Q, we do so by writing $P(p)$ or $Q(q)$. It should be noted that when Q or P is the empty set, then results of this section reduce to those in Sections 6.2 or 6.3, respectively.

The following theorem, the third important consequence of the Regression Lemma, provides such a representation.

Theorem 6.11. Let $\{X_t\}$ be a nondeterministic stationary process, $\{\varepsilon_t\}$ its innovation and $\{P, Q\}$ a partition of the set of integers $\{1, 2, \cdots, p+q\}$. Then, there exist unique scalars $\alpha_i, i \in P, \beta_j, j \in Q$ and a unique stationary process $\{e_{t,p,q}\}$ such that

$$X_t = \sum_{i \in P} \alpha_i X_{t-i} + \varepsilon_t + \sum_{j \in Q} \beta_j \varepsilon_{t-j} + e_{t,p,q}, \tag{6.23}$$

and

$$e_{t,p,q} \in H_{t-(p+q+1)}. \tag{6.24}$$

Proof. To make the idea of the proof clear, let us look at the simple case $p = q = 1,\ P = \{1\}$ and $Q = \{2\}$. Since

$$\hat{X}_0 \in H_{-1} = \bar{sp}\{X_{-1}, H_{-2}\},\ X_{-1} \notin H_{-2},$$

it follows from the Regression Lemma, with $Y = \hat{X}_0, X = X_{-1}$ and $E = H_{-2}$, that there exist a unique scalar α_1 and $e_1 \in H_{-2}$ such that $\hat{X}_0 = \alpha_1 X_{-1} + e_1$. But since $e_1 \in H_{-2} = \bar{sp}\{\varepsilon_{-2}, H_{-3}\},\ \varepsilon_{-2} \notin H_{-3}$, by applying the Regression Lemma, this time with $Y = e_1, X = \varepsilon_2, E = H_{-3}$, it follows that there exist a unique scalar β_2 and $e_{1,1} \in H_{-3}$ such that $e_1 = \beta_2 \in_{-2} + e_{1,1}$. Now, substituting e_1 in $\hat{X}_0$ we obtain

$$X_0 = \alpha_1 X_{-1} + \varepsilon_0 + \beta_2 \varepsilon_{-2} + e_{1,1},$$

by applying U, the (forward) shift operator of the process (Problem 5.7) (b), to both sides of this identity t-times, we arrive at

$$X_t = \alpha_1 X_{t-1} + \varepsilon_t + \beta_2 \varepsilon_{t-2} + \varepsilon_{t,1,1},$$

where $e_{t,1,1} = U^t e_{1,1} \in H_{t-3}$.

For arbitrary p, q, P and Q starting from the integer 1 and ending with $p+q$ the argument proceeds as above; if a given integer, between 1 and $p+q$, belongs to P, the AR decomposition will be used, otherwise the MA should be invoked.

Q.E.D.

Theorem 6.11 and its proof show that the set of vectors $\{X_{t-i}, i \in P\} \bigcup \{\varepsilon_{t-j}, j \in Q\}$, where $\{P, Q\}$ is a partition of $N = \{1, 2, \cdots\}$, could form a basis for the infinite-dimensional space H_{t-1}. The corresponding set of scalars $\{\alpha_i; i \in P\} \bigcup \{\beta_j; j \in Q\}$ which is unique (for a fixed partition of N) is referred to as the *autoregressive and moving average (ARMA) parameters* of a nondeterministic process. It should be noted that a different partition of N does produce a different ARMA parameters and approximant for the process.

A nondeterministic stationary process $\{X_t\}$ is said to be an *autoregressive and moving average* (ARMA) process of order $(p_0, p_0 + q_0)$, $0 \leq p_0, q_0 < \infty$, if

$$e_{t,p_0,q_0} = 0, \ \ t \in \mathbb{Z}, \tag{6.25}$$

in (6.23). In this case,

$$X_t = \sum_{i \in P} \alpha_i X_{t-i} + \varepsilon_t + \sum_{j \in Q} \beta_j \varepsilon_{t-j}, \tag{6.26}$$

where $\{P, Q\}$ is any partition of $\{1, 2, \cdots, p_0 + q_0\}$ with $P = P(p_0)$.

It should be noted that the ARMA$(p, p+q)$ model introduced in (6.26) requires only $p + q$ parameters; in sharp contrast with the standard ARMA$(p, p+q)$ model of Chapter 4, that is,

$$X_t = \sum_{i=1}^{p} \alpha_i X_{t-i} + \varepsilon_t + \sum_{j=1}^{p+q} \beta_j \varepsilon_{t-j}, \tag{6.27}$$

which requires $2p + q$ parameters. Thus, the models in (6.26) are submodels of (6.27) and provide more parsimonious approximants for a stationary time series. The goodness of an ARMA$(p, p+q)$ approximant in (6.23) can be assessed in terms of the magnitude of its remainder $e_{t,p,q}$. Next, conditions are given so that $e_{t,p,q}$ converges to zero in the mean, as p and $q \to \infty$, so that $\{X_t\}$ has ARMA(∞, ∞) representation.

In recent years, due to the development and popularity of automatic model selection criteria such as AIC, BIC and CAT etc., see Choi (1992), there has been considerable interest in fitting ARMA models of successively increasing orders to a given time series data set. In such a model fitting procedure it is presumed, a priori, that the underlying stationary process has an ARMA(∞, ∞) representation with infinite number of parameters; cf. Hannan (1987).

Here, we provide ARMA(∞, ∞) representations for a nondeterministic process; in fact, we show that corresponding to each partition $\{P, Q\}$ of the set of integers $N = \{1, 2, \cdots\}$ there is an ARMA(∞, ∞) representation. This result subsumes the AR, and MA representations of such processes developed earlier in Sections 6.2-6.3. It should be noted that Theorems 6.12 and 6.13 (a) can be also viewed as extensions of the Wold's decomposition and Theorem 6.5. Makagon and Salehi (1992) study nature of solutions of infinite order ARMA models using spectral representations.

Theorem 6.12. Let $\{X_t\}$ be a nondeterministic stationary process and $\{P, Q(\infty)\}$ be a partition of $N = \{1, 2, \cdots\}$ with $P = P(p)$ a finite set. Then,

$$X_t = \sum_{i \in P} \alpha_i X_{t-i} + \varepsilon_t + \sum_{j \in Q(\infty)} \beta_j \varepsilon_{t-j} + e_{t,p},$$

where $\sum_{j \in Q(\infty)} \beta_j^2 < \infty$ and $e_{t,p} \in \bigcap_{q \in Q(\infty)} H_{t-(p+q+1)} = H_{-\infty}(X)$.

Proof. For any $q > 0$, let $Q(q)$ be the set of the first q elements of $Q(\infty)$. Since $\sum_{j \in Q(q)} \beta_j \varepsilon_j$ is orthogonal to $e_{t,p,q} \in H_{t-(p+q+1)}$, we have

$$\sigma^2 \sum_{j \in Q(q)} \beta_j^2 \leq E\left(\sum_{j \in Q(q)} \beta_j \varepsilon_{t-j} + e_{t,p,q} \right)^2 = E\left(X_t - \varepsilon_t - \sum \alpha_i X_{t-i}\right)^2 < \infty.$$

Thus, the sequence $\left\{ \sum_{Q(q)} \beta_j^2 \right\}_q$ is nondecreasing and bounded, so it converges and

$$\sum_{j \in Q(\infty)} \beta_j^2 < \infty,$$

implying that the infinite series $\sum_{j \in Q(\infty)} \beta_j \varepsilon_{t-j}$ is mean-convergent. It also implies the convergence of $e_{t,p,q}$, as $q \to \infty$, to some element $e_{t,p}$; since $e_{t,p,q} \in H_{t-(p+q+1)}(X)$ for each q, it follows that its limit $e_{t,p}$ belongs to $\bigcap_{q \in Q(\infty)} H_{t-(p+q+1)} = H_{-\infty}(X)$. Q.E.D.

Next, we give conditions under which $e_{t,p}$ converges in the mean as $p \to \infty$.

Theorem 6.13. Let $\{X_t\}$ be a nondeterministic stationary process and $\{P(\infty), Q(\infty)\}$ a partition of N such that $\sum_{i \in P(\infty)} \alpha_i X_{t-i}$ is mean-convergent. Then,

(a) $X_t = \sum_{i \in P(\infty)} \alpha_i X_{t-i} + \varepsilon_t + \sum_{j \in Q(\infty)} \beta_j \varepsilon_{t-j} + e_t$, where $\{e_t\}$ with $e_t \in H_{-\infty}(X)$ is a stationary process.

(b) For $\{X_t\}$ purely nondeterministic, we have

$$X_t = \sum_{i \in P(\infty)} \alpha_i X_{t-i} + \varepsilon_t + \sum_{j \in Q(\infty)} \beta_j \varepsilon_{t-j},$$

that is, $\{X_t\}$ has an ARMA(∞, ∞) representation.

It is instructive to compare this theorem with Lemma 6.4 and the ARR in (6.10).

6.5 INFINITE VARIANCE PROCESSES

There is considerable interest in developing a prediction theory for and analyzing time series data from processes with infinite variance. These arise in financial markets, signal processing and network traffic where the distribution of data is asymmetric and heavy-tailed. More examples of non-Gaussian time series with infinite second moments and a version of the iterative model fitting process of Section 2.4.3 can be found in Adler, Feldman and Taqqu (1998).

The notions of covariance and spectral distribution can not be defined properly for such processes, therefore the standard Kolmogorov-Wiener theory of prediction and Box-Jenkins method of time series analysis do not apply directly to this situation. In this section using the Regression Lemma we develop aspects of theory of prediction for infinite variance processes in the time-domain in much the same way as in Sections 6.1–6.4. The results are presented in the same order as for second-order processes. We rely solely on the finite Wold and predictive decompositions of infinite variance processes to solidify and solve certain prediction problems. This underscores the fact that study of convergence of various series in Sections 6.2–6.4 is not necessary for these prediction problems.

6.5.1 Infinite Variance and p-Stationary Processes

We begin with defining *p-th order stationarity* for $0 < p < \infty$, give few examples and introduce its shift operator. For $0 < p < 2$, these processes have infinite variance and stable processes are the most prominent example of this case, see Samoradnitsky and Taqqu (1994); Cambanis, Hardin and Weron (1988); Miamee and Pourahmadi (1988) and Cambanis and Soltani (1984).

Let $(\Omega, \mathcal{F}, P)$ be a probability space. For $0 < p < \infty$, define $L^p = L^p(\Omega, \mathcal{F}, P) = \{X; E|X|^p < \infty\}$ as the set of all random variables with finite p-th moment. Recall that for $p \geq 1, L^p$ is a Banach space with norm $\|X\|_p = (E|X|^p)^{1/p}$, but for $0 < p < 1, L^p$ is a metric space with the metric $d(X, Y) = E|X - Y|^p$. A stochastic process $\{X_t\}$ is said to be *p-stationary* if $X_t \in L^p,\ t \in \mathbb{Z}$ and for all integers $n \geq 1, t_1, \cdots, t_n, h$ and scalars $c_1, \cdots, c_n$,

$$E\left|\sum_{k=1}^{n} c_k X_{t_k+h}\right|^p = E\left|\sum_{k=1}^{n} c_k X_{t_k}\right|^p .$$

It can be seen that strictly stationary processes with finite p-th moment and covariance stationary processes are, respectively, p- and 2-stationary. Conversely, using the Parallelogram law of the Hilbert space L^2, it follows that every 2-stationary process is covariance stationary, see Problem 5.9. Thus, our definition of p-stationarity for $p = 2$ coincides with the notion of covariance stationarity.

The following theorem gives examples and a method of generating p-stationary processes.

Theorem 6.14. (a) Let $\{e_t\}$ be an i.i.d. sequence with $e_t \in L^p$ and $\{c_k\}$ be a sequence of scalars such that the series $\sum_{k=-\infty}^{\infty} c_k e_k$ is convergent in the metric of L^p.

Then, $\{X_t\}$ defined by

$$X_t = \sum_{k=-\infty}^{\infty} c_k e_{t-k},\ t \in \mathbb{Z},$$

is a p-stationary process.

(b) Let $\{e_t\}$ be p-stationary and $\{c_k\}$ be a sequence of scalars such that $\sum_{k=-\infty}^{\infty} c_k e_k$ is convergent in the metric of L^p. Then, $\{X_t\}$ defined by

$$X_t = \sum_{k=-\infty}^{\infty} c_k e_{t-k}, t \in \mathbb{Z},$$

is a p-stationary process.

Proof. It suffices to prove (b). For all integers $n \geq 1, t_1, \cdots, t_n, h$ and scalars $d_1, \cdots, d_n$ we have

$$\begin{aligned}\sum_{j=1}^{n} d_j X_{t_j+h} &= \sum_{k=-\infty}^{\infty} c_k \left(\sum_{j=1}^{n} d_j e_{t_j+h-k} \right) \\ &= \lim_{m\to\infty} \sum_{k=-m}^{m} c_k \left(\sum_{j=1}^{n} d_j e_{t_j+h-k} \right), \\ \sum_{j=1}^{n} d_j X_{t_j} &= \lim_{m\to\infty} \sum_{k=-m}^{m} c_k \left(\sum_{j=1}^{n} d_j e_{t_j-k} \right),\end{aligned}$$

where lim stands for the limit in the metric of L^p. Hence,

$$E\left|\sum_{j=1}^{n} d_j X_{t_j+h}\right|^p = \lim_{m\to\infty} E\left|\sum_{k=-m}^{m} c_k \left(\sum_{j=1}^{n} d_j e_{t_j+h-k} \right)\right|^p = E\left|\sum_{j=1}^{n} d_j X_{t_j}\right|^p,$$

where the second equality holds because of p-stationarity of $\{e_t\}$.

Q.E.D.

For a p-stationary process $X = \{X_t\}$, its time-domain $H(X)$, past and present subspaces $H_t(X), t \in \mathbb{Z}$, and remote past $H_{-\infty}(X)$ are defined as in Section 5.5, with the exception that the closure is taken in the metric of L^p. Consider a linear operator U defined on $L(X) = sp\{X_t; t \in \mathbb{Z}\}$ by

$$U\left(\sum_{k=1}^{n} c_k X_{t_k}\right) = \sum_{k=1}^{n} c_k X_{t_k+1},$$

for all integers $n \geq 1, t_1, \cdots, t_n$ and scalars $c_1, \cdots, c_n$. Since $\{X_t\}$ is p-stationary it follows that U is an *isometry* from $L(X)$ onto $L(X)$. Therefore, using continuity of U it can be extended to a unique *isometric isomorphism* from $H(X)$ onto $H(X)$. This extension of U (also denoted by U) is called the *shift operator* of the p-stationary process $\{X_t\}$. It should be noted that for a 2-stationary process its shift is a unitary operator.

6.5.2 Prediction and Parameterization

For the prediction of p-stationary processes and in the rest of this section we restrict p to the range $1 < p < \infty$. Let $X \in L^p$ and M be a closed subspace. Then, there is a unique element $\hat{X}$ in M which is closest to X, in the sense that

$$E|X - \hat{X}|^p \leq E|X - Y|^p, \text{ for all } Y \in M. \tag{6.28}$$

This element $\hat{X}$ is called the metric projection or simply the *projection* of X onto M. It can be shown that $\hat{X}$ satisfies (6.28), if and only if

$$E(X - \hat{X})^{<p-1>}Y = 0, \text{ for all } Y \in M, \tag{6.29}$$

where for a complex number z, $(z)^{<p-1>} = |z|^{p-2}\bar{z}$ (Cambanis et al., 1988).

For a p-stationary process, the notions of deterministic, nondeterministic, purely nondeterministic, predictors and innovation processes are defined as in Section 5.5. In particular for an integer $\ell \geq 1$, the predictor of $X_{t-1+\ell}$ based on $X_{t-1}, X_{t-2}, \cdots$, denoted by $\hat{X}_{\ell|t-1}$, is the unique projection of $X_{t-1+\ell}$ onto $H_{t-1}(X)$. For $\ell = 1$, we simply use $\hat{X}_t$ to denote this predictor and its error is

$$\varepsilon_t = X_t - \hat{X}_t, \; t \in \mathbb{Z}. \tag{6.30}$$

It is easy to see that $\{X_t\}$ is nondeterministic, if and only if $E|\varepsilon_t|^p > 0$. Unlike the 2-stationary processes, for $p \neq 2$ the innovation $\{\varepsilon_t\}$ is not an orthogonal (white noise) sequence. Perhaps, this is the main source of difficulty in the prediction of p-stationary processes. The next theorem sheds some light on the dependence structure of $\{\varepsilon_t\}$ and $\{\hat{X}_{\ell|t-1}\}_t, \ell \geq 1$, viewed as stochastic processes (Problem 5.8).

Theorem 6.15. Let $\{X_t\}$ be a nondeterministic p-stationary process with the shift operator U, innovation $\{\varepsilon_t\}$ and ℓ-step predictor process $\{\hat{X}_{\ell|t-1}\}$. Then, the processes $\{\varepsilon_t\}_t, \{\hat{X}_{\ell|t-1}\}_t$ are p-stationary with the same shift operator U.

More information about the predictors and parameters of p-stationary processes are obtained through their finite Wold and predictive decompositions developed in the rest of this section. As before, the Regression Lemma is crucial for such decompositions, its statement and proof remain the same-only H should be replaced by L^p and the phrase "Hilbert space" removed. The same applies to the proof of the following important theorem which combines the main results of Theorems 6.3, 6.8 and Corollary 6.9.

Theorem 6.16. Let $\{X_t\}$ be a nondeterministic p-stationary process with innovation $\{\varepsilon_t\}$. Then, for any integer $n \geq 1$,

(a) There are unique scalars $a_1, \cdots, a_n$ and a p-stationary process $\{e_{t,n}\}_t$ with $e_{t,n} \in H_{t-n-1}$ such that

$$X_t = \varepsilon_t + \sum_{k=1}^{n} a_k X_{t-k} + e_{t,n}.$$

(b) There are unique scalars $b_1, \cdots, b_n$ and a p-stationary process $\{V_{t,n}\}_t$ such that

$$X_t = \varepsilon_t + \sum_{k=1}^{n} b_k \varepsilon_{t-k} + V_{t,n}.$$

(c) The sequences $\{a_k\}$ and $\{b_k\}$ are related via

$$b_\ell = \sum_{k=0}^{\ell-1} b_k a_{\ell-k}, \; b_0 = 1, \; \ell = 1, 2 \cdots.$$

In view of Theorem 6.16 to each nondeterministic p-stationary process one can associate two unique sequences of scalars $\{a_k\}$ and $\{b_k\}$, called the autoregressive, and moving average parameters of $\{X_t\}$, respectively. While as a consequence of the orthogonality of the innovation process the MA parameters of 2-stationary processes are square summable, this is not the case for p-stationary processes. All we are able to show is that $|b_k| \leq c2^k, \; k = 1, 2, \cdots$, with c a constant.

Corollary 6.19. Let $\{X_t\}$ be a nondeterministic p-stationary process with $\{\varepsilon_t\}$, $\{b_k\}$ and $\{V_{t,n}\}, n \geq 1$, as in Theorem 6.18. Then,

(a) $V_{t,n} = P_{H_{t-n-1}}\left(X_t - \sum_{k=0}^{n-1} b_k \varepsilon_{t-k}\right)$, where P_M^X is the metric projection of X onto the subspace M.

(b) $|b_k| \leq c2^k, \; k = 1, 2, \cdots$ for some constant c.

Proof. (a) Since $V_{t,n} \in H_{t-n-1}$ and

$$X_t - \sum_{k=0}^{n-1} b_k \varepsilon_{t-k} - V_{t,n} = b_n \varepsilon_{t-n}, \tag{1}$$

in view of (6.29) it suffices to show that

$$E(b_n \varepsilon_{t-n})^{<p-1>} Y = 0, \text{ for all } Y \in H_{t-n-1},$$

but this is evident from the definition of the innovation $\{\varepsilon_t\}$.
(b) It follows from (a) and (1) that

$$||b_n \varepsilon_{t-n}||_p \leq 2 \left\| X_t - \sum_{k=0}^{n-1} b_k \varepsilon_{t-k} \right\|_p \leq 2||X_t||_p + 2 \sum_{k=0}^{n-1} ||b_k \varepsilon_{t-k}||_p.$$

Using $||\varepsilon_t||_p = ||X_t - \hat{X}_t||_p \le ||X_t||_p$, this reduces to

$$|b_n| \le 2\frac{||X_0||_p}{||\varepsilon_0||_p}\left(1 + \sum_{k=0}^{n-1} |b_k|\right), \; n = 1, 2, \cdots$$

from which the result follows.

Q.E.D.

Thus, unlike the 2-stationary case $\{b_k\}$ does not even converge to zero. Nevertheless, with $\{b_k\}$ as in Corollary 6.19 we are able to define a function φ by

$$\varphi(z) = 1 + \sum_{k=1}^{\infty} b_k z^k, \;\; |z| < \frac{1}{2}, \tag{6.31}$$

which is analytic inside the open disc of radius 1/2, called the *transfer function* of the p-stationary process. Since $\varphi(0) = 1 \neq 0$, $\dfrac{1}{\varphi(z)}$ is analytic in a neighborhood of zero and has a Taylor expansion

$$\frac{1}{\varphi(z)} = 1 - \sum_{k=1}^{\infty} a_k z^k, \tag{6.32}$$

where the coefficients a_k's are the AR parameters of $\{X_t\}$. These are the same as those obtained for 2-stationary processes in (5.61).

Fortunately, the coefficients of the multi-step predictors are related to $\{b_k\}$ and $\{a_k\}$ or the Taylor coefficients of φ and φ^{-1} in a simple manner.

Corollary 6.20. Let $\{X_t\}$ be a nondeterministic p-stationary process with innovation $\{\varepsilon_t\}$ and MA parameters $\{b_k\}$. Then, for fixed $\ell \ge 1$ and $n \ge 1$,

(a) There are unique scalars $a_{1,\ell}, \cdots, a_{n,\ell}$ and a p-stationary process $\{e_{t,n,\ell}\}_t$ with $e_{t,n,\ell} \in H_{t-n-1}$ such that

$$\hat{X}_{\ell|t-1} = \sum_{k=1}^{n} a_{k,\ell} X_{t-k} + e_{t,n,\ell}. \tag{6.33}$$

Note: For $\ell = 1, a_{k,1} = a_k, k \ge 1, \; e_{t,n,1} = e_{t,n}$.

(b) There are unique scalars $b_{1,\ell}, \cdots, b_{n,\ell}$ and a p-stationary process $\{V_{t,n,\ell}\}_t$ with $V_{t,n,\ell} \in H_{t-n-1}$ such that

$$\hat{X}_{\ell|t-1} = \sum_{k=1}^{n} b_{k,\ell} \varepsilon_{t-k} + V_{t,n,\ell}.$$

Note: For $\ell = 1, b_{k,1} = b_k, k \ge 1, \;\; V_{t,n,1} = V_{t,n}$.

(c) The sequences $\{a_{k,\ell}\}_k$ and $\{b_{k,\ell}\}_k$ are related via

$$\sum_{k=1}^{m-1} b_k a_{m-k,\ell} = b_{m,\ell}, m = 1, 2, \cdots.$$

Proof. Since $\hat{X}_{\ell|t-1} \in H_{t-1}(X) = \bar{sp}\{X_{t-1}, H_{t-2}\} = \bar{sp}\{\varepsilon_{t-1}, H_{t-2}\}$ proofs of (a), (b) and (c) are exactly as in Theorems 6.3, 6.8 and Corollary 6.9.

Q.E.D.

In view of (c) to find $\{a_{k,\ell}\}_k$ one needs to compute $\{b_{k,\ell}\}_k$ for a fixed $\ell \geq 2$, this seems to be difficult in general. Fortunately, for two important cases of partial interest we are able to show that $\{b_{k,\ell}\}_k$ is obtained by simply shifting $\{b_k\}$, that is $b_{k,\ell} = b_{k+\ell-1}, k \geq 1$.

Corollary 6.21. Let $\{X_t\}$ be a nondeterministic p-stationary process with innovation $\{\varepsilon_t\}$ and MA parameters $\{b_k\}$. If $p = 2$ or $\{\varepsilon_t\}$ is an i.i.d. sequence, then for any $\ell, n \geq 1$. We have

(a) $\hat{X}_{\ell|t-1} = \sum_{k=1}^{n} b_{\ell+k-1}\varepsilon_{t-k} + V_{t-1+\ell,n+\ell-1}, V_{t-1+\ell,n+\ell-1} \in H_{t-n-1}.$

(b) $\hat{X}_{\ell|t-1} = \sum_{k=1}^{n} a_{k,\ell}X_{t-k} + e_{t,n,\ell},\ \ e_{t,n,\ell} \in H_{t-n-1}$, where

$$a_{k,\ell} = \sum_{j=0}^{k-1} b_{\ell+j}d_{k-j}, k \geq 1, b_0 = d_0 = 1. \tag{6.34}$$

Proof. For $m \geq \ell \geq 1$, we have from Theorem 6.16 (b) that

$$X_{t-1+\ell} = \sum_{k=0}^{\ell} b_k\varepsilon_{t-1+\ell-k} + \sum_{k=\ell}^{m} b_k\varepsilon_{t-1+\ell-k} + V_{t-1+\ell,m},$$

with $V_{t-1+\ell,m} \in H_{t-1+\ell-m-1}$. For $p = 2$, since the first sum in the equation above is orthogonal to H_{t-1}, it follows that

$$\hat{X}_{\ell|t-1} = \sum_{k=\ell}^{m} b_k\varepsilon_{t-1+\ell-k} + V_{t-1+\ell,m},$$

or equivalently

$$\hat{X}_{\ell|t-1} = \sum_{k=1}^{m-\ell+1} b_{k+\ell-1}\varepsilon_{t-k} + V_{t-1+\ell,m} = \sum_{k=1}^{n} b_{k+\ell-1}\varepsilon_{t-k} + V_{t-1+\ell,n+\ell-1},$$

where $n = m - \ell + 1$ and $V_{t-1+\ell,n+\ell-1} \in H_{t-n-1}$. Comparing this equation with the one in Corollary 6.20 (b) and since $\{b_{k,\ell}\}_k$ is unique the result follows. The proof is similar when $\{\varepsilon_t\}$ is an i.i.d. sequence. (b) follows from Corollary 6.20 (a), (c).

Q.E.D.

It is known, see Corollary 5.13, that a 2-stationary process is purely nondeterministic, if and only if it is a one-sided moving average of its innovation process. Proof of such a result for general p-stationary processes requires deeper techniques

(Cambanis, et al., 1988; Cheng, Miamee and Pourahmadi, 2000) based on the notion of stochastic processes forming basis for their spans. For 2-stationary processes such results are discussed in Chapter 8 and in greater detail in Rozanov (1967).

Problems

6.1 (a) In Lemma 6.1 when X and Y are random variables with finite second moments, show that a is the minimizer of $Q(c) = E(Y - cX)^2$, if and only if X is orthogonal to E.

(b) In Theorem 6.3 (a) show that $a_1, \cdots, a_p$ are the minimizer of $Q(c_1, \cdots, c_p) = E\left(Y - \sum_{k=1}^{p} c_k X_k\right)^2$, if and only if $X_1, \cdots, X_p$ are orthogonal to E_{p+1}.

(c) In Theorem 6.3, is it reasonable to think of $a_1, \cdots, a_p$ as some kind of statistical parameters and $Y = a_1 X_1 + \cdots + a_p X_p + e_{p+1}$ as a statistical model?

(d) In Theorems 6.3 and 6.4 consider the problem of finding $c_1, \cdots, c_m$ so that $Q_1(c_1, \cdots, c_m) = E\left(X_t - \sum_{k=1}^{m} c_k X_{t-k}\right)^2$ or $Q_2(c_1, \cdots, c_m) = E\left(X_t - \sum_{k=1}^{m} c_k \varepsilon_{t-k}\right)^2$ is minimized. Is the minimizer of $Q_1(\cdot)$ the same as $a_1, \cdots, a_m$? Discuss the same question for $Q_2(\cdot)$. Compare and summarize your findings regarding minimizers of $Q_i(\cdot), i = 1, 2$, and their relationships with the first m AR and MA parameters of a stationary process.

6.2 (a) Show that for a stationary AR(p) process, $a_1, \cdots, a_p$ and σ^2 in (6.5)-(6.6) satisfy the system of linear equations $\Gamma_p x = (0, \cdots, 0, -1)$ where $x = (a_1, \cdots, a_p, -1)/\sigma^2$ and $\Gamma_n = (\gamma_{i-j})_{i,j=0}^{n}, n \geq 1$.

(b) Show that for $n \geq 1$,

$$
\begin{aligned}
\sigma^2 \Gamma_{n-1}^{-1} &= \begin{bmatrix} a_0 & & 0 \\ \vdots & \ddots & \\ a_{n-1} & \cdots & a_0 \end{bmatrix} \begin{bmatrix} a_0 & \cdots & a_{n-1} \\ & \ddots & \vdots \\ 0 & & a_0 \end{bmatrix} \\
&- \begin{bmatrix} a_n & & 0 \\ \vdots & \ddots & \\ a_1 & \cdots & a_n \end{bmatrix} \begin{bmatrix} a_n & \cdots & a_1 \\ & \ddots & \vdots \\ 0 & & a_n \end{bmatrix},
\end{aligned}
$$

where $a_0 = 1$ and $a_i = 0$ for $i > p$.

(c) Show that $(x_1, \cdots, x_n)' \Gamma_{n-1}^{-1} (x_1, \cdots, x_n) = n\sigma^{-2}(a_0, \cdots, a_p)' D(a_0, \cdots, a_p)$, where $D = (d_{i,j})$ is a symmetric $(p+1) \times (p+1)$ matrix. Find $d_{i,j}$.
Hint: Start with $p = 1$.

6.3 Consider the processes in Example 6.1.

(a) What are the moving average parameters of the process in Example 6.1 (a)? Why $|a| \geq 1$ is excluded? What are the predictors when $|a| = 1$?

(b) What are the autoregressive parameters of the process in Example 6.1 (b)? Why $|b| \geq 1$ is excluded? When $|b| = 1$, find the AR parameters of $\{X_t\}$. Does $a_k \to 0$ as $k \to \infty$? What is the implication of the last question so far as ARR of the process is concerned?

6.4 Consider a stationary process $\{X_t\}$ with covariance function

$$\gamma_k = \begin{cases} 5 & , \quad k = 0 \\ -2 & , \quad k = \pm 1, \\ 0 & , \quad |k| > 2, \end{cases}$$

(a) Compute the linear least-squares predictor of X_0 biased on X_{-1} and X_{-2}, and its prediction error variance.

(b) Let $\{e_t\}$ be a white noise with $var(e_t) = 1$. Is $X_t = e_t - 2e_{t-1},\ t \in \mathbb{Z}$, a moving average representation of $\{X_t\}$? How about $X_t = 2e_t - e_{t-1}, t \in \mathbb{Z}$?

6.5 Let $\{X_t\}$ be a nondeterministic stationary process with innovation $\{\varepsilon_t\}$ and covariance $\{\gamma_k\}$.

(a) Find the linear least squares predictor of X_0 based on $\varepsilon_{-1}, \cdots, \varepsilon_{-m}$ and its prediction error variance.

(b) Find the linear least squares predictor of X_0 based on $X_{-1}, \cdots, X_{-m}$ and its prediction error variance.

(c) Find the linear least squares predictor of X_0 based on $\varepsilon_{-1}, \cdots, \varepsilon_{-m}, X_{-1}, \cdots, X_{-m}$ and its prediction error variance.

(d) Discuss the behavior of these predictors and their variances as $m \to \infty$, at least for part (a).

(e) Is there any relationship among the coefficients of the predictor in (c) and the ARMA parameters in (6.26).

6.6 Prove or disprove analogues of Problem 6.2 (a)–(c) for the MA(q) process and its covariance matrix given in (6.16)–(6.17).

6.7 **(a)** Prove Theorems 6.15 and 6.16.

(b) In Theorem 6.16 consider the problem of finding $c_1, \cdots, c_m$ so that

$$\begin{aligned} Q_1(c_1, \cdots, c_m) &= E\left| X_t - \sum_{k=1}^{m} c_k X_{t-k} \right|^p \quad \text{and} \\ Q_2(c_1, \cdots, c_m) &= E\left| X_t - \sum_{k=1}^{m} c_k \varepsilon_{t-k} \right|^p \end{aligned}$$

are minimized. Repeat Problems 6.1 (b)–(d) for this situation.

(c) For $m = 1$ find the minimizer of $Q_2(c_1)$ in terms of moments of $\{X_t\}$.

6.8 Complete the proof of Corollary 6.20.

6.9 Does Corollary 6.21 hold when $\{\varepsilon_t\}$ is a p-stationary martingale difference?

6.10 In Theorem 6.7 (b), (c) write $\hat{X}_{\ell|0} = \sum_{k=0}^{\infty} a_{k,\ell} X_{-k}$, $\hat{X}_{\ell|m} = \sum_{k=0}^{\infty} b_{k,\ell-m} X_{m-k}$. Express $\{a_{k,\ell}\}_k$, $\{b_{k,\ell-m}\}_k$ in terms of $\{a_k\}$ and/or $\{b_k\}$. Is there any relation between $\{a_{k,\ell}\}_k$ and $\{b_{k,\ell-m}\}_k$?

7

Finite Prediction and Partial Correlations

In this chapter the more realistic, practical and computational problem of predicting the future based on the finite past of length n is studied. It is conceivable and we show that for n large components of its solution converge to their infinite-past counterparts and hence to the appropriate parameters of the underlying stationary process. This opens up the possibility of computing and approximating some genuinely infinite-dimensional quantities of interest in prediction theory and time series analysis via their finite-dimensional counterparts. The Durbin-Levinson algorithm which recursively computes the finite predictor coefficients eliminates the need to invert Toeplitz matrices of increasing orders. It also serves as a catalyst to link the finite prediction problem to a host of apparently disparate problems and research areas such as the theory of Szegö's orthogonal polynomials, trigonometric moment problem, digital signal processing and filtering, inverse scattering, interpolation theory of functions and operators. The interplay among these areas is discussed in Landau (1987a), Kailath (1987), Szegö and Grenander (1958) and Pollock (1999).

The thread holding together these diverse areas is the concept of *partial autocorrelation function* (PACF) (the *Schur parameters* or *reflection coefficients*) $\{\pi_k\}$. Unlike the covariance function $\{\gamma_k\}$, the $\{\pi_k\}$ provides an unconstrained reparameterization of the process in the sense that for each k, $|\pi_k| \leq 1$ and otherwise $\pi_1, \pi_2, \cdots$ are free. This feature of $\{\pi_k\}$ is useful in simulating from specific ARMA models, computing maximum likelihood estimators within the causality and invertibility region (Jones, 1980), and in characterizing stationary processes that are (1) nondeterministic, (2) have maximum entropy and (3) their finite predictor coefficients converging to their AR parameters at an exponential rate.

In Section 7.1, the finite prediction problem and its relevant parameters are introduced, and computed either through the matrix inversion or more economically through the Cholesky decomposition of the $n \times n$ Toeplitz matrix $\Gamma_n = (\gamma_{i-j})_{i,j=1}^n$. In Section 7.2, the Durbin-Levinson algorithm is used to compute recursively the coefficients of the finite predictor and hence the PACF of a stationary process. The latter discussed in Section 7.3, provides yet another parameterization of stationary processes that is useful in time series data analysis and in diverse areas of applications. Section 7.4 gives a detailed review of multiple and canonical correlations of finite segments of the process. In Sections 7.5 and 7.6 we characterize nondeterministic and AR(p) processes in terms of the PACF, and study the (rate of) convergence of the finite predictor coefficients to the AR parameter through the Baxter's inequality. Using the isomorphism between the time- and spectral-domains, in two appendixes we show that most problems introduced in this chapter can also be solved more expeditiously using the Szegö's orthogonal polynomials.

7.1 THE FINITE PREDICTION PROBLEM

In this section we introduce the finite predictor and discuss the relevance of Cholesky decomposition of Γ_n to a host of problems in prediction theory and time series analysis.

7.1.1 Computing the Finite Predictor

We introduce few finite prediction problems and show the prevalence of the Toeplitz matrices $\{\Gamma_n\}$ in solving prediction problems. Given $X_1, \cdots, X_n$, and covariances $\gamma_0, \gamma_1, \cdots, \gamma_n$ of a real-valued, zero-mean stationary process, the *finite prediction problem* is that of computing the orthogonal projection of X_{n+1} onto the $sp\{X_1, \cdots, X_n\}$ and its prediction error variance, denoted by $\hat{X}_{n+1}$ and σ_n^2, respectively. Recall that this projection satisfies the *normal equations* (Theorem 5.5),

$$(X_{n+1} - \hat{X}_{n+1}, X_j) = 0, \ 1 \le j \le n.$$

Writing

$$\hat{X}_{n+1} = \sum_{k=1}^{n} a_{k,n} X_{n+1-k}, \tag{7.1}$$

for some scalars $a_{k,n}$, it follows that

$$\sum_{k=1}^{n} a_{k,n}(X_{n+1-k}, X_j) = (X_{n+1}, X_j), \ 1 \le j \le n,$$

or in matrix form

$$\begin{bmatrix} \gamma_0 & \gamma_1 & \cdots & \gamma_{n-1} \\ \gamma_1 & \gamma_0 & \cdots & \gamma_{n-2} \\ \vdots & \vdots & \vdots & \vdots \\ \gamma_{n-1} & \gamma_{n-2} & \cdots & \gamma_0 \end{bmatrix} \begin{bmatrix} a_{1,n} \\ a_{2,n} \\ \vdots \\ a_{n,n} \end{bmatrix} = \begin{bmatrix} \gamma_1 \\ \gamma_2 \\ \vdots \\ \gamma_n \end{bmatrix}, \tag{7.2}$$

and

$$\sigma_n^2 = \mathrm{Var}(X_{n+1} - \hat{X}_{n+1}) = (X_{n+1} - \hat{X}_{n+1}, X_{n+1}) = \gamma_0 - \sum_{k=1}^{n} a_{k,n}\gamma_k. \quad (7.3)$$

The equations (7.2) and (7.3) are called the *Yule-Walker* equations of order n. The invertibility of the matrices $\Gamma_n = (\gamma_{i-j})_{i,j=1}^n, n \geq 1$, is equivalent to the uniqueness of $a_{1,n}, \cdots, a_{n,n}$ in (7.1). Some basic and useful results regarding σ_n^2 and Γ_n are summarized below. To minimize confusion, we use $\hat{X}_{t,n}$ and $\varepsilon_{t,n}$ for the finite predictor and error of predicting X_t based on $X_{t-1}, \cdots, X_{t-n}$.

Lemma 7.1. Let $\{X_t\}$ be a zero-mean real-valued stationary process with the covariance function $\{\gamma_k\}$ and $\sigma_n^2, \Gamma_n, n \geq 1$, be as before. Then,

(a) The sequence $\{\sigma_n^2\}_{n=1}^{\infty}$ is bounded and nonincreasing, that is,

$$\sigma_1^2 \geq \sigma_2^2 \geq \cdots \geq 0.$$

Hence, $\sigma_n^2 \to \sigma^2$ as $n \to \infty$, for some $\sigma^2 \geq 0$.

(b) $||\hat{X}_t - \hat{X}_{t,n}||^2 = \sigma_n^2 - \sigma^2$, and $\lim_{n\to\infty} ||\hat{X}_t - \hat{X}_{t,n}|| = 0$, where $\hat{X}_t$ is the predictor of X_t based on the infinite past $\{X_{t-1}, X_{t-2}, \ldots\}$.

(c) If $\sigma_m^2 = 0$ for some m, then $\sigma_n^2 = 0$, for all $n \geq m$.

(d) $\sigma_n^2 > 0$ for an n, if and only if Γ_{n+1} is invertible. In this case,

$$\sigma_n^2 = \gamma_0 - \boldsymbol{\gamma}'_n \Gamma_n^{-1} \boldsymbol{\gamma}_n = \frac{|\Gamma_{n+1}|}{|\Gamma_n|},$$

with $\boldsymbol{\gamma}_n = (\gamma_1, \cdots, \gamma_n)$ a column vector.

(e) If $\{X_t\}$ is nondeterministic, then $\sigma_n^2 > 0$, Γ_n is invertible for all n and $\sigma^2 > 0$. Moreover,

$$\sigma^2 = \exp\left(\lim_{n\to\infty} \frac{1}{n} \log|\Gamma_n|\right).$$

Proof. (a), (b) and (c) are immediate from the definitions of $\hat{X}_{t,n}, \varepsilon_{t,n}$ and σ_n^2. The first part of (d) follows from (7.2) and (7.3), and the second part follows from partitioning Γ_{n+1} as

$$\Gamma_{n+1} = \begin{bmatrix} \gamma_0 & \boldsymbol{\gamma}'_n \\ \boldsymbol{\gamma}_n & \Gamma_n \end{bmatrix},$$

and using the elementary fact that (see Problem 4.14 (a))

$$|\Gamma_{n+1}| = (\gamma_0 - \boldsymbol{\gamma}'_n \Gamma_n^{-1} \boldsymbol{\gamma}_n)|\Gamma_n|.$$

(e) since $\sigma_n^2 \to \sigma^2$, as $n \to \infty$, it follows from (d) and Problem 7.2 that

$$\log \sigma^2 = \lim_{n\to\infty} \frac{1}{n} \sum_{k=1}^{n} \sigma_k^2 = \lim_{n\to\infty} \frac{1}{n} \sum_{k=1}^{n} (\log |\Gamma_{k+1}| - \log |\Gamma_k|) = \lim_{n\to\infty} \frac{1}{n} \log |\Gamma_n|.$$

Q.E.D.

Remark 7.1. In view of Lemma 7.1 (c), processes $\{X_t\}$ for which $\sigma_n^2 = 0$ for some n, have finitely many linearly independent random variables. We find it convenient to decompose the class Γ of all real-valued covariance sequences $\{\gamma_k\}$ as

$$\Gamma = B(\Gamma) + I(\Gamma), \tag{7.4}$$

where $I(\Gamma)$ — the "interior" — consists of all sequences which are strictly positive definite, while $B(\Gamma)$ - the "boundary" - consists of all sequences where equality in $\sum_{i=1}^{n}\sum_{j=1}^{n} c_i c_j \gamma_{i-j} \geq 0$, is achievable for a non-trivial set $\{c_i\}$. For the time being, we shall work with processes for which $\{\gamma_k\} \in I(\Gamma)$.

7.1.2 The Modified Cholesky Decomposition

Triangular or Cholesky decomposition of the matrix Γ_n is useful in computing the predictor $\hat{X}_{n+1}$, the maximum likelihood estimator of ARMA parameters and relating the coefficients of finite and infinite predictors. In many ways, the role of this matrix decomposition to computational problems of time series data analysis and digital signal processing is similar to the role of the spectral factorization in the conceptual and theoretical problems of prediction theory. However, unlike the latter the Cholesky decomposition of Γ_n follows rather easily from (7.1)) which is a manifestation of the ubiquitous Gram-Schmidt procedure.

Since X_1 has zero-mean, then $\hat{X}_1 = 0$, and for $n \geq 0$ we have

$$\varepsilon_{n+1,n} = X_{n+1} - \hat{X}_{n+1} = X_{n+1} - \sum_{k=1}^{n} a_{k,n} X_{n+1-k}. \tag{7.5}$$

Define the column vectors

$$\boldsymbol{X}_n = (X_1, \cdots, X_n), \hat{\boldsymbol{X}}_n = (\hat{X}_1, \cdots, \hat{X}_n),$$

then (7.5) for $n = 0, 1, 2, \cdots$, written in matrix form becomes

$$\boldsymbol{X}_n - \hat{\boldsymbol{X}}_n = A_n \boldsymbol{X}_n, \tag{7.6}$$

where A_n is a unit lower triangular matrix with $a_{0,n} = 1$ and $-a_{k,n}$ as its diagonal and appropriate subdiagonal elements, respectively.

It is instructive to observe the role of the matrix A_n in (7.6) as a linear transformation sending the data vector $\boldsymbol{X}_n$ to its vector of prediction errors $\boldsymbol{X}_n - \hat{\boldsymbol{X}}_n$.

Fortunately, (7.5)–(7.6) can be solved to express $\boldsymbol{X}_n$ in terms of $\boldsymbol{X}_n - \hat{\boldsymbol{X}}_n$, that is,

$$X_{n+1} = \sum_{k=1}^{n} b_{k,n}(X_{n+1-k} - \hat{X}_{n+1-k}), n = 0, 1, 2, \cdots, \tag{7.7}$$

and in matrix form, we obtain

$$\boldsymbol{X}_n = B_n(\boldsymbol{X}_n - \hat{\boldsymbol{X}}_n), \tag{7.8}$$

where B_n also is a unit lower triangular matrix. Thus, the matrix B_n is seen as a linear transformation sending the vector of prediction errors back to the original data. The elements $b_{k,n}$ of the matrix B_n satisfy the recursions (since $B_n A_n = I_n$):

$$\begin{cases} b_{l,n} = \sum_{k=1}^{l} a_{k,n} b_{l-k,n-k}, & n = 1, 2, \cdots; \; l = 1, 2, \cdots, n, \\ b_{0,n} = 1, & n = 0, 1, 2, \cdots, \end{cases} \tag{7.9}$$

which are analogues of (6.18). These are used in Section 7.6 to establish convergence of finite predictor coefficients to their infinite counterparts and to characterize PND processes.

From (7.8) we get the following *modified Cholesky decompositions* of Γ_n and its inverse:

$$\Gamma_n = \text{cov}(\boldsymbol{X}_n) = B_n \Sigma_n B_n', \; \Gamma_n^{-1} = A_n' \Sigma_n^{-1} A_n, n = 1, 2, \cdots, \tag{7.10}$$

where $\Sigma_n = \text{diag}(\sigma_0^2, \sigma_1^2, \cdots, \sigma_{n-1}^2)$. These decompositions are the stationary analogues of similar results developed in Section 3.5 for nonstationary covariances.

These decompositions are particularly useful for fast numerical evaluation of the Gaussian likelihood function (2.59) of $X_1, \cdots, X_n$, since

$$\begin{cases} |\Gamma_n| = \Pi_{i=0}^{n-1} \sigma_i^2, \\ \boldsymbol{X}_n' \Gamma_n^{-1} \boldsymbol{X}_n = \sum_{i=1}^{n} (X_i - \hat{X}_i)^2 / \sigma_{i-1}^2. \end{cases} \tag{7.11}$$

The set of scalars $\{a_{k,n}\}$ and $\{b_{k,n}\}$ in (7.1) and (7.7) is called *the finite predictor coefficients* of $\{X_t\}$. In general, they are not unique, though the simple condition $\{\gamma_n\} \in I(\Gamma)$ guarantees the uniqueness of $\{a_{k,n}\}$ and hence that of $\{b_{k,n}\}$.

The following remarks regarding the Yule-Walker equations are in order:

Remarks 7.2. (a) Given the time series data $x_1, \cdots, x_n$ from $\{X_t\}$, the equations (7.2)–(7.3) with γ_k's replaced by their estimates $\hat{\gamma}_k$'s is called the *sample Yule-Walker equations*. These are solved in exactly the same manner and by using the same procedures (particularly, the Durbin-Levinson algorithms) developed in this chapter. In fact, solutions of the sample version enjoys the same geometric interpretation when an appropriate inner-product is introduced on the space of extended observations (Problem 7.5).

(b) Often it is desirable to predict X_{n+1} based on the $m(1 \leq m < n)$ immediate past values $X_n, X_{n-1}, \cdots, X_{n-m+1}$. The corresponding normal equations are again (7.2)–(7.3) with n replaced by m.

7.1.3 More Prediction Problems

There are other finite prediction problems whose solutions are related to or equivalent to solving the system (7.2)–(7.3). We introduce three of these here.

The first problem is concerned with computing the *multi-step ahead predictor* of $X_{n+\ell}$ based on $X_1, \cdots, X_n$, denoted by $\hat{X}_{n+\ell,n}$, $\ell \geq 1$, with representation

$$\hat{X}_{n+\ell,n} = \sum_{k=1}^{n} a_{k,n}(\ell) X_{n+1-k}, \tag{7.12}$$

where $a_{k,n}(\ell)$'s are chosen so that $\sigma_n^2(\ell) = E(X_{n+\ell} - \hat{X}_{n+\ell,n})^2$ is a minimum (Problem 7.7). In analogy with (7.1)–(7.2), the coefficients of $\hat{X}_{n+\ell,n}$ satisfy the normal equations

$$\begin{bmatrix} \gamma_0 & \gamma_1 & \cdots & \gamma_{n-1} \\ \gamma_1 & \gamma_0 & \cdots & \gamma_{n-2} \\ \vdots & & & \vdots \\ \gamma_{n-1} & \gamma_{n-2} & \cdots & \gamma_0 \end{bmatrix} \begin{bmatrix} a_{1,n(\ell)} \\ \vdots \\ a_{n,n(\ell)} \end{bmatrix} = \begin{bmatrix} \gamma_\ell \\ \vdots \\ \gamma_{\ell+n-1} \end{bmatrix}, \tag{7.13}$$

with

$$\sigma_n^2(\ell) = \gamma_0 - \sum_{k=1}^{n} a_{k,n}(\ell) \gamma_{\ell+k-1}. \tag{7.14}$$

The second problem is concerned with the *interpolation* of X_0 based on $\{X_{-m}, \cdots, X_{-1}, X_1, \cdots, X_n\}$, where $m, n \geq 0$ are finite integers. Let

$$\hat{X}_{0,m,n} = \sum_{k=1}^{m} c_{k,m} X_{-k} + \sum_{k=1}^{n} d_{k,n} X_k, \tag{7.15}$$

denote the best *linear interpolator* of X_0 based on the given data and $\sigma_{m,n}^2 = E(X_0 - \hat{X}_{0,m,n})^2$ be its *interpolation error*. Note that when $n = 0$, the second sum in (7.15) is absent. To write the normal equations for the coefficients of the interpolator, let us define an $(n + m + 1)$ - dimensional vector x by

$$x = (c_{1,m}, \cdots, c_{m,m}, -1, d_{1,n}, \cdots, d_{n,n}) / \sigma_{m,n}^2. \tag{7.16}$$

From the projection theorem and some algebra it follows that the analogues of (7.1)–(7.2) in matrix form is

$$\Gamma_{m+n+1} x = (0, \cdots, 0, 1, 0, \cdots, 0), \tag{7.17}$$

with 1 at the $(m+1)$-st position. It is instructive to compare this with the Problem 6.2 (a); see Bhansali (1990).

The third problem is that of prediction of a future value X_{n+1} based on the *incomplete past* $\{X_{-m}, \cdots, X_{-1}, X_1, \cdots, X_n\}$, $m, n \geq 0$, denoted by $\tilde{X}_{n+1,m,n}$. It has a representation similar to (7.15), whose coefficients and prediction error variance can be computed using analogues of (7.16)-(7.17).

From (7.2), (7.14) and (7.17), it is evident that fast and efficient inversion of the Toeplitz matrices $\Gamma_n, n \geq 1$, is crucial for solving these problems and implementing their solutions in practice. In the next section, however, we avoid inversion of matrices and develop a recursive method for solving these so-called Toeplitz systems.

7.2 DURBIN-LEVINSON ALGORITHM

In this section the Durbin-Levinson recursive algorithm for solving the system of linear equations (7.2), (7.13) and (7.17) is presented. An important feature of this algorithm is eliminating the need to invert matrices of increasingly large orders. There are several methods of deriving it in the literature relying heavily on Γ_n being Toeplitz or the stationarity of $\{X_t\}$. The presentation below is from Golub and Van Loan (1989) and employs certain symmetry properties of Toeplitz matrices. For an alternative method based on orthogonalization of X_1 and $sp\{X_2, \cdots, X_n\}$, see Problem 7.7 or Brockwell and Davis (1991, p.170). Kailath (1987) provides many other alternatives.

For $n \geq 1$, let $\boldsymbol{a}_n = (a_1, \cdots, a_n)$ denote an n-dimensional column vector. We, often need to reverse the order of elements of $\boldsymbol{a}_n$. This operation is performed by the $n \times n$ (exchange) matrix E_n defined by $E_n = (e_n, \cdots, e_1)$, where $e_1, \cdots, e_n$ are the standard base for R^n. Thus, $E_n \boldsymbol{a}_n = (a_n, \cdots, a_1)$ and E_n has the following simple properties:

(i) $E_n = E_n'$ and $E_n^2 = I_n$, where I_n is the $n \times n$ identity matrix.

(ii) for an $n \times n$ (real) Toeplitz matrix Γ_n,

$$\Gamma_n = E_n \Gamma_n E_n,$$

and when Γ_n is invertible

$$\Gamma_n^{-1} E_n = E_n \Gamma_n^{-1}. \tag{7.18}$$

The following lemma shows the simplicity of the idea behind the celebrated Durbin-Levinson algorithm; see Levinson (1947); Durbin (1960).

Lemma 7.2. Suppose for a fixed $n \geq 1$, the n-th order system

$$\Gamma_n \boldsymbol{y} = (\gamma_1, \cdots, \gamma_n) = \boldsymbol{\gamma}_n, \tag{7.19}$$

is solved and Γ_{n+1} is invertible. Then,

(a) The solution of the $(n+1)$-st order system,

$$\begin{bmatrix} \Gamma_n & E_n\gamma_n \\ \gamma_n' E_n & \gamma_0 \end{bmatrix} \begin{bmatrix} z \\ \alpha \end{bmatrix} = \begin{bmatrix} \gamma_n \\ \gamma_{n+1} \end{bmatrix}, \tag{7.20}$$

is given by

$$\begin{cases} \alpha = (\gamma_0 - \gamma_n' \boldsymbol{y})^{-1}(\gamma_{n+1} - \gamma_n' E_n \boldsymbol{y}), \\ z = \boldsymbol{y} - \alpha E_n \boldsymbol{y}. \end{cases}$$

(b) If $\Gamma_n \boldsymbol{x} = \boldsymbol{b} = (b_1, \cdots, b_n)$ and (7.19) is solved, then

$$\begin{bmatrix} \Gamma_n & E_n\gamma \\ \gamma_n' E_n & \gamma_0 \end{bmatrix} \begin{bmatrix} \boldsymbol{\nu} \\ \beta \end{bmatrix} = \begin{bmatrix} \boldsymbol{b} \\ b_{n+1} \end{bmatrix},$$

is solved by

$$\begin{cases} \beta &= (b_{n+1} - \gamma_n' E_n \boldsymbol{x})/(\gamma_0 - \gamma_n' E_n \boldsymbol{x}), \\ \boldsymbol{\nu} &= \boldsymbol{x} - \beta E_n \boldsymbol{y}. \end{cases}$$

Proof. (a) From (7.20) we have,

$$\begin{cases} \Gamma_n z + \alpha E_n \gamma_n = \gamma_n, \\ \gamma_n' E_n z + \gamma_0 \alpha = \gamma_{n+1}, \end{cases}$$

which in conjunction with (7.19) implies

$$\begin{cases} z = \Gamma_n^{-1}\gamma_n - \alpha \Gamma_n^{-1} E_n \gamma_n = \boldsymbol{y} - \alpha \Gamma_n^{-1} E_n \gamma_n, \\ \gamma_0 \alpha = \gamma_{n+1} - \gamma_n' E_n z. \end{cases} \tag{7.21}$$

Since Γ_n is invertible we find from (7.18) and the first equation in (7.21) that

$$z = \boldsymbol{y} - \alpha E_n \boldsymbol{y},$$

as desired. Substituting this in the second equation in (7.21) we obtain

$$\gamma_0 \alpha = \gamma_{n+1} - \gamma_n' E_n \boldsymbol{y} + \alpha \gamma_n' E_n^2 \boldsymbol{y}. \tag{7.22}$$

Since $E_n^2 = I$, from (7.22) we get

$$(\gamma_0 - \gamma_n' y)\alpha = \gamma_{n+1} - \gamma_n' E_n y,$$

and the result follows. Proof of (b) is similar to (a).

Q.E.D.

To obtain the Durbin-Levinson algorithm from Lemma 7.2, replace $\boldsymbol{y}$ by $\boldsymbol{a}_n$, $\begin{bmatrix} z \\ \alpha \end{bmatrix}$ by $\boldsymbol{a}_{n+1}$ and take $\{\gamma_k\} \in I(\Gamma)$.

Theorem 7.3. (Durbin-Levinson Algorithm) Let $\{X_t\}$ be a real-valued zero-mean stationary process with the covariance function $\{\gamma_k\} \in I(\Gamma)$. Then, given that $\boldsymbol{a}_n = (a_{1,n}, \cdots, a_{n,n})$ solves (7.2), the $\boldsymbol{a}_{n+1}$ is obtained using

(i) $a_{n+1,n+1} = (\gamma_0 - \boldsymbol{\gamma}'_n \boldsymbol{a}_n)^{-1}(\gamma_{n+1} - \boldsymbol{\gamma}'_n E_n \boldsymbol{a}_n),\ n = 1, 2, \cdots$.

(ii) $\boldsymbol{a}_{n,n+1} = \boldsymbol{a}_n - a_{n+1,n+1} E_n \boldsymbol{a}_n,\ n = 1, 2, \cdots$, where $\boldsymbol{a}_{n,n+1}$, is the vector of first n elements of $\boldsymbol{a}_{n+1}$.

While the solution of a general linear system of order n requires $O(n^3)$ operations, that is, the number of multiplications required is bounded by a multiple of n^3, the Durbin-Levinson algorithm taking advantage of the Toeplitz structure of the matrix Γ_n in (7.2) solves such a system in $O(n^2)$ operations, which is considerably faster when n is large.

A recursive formula for computing σ_n^2 is given next. It indicates the distinguished role of the sequence $\{a_{n,n}\}$ in the computational problems related to the process $\{X_t\}$ in Theorem 7.3 (ii).

Corollary 7.4. With assumption and notation as above, for $n \geq 1$,

(i) $\sigma_n^2 = (1 - a_{n,n}^2)\sigma_{n-1}^2$, $\sigma_0^2 = \gamma_0$.

(ii) $\sigma_n^2 = \gamma_0 \Pi_{k=1}^n (1 - a_{k,k}^2)$ and $|a_{n,n}| < 1$.

Proof. Let the column vector $\boldsymbol{a}_{n-1,n}$ denote the vector of the first $n - 1$ elements of $\boldsymbol{a}_n$. We get from (7.3) that

$$\begin{aligned}
\sigma_n^2 &= \gamma_0 - \boldsymbol{\gamma}'_n [\boldsymbol{a}_{n-1,n} a_{n,n}] = \gamma_0 - \boldsymbol{\gamma}'_{n-1} \boldsymbol{a}_{n-1,n} - \gamma_n a_{n,n} \\
&= \gamma_0 - \boldsymbol{\gamma}'_{n-1}(\boldsymbol{a}_{n-1} - a_{n,n} E_{n-1} \boldsymbol{a}_{n-1}) - \gamma_n a_{n,n} \\
&= \gamma_0 - \boldsymbol{\gamma}'_n \boldsymbol{a}_{n-1} + a_{n,n} \boldsymbol{\gamma}'_n E_{n-1} \boldsymbol{a}_{n-1}) - \gamma_n a_{n,n} \\
&= \sigma_{n-1}^2 - a_{n,n}(\gamma_n - \boldsymbol{\gamma}'_{n-1} E_{n-1} \boldsymbol{a}_{n-1}) \\
&= \sigma_{n-1}^2 - a_{n,n}^2 \sigma_{n-1}^2 = (1 - a_{n,n}^2)\sigma_{n-1}^2,
\end{aligned}$$

where in the third equality we have replaced $\boldsymbol{a}_{n-1,n}$ by its value from Theorem 7.3 (ii) and in the fifth equality we have replaced the quantity in the parenthesis by its value from Theorem 7.3 (i). (ii) is immediate from (i).

Q.E.D.

Now we apply the Durbin-Levinson algorithm to few concrete cases.

Example 7.1. (a) Let $\{X_t\}$ be a zero-mean, real-valued stationary process with the covariance function $\gamma_k = a^{|k|}, |a| < 1, k \in \mathbb{Z}$. Then, $a_{1,1} = \frac{\gamma_1}{\gamma_0} = a$, and from Theorem 7.3 we find that

$$a_{2,2} = (1 - a^2)^{-1}(a^2 - a^2) = 0,\ a_{1,2} = a_{1,1} - a_{2,2} a_{1,1} = a,$$

and finally, $\boldsymbol{a}_n = (a, 0, \cdots, 0)$, for $n \geq 1$. Thus, $\hat{X}_{n+1,n} = aX_n, \sigma_n^2 = 1 - a^2$ for all $n \geq 1$.

(b) Let $\{X_t\}$ be a stationary process with the covariance function

$$\gamma_0 = 2, \gamma_{\pm 1} = -1, \ \gamma_k = 0, \ \text{for } |k| \geq 2.$$

Here, $a_{1,1} = \frac{\gamma_1}{\gamma_0} = -\frac{1}{2}$, and from Theorem 7.3 (i) $a_{n+1,n+1} = (2+a_{1,n})^{-1} a_{n,n}, n = 1, 2, \cdots$. and for $n = 2, 3, \cdots$ we obtain

$$\begin{aligned} \boldsymbol{a}_2 &= (a_{1,2}, a_{2,2}) = \left(-\frac{2}{3}, -\frac{1}{3}\right), \\ \boldsymbol{a}_3 &= (a_{1,3}, a_{2,3} a_{3,3}) = \left(-\frac{3}{4}, -\frac{1}{2}, -\frac{1}{4}\right). \\ \boldsymbol{a}_4 &= \left(-\frac{4}{5}, -\frac{3}{5}, -\frac{2}{5}, -\frac{1}{5}\right). \end{aligned}$$

In general, using induction

$$a_{k,n} = -\left(1 - \frac{k}{n+1}\right), n = 1, 2, \cdots, 1 \leq k \leq n. \tag{7.23}$$

In this case, the predictor of X_{n+1} based on the n observations $X_1, \cdots, X_n$ is given by

$$\hat{X}_{n+1,n} = -\sum_{k=1}^{n} \left(1 - \frac{k}{n+1}\right) X_k, \tag{7.24}$$

with prediction error variance

$$\sigma_n^2 = 2\left(1 - \frac{1}{2^2}\right)\left(1 - \frac{1}{3^2}\right) \cdots \left(1 - \frac{1}{(n+1)^2}\right), n = 1, 2, \cdots. \tag{7.25}$$

This example and the representation (7.24) of the finite predictor are referred to in the sequel when discussing convergence of finite predictors as $n \to \infty$.

7.3 PARTIAL CORRELATIONS

In this section we show that the partial correlation function $\{\pi_n\}$ provides essentially an unconstrained reparameterization of the covariance function $\{\gamma_n\}$.

To motivate the notion of *partial correlation* we note that if the error of prediction of X_n based on $X_1, \cdots, X_{n-1}$, is reduced considerably by the inclusion of X_0, then of course, X_0 should be included in its prediction. The reduction in prediction error is measured by $a_{n,n}$ in Corollary 7.4 and is called the *partial correlation (coefficient)* between X_0 and X_n. It is, indeed, the correlation coefficient between X_0 and X_n after eliminating the effects of the intervening variables $X_1, \cdots, X_{n-1}$. More precisely, let $\hat{X}_{n,n-1}$ be the linear predictor of X_n based on $X_1, \cdots, X_{n-1}$ and set $\varepsilon_{n,n-1} = X_n - \hat{X}_{n-1,n}$. Similarly, define $\tilde{X}_{0,n-1}$ as the *(backward) predictor* of X_0 based on $X_1, \cdots, X_{n-1}$, and set

$$\tilde{\varepsilon}_{0,n-1} = X_0 - \tilde{X}_{0,n-1}. \tag{7.26}$$

From stationarity of $\{X_t\}$ and the definition of σ_n^2, we have for $n \geq 0$ that

$$\sigma_n^2 = \text{Var}(\varepsilon_{n+1,n}) = \text{Var}(\tilde{\varepsilon}_{0,n}). \tag{7.27}$$

Then, the *partial autocorrelation function* (PACF) of a mean-zero stationary process $\{X_t\}$, denoted by $\{\pi_n\}$, is

$$\pi_n = \text{Corr}(\varepsilon_{n,n-1}, \tilde{\varepsilon}_{0,n-1}), n = 1, 2, \cdots.$$

We show that the PACF $\{\pi_n\}$ is precisely the sequence $\{a_{n,n}\}$ appearing in the Durbin-Levinson algorithm.

Lemma 7.8. For $\{X_t\}$ a mean-zero real-valued stationary process with $\{\gamma_k\} \in I(\Gamma)$ we have

$$\pi_n = a_{n,n}, \; n = 1, 2, \cdots.$$

Proof. Since $\text{Cov}(\tilde{X}_{0,n-1}, X_n - \hat{X}_{n,n-1}) = 0$, from the definition of π_n and Theorem 7.3 (i), it follows that

$$\begin{aligned}
\pi_n &= \sigma_{n-1}^{-2}\text{Cov}(X_0, X_n - \hat{X}_{n+1,n}) = \sigma_{n-1}^{-2}\left[\gamma_n - \sum_{k=1}^{n} a_{k,n}\text{Cov}(X_0, X_k)\right] \\
&= \sigma_{n-1}^{-2}\left[\gamma_n - \sum_{k=1}^{n} a_{k,n}\gamma_{n-k}\right] = \sigma_{n-1}^{-2}[\gamma_n - \boldsymbol{\gamma}_{n-1}' E_n \boldsymbol{a}_{n-1}] = a_{n,n}.
\end{aligned}$$

Q.E.D.

Now that π_n is defined and can be computed recursively, we establish a one-to-one correspondence between $\{\gamma_k\}$ and $\{\pi_n\}$. This shows that $\{\pi_n\}$ provides an alternative, albeit simpler and unconstrained, parameterization of stationary processes than the covariance function (Ramsey, 1974). This is done by showing that to each sequence $\{\pi_n\}$ one can associate a positive-definite sequence $\{\gamma_k\}$ and vice versa.

Given a covariance function $\{\gamma_n\} \in I(\Gamma)$, the Durbin-Levinson algorithm generates its partial correlation function $\{\pi_n\}_{n=1}^{\infty}$ with $|\pi_n| < 1$, for all n. Conversely, given any such sequence $\{\pi_n\}$, one can define via Theorem 7.3, a unique positive-definite sequence $\{\gamma_k\}$ by

$$\gamma_0 = 1, \gamma_1 = \pi_1, \gamma_{n+1} = \sigma_n^2 \pi_{n+1} + \boldsymbol{\gamma}_n' E_n \boldsymbol{a}_n, \; n \geq 1. \tag{7.28}$$

From Corollary 7.4 (ii) and Lemma 7.1 (d) it follows that, for all $n \geq 1$,

$$|\Gamma_{n+1}| = \gamma_0 \Pi_{k=1}^{n}(1 - \pi_k^2)^{n+1-k} > 0,$$

(since $|\pi_k| < 1$) thus $\{\gamma_k\}$ is positive-definite.

To prove a general version of this result, we need the class S of all real-valued sequences $s = \{s_k, k = 0, 1, \cdots\}$ which satisfy

(a) $|s_k| \leq 1$ for $k = 0, 1, \cdots$, and

(b) $|s_k| = 1$ implies $s_{k+1} = s_k$.

As in (7.4) it is convenient to decompose S into

$$S = B(S) + I(S),$$

where $I(S)$ consists of all $s \in S$ for which (a) holds as a strict inequality for all k and $B(S)$ consists of sequences for which $|s_k| = 1$ for some k.

Theorem 7.9. There is a one-to-one correspondence between the partial correlation function $\{\pi_n\}_{n=1}^{\infty}$ and the covariance function $\{\gamma_k\}$ of stationary processes.

Proof. The one-to-one correspondence among elements of $I(\Gamma)$ and $I(S)$ were established earlier. Now let $\{\gamma_k\} \in B(\Gamma)$, then there exists a positive integer p for which $|\Gamma_k| = 0$ for all $k > p$ and $|\Gamma_k| > 0$ for all $k \le p$. It follows from Theorem 7.3 and Corollary 7.4 that there exists a unique sequence $\{\pi_k\}_{k=1}^{p}$ with $|\pi_k| < 1$ for $k < p$ and with $|\pi_p| = 1$. Letting $\pi_{p+k} = \pi_k, k = 1, 2, \cdots$, we arrive at a full sequence $\{\pi_k\} \in B(S)$. Conversely, let $\{\pi_k\} \in B(S)$. There exists a p for which $|\pi_k| < 1$ for all $k < p$ and $|\pi_k| = 1$ for all $k \ge p$. From (7.28), there is a unique solution $\{1, \gamma_1, \cdots, \gamma_p\}$ with $|\Gamma_k| > 0$ for $k \le p$ and $|\Gamma_{p+1}| = 0$. Let $(X_1, \cdots, X_{p+1})$ be a mean-zero random vector with the covariance matrix Γ_{p+1} (of rank p). Since, the residuals from regressions of X_1 on $\{X_2, \cdots, X_{p+1}\}$ and X_{p+1} on $\{X_1, \cdots, X_p\}$ are zero, we get

$$X_t = \sum_{j=1}^{p} \alpha_j X_{t-j}, \quad t = p+1, \tag{7.29}$$

$$X_t = \sum_{j=1}^{p} \beta_j X_{t+j}, \quad t = 1, \tag{7.30}$$

extending the sequence $X_1, \cdots, X_{p+1}$ according to the difference equation (7.29) for $t = p+2, p+3, \cdots$ and (7.30) for $t = 0, -1, -2, \cdots$, we obtain a stationary time series for which $\{\pi_k\}$ is its PACF. Clearly, the corresponding $\{\gamma_k\}$ is in $B(\Gamma)$.

Q.E.D.

In view of this theorem the PACF $\{\pi_n\}$ provides a reparameterization of stationary processes so that each π_n is free to vary over the interval $[-1, 1]$ independently of the others. This result was predated by Theorem 7.27 in the literature of orthogonal polynomials (Szegö and Grenander, 1958). In Section 7.5, this reparameterization is used to show that a stationary process is nondeterministic if and only if the sequence $\{\pi_n\}$ is square-summable. Readers more interested in such characterization results may proceed directly to Section 7.5. More practical applications of this reparameterization to simulation of ARMA models is discussed by Jones (1987) and to Bayesian computation of certain posteriors by Marriott and Smith (1992).

7.4 MULTIPLE AND CANONICAL CORRELATIONS

In this section we define multiple and canonical correlations for finite segments of a stationary process and indicate their roles in assessing various predictors.

7.4.1 Multiple Correlation

A different way to assess the accuracy of predictor of X_{n+1} based on $X_1, \cdots, X_n$ is through the *multiple correlation coefficient,* in symbols R_n^2, defined as the maximum correlation between X_{n+1} and all linear functions $\sum_{k=1}^{n} c_k X_k$ of the finite past. More precisely, for an $n \geq 1$, let $\boldsymbol{c} = (c_1, \cdots, c_n)$, then

$$R_n^2 = \max_{\boldsymbol{c} \neq 0} \frac{\text{Cov}\left(X_{n+1}, \sum_1^n c_k X_k\right)}{\sqrt{\text{Var}(X_{n+1})\text{Var}\left(\sum_1^n c_k X_k\right)}} = \max_{\boldsymbol{c} \neq 0} \frac{\boldsymbol{c}'\boldsymbol{\gamma}_n}{\sqrt{\gamma_0 \boldsymbol{c}'\Gamma_n \boldsymbol{c}}}.$$

We show the maximum

$$R_n = \sqrt{\boldsymbol{\gamma}_n'\Gamma_n^{-1}\boldsymbol{\gamma}_n/\gamma_0}, \tag{7.31}$$

is attained when $\boldsymbol{c} = (a_{1,n}, \cdots, a_{n,n}) = \boldsymbol{a}_n$.

To establish this, note that with $\boldsymbol{u} = \Gamma_n^{\frac{1}{2}}\boldsymbol{c}, \boldsymbol{v} = \Gamma_n^{-\frac{1}{2}}\boldsymbol{\gamma}_n$, and using the Cauchy-Schwarz inequality we have

$$\begin{aligned}
\frac{\boldsymbol{c}'\boldsymbol{\gamma}_n}{(\gamma_0 \boldsymbol{c}'\Gamma_n \boldsymbol{c})^{\frac{1}{2}}} &= \frac{\boldsymbol{c}'\Gamma_n^{\frac{1}{2}}\Gamma_n^{-\frac{1}{2}}\boldsymbol{\gamma}_n}{(\gamma_0 \boldsymbol{c}'\Gamma_n \boldsymbol{c})^{\frac{1}{2}}} = \frac{\boldsymbol{u}'\boldsymbol{v}}{(\gamma_0 \boldsymbol{c}'\Gamma_n \boldsymbol{c})^{\frac{1}{2}}} \\
&\leq \frac{(\boldsymbol{u}'\boldsymbol{u})^{\frac{1}{2}}(\boldsymbol{v}'\boldsymbol{v})^{\frac{1}{2}}}{(\gamma_0 \boldsymbol{c}'\Gamma_n \boldsymbol{c})^{\frac{1}{2}}} \\
&= (\boldsymbol{c}'\Gamma_n \boldsymbol{c})^{\frac{1}{2}}(\boldsymbol{\gamma}_n'\Gamma_n^{-1}\boldsymbol{\gamma}_n)^{\frac{1}{2}}/(\gamma_0 \boldsymbol{c}'\Gamma_n \boldsymbol{c})^{\frac{1}{2}} = (\boldsymbol{\gamma}_n'\Gamma_n^{-1}\boldsymbol{\gamma}_n/\gamma_0)^{\frac{1}{2}},
\end{aligned}$$

with equality if and only if $\boldsymbol{u} = \boldsymbol{v}$ or $\boldsymbol{c} = \Gamma_n^{-1}\boldsymbol{\gamma}_n$, which gives (7.31).

The importance of R_n in measuring the accuracy of prediction of X_{n+1} based on $X_1, \cdots, X_n$ and its relation with partial correlation follow from Lemma 7.1 (d) which yields

$$\sigma_n^2 = \gamma_0(1 - \boldsymbol{\gamma}_n'\Gamma_n^{-1}\boldsymbol{\gamma}_n/\gamma_0) = \gamma_0(1 - R_n^2), \tag{7.32}$$

or

$$R_n^2 = 1 - \Pi_{k=1}^n(1 - \pi_k^2). \tag{7.33}$$

Thus, $\{\pi_k\}$ can be used to compute $\{R_n\}$.

7.4.2 Canonical Correlations

The canonical correlation analysis between past and future of a time series has a broader goal than the multiple or partial correlation analysis. It seeks to identify and quantify the dependence between the *finite past* $\{X_{-m}, \cdots, X_{-1}\}$ and *finite future* $\{X_0, X_1, \cdots, X_{n-1}\}$, $m, n \geq 1$, by studying the correlation between linear combination of random variables in the past and future. This is done sequentially by first finding the pair of linear combinations having the largest correlation, and then finding successively other pair of linear combinations having the largest correlation among all pairs orthogonal to the earlier selected pairs. The pairs of selected linear combinations are called *canonical variables* and their correlations are called *canonical correlations*; see Anderson (1984). The latter measures the strength of dependence between past and future, and the former reduces high-dimensional relationships among many random variables into a few carefully selected and significant linear combinations. We present a method for computing canonical correlations and variables for general zero-mean random vectors $X = (X_1, \cdots, X_m)$ and $Y = (Y_1, \cdots, Y_n)$ with covariances

$$\text{Cov}(X) = \Sigma_{11}, \;\; \text{Cov}(Y) = \Sigma_{22} \text{ and } \text{Cov}(X, Y) = \Sigma_{12}.$$

This general setup is handy for the applications in time series analysis where m, n and hence mn become large with the passage of time. The $m \times n$ covariance matrix Σ_{12} contains all covariances between components of X and Y, when mn is large interpreting elements of Σ_{12} collectively is nearly impossible. In this setup the goal of canonical correlation analysis is to summarize the information in Σ_{12} in terms of few linear combinations. Let

$$U = \boldsymbol{a}'\boldsymbol{X}, \;\; V = \boldsymbol{b}'\boldsymbol{Y}$$

for some pairs of coefficient vectors $\boldsymbol{a}$ and $\boldsymbol{b}$. Then

$$\text{Var}(U) = \boldsymbol{a}'\Sigma_{11}\boldsymbol{a}, \;\; \text{Var}(V) = \boldsymbol{b}'\Sigma_{22}\boldsymbol{b}, \; \text{Cov}(U, V) = \boldsymbol{a}'\Sigma_{12}\boldsymbol{b}.$$

Our goal is to find vectors $\boldsymbol{a}$ and $\boldsymbol{b}$ so that

$$\text{Corr}(U, V) = \frac{\boldsymbol{a}'\Sigma_{12}\boldsymbol{b}}{\sqrt{\boldsymbol{a}'\Sigma_{11}\boldsymbol{a}}\sqrt{\boldsymbol{b}'\Sigma_{22}\boldsymbol{b}}}, \tag{7.34}$$

is as large as possible. Let us arrange the multiple solutions of this optimization problem as follows:

The *first pair of canonical variables* are the pair (U_1, V_1), having unit variances, which maximizes the correlation in (7.34). The *kth pair of canonical variables* are the pair (U_k, V_k), having unit variances, which maximize the correlation (7.34) among all choices which are orthogonal to $(U_j, V_j), 1 \leq j \leq k-1, k = 2, \cdots, min\,\{m, n\}$. The correlation ρ_k between the kth pair (U_k, V_k) is called the *kth canonical correlation* between X and Y. It is evident that $\rho_1 \geq \rho_2 \geq \cdots \geq \rho_m \geq 0$. Great simplification occurs when the first few canonical correlations are positive and the rest are zero.

Usually the first canonical correlation, in symbols ρ is simply referred to as the *canonical correlation* or *maximal correlation*.

Next, we give the details for computing the canonical correlations and variates when Σ_{11} and Σ_{22} are invertible and $m \le n$. Let $\rho_1 \ge \rho_2 \ge \cdots \ge \rho_m$ and $\boldsymbol{e}_1, \cdots, \boldsymbol{e}_m$ be the eigenvalues and the corresponding normalized eigenvectors of the $m \times m$ matrix $M = \Sigma_{11}^{-1/2}\Sigma_{12}\Sigma_{22}^{-1}\Sigma_{12}'\Sigma_{11}^{-1/2}$. (It can be shown that $\rho_1, \cdots, \rho_m$ are also the m largest eigenvalues of the $n \times n$ matrix $N = \Sigma_{22}^{-1/2}\Sigma_{12}'\Sigma_{11}^{-1}\Sigma_{12}\Sigma_{22}^{-1/2}$ with the corresponding $n \times 1$ normalized eigenvectors $\boldsymbol{h}_1, \cdots, \boldsymbol{h}_m$.) We show that

$$\max_{\boldsymbol{a},\boldsymbol{b}} \operatorname{Corr}(U, V) = \rho_1 \tag{7.35}$$

is attained for

$$U_1 = \boldsymbol{e}_1'\Sigma_{11}^{-1/2}X, \; V_1 = \boldsymbol{h}_1'\Sigma_{22}^{-1/2}Y, \tag{7.36}$$

and the kth pair of canonical variates, $k = 2, \cdots, m$,

$$U_k = \boldsymbol{e}_k'\Sigma_{11}^{-1/2}X, \; V_k = \boldsymbol{h}_k'\Sigma_{22}^{-1/2}Y, \tag{7.37}$$

maximizes $corr(U_k, V_k)$ among those linear combinations orthogonal to $U_j, V_j, j = 1, \cdots, k-1$, i.e. $\rho_k = \operatorname{Corr}(U_k, V_k)$. Proof of these assertions is based on the following useful result.

Lemma 7.10. (Hilbert-Courant). Let A be an $m \times m$ positive definite matrix with eigenvalues $\lambda_1 \ge \lambda_2 \ge \cdots \ge \lambda_m > 0$ and corresponding normalized eigenvectors $\boldsymbol{e}_1, \cdots, \boldsymbol{e}_m$. Then,

(a) $\displaystyle\max_{\boldsymbol{x}\neq 0} \frac{\boldsymbol{x}'A\boldsymbol{x}}{\boldsymbol{x}'\boldsymbol{x}} = \lambda_1$ is attained when $\boldsymbol{x} = \boldsymbol{e}_1$,

$\displaystyle\min_{\boldsymbol{x}\neq 0} \frac{\boldsymbol{x}'A\boldsymbol{x}}{\boldsymbol{x}'\boldsymbol{x}} = \lambda_m$ is attained when $\boldsymbol{x} = \boldsymbol{e}_m$.

(b) $\displaystyle\max_{\boldsymbol{x}\perp \boldsymbol{e}_1,\cdots,\boldsymbol{e}_{k-1}} \frac{\boldsymbol{x}'A\boldsymbol{x}}{\boldsymbol{x}'\boldsymbol{x}} = \lambda_k$ is attained when $\boldsymbol{x} = \boldsymbol{e}_k$, $k = 2, \cdots, m$.

Proof. (a) Let P be the orthogonal matrix with columns $\boldsymbol{e}_1, \cdots, \boldsymbol{e}_m$ and $\Lambda = \operatorname{diag}(\lambda_1, \cdots, \lambda_m)$. For any $\boldsymbol{x} \neq 0$, set $\boldsymbol{y} = P'\boldsymbol{x}$, then

$$\frac{\boldsymbol{x}'A\boldsymbol{x}}{\boldsymbol{x}'\boldsymbol{x}} = \frac{\boldsymbol{y}'\Lambda\boldsymbol{y}}{\boldsymbol{y}'\boldsymbol{y}} = \frac{\sum_{i=1}^{m}\lambda_i y_i^2}{\sum_{i=1}^{m} y_i^2} \le \lambda_1. \tag{7.38}$$

But, for $\boldsymbol{x} = \boldsymbol{e}_1, \boldsymbol{y} = P'\boldsymbol{e}_1 = (1, 0, \cdots, 0)'$ and $\dfrac{\boldsymbol{y}'\Lambda\boldsymbol{y}}{\boldsymbol{y}'\boldsymbol{y}} = \lambda_1$, showing that the upper bound in (7.38) is attained. A similar argument gives the proof of the second assertion.

(b) Since $\boldsymbol{x} \perp \boldsymbol{e}_1, \cdots, \boldsymbol{e}_{k-1}$ and $\boldsymbol{x} = P\boldsymbol{y} = y_1\boldsymbol{e}_1 + \cdots + y_m\boldsymbol{e}_m$, it follows that

$$0 = \boldsymbol{e}_i'\boldsymbol{x} = y_1\boldsymbol{e}_i'\boldsymbol{e}_1 + y_2\boldsymbol{e}_i'\boldsymbol{e}_2 + \cdots + y_m\boldsymbol{e}_i'\boldsymbol{e}_m = y_i, \quad i \le k-1.$$

Thus, from the left-hand side of (7.38) we get

$$\frac{\boldsymbol{x}'A\boldsymbol{x}}{\boldsymbol{x}'\boldsymbol{x}} = \frac{\sum_{i=k}^{m} \lambda_i y_i^2}{\sum_{i=k}^{m} y_i^2},$$

taking $y_k = 1,\ y_{k+1} = \cdots = y_m = 0$ gives the desired result.

Q.E.D.

To establish (7.35)-(7.36), set $\boldsymbol{c} = \Sigma_{11}^{1/2}\boldsymbol{a},\ \boldsymbol{d} = \Sigma_{22}^{1/2}\boldsymbol{b}$, since $\mathrm{Var}(\boldsymbol{a}'\boldsymbol{X}) = \boldsymbol{c}'\boldsymbol{c} = 1 = \boldsymbol{d}'\boldsymbol{d} = \mathrm{Var}(\boldsymbol{b}'\boldsymbol{Y})$, from (7.34) and the Cauchy-Schwarz inequality we get

$$\mathrm{Corr}(\boldsymbol{a}'\boldsymbol{X},\ \boldsymbol{b}'\boldsymbol{Y}) = \boldsymbol{c}'\Sigma_{11}^{-1/2}\Sigma_{12}\Sigma_{22}^{-1/2}\,\boldsymbol{d} \le (\boldsymbol{c}'M\boldsymbol{c})^{1/2}(\boldsymbol{d}'\boldsymbol{d})^{1/2}, \tag{7.39}$$

where M is the $m \times m$ positive-definite matrix defined earlier. Applying Lemma 7.10 (a) to M yields

$$\boldsymbol{c}'M\boldsymbol{c} \le \lambda_1\boldsymbol{c}'\boldsymbol{c}, \tag{7.40}$$

where λ_1 is the largest eigenvalue of M. Equality occurs in (7.40) for $\boldsymbol{c} = \boldsymbol{e}_1$, the corresponding normalized eigenvectors of λ_1, and in (7.39) for $\boldsymbol{d}$ proportional to $\Sigma_{22}^{-1/2}\Sigma_{12}'\Sigma_{11}^{-1/2}\boldsymbol{e}_1$. Thus

$$\max_{\boldsymbol{a},\boldsymbol{b}} \mathrm{Corr}(\boldsymbol{a}'\boldsymbol{X}, \boldsymbol{b}'\boldsymbol{Y}) = \lambda_1,$$

with equality occurring for $\boldsymbol{a} = \Sigma_{11}^{-1/2}\boldsymbol{c} = \Sigma_{11}^{-1/2}\boldsymbol{e}_1$ and with $\boldsymbol{b} = \Sigma_{22}^{-1/2}\Sigma_{22}^{-1/2}\Sigma_{12}'\Sigma_{11}^{-1/2}\boldsymbol{e}_1 = \Sigma_{22}^{-1/2}\boldsymbol{h}_1$. The last equality follows by multiplying both sides of $M\boldsymbol{e}_1 = \lambda_1\boldsymbol{e}_1$ by $\Sigma_{22}^{-1/2}\Sigma_{12}'\Sigma_{11}^{-1/2}$ and identifying $(\lambda_1, \boldsymbol{h}_1)$ as an eigenvalue-eigenvector pair for the $n \times n$ matrix N introduced earlier. This shows that $U_1 = \boldsymbol{e}_1'\Sigma_{11}^{-1/2}\boldsymbol{X}, V_1 = h_1'\Sigma_{22}^{-1/2}\boldsymbol{Y}$ is the first pair of canonical variables with correlation $\rho_1 = \lambda_1$.

At the kth stage, we require $\boldsymbol{c} \perp \boldsymbol{e}_1, \boldsymbol{e}_2, \cdots, \boldsymbol{e}_{k-1}$. Lemma 7.10 (b) then yields

$$\boldsymbol{c}'M\boldsymbol{c} \le \lambda_k\boldsymbol{c}'\boldsymbol{c},$$

and from (7.39)

$$\mathrm{Corr}(\boldsymbol{a}'\boldsymbol{X}, \boldsymbol{b}'\boldsymbol{Y}) \le \lambda_k,$$

with equality from $\boldsymbol{c} = \boldsymbol{e}_k$ or $\boldsymbol{a} = \Sigma_{11}^{-1/2}\boldsymbol{e}_k$ and $\boldsymbol{b} = \Sigma_{22}^{-1/2}\boldsymbol{h}_k$. Thus $U_k = \boldsymbol{e}_k'\Sigma_{11}^{-1/2}\boldsymbol{X}$ and $V_k = \boldsymbol{h}_k'\Sigma_{22}^{-1/2}\boldsymbol{Y}$ are the kth canonical pair of random variables with correlation $\rho_k = \lambda_k$ as announced in (7.37).

Computation of $\Sigma_{ii}^{-1/2}$, $i = 1, 2$, appearing in (7.37) is not necessary in practice. Furthermore, to ease the computational burden, we may use the equations

$$\begin{cases} |\Sigma_{12}\Sigma_{22}^{-1}\Sigma_{12}' - \rho\Sigma_{11}| = 0, \\ \Sigma_{12}\Sigma_{22}^{-1}\Sigma_{12}'\boldsymbol{a} = \rho\Sigma_{11}\boldsymbol{a}, \\ \Sigma_{12}'\Sigma_{11}^{-1}\Sigma_{12}\boldsymbol{b} = \rho\Sigma_{22}\boldsymbol{b}, \end{cases} \tag{7.41}$$

to obtain the canonical correlations and the coefficient vectors $\boldsymbol{a}$ and $\boldsymbol{b}$.

For a stationary process with $\boldsymbol{X} = (\boldsymbol{X}_{-n}, \cdots, \boldsymbol{X}_{-1})$, $\boldsymbol{Y} = (\boldsymbol{X}_0, \cdots, \boldsymbol{X}_{n-1})$ we have $\Sigma_{11} = \Sigma_{22} = \Gamma_n$, so that solving (7.41) requires inversion of only one matrix. In Chapter 8, we show that for n large the matrix inversion can be avoided and replaced by the Cholesky decomposition of the covariance matrix of the stationary process.

7.5 CHARACTERIZATION OF NONDETERMINISTIC AND AR(p) PROCESSES

What are the necessary and sufficient conditions for a stationary process to be nondeterministic? Purely nondeterministic? ARMA(p, q)? AR(p)? In general, characterization of these properties in terms of the covariance function $\{\gamma_k\}$ is not easy. Exceptions to this rule, however, is the characterization of ARMA(p, q) processes in Section 4.5 which assumes a priority that the process is nondeterministic. As yet, unfortunately, there is no widely known characterization of nondeterministic processes in terms of covariance functions.

In this section, the PACF $\{\pi_n\}$ is used to characterize nondeterministic and AR(p) processes in the time-domain in a rather simple manner. The key idea is the observation that $\lim_n \sigma_n^2$ exists and from Corollary 7.4 (ii) the *innovation variance* can be expressed in terms of the PACF:

$$\sigma^2 = \gamma_0 \Pi_{k=1}^{\infty}(1 - \pi_k^2). \tag{7.42}$$

This is a useful alternative to the Kolmogorov-Szegö formula expressing the innovation variance in term of logarithm of the spectral density function. Now we use (7.42) to characterize nondeterministic processes in terms of square summability of $\{\pi_n\}$; see Degerine (1982).

Theorem 7.11. A stationary process with $\{\gamma_k\} \in I(\Gamma)$ and PACF $\{\pi_n\}$ is nondeterministic, if and only if

$$\sum_{n=1}^{\infty} \pi_n^2 < \infty.$$

Proof. A stationary process is nondeterministic, if and only if $\sigma^2 > 0$. In view of (7.42), this is equivalent to

$$\Pi_{k=1}^{\infty}(1 - \pi_k^2) > 0. \tag{7.43}$$

Since $0 \leq |\pi_k| < 1$ for all k, it follows that (7.43) is equivalent to $\sum_{k=1}^{\infty} |\pi_k|^2 < \infty$, see Problem 7.2.

Q.E.D.

Now we use $\{\pi_n\}$ to give two distinct characterizations of finite autoregressions.

Theorem 7.12. A stationary process with the PACF $\{\pi_n\}$ is a finite order autoregressive process, if and only if $\pi_n \neq 0$, for finitely many values of n.

Proof. Suppose $\{X_t\}$ is a causal AR(p) process for some finite p, that is, there are unique $a_1, \cdots, a_p$ and a white noise $\{\varepsilon_t\}$ such that $X_t = \sum_{k=1}^{p} a_k X_{t-k} + \varepsilon_t$. Then, for $n \geq p$ we have

$$\hat{X}_{t,n} = \sum_{k=1}^{p} a_k X_{t-k} = \hat{X}_t \text{ and } \sigma_n^2 = \sigma_p^2.$$

Now it follows from Corollary 7.4 (i) that $1 - \pi_n^2 = 1$ or $\pi_n = 0$ for $n \geq p$.

Conversely, suppose $\{\pi_n\}$ has finitely many non-zero elements. Then $\{X_t\}$ is nondeterministic by Theorem 7.11. Let $p = \max\{n \geq 1; \pi_n \neq 0\}$, note that for all $n \geq p$, $\boldsymbol{a}_n = (a_{1,p}, a_{2,p}, \cdots, a_{p,p}, 0, \cdots, 0)$, see Theorem 7.3 (ii), and $\hat{X}_{t,n} = \sum_{k=1}^{n} a_{k,n} X_{t-k} = \sum_{k=1}^{p} a_{k,p} X_{t-k}$. Thus, $\hat{X}_t = \lim_{n\to\infty} \hat{X}_{t,n} = \sum_{k=1}^{p} a_{k,p} X_{t-k}$, which shows that $\{X_t\}$ is an AR(p) process with the innovation $\varepsilon_t = X_t - \hat{X}_t$.

Q.E.D.

Our second characterization of AR(p) processes is in terms of their maximum entropy, see Burg (1967) and Landau (1987a). It shows that, among all Gaussian stationary processes sharing the same autocovariances $\gamma_0, \cdots, \gamma_p$, AR($p$) processes have the largest one-step ahead prediction error variance.

Theorem 7.13. Among all stationary processes with the same autocovariances $\gamma_0, \cdots, \gamma_p$, an AR($p$) process has the largest one-step ahead prediction error variance.

Proof. Since knowing $\{\gamma_0, \gamma_1, \cdots, \gamma_p\}$ is equivalent to knowing $\{\gamma_0, \pi_1, \cdots, \pi_p\}$, see Theorem 7.9, and

$$\sigma^2 = \gamma_0 \Pi_{k=1}^{\infty} (1 - \pi_k^2) \leq \gamma_0 \Pi_{k=1}^{p} (1 - \pi_k^2),$$

with equality, if and only if $\pi_k = 0$ for $k > p$. It follows from Theorem 7.12 that $\{X_t\}$ is an AR(p) process.

Q.E.D.

For more information and discussion on the maximum entropy principle see Section A7.8.

7.6 CONVERGENCE OF PREDICTORS

In this section we show that convergence of the finite predictors to their infinite counterpart, which is generally expected, leads to some unexpected results on the nature of the process, convergence of finite predictor coefficients and autoregressive representation of the process.

7.6.1 Convergence of Predictor Coefficients

This section shows that the finite predictor coefficients $a_{k,n}$, and $b_{k,n}$ converge to the AR and MA parameters as expected. But, it is surprising that the norm convergence of $\{b_{k,n}\}$ to $\{b_k\}$ is equivalent to $\{X_t\}$ being purely nondeterministic, and the norm convergence of $\{a_{k,n}\}$ to $\{a_k\}$ is related to the question of $\{X_t\}$ having *autoregressive representation* (ARR). The rates of convergence of these coefficients (Baxter's inequality), however, play important roles in deriving asymptotic distributions of estimators obtained from the method of autoregressive model fitting. The next three largely ignored results due to Degerine (1982, 1990, 1994) deserve better recognition.

Theorem 7.14. Let $\{X_t\}$ be a nondeterministic stationary process with AR and MA parameters $\{a_k\}$ and $\{b_k\}$, respectively. Then,

(a) $\lim_{n\to\infty} b_{k,n} = b_k,\ k = 1, 2, \cdots.$

(b) $\lim_{n\to\infty} a_{k,n} = a_k,\ k = 1, 2, \cdots.$

Proof. From (7.7), replacing k by $n - k$ and using stationarity, we get

$$\sigma_{n-k}^2 b_{k,n} = (X_{n+1}, X_{n-k+1} - \hat{X}_{n-k+1}) = (X_t, X_{t-k} - \hat{X}_{t-k,n}), \tag{1}$$

where $\hat{X}_{t-k,n}$ is the projection of X_{t-k} onto $sp\{X_{t-k-1}, \cdots, X_{t-k-n}\}$. Since $X_{t-k} - \hat{X}_{t-k}, n$ converges to ε_{t-k} in norm as $n \to \infty$, and from Theorem 6.8 (c), we have

$$(X_t, \varepsilon_{t-k}) = \sigma^2 b_k,$$

it follows from (1) that

$$\lim_{n\to\infty} \sigma_{n-k}^2 b_{k,n} = \sigma^2 b_k.$$

Now the result follows from $\lim_{n\to\infty} \sigma_{n-k}^2 = \sigma^2 > 0$.

We prove (b) by induction, note that $a_{0,n} = a_0 = 1$. Suppose that (b) is true for all $k \leq l - 1$, where l is an integer, then we need to show that $\lim_{n\to\infty} a_{l,n} = a_l$.

From (7.9), by rearranging the terms we get

$$a_{l,n} = b_{l,n} - \sum_{k=1}^{l-1} a_{k,n} b_{l-k,n-k},$$

from (a) and the assumption of the induction it follows that $\lim_{n\to\infty} a_{l,n}$ exists. Moreover, from (6.18) we obtain

$$\lim_{n\to\infty} a_{l,n} = b_l - \sum_{k=1}^{l-1} a_k b_{l-k} = a_l.$$

Q.E.D.

Norm-convergence of the sequences $\{a_{k,n}\}$, $\{b_{k,n}\}$ are important in proving limit theorems for the estimators of parameters of a stationary time series (Hannan, 1987) as well as characterizing PND processes. The following lemma plays an important role in establishing the norm convergence of $\boldsymbol{b}_n = (1, b_{1,n}, \cdots, b_{n,n}, 0, 0, \cdots)$ to $\boldsymbol{b} = (1, b_1, b_2, \cdots)$.

Lemma 7.15. Let $\{X_t\}$ be a nondeterministic stationary process with the Wold decomposition

$$X_t = \sum_{k=0}^{\infty} b_k \varepsilon_{t-k} + V_t,$$

and the *finite innovation representation*

$$X_t = \sum_{k=0}^{n} b_{k,n} \varepsilon_{t-k,n-k},$$

where

$$\varepsilon_{t-k,n} = X_{t-k} - \hat{X}_{t-k,n}, 0 \le k \le n.$$

Then, for $t \in \mathbb{Z}$,

$$\sum_{k=0}^{n} (b_{k,n} - b_k)\varepsilon_{t-k,n-k} \to V_t, \text{ as } n \to \infty.$$

Proof. For integers $m < n$, using the two representations for X_t given above we have

$$\begin{aligned} V_t - \sum_{k=0}^{n} (b_{k,n} - b_k)\varepsilon_{t-k,n-k} &= \sum_{k=0}^{m} b_k(\varepsilon_{t-k,n-k} - \varepsilon_{t-k}) \\ &\quad + \sum_{k=m+1}^{n} b_k \varepsilon_{t-k,n-k} - \sum_{k=m+1}^{\infty} b_k \varepsilon_{t-k} \\ &= I_1 + I_2 + I_3. \end{aligned}$$

Now,

$$||I_1|| \le \sum_{k=0}^{m} |b_k| ||\varepsilon_{t-k,n-k} - \varepsilon_{t-k}||,$$

$$||I_2||^2 = \sum_{k=m+1}^{n} b_k^2 \sigma_{n-k}^2 \le \sigma_0^2 \sum_{k=m+1}^{\infty} b_k^2,$$

and

$$||I_3||^2 = \sigma^2 \sum_{k=m+1}^{\infty} b_k^2 \le \sigma_0^2 \sum_{k=m+1}^{\infty} b_k^2.$$

For m fixed, $||I_1|| \to 0$ as $n \to \infty$, cf. Lemma 7.1(b). Since $\sum_{k=0}^{\infty} b_k^2 < \infty$, for any $\epsilon > 0$ we can choose m (large enough) so that

$$\left(\sigma_0^2 \sum_{k=m+1}^{\infty} b_k^2 \right)^{\frac{1}{2}} < \frac{\epsilon}{3},$$

and

$$||I_i|| < \frac{\epsilon}{3}, \; i = 1, 2, 3.$$

Therefore the desired result follows from $||I_1|| + ||I_2|| + ||I_3|| < \epsilon$.

Q.E.D.

It is instructive to note that the statement of the lemma provides a recipe for capturing V_t, from the remote past, in terms of finite prediction errors, finite predictor coefficients and MA parameters of the process.

Theorem 7.19. A nondeterministic stationary process $\{X_t\}$ is purely nondeterministic, if and only if

$$\lim_{n\to\infty} \sum_{k=0}^{\infty} (b_k - b_{k,n})^2 = \lim_{n\to\infty} ||\boldsymbol{b}_n - \boldsymbol{b}||_2^2 = 0.$$

Proof. In view of Lemma 7.18, it suffices to show that l^2-convergence of $\boldsymbol{b}_n$ to $\boldsymbol{b}$ is equivalent to the norm convergence of $R_{t,n} = \sum_{k=0}^{n} (b_{k,n} - b_k)\varepsilon_{t-k,n-k}$ to zero. For this note that

$$||R_{t,n}||^2 = \sum_{k=0}^{n} (b_k - b_{k,n})^2 \sigma_{n-k}^2 \le \sigma_0^2 \sum_{k=0}^{n} (b_k - b_{k,n})^2 \le \sigma_0^2 ||\boldsymbol{b}_n - \boldsymbol{b}||_2^2,$$

and conversely

$$\sum_{k=0}^{\infty} (b_k - b_{k,n})^2 \le \sum_{k=0}^{n} (b_k - b_{k,n})^2 \frac{\sigma_{n-k}^2}{\sigma^2} + \sum_{k=n+1}^{\infty} b_k^2 = \frac{1}{\sigma^2} ||R_{t,n}||^2 + \sum_{k=n+1}^{\infty} b_k^2,$$

from which the desired results follows.

7.6.2 Rates of Convergence

It is much harder to establish the norm convergence of $\boldsymbol{a}_n = (a_{1,n}, \cdots, a_{n,n}, 0, 0, \cdots)'$ to $\boldsymbol{a} = (a_1, a_2, \cdots)$; see Problem 6.3, since $\boldsymbol{a}$ is not always in l^2. The following results shed some light on the nature of the conditions and the importance of the size of $t_n = \left\|\sum_{n+1}^{\infty} a_k X_{t-k}\right\|$ in controlling the rate and mode of convergence of some quantities of interest in time series analysis and prediction theory.

Lemma 7.20. Let $\{X_t\}$ be a PND stationary process with ARR. Then, for any $n \geq 1$,

(a) $||\hat{X}_{t,n} - \hat{X}_t||^2 = ||\varepsilon_{t,n} - \varepsilon_t||^2 = \sigma_n^2 - \sigma^2 \leq t_n^2$

(b) $\sum_{k=1}^{n} |a_{k,n} - a_k|^2 \leq 4\lambda_{1,n}^{-1} t_n^2$, where $\lambda_{1,n}$ is the smallest eigenvalue of Γ_n.

(c) For any $1 \leq j < n$, $\sum_{k=1}^{n} |a_{k,n} - a_{k,n-j}|^2 \leq 4\lambda_{1,n}^{-1} t_n^2$, where $a_{k,n-j} = 0$ for $k > n - j$.

(d) For a fixed j, $1 \leq j < n$, $b_{\ell,n} - b_{\ell,n-j} = O(t_n)$ and $b_{\ell,n} - b_\ell = O(t_n)$.

Note: We write $a_n = 0(b_n)$ if the sequence $\{a_n/b_n\}$ is bounded.

Proof. (a) Note that

$$X_{n+1} = \sum_{i=1}^{n} a_i X_{n+1-i} + \sum_{i=n+1}^{\infty} a_i X_{n+1-i} + \varepsilon_{n+1}.$$

Hence from the definition of σ_n^2 it follows that

$$\begin{aligned} \sigma_n^2 - \sigma^2 &\leq \left\| X_{n+1} - \sum_{i=1}^{n} a_i X_{n+1-i} \right\|^2 - \sigma^2 \\ &= \left\| \varepsilon_{n+1} - \sum_{i=n+1}^{\infty} a_i X_{n+1-i} \right\|^2 - \sigma^2 \\ &= \left\| \sum_{i=n+1}^{\infty} a_i X_{n+1-i} \right\|^2 \end{aligned}$$

as desired. The last equality is a consequence of ε_{n+1} and $\sum_{i=n+1}^{\infty} a_i X_{n+1-i}$ being orthogonal.

(b) Let T_n stand for $\sum_{k=n+1}^{\infty} a_k X_{t-k}$. Then, using the fact that $x'\Gamma_n x \geq \lambda_{j,n} x'x$ for any vector x, we obtain

$$\begin{aligned}
\lambda_{1,n} \sum_{k=1}^{n} |a_{k,n} - a_k|^2 &\leq \left\| \sum_{k=1}^{n} (a_{k,n} - a_k) X_{t-k} \right\|^2 \\
&= \left\| \hat{X}_{t,n} - \left(\sum_{k=1}^{n} a_k X_{t-k} + T_n - T_n + \varepsilon_t - \varepsilon_t \right) \right\|^2 \\
&= \|\hat{X}_{t,n} - X_t + \varepsilon_t + T_n\|^2 \\
&= \|\hat{X}_{t,n} - \hat{X}_t + T_n\|^2 = \|\hat{X}_{t,n} - \hat{X}_t\|^2 + \|T_n\|^2 \\
&\quad +2(\hat{X}_{t,n} - \hat{X}_t, T_n) \leq 2t_n^2 + 2\|\hat{X}_{t,n} - \hat{X}_t\|\|T_n\| \\
&= 4t_n^2.
\end{aligned}$$

(c) Using arguments similar to (b) we have

$$\begin{aligned}
\lambda_{1,n} \sum_{k=1}^{n} |a_{k,n} - a_{k,n-j}|^2 &\leq \|\hat{X}_{t,n} - \hat{X}_{t,n-j}\|^2 \\
&= \|-X_t + \hat{X}_{t,n} + \varepsilon_t + X_t - \hat{X}_{t,n-j} - \varepsilon_t\|^2 = \|\hat{X}_t + \hat{X}_t - \hat{X}_{t,n-j}\|^2 \\
&= \|\hat{X}_{t,n} - \hat{X}_t\|^2 + \|\hat{X}_t - \hat{X}_{t,n-j}\|^2 + 2(\hat{X}_{t,n} - \hat{X}_t, \hat{X}_{t,n-j}) \leq 4t_n^2,
\end{aligned}$$

where the last inequality follows from the Cauchy-Schwartz inequality and (a).

(d) Since $b_{1,n} = a_{1,n}$, using (c) it follows that for a fixed $j, 1 \leq j < n, b_{1,n} - b_{1,n-j} = a_{1,n} - a_{1,n-j} = O(t_n)$. From (7.9) we have $b_{\ell,n} = \sum_{k=1}^{\ell} a_{k,n} b_{\ell-k,n-k}$, and $b_{\ell,n-j} = \sum_{k=1}^{\ell} a_{k,n-j} b_{\ell-k,n-k}$. Subtracting these we obtain

$$\begin{aligned}
b_{\ell,n} - b_{\ell,n-j} &= \sum_{k=1}^{\ell} (a_{k,n} b_{\ell-k,n-k} - a_{k,n-j} b_{\ell-k,n-k}) \\
&= \sum_{k=1}^{\ell} \{a_{k,n}(b_{\ell-k,n-k} - b_{\ell-k,n-j-k}) \\
&\quad +(a_{k,n} - a_{k,n-j}) b_{\ell-k,n-j-k}\}.
\end{aligned}$$

Now using (c) and the fact that $b_{1,n} - b_{1,n-j} = O(t_n)$, the corresponding result for $b_{\ell,n} - b_{\ell,n-j}$ follows by induction.

For a proof of the last half of (d), we use (7.9) and $b_\ell = \sum_{k=1}^{\ell} a_k b_{\ell-k}$ to obtain

$$\begin{aligned} b_{\ell,n} - b_\ell &= \sum_{k=1}^{\ell} a_{k,n} b_{\ell-k,n-k} - \sum_{k=1}^{\ell} a_k b_{\ell-k} \\ &= \sum_{k=1}^{\ell} (a_{k,n} - a_k) b_{\ell-k,n-k} + \sum_{k=1}^{\ell} a_k (b_{\ell-k,n-k} - b_{\ell-k}) \\ &= \sum_{k=1}^{\ell} (a_{k,n} - a_k) b_{\ell-k,n-k} + \sum_{k=1}^{\ell} a_k (b_{\ell-k,n-k} - b_{\ell-k,n}) \\ &\quad + \sum_{k=1}^{\ell} a_k (b_{\ell-k,n} - b_{\ell-k}). \end{aligned}$$

The first and second sums are $O(t_n)$ by (b) and the first half of (d). Since $b_1 = a_1$ and $b_{1,n} = a_{1,n}$, the third sum is $O(t_n)$ by the same inductive argument used earlier. Q.E.D.

The covariance function (spectrum) of a stationary process is said to be *bounded* if there are constants c and d with $0 < c \leq d < \infty$ so that for all n

$$cI_n \leq \Gamma_n \leq dI_n \tag{7.44}$$

This condition is equivalent to the *boundedness* of the spectral density function of the process, namely, $c \leq f(\lambda) \leq d$.

Theorem 7.21. Let $\{X_t\}$ be a stationary process with the boundedness property (7.44) and AR parameters $\{a_k\}$. Then,

(a) $\boldsymbol{a}_n$ converges to $\boldsymbol{a}$ in the ℓ^2-norm, that is,

$$\lim_{n\to\infty} \sum_{k=1}^{n} (a_{k,n} - a_k)^2 = 0.$$

(b) If $\sum_{k=1}^{\infty} |a_k| < \infty$, then

$$\max_{1<k\leq n} |a_{k,n} - a_k|^2 \leq \sum_{k=1}^{n} (a_{k,n} - a_k)^2 \leq \frac{\gamma_0}{c} \sum_{k=n+1}^{\infty} |a_k|. \tag{7.45}$$

Proof of this theorem is immediate from Lemma 7.20 (b), (c) and (7.44). Note that (4.45) gives bounds for the ℓ^2-norm of $\boldsymbol{a}_n - \boldsymbol{a}$. A sharper ($\ell^1-$) version of this inequality is needed in proving certain asymptotic results of interest in the method of autoregressive model fitting, see Berk (1974). This result commonly referred to as the *Baxter's inequality*, Baxter (1962), holds under additional conditions on $\{\Gamma_n\}$, that are better stated in terms of the smoothness of the spectral density function of the process. We recall that a positive continuous function defined on a compact set is bounded in the above sense.

Theorem 7.22. Let $\{X_t\}$ be a stationary process with a positive and continuous spectral density function defined on $(-\pi, \pi]$. Then there exists an integer N and a constant c such that for all $n \geq N$,

$$\sum_{k=1}^{n} |a_{k,n} - a_k| \leq c \sum_{k=n+1}^{\infty} |a_k|. \tag{7.46}$$

The theory of orthogonal polynomials is used in proving the Baxter's inequality and obtaining rates of convergence of t_n stated in the next two theorems. We do not present their lengthy proofs in this book.

Example 7.2. To get a feel for the nature of convergence of $a_{k,n}$ to a_k, we take another look at Example 7.1 (a), (b). For the first example we have

$$\begin{aligned} a_1 &= a, \;\; a_k = 0, \;\; k \geq 1, \\ a_{1,n} &= a, \;\; a_{k,n} = 0, \;\; n \geq 2, \;\; n \geq k > 1. \end{aligned}$$

In this case, $a_{k,n} - a_k = 0$, for $n \geq 1$, $1 \leq k \leq n$, and $\max_{1 \leq k \leq n} |a_{k,n} - a_k| = 0$, showing that $a_{k,n}$ converges to a_k *exponentially and uniformly in* k as $n \to \infty$. While for the second example, we have

$$\begin{aligned} a_k &= -1, \; k \geq 1, \\ a_{k,n} &= 1 - \frac{k}{n+1}, \;\; n \geq 1, \; 1 \leq k \leq n, \\ a_{k,n} - a_k &= \frac{k}{n+1} \to 0 \text{ as } n \to \infty (\text{for a fixed } k). \end{aligned}$$

In this case $a_{k,n}$ converges to a_k rather slowly, and the convergence is not uniform in k. Indeed,

$$\max_{1 \leq k \leq n} |a_{k,n} - a_k| = \frac{n}{n+1} \to 1 \text{ as } n \to \infty.$$

The difference in these two examples can be traced to the magnitude of π_n or summability of the series $\sum_{n=1}^{\infty} |\pi_n|$. For the first example, $\sum_{n=1}^{\infty} |\pi_n| = |a| < \infty$, while for the second

$$\sum_{n=1}^{\infty} |\pi_n| = \sum_{n=1}^{\infty} \frac{1}{n+1} = \infty.$$

Another result in Baxter (1962), gives a necessary and sufficient condition for the exponential rate of convergence of $a_{k,n}$ to a_k, in terms of summability of the partial correlations. Its proof employs advanced results from the theory of orthogonal polynomials and is omitted.

Theorem 7.23. Let $\{a_{k,n}\}$ and $\{a_k\}$ be as before. Then, for each k, $a_{k,n}$ converges to a_k exponentially as $n \to \infty$, if and only if

$$\sum_{n=1}^{\infty} |\pi_n| < \infty.$$

Results on the polynomial and exponential rate of convergence of t_n, the tail of the predictor series are given next under additional smoothness conditions than those in Theorem 7.22. For more results of this type, see Ibragimov and Rozanov (1978).

Theorem 7.24. Let $\{X_t\}$ be a stationary process with the spectral density f. Then,

(a) $t_n = O(\delta^n)$, for some $0 \leq \delta < 1$, if and only if f coincides in $(-\pi, \pi]$ almost everywhere with a strictly positive analytic function (Rosenblatt and Grenander, 1954).

(b) $t_n = O(n^{-2p}), p > \frac{1}{2}, p$ non-integer, if and only if f coincides almost everywhere with a strictly positive continuous function which has an absolutely continuous derivative $f^{(r-1)}, r = [p]$, and an rth derivative $f^{(r)} \in L^2(-\pi, \pi]$ satisfying a Hölder condition of order $\alpha = p - [p]$. The latter means that

$$\sup_{t \leq h} \left(\int_{-\pi}^{\pi} |f^{(r)}(\lambda + t) - f^{(r)}(\lambda)|^2 d\lambda \right)^{\frac{1}{2}} = O(h^{\alpha}).$$

As noted earlier proofs of the previous three theorems rely on the theory of orthogonal polynomials, for a direct proof see Cheng and Pourahmadi (1993) and references therein. A brief introduction to this theory with a geometric flavor is provided next as an appendix.

A7.7 SZEGÖ's ORTHOGONAL POLYNOMIALS

Szegö's orthogonal polynomials have lives of their own with applications in pure and applied mathematics, physical and engineering sciences. Interestingly, some of the concepts and techniques developed in connection with these polynomials predate those developed in prediction theory and time series analysis. Szegö's orthogonal polynomials obtained by applying the Gram-Schmidt process to the exponentials $1, e^{i\lambda}, e^{2i\lambda}, \cdots$, in $L^2(F)$ are the isomorphs of the finite prediction errors of a stationary process with the spectral measure F. Nevertheless, there are certain questions of analytic nature that are better understood and can be answered more succinctly in the spectral-domain using the orthogonal polynomials; see Theorems 7.22–7.24. Two other notable examples are the nature of zeros of the characteristic polynomials of AR models fitted by solving the Yule-Walker equations and solution of the maximum entropy problem given a finite number of covariances.

Our presentation follows closely Landau's (1987a) which relies heavily on orthogonal decomposition of a sequence of increasing finite-dimensional subspaces generated by progressive segments of a covariance function. Interestingly, this is reminiscent of the approach in Section 5.2 employed to develop the Wold decomposition and is bound to reduce some of the analytic intricacies of the subject and to illuminate the connection with other apparently unrelated areas. This extended appendix should serve as a suitable preparation for the more advanced topics on Hilbert spaces and function theory developed in Chapters 9 and 10.

A7.7.1 An Inner Product for Polynomials

In this section a natural inner product is attached to a complex-valued covariance function.

The positive-definiteness can be thought of as defining an inner product on the linear space of finite sequences $a = (a_0, a_1, \cdots, a_n)$ or equivalently on the trigonometric polynomials $A(z) = \sum_{k=0}^{n} a_k z^k = \sum_{k=0}^{n} a_k e^{ik\lambda}, |z| = 1$, of degree n. Let $\mathcal{P}_n$ denote the $(n+1)$-dimensional space of such polynomials. More precisely, for fixed $n > 0$ assume $\gamma_0 = 1, \gamma_1, \cdots \gamma_n$ are given so that $\Gamma_{n+1} = (\gamma_{i-j})_{i,j=0}^{n}$ with $\gamma_{-k} = \bar{\gamma}_k$ is positive-definite. Then, introduce an inner product in $\mathcal{P}_n$ by setting, for $B = B(z) = \sum_{k=0}^{n} b_k z^k$,

$$\begin{cases} [A, B] = (a, \Gamma_{n+1} b) = (\Gamma_{n+1} a, b) = \overline{[B, A]} = \sum_{j,k} a_j \bar{b}_k \gamma_{j-k}, \\ \| A \|^2 = [A, A], \end{cases} \tag{7.47}$$

where $(\cdot, \cdot)$ is the usual inner product in $\mathbb{C}^{n+1}$. In the rest of this section for notational simplicity and compatibility with other subscripts we replace $n+1$ in Γ_{n+1} by n and bear in mind that Γ_n is, indeed, an $(n+1) \times (n+1)$ matrix. We refer to a as the *coefficient vector* of A, to a_n as the *leading coefficient* of A, and will say that A and B are *orthogonal* whenever $[A, B] = 0$.

The main advantage of defining this inner product in terms of trigonometric polynomials, rather than the vectors themselves, is that the Toeplitz nature of Γ_n is now succinctly expressed by

$$[zA, zB] = [A, B], \tag{7.48}$$

for $A, B \in \mathcal{P}_{n-1}$, which is a reincarnation of the definition of *stationarity*. An important consequence of (7.48) is that as progressively more covariances are given (increasing n), the corresponding Γ_n could define the scalar product on the increasing family $\{\mathcal{P}_n\}$ without *altering* it where previously specified. To this end, one must choose an orthogonal basis for $\mathcal{P}_n$ to span subspaces of polynomials of successively increasing degree. A natural candidate to generate such polynomials is the Gram-Schmidt process applied to $1, z, \cdots, z^n$, in the following inductive form. For connection with the Yule-Walker equations; see (7.17) and (7.19).

Suppose the basis for $\mathcal{P}_{n-1}$ are found and let $P_n(z) \in \mathcal{P}_n$ have coefficient vector $\boldsymbol{a}_n$ satisfying

$$\Gamma_n \boldsymbol{a}_n = (0, \cdots, 0, 1). \tag{7.49}$$

Since any polynomial $Q_{n-1} \in \mathcal{P}_{n-1} \subset \mathcal{P}_n$ has coefficient vector of the form $(*, \cdots, *, 0)$ where $*$ stands for a possibly non-zero entry, we have $[Q_{n-1}, P_n] = 0$ by (7.47). Thus, P_n is a scalar multiple of the desired n^{th} basis element. To normalize it, using (7.47) and (7.49) we obtain

$$\| P_n \|^2 = (\boldsymbol{a}_n, \Gamma_n \boldsymbol{a}_n) = \sigma_n^{-2} > 0, \tag{7.50}$$

so that $\sigma_n^{-1} P_n(z)$ which has degree n and leading coefficient σ_n^{-1} is the nth (normalized) orthogonal polynomial. The following analytic property of P_n derived from orthogonality and the fundamental theorem of algebra is noteworthy.

Lemma 7.25. All the zeros of $P_n(z)$ lie in $|z| < 1$, i.e. inside the unit circle.

Proof. Suppose α to be a zero of P_n, so that $P_n(z) = (z - \alpha)Q_{n-1}(z)$, or

$$P_n(z) + \alpha Q_{n-1}(z) = z Q_{n-1}(z),$$

for some $Q_{n-1} \in \mathcal{P}_{n-1}$. Since $P_n \perp Q_{n-1}$, taking norms of both sides and using (7.48), we arrive at

$$\| P_n \|^2 + |\alpha|^2 \| Q_{n-1} \|^2 = \| z Q_{n-1} \|^2 = \| Q_{n-1} \|^2,$$

from which $1 - |\alpha|^2 = \dfrac{\| P_n \|^2}{\| Q_{n-1} \|^2} > 0$ or $|\alpha| < 1$.

Q.E.D.

Thus far we know how to find $P_n(z)$ by solving the linear system (7.49). Our next major goal is to avoid inverting Γ_n in (7.49) by developing a recursion for $\{P_n\}$. Doing this, however, requires another sequence of orthogonal polynomials introduced in the next section.

A7.7.2 Evaluation Polynomials

In this section, we show that the operation which assigns to $A_n(z) \in \mathcal{P}_n$ its value at a fixed $\alpha \in \mathbb{C}$ gives rise to a unique sequence of orthogonal polynomials $\{E_n^\alpha\}$, $E_n^\alpha \in \mathcal{P}_n$. The polynomial E_n^α, called the *evaluation polynomial* for α or the *reproducing kernel of* $\mathcal{P}_n$, plays vital roles in the sequel in studying $\{P_n\}$ and solving many analytic problems related to stationary processes.

The starting point is the observation that *point evaluation* at $\alpha \in \mathbb{C}$ is a *bounded linear functional*, see Section 9.7.2, on the inner product space $\mathcal{P}_n$, namely, for $A_n, B_n \in \mathcal{P}_n$,

(i) $(A_n + B_n)(\alpha) = A_n(\alpha) + B_n(\alpha)$,

(ii) $|A_n(\alpha)| \leq \| A_n \|_2$.

Thus, using the Riesz representation theorem, it can be represented as the inner product

$$A_n(\alpha) = [A_n, E_n^\alpha], \tag{7.51}$$

where E_n^α is a unique element of $\mathcal{P}_n$. Some useful properties of E_n^α are immediate from (7.51).

Lemma 7.26. The evaluation polynomial E_n^α has the following properties:

(a) $\overline{E_n^\alpha(\beta)} = E_n^\beta(\alpha)$ and $E_n^\alpha(\alpha) = \| E_n^\alpha \|^2$.

(b) $E_n^\alpha(z) = \sum_{k=0}^{n} \sigma_k^2 \overline{P_k(\alpha)} P_k(z)$.

(c) $\| E_n^\alpha \|^{-2} = \inf_{Q_n(\alpha)=1} \| Q_n(\alpha) \|^2$

Proof. (a) From (7.51) we have

$$\begin{aligned} E_n^\alpha(\beta) &= [E_n^\alpha, E_n^\beta] = \overline{[E_n^\beta, E_n^\alpha]} = \overline{E_n^\beta(\alpha)}, \\ \| E_n^\alpha \|^2 &= [E_n^\alpha, E_n^\alpha] = E_n^\alpha(\alpha). \end{aligned}$$

(b) Let $E_n^\alpha = \sum_{k=0}^{n} a_k P_k$ be the decomposition of E_n^α in the orthogonal basis formed by $\{P_k\}$. Then, by taking the inner product of both sides with P_j and using (7.50), (7.47) and (7.51) we obtain

$$a_j \| P_j \|^2 = a_j \sigma_j^{-2} = [E_n^\alpha, P_j] = \overline{[P_j, E_n^\alpha]} = \overline{P_j(\alpha)}.$$

(c) Let $1 = Q_n(\alpha) = [Q_n, E_n^\alpha]$. Then, by the Cauchy-Schwarz inequality

$$1 = |[Q_n, E_n^\alpha]|^2 \leq \| Q_n \|^2 \, \| E_n^\alpha \|^2,$$

and equality is attained only if Q_n is a multiple of E_n^α, that is, for $Q_n(z) = \dfrac{E_n^\alpha(z)}{E_n^\alpha(\alpha)}$.

Q.E.D.

Now, let us single out $E_n^0(z)$, the evaluation polynomial at $\alpha = 0$ and show its special connection to $P_n(z)$. To this end, since $Q_n(0)$ is the constant term of $Q_n(z)$, (7.51) and (7.49) imply that the coefficient vector $\boldsymbol{b}_n$ of $E_n^0(z)$ satisfies

$$\Gamma_n \boldsymbol{b}_n = (1, 0, \cdots, 0). \tag{7.52}$$

Since Γ_n is Toeplitz and Hermitian, we find upon comparing (7.52) and (7.49) that $\boldsymbol{b}_n$ is obtainable from $\boldsymbol{a}_n$ by writing the components of $\bar{\boldsymbol{a}}_n$ in reverse order. Consequently, $P_n(z)$ determines $E_n^0(z)$ and vice versa. One can express this equivalency by

$$\begin{cases} E_n^0(z) &= z^n \bar{P}_n(1/z), \\ P_n(z) &= z^n \bar{E}_n^0(1/z). \end{cases} \tag{7.53}$$

From Lemma 7.26 (a) and (7.53) we have

$$\begin{cases} \| E_n^0 \|^2 = E_n^0(0) = \sigma_n^{-2} = \| P_n \|^2, \\ |E_n^0(e^{i\lambda})|^2 = |P_n(e^{i\lambda})|^2. \end{cases} \tag{7.54}$$

The importance of $E_n^0(z)$ is well-known in the literature of orthogonal polynomials (Szegö and Grenander, 1958) where it is often denoted by $P^*(z)$. The advantage of the derivation through (7.51) is its characterization by its action in the inner product space, rather than by its analytic form (7.53). Next, we show that the map from $(\gamma_0, \cdots, \gamma_n)$ to $E_n^0(z)$ is one-to-one.

Theorem 7.27. The evaluation polynomial $E_n^0(z)$ determines $(\gamma_0, \cdots, \gamma_n)$.

Proof. It is evident from (7.53) that $E_n^0(z)$ determines $P_n(z)$ and from Lemma 7.26 (b) that it also determines

$$E_n^0(z) - \overline{P_n(0)} P_n(z) \sigma_n^2 = E_{n-1}^0(z).$$

Proceeding in this manner, it is seen that $E_n^0(z)$ determines all the $P_k(z), k \leq n$. Thus, for $j \leq n$ the γ_j is determined from the expansion $z^j = \sum_{k=0}^{j} a_{j,k} P_k(z)$ by

$$\gamma_j = [z^j, 1] = [\Sigma a_{j,k} P_k, P_0] = a_{j,0}.$$

Q.E.D.

Theorem 7.9 due to Ramsey (1974), though proved much later, is a time-domain version of the above result.

A7.7.3 Durbin-Levinson, Burg and Schur Algorithms

In this section, using the basic shift-invariant property (7.48) of the inner product we derive recursions for computing the coefficient vector of $P_n(z)$ and introduce the Schur's parameters and other relevant quantities of interest in time series analysis, digital signal processing, and numerical linear algebra.

Durbin-Levinson Algorithm: We begin with the prediction-theoretic meanings of P_n, E_n^0 and σ_n^2. Since $\sigma_n^2 P_n(z)$ has leading coefficient 1 and is orthogonal to $\mathcal{P}_{n-1}$, it follows that $\| \sigma_n^2 P_n \| = \sigma_n$ measures the distance of z^n from $\mathcal{P}_{n-1}$ and is positive, if and only if Γ_n is positive-definite. From the isomorphism between $\{X_t\}$ and $\{e^{it\lambda}\}$, it is evident that $z^n - \sigma_n^2 P_n$ corresponds to the (forward) prediction error of length n and $1 - \sigma_n^2 E_n^0$ is the isomorph of the similar (backward) prediction error.

From (7.48), $\sigma_n^{-2} z P_n(z) \in \mathcal{P}_{n+1}$ is orthogonal to all polynomials of the form $z A_{n-1}$, so it needs a one-dimensional adjustment to be orthogonal to all of $\mathcal{P}_n$ and hence producing P_{n+1}. Specifically, $Q(z) = \sigma_{n+1}^2 P_{n+1}(z) - \sigma_n^2 z P_n(z) \in \mathcal{P}_n$, because their leading terms cancel and $[zA_{n-1}, Q] = 0$. Thus, Q and E_n^0 being orthogonal to the same $(n-1)$-dimensional subspace of $\mathcal{P}_n$ are proportional, namely,

$$\sigma_{n+1}^2 P_{n+1}(z) = \sigma_n^2 z P_n(z) + \pi_n \sigma_n^2 E_n^0(z), \tag{7.55}$$

for a constant π_n. It turns out π_n has considerable statistical and physical interpretations as soon as we recognize (7.55) as the *Durbin-Levinson algorithm* for generating $\{P_n\}$ recursively. It also leads to computational improvements in the inversion of Toeplitz matrices $\{\Gamma_n\}$ and other quantities of interest in time series analysis and digital signal processing.

As for the meaning of π_n, evaluating (7.55) at $z = 0$ and using (7.54) one obtains

$$\pi_n = \sigma_{n+1}^2 P_{n+1}(0) = \frac{P_{n+1}(0)}{\| E_{n+1}^0 \|^2} = \left[\frac{P_{n+1}}{\| P_{n+1} \|}, \frac{E_{n+1}^0}{\| E_{n+1}^0 \|} \right],$$

so that π_n is, indeed, the correlation coefficient between the normalized forward and backward prediction errors of length $n+1$. Alternatively, forming the inner product of (7.55) with E_n^0 and noting from (7.48) that $\| zP_n \|^2 = \| P_n \|^2 = \sigma_n^{-2}$, one obtains

$$\pi_n = -\left[\sigma_n^2 z P_n(z), E_n^0\right] = -\left[\frac{zP_n(z)}{\| zP_n \|}, \frac{E_n^0}{\| E_n^0 \|} \right], \tag{7.56}$$

so that $-\pi_n$ is also the correlation coefficient between the normalized backward prediction error of length n and forward prediction error of length n, advanced by one step.

To further understand $\{\pi_n\}$ and the range of is values, let us rewrite (7.55) as

$$\sigma_{n+1}^2 P_{n+1}(z) - \pi_n \sigma_n^2 E_n^0(z) = \sigma_n^2 z P_n(z),$$

in which the components on the left are orthogonal. Then, on taking norms and using (7.48) and (7.54), we find that

$$\sigma_{n+1}^2 + \sigma_n^2 |\pi_n|^2 = \sigma_n^2,$$

or

$$\sigma_{n+1}^2 = \sigma_n^2 (1 - |\pi_n|^2), \tag{7.57}$$

which is the result in Corollary 7.4 (i).

Suppose that Γ_{n+1} is positive-definite so that P_{n+1} is well-defined, then it follows from (7.57) that $|\pi_n| < 1$. Conversely, for each π_n with $|\pi_n| < 1$, one can define σ_{n+1}^2 by (7.57) and P_{n+1} by (7.55). Such P_{n+1} is automatically orthogonal to all polynomials of the form $zA_{n-1}(z)$, to be orthogonal to all of $\mathcal{P}_n$ it needs only be orthogonal to 1, a condition which determines γ_{n+1}, see proof of Theorem 7.27. Using this γ_{n+1}, the matrix Γ_{n+1} will be positive-definite and will generate P_{n+1} as the $(n+1)-st$ orthogonal polynomial. Thus, the choice of $\{\pi_0, \ldots, \pi_n, \ldots\}$ with $|\pi_k| < 1$ corresponds in a 1-to-1 way to positive-definite Toeplitz matrices $\{\Gamma_n\}$.

Finally, using (7.53), one can also write (7.55) as

$$\sigma_{n+1}^2 E_{n+1}^0(z) = \sigma_n^2 E_n^0(z) + \bar{\pi}_n \sigma_n^2 z P_n(z), \tag{7.58}$$

so that these two recursions together represent a way of generating the successive forward and backward predictors by means of the constants $\{\pi_n\}$. These equations have been viewed as modeling wave propagation in a stratified medium, in which the wave is partially transmitted and partially reflected at each layer boundary. As such, they have applications in geophysics, inverse scattering, modeling vocal tract and lossless transmission lines (Landau, 1987a,b; Kailath 1987).

Burg's Algorithm: The above construction of Γ_{n+1} from $\{\pi_0, \cdots, \pi_n\}$ combined with (7.56) or its equivalent

$$\pi_n = \frac{-2[zP_n, E_n^0]}{\| zP_n \|^2 + \| E_n^0 \|^2} \tag{7.59}$$

is at the heart of *Burg's algorithm* for generating the orthogonal polynomials from the sample data $x_1, \cdots, x_n$. To this end, if $\pi_0, \cdots, \pi_{n-1}$ are known (estimated), then P_{n+1} and E_{n+1}^0 are determined using (7.58) and (7.55) with π_n estimated using (7.59) by averaging over all the consecutive segments of lengths $n+1$ which the data may contain. Due to practical importance of this algorithm, we make a digression to motivate and explain it using the data.

The (forward) least-squares and Yule-Walker estimates of the AR parameters obtained by minimizing the sum of prediction errors in Sections 2.5 and 7.1 are not *time-reversible*, in that reversing the order of the data sequence changes the values of the estimators. This can be remedied by using the *forward-backward least-squares* or the *Burg's estimator* by minimizing the sum of forward and backward prediction errors, see Section 7.3, as a function of $a_{p,p}$:

$$Q_p(a_{p,p}) = \sum_{t=p+1}^{n} \left(X_t - \sum_{i=1}^{p} a_{i,p} X_{t-i} \right)^2 + \sum_{t=1}^{n-p} \left(X_t - \sum_{i=1}^{p} a_{i,p} X_{t+i} \right)^2 . \tag{7.60}$$

Burg's algorithm proceeds by minimizing successively, $Q_1(a_{1,1}), Q_2(a_{2,2}), \cdots$, $Q_p(a_{p,p})$. Note that after fitting a model of order $p-1$, by the Durbin-Levinson recursion (Theorem 7.3 (ii)), we have

$$a_{j,p} = a_{j,p-1} - a_{p,p} a_{p-j,p-1}, j = 1, \cdots, p-1,$$

so that all parameters of order p are determined by $a_{p,p}$ and those of order $p-1$. Thus, $Q_p(\cdot)$ as a function of $a_{p,p}$ when minimized gives the following analogue of (7.59),

$$\hat{a}_{p,p} = -\frac{2B_p}{A_p + C_p}, \tag{7.61}$$

where

$$\begin{cases} A_p = \sum_{i=1}^{p}\sum_{j=1}^{p} \hat{a}_{p-i,p-1}\hat{a}_{p-j,p-1}d_{ij}, \\ B_p = \sum_{i=0}^{p-1}\sum_{j=1}^{p} \hat{a}_{i,p-1}\hat{a}_{p-j,p-1}d_{ij}, \\ C_p = \sum_{i=0}^{p-1}\sum_{j=0}^{p-1} \hat{a}_{i,p-1}\hat{a}_{j,p-1}d_{ij}, (\hat{a}_{0,p-1} = -1), \end{cases} \tag{7.62}$$

and d_{ij}'s are as in (4.68).

In the same vein, the Kay's method (Landau, 1987a) proceeds by maximizing the 'partial log-likelihood' of the AR(p) model with respect to $a_{p,p}$:

$$-2L_p(a_{p,p}) = n\log 2\pi\sigma^2 + \log \Pi_{i=1}^{p}(1-a_{i,i}^2)^{-1} + \frac{n}{\sigma^2}\sum_{i=0}^{p}\sum_{j=0}^{p} a_{i,p}a_{j,p}d_{ij}.$$

It can be shown that Kay's method is equivalent to minimizing

$$K_p(a_{p,p}) = n\log(A_p a_{p,p}^2 + 2B_p a_{p,p} + C_p) - p\log(1-a_{p,p}^2)$$

with respect to $a_{p,p} \in (-1, 1)$, where A_p, B_p and C_p are as above.

Schur's Algorithm: The Schur algorithm is closely tied with checking the positive-definiteness of a function $\{\gamma_k\}$ on the one hand, and efficient computation of $\{\pi_n\}$ and inversion of Toeplitz matrices on the other. It has numerous important applications in electrical engineering (Kailath, 1987) and consequences in function and operator theory (Gohberg, 1986).

It all began with Caratheodory and Toeplitz associating to a sequence $\gamma_0, \gamma_1, \cdots$, with $\gamma_0 = 1$ the function

$$\gamma(z) = 1 + 2\sum_{k=1}^{\infty}\gamma_k z^k,$$

and showing that $\{\gamma_k\}$ is positive-definite, if and only if $\gamma(z)$ is analytic and $Re\gamma(z) \geq 0$ for $|z| < 1$. This characterization was further illuminated by Schur in 1917 using the function

$$g(z) = \frac{\gamma(z)-1}{\gamma(z)+1}.$$

Evidently, $g(z)$ is analytic where $\gamma(z)$ is, and in $|z| < 1$,

$$g(z) \leq 1, \text{ if and only if } Re\gamma(z) \geq 0.$$

Thus, a positive-definiteness test for $\{\gamma_k\}$ is equivalent to a boundedness test for $g(z)$. Schur devised the following simple recursive test for checking boundedness of $g(\cdot)$.

Since $g(0) = 0$, by Schwarz's lemma $|g(z)| \leq |z|$, hence for $g_1(z) = g(z)/z$ we have $|g_1(z)| \leq 1$ and $\dfrac{g_1(z) - g_1(0)}{1 - g_1(z)\overline{g_1(0)}}$ takes $|z| \leq 1$ into itself and 0 onto 0, just like $g(z)$. By repeating this process, we get

$$g_{n+1}(z) = \frac{g_n(z) - g_n(0)}{z(1 - g_n(z)\overline{g_n(0)})}, n \geq 1, \tag{7.63}$$

and the quantities $\{g_n(0)\}$. Using this sequence of functions it can be proved that $g(z)$ is bounded, if and only if $|g_n(0)| < 1$, for all $n \geq 0$. It turns out that $\{g_n(0)\}$ is precisely the sequence $\{\pi_n\}$ encountered in (7.55) and later. To see this, consider the quotient $B_n(z) = \dfrac{P_n(z)}{E_n^0(z)}$ with $B_0(z) = 1$. From (7.55) and (7.58) it follows that

$$B_{n+1}(z) = \frac{zB_n(z) + \pi_n}{1 + \bar{\pi}_n z B_n(z)},$$

or

$$B_n(z) = \frac{B_{n+1}(z) - B_{n+1}(0)}{z(1 - B_{n+1}(z)\overline{B_{n+1}(0)})},$$

which is the Schur sequence in the reverse order.

A7.8 TRUNCATED MOMENT PROBLEM

Often, in practice from design considerations or the available data one knows or can estimate the first $m+1$ covariances $\gamma_0, \gamma_1, \cdots, \gamma_m$ of a stationary process. We call any spectral distribution F with

$$\gamma_k = \frac{1}{2\pi}\int_{-\pi}^{\pi} e^{ik\lambda} dF(\lambda), 0 \leq k \leq m, \tag{7.64}$$

a spectrum consistent with the data. Let us denote by $\mathcal{F}_m$ the collection of all such distributions and note that it consists of all solutions to the *truncated moment problem* (7.64) in which (2.36) holds only for $\gamma_0, \cdots, \gamma_m$. How should one select an appropriate distribution from $\mathcal{F}_m$ on the strength of the available information?

There are various optimality criteria to guide selection giving prominence to certain members of $\mathcal{F}_m$. In this section, we discuss two apparently distinct criteria that lead to the same extension of $\gamma_0, \cdots, \gamma_m$ to a complete sequence of covariances corresponding to an AR(m) process.

A7.8.1 Worst Linear Prediction

In this section we show that $\mathcal{F}_m$ contains a member with the spectral density

$$f_m(\lambda) = \frac{E_m^0(0)}{|E_m^0(e^{i\lambda})|^2} = \frac{1}{\sigma_m^2 |P_m(e^{i\lambda})|^2}, \tag{7.65}$$

corresponding to an AR(m) process with the largest one-step ahead prediction error variance.

Lemma 7.28. Let F_m be the spectral distribution corresponding to (7.65). Then,

(a) $F_m \in \mathcal{F}_m$.

(b) With $\{c_k\}$ denoting the covariances corresponding to F_m, that is,

$$c_k = \int_{-\pi}^{\pi} e^{ik\lambda} dF_m(\lambda), k \geq 0,$$

and for the inner product defined by $\{c_k\}$ in (7.47) we have

$$E_n^0(z) = E_m^0(z), n \geq m. \tag{7.66}$$

Proof. We need to verify that $c_k = \gamma_k, 0 \leq k \leq m$. But this follows from (b), for if $c_0, \cdots, c_m$ and $\gamma_0, \cdots, \gamma_m$ give rise to the same $E_m^0(z)$, then they must coincide by Theorem 7.27. (b) Using the integral representation of the inner product in $L^2(F_m)$ and the definition of f_m, we have for any polynomial Q_n,

$$\begin{aligned}
[Q_n, E_m^0] &= \frac{1}{2\pi} \int_{-\pi}^{\pi} Q_n(e^{i\lambda}) \overline{E_m^0(e^{i\lambda})} f_m(\lambda) d\lambda \\
&= \frac{1}{2\pi} \int_{-\pi}^{\pi} \frac{Q_n(e^{i\lambda})}{E_m^0(e^{i\lambda})} E_m^0(0) d\lambda \\
&= \frac{1}{2\pi i} \int_{|z|=1} \frac{Q_n(z) E_m^0(0)}{E_m^0(z)} \frac{dz}{z}.
\end{aligned}$$

By (7.53), the zeros of $E_m^0(z)$ are given by $1/\bar{\alpha}_i$, with α_i the zeros of $P_m(z)$ which lie in $|z| > 1$ (Lemma 7.25). Thus, the only singularity of the integrand is the pole at $z = 0$, so that by Cauchy's formula, $[Q_n, E_m^0] = Q_n(0)$ for each n and therefore E_m^0 coincides with E_n^0.

Q.E.D.

From (7.53) and using (7.66) it is immediate that

$$P_n(z) = z^{n-m} P_m(z), n \geq m. \tag{7.67}$$

The distinguished role of f_m with regard to prediction error variance is discussed next. To this end, from (2.49), σ_m^2 the finite prediction error variance of X_m based on $X_{m-1}, \cdots, X_0$ is given by

$$\sigma_m^2(F) = \inf_{c_k} \frac{1}{2\pi} \int_{-\pi}^{\pi} \left| 1 - \sum_{k=1}^{m} c_k e^{ik\lambda} \right|^2 dF(\lambda).$$

On expanding the integrand, it is evident that $\sigma_m^2(F)$ depends only on $\gamma_0, \cdots, \gamma_m$ and therefore is unchanged should F be replaced by any other member of $\mathcal{F}_m$. To compute σ_m^2, we note that $Q_m(z) = 1 - \sum_{k=1}^{m} a_k z^k \in \mathcal{P}_m$ with $Q_m(0) = 1$, and conversely any $Q_m(z)$ with $Q_m(0) = 1$ has the above form. Thus,

$$\sigma_m^2(F) = \inf_{Q_m(0)=1} \| Q_m \|^2 = \frac{1}{E_m^0(0)}, \tag{7.68}$$

where the second equality follows from Lemma 7.26 (c) and (a).

Let $\sigma^2(F)$ be the prediction error variance based on the infinite past corresponding to a spectral distribution F in $\mathcal{F}_m$. By monotonicity of $\{\sigma_m^2(F)\}$ we have

$$0 \leq \sigma^2(F) \leq \sigma_m^2(F) = 1/E_m^0(0). \tag{7.69}$$

The extreme distributions of $\mathcal{F}_m$ having simply $m+1$ jumps yield $\sigma_{m+1}^2(F) = 0 = \sigma^2(F)$ corresponding to perfect predictability. On the other hand, for the distribution corresponding to f_m in (7.65), for $n > m$ and from Lemma 7.28 (b) we have

$$\sigma_n^2(F_m) = 1/E_n^0(0) = 1/E_m^0(0) = \sigma_m^2(F_m),$$

not depending on n, and

$$\sigma^2(F_m) = \sigma_m^2(F_m).$$

Thus, it enjoys the opposite property, in that, it produces the worst prediction from the infinite past, consistent with the given correlations $\gamma_0, \cdots, \gamma_m$. This can be explained by (7.67) indicating that the spectrum f_m corresponds to a Markovian or an AR(m) process in which the entire past influences the present through only the proceeding m lags.

A7.8.2 Maximum Entropy

In this section the notion of entropy is introduced as a measure of randomness. Thus, intuitively larger entropy corresponds to less information. Using the entropy maximization principle, the spectrum f_m in (7.65) also appears as the solution to the truncated moment problem.

The *entropy* of the random vector $X = (X_1, \cdots, X_n)$ with the probability density $f_X(x_1, \cdots, x_n)$ is defined by

$$H_n(X) = -\int \cdots \int f_X(x_1, \cdots, x_n) \log f_X(x_1, \cdots, x_n) dx_1 \cdots dx_n.$$

The *entropy rate* of a strictly stationary process $X = \{X_t\}$ is defined as the limit of $\frac{1}{n} H_n(X)$, namely,

$$H(X) = \lim_{n\to\infty} \sup \frac{H_n(X)}{n}.$$

As usual, for zero-mean Gaussian vectors H_n is easy to compute, and moreover they maximize the entropy among the class of all processes with the same covariances

(Landau, 1987a; Cover and Thomas, 1990). To see the latter, let $Y = \{Y_t\}$ be a Gaussian process having the same covariance as $X = \{X_t\}$. Since $\log f_Y$ is quadratic, $\int f_X \log f_Y = \int f_Y \log f_Y$, and from the basic property

$$\int f_X \log \frac{f_Y}{f_X} \leq 0,$$

valid for any two densities by convexity of the logarithm it follows that for all $n \geq 1$,

$$H_n(X) \leq H_n(Y). \tag{7.70}$$

The *entropy maximization principle* seeks to maximize the entropy rate $H(X)$ of a process subject to the prescribed values of the $m+1$ covariances $\gamma_0, \cdots, \gamma_m$.

For a Gaussian stationary process, a direct computation of H_n gives

$$H_n(Y) = \frac{n}{2} \log 2\pi e |\Gamma_{n-1}|^{1/n}. \tag{7.71}$$

From Lemma 7.1 (c) and monotonicity of σ_m^2, we have for all $n \geq m$,

$$\frac{|\Gamma_{n+1}|}{|\Gamma_n|} \leq \sigma_m^2,$$

or

$$|\Gamma_{n+1}| \leq \sigma_m^{2(n+1-m)} |\Gamma_m|. \tag{7.72}$$

From (7.70)–(7.72) we obtain

$$H_n(X) \leq H_n(Y) \leq H_n(Y^{(m)}),$$

and therefore (Landau, 1987a, p. 58) we have

$$H(X) \leq H(Y) \leq H(Y^{(m)}),$$

where $\{Y_t^{(m)}\}$ is a zero-mean Gaussian AR(m) process with the first $m+1$ covariances $\gamma_0, \cdots, \gamma_m$, which attains the maximum entropy rate subject to the given information.

The preceding discussion reveals that an extremal problem involving entropy and a geometric one dealing with prediction and best approximation in a Euclidean space lead to the same solution. Traditionally, the connection between prediction and entropy maximization is established using the considerably deeper Kolmogorov-Szegö Theorem VII (b). In the presentation here, the relevant quantities are already linked in the finite-dimension. Indeed, since $\sigma_m^2 = 1/E_m^0(0)$ and $f_m(\lambda) = E_m^0(0)|E_m^0(z)|^{-2}$, the Kolmogorov-Szegö formula in this setup follows trivially by applying the mean-value theorem to the harmonic function $\log E_m^0(0)|E_m^0(z)|^{-2}$, see Landau (1987a, p. 58).

Problems

7.1 For the three covariance functions in Example 5.1 solve the equations (7.2)–(7.3) for $n = 1, 2, 3$.

7.2 **(a)** Show that if $\{a_n\}$ converges to a, then

$$\lim_{n\to\infty} \frac{a_1 + a_2 + \cdots + a_n}{n} = a.$$

(b) Provide the necessary details of the proof of Lemma 7.1.

(c) Suppose $0 \leq u_n < 1$. Show that

$$\Pi_{n=1}^{\infty}(1 - u_n) > 0 \text{ if and only if } \sum_{n=1}^{\infty} u_n < \infty.$$

7.3 **(a)** For the (finite order) harmonic process in Problem 5.5, find the $a_{k,n}, b_{k,n}$ and the PACF for $m = 0, 1, 2$. For m arbitrary, what is the value of π_n for $n > m$?

(b) For the MA(1) process $X_t = \varepsilon_t + \cdot 6\varepsilon_{t-1}$, $\text{Var}(\varepsilon_t) = 1$, find

$$\{a_{k,n}\} \text{ and } \{b_{k,n}\} \text{ for } n = 1, 2, 3.$$

(c) For the ARMA(1,1) process $X_t = \alpha X_{t-1} + \varepsilon_t + \beta\varepsilon_{t-1}$, $\text{Var}(\varepsilon_t) = 1$, $|\alpha| < |\beta| < 1, \alpha \neq \beta$, find $\{a_{k,n}\}$ and $\{b_{k,n}\}$ for $n = 1, 2$.

7.4 **(a)** Show that the Gram-Schmidt procedure applied to $\{X_n\}_{n=0}^{\infty}$ produces the sequence $\{\varepsilon_{n+1,n}\}_{n=0}^{\infty}$ in (7.5).

(b) Apply the Gram-Schmidt procedure to $\{X_n - \hat{X}_n\}_{n=0}^{\infty}$ and write the resulting orthogonal sequence in terms of the innovation $\{\varepsilon_t\}$, here $\hat{X}_n$ is the predictor of X_n based on the infinite past $\{X_t; t \leq -1\}$.

7.5 Extend the time series data $x_1, \cdots, x_n$ from a zero-mean process $\{X_t\}$ by defining $x_t = 0$ for $t \leq 0$ or $t > n$. Let us write $\mathbf{x}_t$ for the sequence $\{x_t; t = 0, \pm, \cdots\}$ or $\mathbf{x}_t = (\cdots, 0, x_1, \cdots, x_n, 0, \cdots)$ and the shifted sequence $\{x_{t-j}; t = 0, \pm 1, \cdots\}$ will be denoted by $\mathbf{x}_{t-j}$. The space S of real sequences with finitely many non-zero components is an inner-product space with the following inner product

$$< u, v >= \frac{1}{n}\sum_{t=1}^{n} u_t v_t, u, v \in S.$$

(a) Show that

$$< \mathbf{x}_{t+k}, \mathbf{x}_t >= \frac{1}{n}\sum_{t=1}^{n-k} x_t x_{t+k} = \hat{\gamma}_k, k = 0, 1, 2, \cdots.$$

(b) Write the normal equations for predicting $\mathbf{x}_{n+1}$ based on $\mathbf{x}_n, \cdots, \mathbf{x}_1$, and compare them with the sample Yule-Walker equations in Remark 7.2 (a). Develop the Durbin - Levinson algorithm for this case.

(c) (Data taper) Consider the sequences

$$\mathbf{x}_t = (\cdots, 0, h_1 x_n, \cdots, h_n x_n, 0, \cdots),$$

where h_t's are scalars with $\sum_{t=1}^{n} h_t^2 = n$, called a *data taper*. Repeat (a) and (b) above for a tapered time series; see Koopmans (1974, Chap. 9).

7.6 (a) In (7.15) for $n = m$, show that $c_{k,n} = d_{k,n}$, for all k, n, and $\mathbf{c}_n = (c_{1,n}, \cdots, c_{n,n})$ satisfies the equation

$$(\Gamma_n + H_n)\mathbf{c}_n = \gamma_n,$$

where the $n \times n$ matrix $H_n = (\gamma_{k+j})_{k,j=1}^{n}$ is a Hankel matrix.

(b) For $m = n = 2$, solve the normal equations in (a) for an ARMA(1,1) process. Write the interpolator formula (7.15) and compute the variance of the interpolation error $\sigma_{2,2}^2$.

(c) Find analogue of (7.16)–(7.17) for the predictor of X_{n+1} based on the incomplete past. What happens when $n = 0$?

7.7 (a) Derive the results in Theorem 7.3 by orthogonalizing X_1 and $sp\{X_2, \cdots, X_n\}$.

(b) Suppose that (7.2) is solved recursively using the Durbin-Levinson algorithm. Then, show that for any $\ell \geq 2$ the coefficients $a_{k,n}(\ell)$ and $\sigma^2(\ell)$ in (7.13)–(7.14) can be computed recursively via

$$a_{1,1}(\ell) = \rho(\ell), \; \sigma_0^2(\ell) = \gamma(0),$$

for $n \geq 2$,

$$a_{n,n}(\ell) = \left[\gamma(n + \ell - 1) - \sum_{j=1}^{n-1} a_{n-1,j}(\ell)\gamma_{n+\ell-j-1} \right]^{-1} \sigma_{n-1}(1),$$

$$\begin{bmatrix} a_{1,n}(\ell) \\ \vdots \\ a_{n-1,n}(\ell) \end{bmatrix} = \begin{bmatrix} a_{1,n-1}(\ell) \\ \vdots \\ a_{n-1,n-1}(\ell) \end{bmatrix} - a_{n,n}(\ell) \begin{bmatrix} a_{n-1,n-1}(1) \\ \vdots \\ a_{1,n-1}(1) \end{bmatrix},$$

$$\sigma_n^2(\ell) = \sigma_{n-1}^2(\ell)[1 - \alpha^2(n, \ell)],$$

where

$$\alpha(n, \ell) = \frac{\sigma_n^2(1)}{\sigma_n^2(\ell)} a_{n,n}(\ell).$$

Furthermore,

$$\sigma^2(\ell) = \lim_{n\to\infty} \sigma_n^2(\ell) = \gamma(0)\Pi_{n=1}^{\infty}[1 - \alpha^2(n,\ell)].$$

For $\ell = 1, \{\alpha(n,1)\}$ is precisely the PACF of $\{X_t\}$. What is the statistical interpretation of $\alpha(n,\ell)$, for $\ell \geq 2$?

(c) Compute $\{a_{k,n}(\ell)\}$ for $n = 2, \ell = 3$ and $\{X_t\}$ an ARMA(1,1) process.

(d) Compute the canonical correlations between $\boldsymbol{X} = (X_{-2}, X_{-1})$ and $\boldsymbol{Y} = (X_0, X_1)$ for the MA(1) process $X_t = \varepsilon_t + 0.6\varepsilon_{t-1}$, $\text{Var}(\varepsilon_t) = 1$.

7.8 Let $\{\varepsilon_t\}$ be an i.i.d. sequence of random variables with mean zero, $\sigma^2 = E\varepsilon_t^2$ and finite fourth moment, that is, $E\varepsilon_t^4 < \infty$. The sequences $\{c_n\}$ and $\{d_n\}$ are in ℓ^2 with $C^2 = \sum_{k=0}^{\infty} c_k^2$ and $D^2 = \sum_{k=0}^{\infty} d_k^2$.

(a) Define the MA(∞) processes $\{U_t\}$ and $\{V_t\}$ as

$$U_t = \sum_{k=0}^{\infty} c_k \varepsilon_{t-k}, V_t = \sum_{k=0}^{\infty} d_k \varepsilon_{t-k}, t = 0, \pm 1, \cdots,$$

show that for any $m \geq 1$,

$$E\left(\sum_{i=1}^{m} U_t V_t\right)^2 \leq 3m^2 C^2 D^2 E\varepsilon_t^4.$$

(b) For a PND process $\{X_t\}$ with AR parameters $\{a_n\}$, innovation $\{\varepsilon_t\}$ with the above properties, and $\lambda_{1,n} \leq K$, set

$$\varepsilon_t = \sum_{i=0}^{\infty} a_i X_{t-i}, \varepsilon_{t,k} = \sum_{i=0}^{k} a_i X_{t-i},$$

and show that

$$E\left(\sum_{t=k+1}^{n} U_t(\varepsilon_{t,k} - \varepsilon_t)\right)^2 \leq 6\pi E(\varepsilon_t/\sigma)^4 K E\varepsilon^2 (n-k)^2 (a_{k+1}^2 + a_{k+2}^2 + \cdots)$$

Hint: Take $V_t = \varepsilon_{t,k} - \varepsilon_t = a_{k+1}X_{t-k-1} + a_{k+2}X_{t-k-1} + \cdots$
$= d_{k+1}\varepsilon_{t-k-1} + d_{k+2}\varepsilon_{t-k-2} + \cdots$
and use (a); see Berk (1974).

8

Missing Values: Past and Future

In this chapter, we show how the future observations and their dependence on the past can be used to update predictors, interpolate missing values, estimate impacts of interventions and outliers, and compute the maximal correlation (angle) between past and future. These practical issues of interest in time series analysis are dealt with in the first half of the chapter, they serve as great motivations for the more theoretical topics such as the Kolmogorov's interpolation problem, the Helson-Szegö's problem of positivity of the angle between past and future (canonical correlations), characterization of ARMA models and strongly mixing Gaussian stationary sequences. Heuristically, one expects to estimate the impact of abrupt changes by comparing forecasts made using data prior to the suspected change with data actually occurring after the change. For specific changes one may need to rely on suitable functions of the forecast errors. Fortunately, this intuition is given credence in this chapter as we rely solely on the multi-step forecast errors $\{X_k - \hat{X}_k;\ 0 < k \leq m\}$ and their orthogonality to the past $\{X_t; t \leq 0\}$. This serves as a catalyst to link together several apparently distinct topics of practical and theoretical interest.

In Section 8.1, the best predictable linear combination of future values $X_1, \cdots, X_m$ are found and shown to be related to computing the canonical variables and correlations between the infinite past and finite future. Section 8.2 deals with the ubiquitous problem of incomplete and/or contaminated data, the interpolation of a single missing value as the most generic problem in this category is discussed in detail. Computing the interpolator and its error when several values are missing is discussed in Section 8.3. We show in Section 8.4 that computing predictors of future values based on incomplete past and estimators of impacts of interventions are related to computing the interpolators of certain missing values. The notions of *minimality*, *inverse process* and their roles in time series model identification are discussed in Section 8.5. A time-domain characterization of the minimality in terms of the square-summability of the AR parameters is given. Computing the angle or canonical correlation between past and future, conditions for positivity of the angle of a stationary process, convergence of the series representations of its predictor and interpolator and characterization of ARMA models in terms of canonical correlations are discussed in Sections 8.6. Section 8.7 provides a finite-dimensional method for approximating canonical correlations.

8.1 PREDICTION OF LINEAR FUNCTIONS OF FUTURE

In this section, we compute the predictor and the prediction error variance of the linear form

$$X = \sum_{r=1}^{m} c_r X_r, \tag{8.1}$$

where m is a finite integer and $c_1, \cdots, c_m$ are constants. In practice, the need for prediction of such linear forms arise when the forecaster is interested in the forecast of a linear combination of m future values. For example, if sales are recorded monthly, the forecaster might be interested in the next year's total sales ($m = 12$). Prediction of X in (8.1) is also important from theoretical point of views as indicated in the following sections.

Let $H_0 = \overline{sp}\{X_t;\ t \leq 0\}$. From (8.1) and the linearity of the projection it follows that

$$\begin{cases} \hat{X} = P_{H_0} X = \sum_{r=1}^{m} c_r \hat{X}_r, \\ X - \hat{X} = \sum_{r=1}^{m} c_r (X_r - \hat{X}_r), \\ \text{Var}(X - \hat{X}) = \sum_{r=1}^{m} \sum_{s=1}^{m} g_{rs} c_r c_s = \mathbf{c}' G_m \mathbf{c}, \end{cases} \tag{8.2}$$

where $g_{rs} = \sigma^2 \sum_{k=0}^{m-1} b_{r+k-m} b_{s+k-m}, G_m = (g_{rs})_{r,s=1}^{m}$ and $\mathbf{c} = (c_1, \cdots, c_m)$. The matrix G_m appearing in (8.2) is, in fact, the covariance matrix of the vector of prediction errors $(X_1 - \hat{X}_1, \cdots, X_m - \hat{X}_m)$. It can also be identified as the upper left $(m \times m)$ submatrix of the infinite matrix

$$G = \sigma^2 T T', \tag{8.3}$$

where T is $T = (b_{i-j})_{i,\, j=0}^{\infty}$.

A barometer of predictability of X can be defined in terms of the ratio of the prediction error variance of X to its variance. More specifically, we use the quantity $\lambda(X) = 1 - \dfrac{\text{Var}(X - \hat{X})}{\text{Var}(X)}$, as a *measure of predictability* of X (Yaglom, 1963). Note that larger values of $\lambda(X)$ indicate that X is more predictable. Since

$$\text{Var}(X) = \sum_{r=1}^{m}\sum_{s=1}^{m} c_r c_s \gamma_{r-s} = \boldsymbol{c}'\Gamma_m \boldsymbol{c},$$

we have

$$\lambda(X) = 1 - \frac{\boldsymbol{c}' G_m \boldsymbol{c}}{\boldsymbol{c}'\Gamma_m \boldsymbol{c}}. \tag{8.4}$$

For a process $\{X_t\}$ we are interested in the *most predictable linear function* of the future values. That is, finding $c_1, \cdots, c_m$, so that $\lambda(X)$ in (8.4)is maximized. This problem, however, is closely related to finding the canonical variables and correlations between the infinite past $\{X_t; t \leq 0\}$ and the finite future $\{X_1, \cdots, X_m\}$ studied in Chapter 7.

Theorem 8.1. Let $\{X_t\}$ be a nondeterministic stationary process and $X = \sum_{r=1}^{m} c_r X_r$, where $m \geq 1$ is fixed. Let $\lambda_1 < \cdots < \lambda_k, (k \leq m)$, be the distinct roots of the determinantal equation

$$\det(G_m - \lambda \Gamma_m) = 0,$$

and $\boldsymbol{c}_1, \cdots, \boldsymbol{c}_m$ be the corresponding orthonormalized eigenvectors, that is,

$$\boldsymbol{c}_i' \Gamma_m \boldsymbol{c}_j = \delta_{i,j}, i, j = 1, 2, \cdots, m.$$

Then, $\boldsymbol{U}_1 = \boldsymbol{c}_1' \boldsymbol{X}_m = \sum_{r=1}^{m} c_{1,r} X_r$ is the best predictable linear function of the finite future with the measure of predictability

$$\lambda(\boldsymbol{U}_1) = 1 - \lambda_1,$$

and, in general, the measure of predictability of $\boldsymbol{U}_k = \boldsymbol{c}_k' \boldsymbol{X}_m$ is $1 - \lambda_k$.

The next two examples illustrate the computation of the most predictable components of a stationary process.

Example 8.1. For a stationary process with the moving average representation

$$X_t = \varepsilon_t + \varepsilon_{t-1}, \ \text{Var}(\varepsilon_t) = 1,$$

and for $m = 2$, we have

$$\Gamma_2 = \begin{pmatrix} 2 & 1 \\ 1 & 2 \end{pmatrix}, G_2 = \begin{pmatrix} 1 & 1 \\ 1 & 2 \end{pmatrix},$$
$$\det(G_2 - \lambda \Gamma_2) = (1 - \lambda)(1 - 3\lambda) = 0.$$

Thus, $\lambda_1 = \frac{1}{3}, \lambda_2 = 1$, and the corresponding orthonormalized eigenvectors are

$$\boldsymbol{c}_1 = \left(-\frac{2}{\sqrt{6}}, \frac{1}{\sqrt{6}}\right), \boldsymbol{c}_2 = \left(0, \frac{1}{\sqrt{2}}\right).$$

Thus

$$\boldsymbol{U}_1 = -\frac{2}{\sqrt{6}}X_1 + \frac{1}{\sqrt{6}}X_2, \lambda(\boldsymbol{U}_1) = \frac{2}{3},$$

is the best predictable linear function of the future values X_{t+1} and X_{t+2}, and

$$\boldsymbol{U}_2 = \frac{1}{\sqrt{2}}X_2, \ \lambda(\boldsymbol{U}_2) = 0,$$

is the second best predictable linear function of the same future values. Note that $\boldsymbol{U}_2$ is actually uncorrelated with the past of the process.

For an arbitrary m, by expanding $|G_m - \lambda\Gamma_m|$ by the first row, we have

$$|G_m - \lambda\Gamma_m| = (1 - 2\lambda)D_m - (1 - \lambda)^2 D_{m-1}, \tag{8.5}$$

where D_m is the determinant of an $(m-1) \times (m-1)$ matrix with $2 - \lambda$ as diagonal elements, $1 - \lambda$ on diagonals immediately above and below the main diagonal and zeros elsewhere; see Problem 8.1.

Example 8.2. For the various processes (models) of Problem 4.10, the best predictable linear functions of $m = 8$ future values and their measures of predictability are given in the following table. For simplicity, we have taken $\sigma_e^2 = 1$.

Model	λ_1	Best Predictable Linear Functional
1	.1434	$X_1 - .36X_2$
2	.01	$X_1 - .38X_2$
3	.1359	$X_1 - .29X_2 - .14X_3 + .19X_4 - .19X_5$
4	.085	$X_1 - .48X_2 + .08X_3$
5	.1524	$X_1 - .30X_2 - .05X_3$
6	.1324	$X_1 - .41X_2 + .13X_3$

Note that for model 2, λ_1 is significantly different from that for the other models, this difference persists even for larger m. Model 2, in fact, represents a nonstationary process. In spite of this, its best predictable linear function is almost the same as that for model 1.

8.2 THE MISSING VALUE PROBLEM

An important problem in time series data analysis is that of estimating missing data or values of the series that for some reasons can not be observed completely or measured accurately. It should be noted that there are situations where a priori the time series is observed completely, but prudent statistical analysis of the data may require deletion of some of the (suspect) values. This happens, for example, in outlier detection and cross-validation.

8.2.1 A Single Missing Value

In this section we solve the *missing value problem* in time series analysis. The simplest form of this problem is that of finding the best linear predictor of X_0 and its prediction error variance based on the knowledge of

$$\{X_t; t \leq m, t \neq 0\},\ 0 \leq m < \infty,$$

and the covariance function $\{\gamma_k\}$ of the process. Note that when $m = 0$, this reduces to the one-step ahead prediction problem studied in Chapters 5 and 6, and the limiting case $m = \infty$ is the *interpolation problem* discussed in Section 8.5.

Let $H_{m,0}$ denote the space of linear predictors for X_0, that is

$$H_{m,0} = \overline{sp}\{X_t; t \leq m, t \neq 0\}.$$

Of course, the orthogonal projection of X_0 onto $H_{m,0}$, denoted by $\hat{X}'_{0,m}$ gives the best linear predictor of X_0, but computing this predictor is not easy. Although we can compute projections of X_0 onto $H_{-1} = \overline{sp}\{X_t; t \leq -1\}$ and $sp\{X_1, \cdots, X_m\}$ separately, it is difficult to compute the projection of X_0 on $H_{m,0}$ because H_{-1} and $sp\{X_1, \cdots, X_m\}$ are not orthogonal.

Our approach consists of writing $H_{m,0}$ as the orthogonal sum of H_{-1} and the m-dimensional subspace $sp\{X_k - \hat{X}_k; 1 \leq k \leq m\}$. This simple and fruitful idea is exploited throughout this chapter to reduce many diverse least-squares problems involving future of a process to the usual multi-step predictions plus a finite-dimensional regression problem.

Theorem 8.2. Let $\{X_t\}$ be a nondeterministic stationary process with the AR parameters $\{a_k\}$ and innovation $\{\varepsilon_t\}$. Then, with notation as above,

(a) The linear least-squares predictor of X_0 based on $\{X_t; t \leq m, t \neq 0\}$ is given by

$$\hat{X}'_{0,m} = \hat{X}_0 + \sum_{k=1}^{m} \beta_{k,m}(X_k - \hat{X}_k), \tag{8.6}$$

where

$$\beta_{k,m} = \left(1 + \sum_{i=1}^{m} a_i^2\right)^{-1} \left(a_k - \sum_{i=1}^{m-k} a_i a_{i+k}\right),\ k = 1, 2, \cdots, m \tag{8.7}$$

(b) the prediction error has a (forward) moving average representation:

$$X_0 - \hat{X}'_{0,m} = \left(1 + \sum_{i=1}^{m} a_i^2\right)^{-1} (\varepsilon_0 - a_1\varepsilon_1 - \cdots - a_m\varepsilon_m), \tag{8.8}$$

with variance

$$\sigma_m'^2 = \mathrm{Var}(X_0 - \hat{X}'_{0,m}) = \sigma^2 \left(1 + \sum_{i=1}^{m} a_i^2\right)^{-1}. \tag{8.9}$$

Note that the coefficients $\beta'_{k,m}$ resembles the ACF of a finite MA process. We discuss this point further toward the end of the present section. The autoregressive parameters of $\{X_t\}$ play a dominant role in these formulas and determine the amount of reduction in variance when predicting X_0 using the additional data $X_1, \cdots, X_m$. Proof of the theorem is broken into several lemmas that are of independent interest and used in later sections (Pourahmadi, 1989).

Lemma 8.3. Let $\{X_t\}$ be a nondeterministic stationary process and let $\hat{X}_k$ be the predictor of X_k based on $H_{-1}, k \geq 1$. Then,

$$H_{m,0} = H_{-1} \oplus sp\{X_k - \hat{X}_k; 1 \leq k \leq m\}.$$

Proof. Obviously, $H_{-1} \oplus sp\{X_k - \hat{X}_k; 1 \leq k \leq m\} \subseteq H_{m,0}$. To show the other inclusion, it is enough to note that any $X \in H_{m,0}$ can be written as $X = Y + \sum_{k=1}^{m} c_k X_k$, where $Y \in H_{-1}$, thus

$$X = \left(Y + \sum_{k=1}^{m} c_k \hat{X}_k\right) + \sum_{k=1}^{m} c_k (X_k - \hat{X}_k) \in H_{-1} \oplus sp\{X_k - \hat{X}_k; 1 \leq k \leq m\}.$$

Q.E.D.

Since H_{-1} and $sp\{X_k - \hat{X}_k; 1 \leq k \leq m\}$ are orthogonal subspaces, it follows immediately that $\hat{X}'_{0,m} = \hat{X}_0 + P^{X_0}_{sp\{X_k - \hat{X}_k; 1 \leq k \leq m\}}$. To obtain (8.7)–(8.9), we have to compute the projection of X_0 onto the indicated m-dimensional subspace, this is done in the following lemmas after introducing some notation and conventions that are used throughout this chapter.

For a nondeterministic stationary process $\{X_t\}$ with AR and MA parameters $\{a_k\}$, $\{b_k\}$ and given integer $m \geq 1$, let

$$\boldsymbol{b} = (b_1, \cdots, b_m),\ \boldsymbol{a} = (a_1, \cdots, a_m).$$

Only when it is necessary we use $\boldsymbol{b}_m$ to indicate that $\boldsymbol{b}$ is an m-vector. For a given vector $\boldsymbol{x} = (x_1, \cdots, x_m)$, $\boldsymbol{x}_r$ stands for the vector whose components are in reverse order, that is, $\boldsymbol{x}_r = (x_m, \cdots, x_1)$.

Lemma 8.4. For a nondeterministic stationary process with MA parameters $\{b_k\}$ we have

$$P^{X_0}_{\overline{sp}\{X_k - \hat{X}_k; 1 \leq k \leq m\}} = \sum_{k=1}^{m} \beta_{k,m}(X_k - \hat{X}_k),$$

where the vector $\boldsymbol{\beta} = (\beta_{1,m}, \cdots, \beta_{m,m})$ satisfies the *normal equations*

$$A_m \boldsymbol{\beta} = \boldsymbol{b},$$

with

$$A_m = T_m T_m' + \boldsymbol{b}\boldsymbol{b}'$$

and $T_m = (b_{i-j})_{i,j=0}^{m-1}$ is an $m \times m$ lower triangular matrix.

Proof. From (5.57),

$$\begin{bmatrix} X_1 - \hat{X}_1 \\ \vdots \\ X_m - \hat{X}_m \end{bmatrix} = \begin{bmatrix} b_1 & 1 & 0 & \cdot & \cdot & \cdot & 0 \\ b_2 & b_1 & 1 & \cdot & \cdot & \cdot & 0 \\ \vdots & \vdots & \vdots & \vdots & \vdots & \vdots & \vdots \\ b_m & b_{m-1} & \cdot & \cdot & \cdot & \cdot & 1 \end{bmatrix} \begin{bmatrix} \varepsilon_0 \\ \vdots \\ \varepsilon_m \end{bmatrix}$$

$$= \begin{bmatrix} \boldsymbol{b} \vdots T_m \end{bmatrix} \begin{bmatrix} \varepsilon_0 \\ \vdots \\ \varepsilon_m \end{bmatrix}, \tag{8.10}$$

thus

$$[\mathrm{Cov}(X_k - \hat{X}_k, X_l - \hat{X}_l)]_{k,l=1}^m = \sigma^2 A_m,$$

and using the Wold decomposition of $\{X_t\}$, we get

$$[\mathrm{Cov}(X_k - \hat{X}_k, X_0)]_{k=1}^m = \sigma^2\, \boldsymbol{b},$$

from which the result follows.

Q.E.D.

Next, we express $\boldsymbol{\beta}$ in terms of the AR parameters of $\{X_t\}$ and establish (8.7)–(8.9) in the following two lemmas.

Lemma 8.5. For a nondeterministic stationary process with AR parameters $\{a_k\}$, MA parameters $\{b_k\}$, let $G_m = T_m T_m'$, $\boldsymbol{d} = G_m^{-1}\boldsymbol{b}$, then

(i) $\boldsymbol{a} = T_m^{-1}\boldsymbol{b}, \boldsymbol{d} = T_m'^{-1}\, \boldsymbol{a}$,

(ii) $\boldsymbol{b}' G_m^{-1} \boldsymbol{b} = \sum_{k=1}^{m} a_k^2 = \boldsymbol{a}'\boldsymbol{a}$,

(iii) $d_k = a_k - \sum_{i=1}^{m-k} a_i a_{i+k}, k = 1, 2, \cdots, m.$

(iv) $A_m^{-1} = G_m^{-1} - (1 + \boldsymbol{a}'\boldsymbol{a})^{-1}\boldsymbol{d}\boldsymbol{d}'$,

(v) $\boldsymbol{\beta} = \left(1 + \sum_{k=1}^{m} a_k^2\right)^{-1} \boldsymbol{d}.$

Proof. (i) is essentially the same as (6.18). (ii) is immediate from (i) and the definition of G_m. (iii) follows by solving the system of linear equations $G_m \boldsymbol{d} = \boldsymbol{b}$ and using (i). (iv) can be checked directly. (v) follows from (iv) and some algebra. Q.E.D.

The next lemma provides an MA representation for the prediction error.

Lemma 8.6. For a nondeterministic stationary process with AR parameters $\{a_k\}$ and innovation $\{\varepsilon_t\}$, we have

(i) $X_0 - \hat{X}'_{0,m} = (1 + \boldsymbol{a}'\boldsymbol{a})^{-1}(\varepsilon_0 - a_1\varepsilon_1 - \cdots - a_m\varepsilon_m)$.

(ii) $\mathrm{Var}(X_0 - \hat{X}'_{0,m}) = \sigma^2 \left(1 + \sum_{k=1}^{m} a_k^2\right)^{-1}$.

Proof. From (8.10) and the representation of $\hat{X}'_{0,m}$ we obtain

$$\begin{aligned} X_0 - \hat{X}'_{0,m} &= X_0 - \hat{X}_0 - \sum_{k=1}^{m} \beta_{k,m}(X_k - \hat{X}_k) \\ &= (1, -\beta_{1,m}, \cdots, -\beta_{m,m})'(X_0 - \hat{X}_0, \cdots X_m - \hat{X}_m) \\ &= (1, -\boldsymbol{\beta}') \begin{pmatrix} 1 & 0 \\ \boldsymbol{b} & T_m \end{pmatrix} \boldsymbol{\varepsilon}_{m+1} = (1 - \boldsymbol{\beta}'\boldsymbol{b}, -\boldsymbol{\beta}'T_m)\boldsymbol{\varepsilon}_{m+1}, \end{aligned}$$

where $\boldsymbol{\varepsilon}_{m+1} = (\varepsilon_0, \varepsilon_1, \cdots, \varepsilon_m)$. From Lemma 8.5 (iv), (iii) and (i) we obtain

$$\begin{aligned} \boldsymbol{\beta}'\boldsymbol{b} &= (1 + \boldsymbol{a}'\boldsymbol{a})^{-1}\boldsymbol{d}'\boldsymbol{b} = (1 + \boldsymbol{a}'\boldsymbol{a})^{-1}\boldsymbol{a}'T_m^{-1}\boldsymbol{b} = (1 + \boldsymbol{a}'\boldsymbol{a})^{-1}\boldsymbol{a}'\boldsymbol{a}, \\ \boldsymbol{\beta}'T_m &= (1 + \boldsymbol{a}'\boldsymbol{a})^{-1}\boldsymbol{d}'T_m = (1 + \boldsymbol{a}'\boldsymbol{a})^{-1}\boldsymbol{a}', \end{aligned}$$

from which the desired result follows. (ii) is immediate from (i).

Q.E.D.

Example 8.3. We illustrate (8.6)–(8.9) for stationary AR(p) and MA(1) processes.

(i) Let $X_t = \sum_{i=1}^{p} a_i X_{t-i} + \varepsilon_t, 1 \le p < \infty,\ \mathrm{Var}(\varepsilon_t) = \sigma^2$, then from Lemma 8.5 (v) we have

$$\begin{cases} \beta_{k,m} &= \left(1 + \sum_{i=1}^{m} a_i^2\right)^{-1} \left(a_k - \sum_{i=1}^{m-k} a_i a_{i+k}\right), k = 1, \cdots, m, \\ \beta_{k,m} &= 0, \text{ for } p < k \le m. \end{cases}$$

since $a_i = 0$ for $i > p$. Thus if $m \ge p$, we obtain the simple formula,

$$\hat{X}_{0,m} = -\sum_{k=1}^{p} \beta_{k,m}(X_k + X_{-k}),$$

for the predictor of X_0, and the prediction error variance is given by

$$\text{Var}(X_0 - \hat{X}'_{0,m}) = \sigma^2 \left(1 + \sum_{k=1}^{p} a_k^2\right)^{-1}.$$

For $p = 1$ and any $m \geq 1$ we obtain the familiar formulas,

$$\hat{X}'_{0,m} = \frac{a_1}{1 + a_1^2}(X_1 + X_{-1}), \text{ Var } (X_0 - \hat{X}'_{0,m}) = \frac{\sigma^2}{1 + a_1^2}.$$

(ii) Let $X_t = \varepsilon_t - b\varepsilon_{t-1}, |b| \leq 1, \text{Var}(\varepsilon_t) = \sigma^2$. In this case

$$a_k = b^k, k = 1, 2, \cdots,$$

and from (8.7) for a given m we have

$$\beta_{k,m} = b^k \frac{1 + \sum_{i=1}^{m-k} b^{2i}}{1 + \sum_{i=1}^{m} b^{2i}}, 1 \leq k \leq m,$$

which is nonzero for all k. Thus, unlike the AR(p) process $\hat{X}'_{0,m}$ includes every $X_m, 1 \leq k \leq m$. Moreover,

$$\text{Var}(X_0 - \hat{X}'_{0,m}) = \sigma^2 \left(1 + \sum_{i=1}^{m} b^{2i}\right)^{-1}.$$

Note that for $|b| < 1$,

$$\lim_{m \to \infty} \beta_{k,m} = b^k, \text{ for } all\ k,$$

and

$$\lim_{m \to \infty} \text{Var}(X_0 - \hat{X}'_{0,m}) = \sigma^2(1 - b^2).$$

However, when $b^2 = 1$, then

$$\lim_{m \to \infty} \beta_{k,m} = \begin{cases} 1 & \text{if } b = 1, \\ (-1)^k & \text{if } b = -1, \end{cases}$$

and

$$\lim_{m \to \infty} \text{Var}(X_0 - \hat{X}'_{0,m}) = 0.$$

The differences in the behavior of $\lim_{m\to\infty} \beta_{k,m}$ and $\lim_{m\to\infty} \text{Var}(X_0 - \hat{X}'_{0,m})$ when $|b| < 1$ and $|b| = 1$ are not incidental. They are simple examples of a more general phenomenon discussed in Section 8.5. For $m \geq 0$, the *m-innovation* or *generalized innovation* of a process $\{X_t\}$ is defined by

$$\varepsilon_{t,m} = X_t - \hat{X}'_{t,m},$$

where $\hat{X}'_{t,m}$ is the predictor of X_t based on the data $\{X_s; s \leq t + m, s \neq t\}$. Note that for $m = 0, \varepsilon_{t,0}$ is precisely the innovation ε_t or the one-step ahead prediction error. For a nondeterministic stationary process $\{X_t\}$, it follows from Lemma 8.6 (i) that its m-innovation can be expressed in terms of $\{\varepsilon_t\}$ via

$$\varepsilon_{t,m} = (1 + \boldsymbol{a}'\boldsymbol{a})^{-1} \left(\varepsilon_t - \sum_{k=1}^{m} a_k \varepsilon_{t+k} \right). \tag{8.11}$$

Thus, the m-innovation is a (forward) MA(m) process. In particular for the AR(p) process of Example 8.3 (i), and any fixed $m \geq p$ the m-innovation process $\{\varepsilon_{t,m}\}$ is a (forward) MA(p) process. The process is not an orthogonal process, but is *bioorthogonal* to $\{X_t\}$, that is,

$$\text{Cov}(X_s, \varepsilon_{t,m}) = \sigma'^2_m \delta_{s,t}, \tag{8.12}$$

with σ'^2_m given in (8.9).

8.2.2 Several Missing Values

This section is devoted to the more general problem of estimation of several missing values with no restriction on the pattern of missingness. All the ideas and techniques of the previous section are employed here, but the notation is more complicated.

The set of integers

$$M = \{n_1, \cdots, n_m\}, n_1 = 0 < n_2 < \cdots < n_m,$$

is used to denote the index of the missing values, $m = card(M)$ represents the total number of missing values. The set of integers

$$K = \{k_1, \cdots, k_r\}, 0 < k_1, \cdots, < k_r < \infty,$$

denotes the index of known values after X_0 - the first missing value - thus $r = card(K)$ is the number of known values after X_0. For example, for one missing value, $M = \{0\}$ and $K = \{1, 2, \cdots, r\}$. We use X_M to denote the vector of missing values:

$$X_M = (X_{n_1}, \cdots, X_{n_m}),$$

and X_K the vector of known values after X_0: $X_K = (X_{k_1}, \cdots, X_{k_r})$. For a vector X, $\hat{X}$ denotes the vector of predictors of the entries of X based on the infinite past

$X_{-1}, X_{-2}, \cdots, \hat{X}'$ will denote the vector of interpolators of the entries of X based on all the observed values $\{X_t; t \leq -1 \text{ or } t = k_1, \cdots, k_r\}$.

The analogue of the matrix A_m in Lemma 8.4 is denoted by A_K, in fact,

$$A_K = \text{Cov}(X_K - \hat{X}_K) = [\text{Cov}(X_{k_i} - \hat{X}_{k_i}, X_{k_j} - \hat{X}_{k_j})]_{i,j=1,\cdots,r}, \tag{8.13}$$

is the $r \times r$ principal submatrix of the (infinite) matrix $G = \sigma^2 TT'$, obtained by choosing rows and columns of G with indices in K.

For interpolating an individual missing value X_{n_j}, the analogues of $\boldsymbol{b}$ and $\boldsymbol{\beta}$ in Lemma 8.4 are denoted by $\boldsymbol{b}_j$ and $\boldsymbol{\beta}_j$, respectively, the $r \times m$ matrices

$$B = [\boldsymbol{b}_1, \cdots, \boldsymbol{b}_m], \ \boldsymbol{\beta} = [\boldsymbol{\beta}_1, \cdots, \boldsymbol{\beta}_m]$$

are needed in the interpolation of all missing values. The matrix B is identified as

$$B = \text{Cov}(X_K - \hat{X}_K, X_M) = \text{Cov}(X_K - \hat{X}_K, X_M - \hat{X}_M), \tag{8.14}$$

which is the $r \times m$ submatrix of G whose row indices belong to K and column indices to M. The matrix A_K does not depend (directly) on M or the pattern of missing values, while B and $\boldsymbol{\beta}$ do depend on M as well as K; to keep the notation simple we suppress the dependence of the entries of B and β on M or m as much as possible.

Theorem 8.7. Let $\{X_t\}$ be a nondeterministic stationary process. Then, with notation as above, we have

(a) For any $1 \leq j \leq m$,

$$\hat{X}'_{n_j} = \hat{X}_{n_j} + \sum_{i=1}^{r} \beta_{i,j}(X_{k_i} - \hat{X}_{k_i}),$$

where the vector

$$\boldsymbol{\beta}_j = (\beta_{1,j}, \cdots, \beta_{r,j})$$

is the solution of the *normal equations*

$$A_K \boldsymbol{\beta}_j = \boldsymbol{b}_j.$$

Using the vector form, the interpolators of all missing values can be written compactly as

$$\hat{X}'_M = \hat{X}_M + \boldsymbol{\beta}'(X_K - \hat{X}_K),$$

where the $r \times m$ matrix $\boldsymbol{\beta}$ satisfies the *normal equations*

$$A_K \boldsymbol{\beta} = B.$$

(b) $\text{Cov}(X_M - \hat{X}'_M) = \text{Cov}(X_M - \hat{X}_M) - \sigma^2 B' A_K^{-1} B.$

This shows how the additional observations in X_K will improve the predictors of the missing values. To use it one needs to compute the multi-step predictions of $X_l, l = 0, 1, \cdots, m+r-1$, and invert the $r \times r$ matrix A_K. When r is large this is a computationally difficult task. Furthermore, when more than one value is missing and M is arbitrary, simple formulas for $\hat{X}'_M$ and $\text{Cov}(X_M - \hat{X}'_M)$ might be hard to find. However, for consecutive indices, namely,

$$M = \{0, 1, \cdots, m-1\},\ K = \{m, m+1, \cdots, m+r-1\},$$

the A_K and B assume simpler forms, and the required computation is reduced considerably. To illustrate this, let us define the $r \times m$ matrix

$$B_{r,m} = (b_{m+i-j})_{\substack{i=0,\cdots,r-1,\\ j=0,\cdots,m-1}}$$

then it follows from (8.13) and (8.14) that

$$A_K = B_{r,m}B'_{r,m} + G_r, \tag{8.15}$$

where $G_r = T_r T'_r$ is as before, and

$$B = B_{r,m}T_m. \tag{8.16}$$

Note that when $m = 1$, then $B_{r,1} = B = \boldsymbol{b}$ as in Lemma 8.4.

Owing to the simple form of A_K in (8.15), it can be shown that.

$$A_K^{-1} = G_r^{-1} - G_r^{-1}B_{r,m}(I + B'_{r,m}G_r^{-1}B_{r,m})^{-1}B'_{r,m}G_r^{-1}. \tag{8.17}$$

Let A, D be the $r \times m$ matrices satisfying the equations

$$T'_r A = B_{r,m}, T_r D = A. \tag{8.18}$$

From Theorem 8.7, (8.17), and (8.18), it follows that

$$\boldsymbol{\beta} = D(I_m + A'A)^{-1}T_m, \tag{8.19}$$

and

$$\text{Cov}(X_M - \hat{X}'_M) = \sigma^2 T'_m(I_m + A'A)^{-1}T_m. \tag{8.20}$$

It is important to note that the $r \times m$ matrices A and D defined in (8.18) are analogues of the $r \times 1$ vectors $\boldsymbol{a}$ and $\boldsymbol{d}$ defined in Lemma 8.4, in fact, it is not hard to show that the last column of A is the first r autoregressive coefficients of X_0 based on the infinite past $X_{-1}, X_{-2}, \cdots$; the $(m-1)$st column of A is the first r autoregressive coefficients of the predictor of X_1 based on $X_{-1}, X_{-2}, \cdots$, etc. Also the elements of the matrix D can be expressed in terms of the elements of the matrix A just as in the Lemma 8.5. This shows that computation of A and D, in (8.18), do not require any matrix inversion, the only matrix inversion required is that of $I_m + A'A$ in computing $\boldsymbol{\beta}$ and $\text{Cov}(X_M - \hat{X}'_M)$, in (8.19) and (8.20).

8.3 PREDICTION USING INCOMPLETE PAST

In this section we compute predictors of future values and their errors when the infinite past is not known completely and few of its values are missing. For simplicity, we focus on predicting X_{m+1} based on $\{X_t; t \leq m, t \neq 0\}$, i.e. when only X_0 from the past is missing (Pourahmadi, 1992; Pourahmadi and Soofi, 2000). Surprisingly the method of solution and essentially all formulas are similar to those in Section 8.2. Extension to more general settings is possible, but is not pursued here.

Theorem 8.8. Let $\{X_t\}$ be a nondeterministic stationary process with the AR parameters $\{a_k\}$ and innovations $\{\varepsilon_t\}$. For $m \geq 0$, let $\hat{X}_{m+1,0}$ denote the predictor of X_{m+1} based on the incomplete past $\{X_t; t \leq m,\ t \neq 0\}$. Then

(a) The linear least squares predictor of X_{m+1} based on the incomplete past $\{X_t; t \leq m, t \neq 0\}$ is given by

$$\hat{X}_{m+1,0} = \hat{X}_{m+1} + \sum_{k=1}^{m} \beta_{k,m}(X_k - \hat{X}_k) = \hat{X}_{m+1} + \boldsymbol{\beta}'(X - \hat{X}), \tag{8.21}$$

where $\boldsymbol{\beta} = (\beta_{1,m}, \cdots, \beta_{m,m})$ satisfies the normal equations

$$A_m \boldsymbol{\beta} = \mathbf{c} \tag{8.22}$$

with A_m as in Lemma 8.4, and

$$\mathbf{c} = \sigma^{-2}\text{Cov}(X_{m+1} - \hat{X}_{m+1}, X - \hat{X}) = b_{m+1}\boldsymbol{b} + T_m \boldsymbol{b}_r, \tag{8.23}$$

where $\boldsymbol{b}_r = (b_m, b_{m-1}, \cdots, b_1)$.

(b) The prediction error has the MA$(m+1)$ representation:

$$X_{m+1} - \hat{X}_{m+1,0} = \varepsilon_{m+1} + \alpha \sum_{k=0}^{m} a_k \varepsilon_k,$$

where

$$\alpha = a_{m+1}\left(\sum_{k=0}^{m} a^2\right)^{-1}.$$

(c) $\text{Var}(X_{m+1} - \hat{X}_{m+1,0}) = \sigma^2 \left(\sum_{k=0}^{m} a_k^2\right)\left(\sum_{k=0}^{m-1} a_k^2\right)^{-1}.$

Proof. (a) The representation (8.21) and the normal equations (8.22) follow from the same lines of arguments used in Lemmas 8.3 and (8.8). The following identities are needed in obtaining the representation for the prediction error. From Lemma 8.5 (i) and (6.18) we have

$$\begin{cases} T_m^{-1}\boldsymbol{c} &= (b_{m+1}T_m^{-1}\boldsymbol{b} + \boldsymbol{b}_r) = b_{m+1}\boldsymbol{a} + \boldsymbol{b}_r, \\ \boldsymbol{d}'\boldsymbol{c} &= \boldsymbol{a}'T_m^{-1}(b_{m+1}\boldsymbol{b} + T_m\boldsymbol{b}_r) = \boldsymbol{a}'(b_{m+1}\boldsymbol{a} + \boldsymbol{b}_r), \\ \boldsymbol{a}'\boldsymbol{b}_r &= b_{m+1} - a_{m+1}. \end{cases} \tag{8.24}$$

Also from (8.22) and Lemma 8.5 (iv) we obtain

$$\begin{aligned} \boldsymbol{\beta} &= [G^{-1} - (1+\boldsymbol{a}'\boldsymbol{a})^{-1}\boldsymbol{d}\boldsymbol{d}']\boldsymbol{c} = [T'^{-1} - (1+\boldsymbol{a}'\boldsymbol{a})^{-1}\boldsymbol{d}\boldsymbol{a}'](b_{m+1}\boldsymbol{a} + \boldsymbol{b}_r) \\ &= T'^{-1}[I - (1+\boldsymbol{a}'\boldsymbol{a})^{-1}\boldsymbol{a}\boldsymbol{a}'](b_{m+1}\boldsymbol{a} + \boldsymbol{b}_r). \end{aligned} \tag{8.25}$$

Next, we express $X_{m+1} - \hat{X}_{m+1,0}$ in terms of the innovation $\{\varepsilon_t\}$ as follows:

$$\begin{aligned} X_{m+1} - \hat{X}_{m+1,0} &= X_{m+1} - \hat{X}_{m+1} - \sum_{k=1}^{m} \beta_{k,m}(X_k - \hat{X}_k) \\ &= (-\beta_{1,m}, \cdots, -\beta_{m,m}, 1) \left[\begin{array}{c|cccc|c} b_1 & 1 & 0 & \cdots & & 0 \\ b_2 & b_1 & 1 & \cdots & & 0 \\ \vdots & & & & \vdots & \\ \cdots & \cdots & \cdots & \cdots & & \cdots \\ b_{m+1} & b_m & b_{m-1} & b_1 & & 1 \end{array}\right] \begin{bmatrix} \varepsilon_0 \\ \vdots \\ \varepsilon_m \\ \varepsilon_{m+1} \end{bmatrix} \\ &= (b_{m+1} - \boldsymbol{\beta}'\boldsymbol{b}_r,\ \boldsymbol{b}_r' - \boldsymbol{\beta}'T_m,\ 1)\boldsymbol{\varepsilon}_{m+2}. \end{aligned} \tag{8.26}$$

From the last identity in (8.24), (8.25) and some algebra it follows that

$$\begin{aligned} \boldsymbol{\beta}'\boldsymbol{b} &= (1+\boldsymbol{a}'\boldsymbol{a})^{-1}(b_{m+1}\boldsymbol{a}'\boldsymbol{a} + \boldsymbol{a}'\boldsymbol{b}_r), \\ b_{m+1} - \boldsymbol{\beta}'\boldsymbol{b} &= (1+\boldsymbol{a}'\boldsymbol{a})^{-1}a_{m+1} \\ \boldsymbol{b}_r' - \boldsymbol{\beta}'T_m &= a_{m+1}(1+\boldsymbol{a}'\boldsymbol{a})^{-1}\boldsymbol{a}' = \alpha\boldsymbol{a}'. \end{aligned}$$

Thus,

$$X_{m+1} - \hat{X}_{m+1,0} = \alpha\varepsilon_0 + \alpha\sum_{k=1}^{m} a_k\varepsilon_k + \varepsilon_{m+1}, \tag{8.27}$$

from which (c) is immediate.

Q.E.D.

Note that the increase in variance due to X_0 being missing or excluded is

$$\mathrm{Var}(X_{m+1} - \hat{X}_{m+1,0}) - \sigma^2 = \sigma^2 a_{m+1}^2 \left(\sum_{k=0}^{m} a_k^2\right)^{-1} = \sigma^2 a_{m+1}\alpha. \tag{8.28}$$

For the AR(p) process in Example 8.3 (i) this increase is zero when $m > p$, as expected. But for the MA(1) process in Example 8.3 (ii) this increase is never zero. For any stationary process one expects the variance in (8.28) to converge to zero as $m \to \infty$. But, this is not true in general. What are the necessary and sufficient conditions for this to happen?

8.4 INTERVENTION ANALYSIS

Stochastic processes representing real-life systems and phenomena are frequently affected by policy changes and other external events usually referred to as *interventions*. In this section we develop methods for estimating *impacts* of such interventions.

Intervention analysis developed by Box and Tiao (1975, 1976) has been used extensively in various areas of applications, a slight modification of it is used in detecting various types of outliers and structural disturbances in time series. In the Box-Tiao approach, ARMA-like difference equation models are used to model the possible dynamic characteristics of the intervention and the noise. Here, we propose a nonparametric and more general approach to intervention analysis. It is motivated by the simple and intuitively appealing fact that the multi-step ahead prediction errors of the post-intervention measurements contain all the relevant information regarding the interventions and their impacts. Consequently, the matrix G_m in (8.2) and its inverse play a major role in relating the estimators of impacts of interventions to quantities appeared earlier in solving the incomplete data problems; see Sections 8.2–8.3.

To begin, suppose the first intervention occurs at a known time, say, $t = 1$. Therefore, the post-intervention data $X_1, \cdots, X_m$ cannot be observed directly, instead one observes

$$Y_t = f_t + X_t, \;\; 1 \leq t \leq m, \tag{8.29}$$

where f_t is the impact of possibly many interventions on the level of the time series at time t. Let $f = (f_1, \cdots, f_m)$ and $\hat{X} = (\hat{X}_1, \cdots, \hat{X}_m)$ where $\hat{X}_\ell, \; 1 \leq \ell \leq m$, is the ℓ-step ahead predictor of X_ℓ based on the pre-intervention data $\{X_t; t \leq 0\}$. The *ℓ-step ahead prediction errors* of the post-intervention measurements defined by

$$\hat{f}_\ell = Y_\ell - \hat{X}_\ell, \;\; \hat{f} = (\hat{f}_1, \cdots, \hat{f}_m), \tag{8.30}$$

provides an estimator of f. To study the statistical properties of the estimator we express $\hat{f}$ in terms of the components of the series $\{X_t\}$, whose statistical properties are assumed known prior to the intervention. From (8.29)–(8.30) and (8.10) we have

$$\hat{f} = f + X - \hat{X} = f + T_m \varepsilon, \tag{8.31}$$

where $\varepsilon = (\varepsilon_1, \cdots, \varepsilon_m)$ with $E(\varepsilon) = 0$. Then, $E(\hat{f}) = f$, i.e. $\hat{f}$ is an unbiased estimator of f with the covariance matrix

$$\text{Cov}(\hat{f}) = \sigma^2 T_m T_m' = G_m. \tag{8.32}$$

It should be noted that $\hat{f}$ is a *nonparametric estimator* of f in the sense that no model for f is assumed yet. In fact, a timeplot of the entries of $\hat{f}$ could provide valuable information about the nature and possibly the number of interventions and their impacts. In other words, such timeplots can be used to postulate appropriate models for f.

The parametric models for the impact of interventions as postulated in Box and Tiao (1975) are equivalent to assuming

$$f_t = \beta_1 z_{t,1} + \cdots + \beta_p z_{t,p}, \quad 1 \le t \le m, \tag{8.33}$$

for some known variables $z_{t,i}, 1 \le i \le p$ and unknown parameters $\beta_1, \cdots, \beta_p$ which assess the impact of interventions. Rewriting this in matrix form and from (8.31) we obtain

$$\hat{f} = Z\beta + T_n\varepsilon, \tag{8.34}$$

where $\beta = (\beta_1, \cdots, \beta_p)$ and $Z = (z_{t,i})$ is the $m \times p$ "design" matrix.

The generalized least squares estimator of $\boldsymbol{\beta}$ in (8.34) is given by

$$\hat{\boldsymbol{\beta}} = (Z'G_m^{-1}Z)^{-1}Z'G_m^{-1}\hat{f}. \tag{8.35}$$

Expressing $\hat{\boldsymbol{\beta}}$ in terms of the parameters and innovations of $\{X_t\}$ makes the statistical properties of $\hat{\boldsymbol{\beta}}$ more transparent. Let us define the auxiliary matrix U by

$$U = T_m^{-1}Z. \tag{8.36}$$

Then, using (8.34)–(8.36) we obtain

$$\hat{\boldsymbol{\beta}} = \boldsymbol{\beta} + (U'U)U'\varepsilon,$$

consequently, $\hat{\boldsymbol{\beta}}$ is an unbiased estimator of $\boldsymbol{\beta}$ with

$$\text{Cov}(\hat{\boldsymbol{\beta}}) = \sigma^2(U'U)^{-1}.$$

The following examples (Tsay, 1988) demonstrate the versatility of intervention analysis and clarify the relationship among the above formulas and those obtained for the interpolation of missing values and m-innovations introduced in (8.11).

Example 8.4. (a) *Additive Outlier (AO)*. The presence of an additive outlier at time $t = 1$, that is, when X_1 is replaced by $X_1 + \beta$ is a special case of (8.29) or (8.33) with

$$f_1 = \beta, \;\; f_t = 0, \;\; 2 \le t \le m, \text{ or } Z = (1, 0, \cdots, 0), p = 1.$$

Thus,

$$\begin{aligned} U &= T_m^{-1}Z = (1, -a_1, \cdots, -a_{m-1}),\; U'U = \sum_{k=0}^{m-1} a_k^2, \\ \hat{\beta} &= \beta + \left(\sum_{k=0}^{m-1} a_k^2\right)^{-1} \left(\varepsilon_1 - \sum_{k=1}^{m-1} a_k \varepsilon_{1+k}\right), \\ \text{Var}(\hat{\beta}) &= \sigma^2 \left(\sum_{k=0}^{m-1} a_k^2\right)^{-1}. \end{aligned}$$

(b) *Innovation Outlier (IO)*. An innovation outlier of size β at time $t = 1$, occurs if ε_1 is replaced by $\varepsilon_1 + \beta$. It can be shown that its effect propagates so that

$$f_t = \beta b_{t-1}, \quad 1 \le t \le m, \quad Z = (1, b_1, \cdots, b_{m-1}), p = 1.$$

Then,

$$U = T_m^{-1} Z = (1, 0, \cdots, 0), \quad \hat{\beta} = \beta + \varepsilon_1 \text{ and } \mathrm{Var}(\hat{\beta}) = \sigma^2.$$

(c) *Level Change.* When the mean of a series shifts to a new level due to some interventions, then

$$\begin{aligned}
f_t &= \beta, \ 1 \le t \le m, \quad Z = (1, 1, \cdots, 1), p = 1 \\
U &= \left(1, 1 - a_1, 1 - a_1 - a_2, \cdots, 1 - \sum_{k=1}^{m-1} a_k\right) = (1, u_1, \cdots, u_{m-1}), \\
\hat{\beta} &= \beta + \left(\sum_{k=0}^{m-1} u_k^2\right)^{-1} \left(\varepsilon_1 - \sum_{k=1}^{m-1} u_k \varepsilon_{1+k}\right) \text{ and } \mathrm{Var}(\hat{\beta}) = \sigma^2 \left(\sum_{k=0}^{m-1} u_k^2\right)^{-1}.
\end{aligned}$$

(d) *Transient Change.* When the mean of the series changes slowly with time due to some interventions, then

$$\begin{aligned}
f_t &= \beta\delta^{t-1}, 1 \le t \le m, \ |\delta| \le 1, \ Z = (1, \delta, \cdots, \delta^{m-1}), p = 1. \\
U &= \left(1, 1 - a_1\delta, \cdots, 1 - \sum_{k=1}^{m-1} a_k \delta^k\right) = (1, u_1, \cdots, u_{m-1}), \\
\hat{\beta} &= \beta + \left(\sum_{k=0}^{m-1} u_k^2\right)^{-1} \left(\varepsilon_1 - \sum_{k=1}^{m-1} u_k \varepsilon_{1+k}\right) \text{ and } \mathrm{Var}(\hat{\beta}) = \sigma^2 \left(\sum_{k=0}^{m-1} u_k^2\right)^{-1}.
\end{aligned}$$

(d) IO at time $t = 1$ and AO at time $t = 2$: This is a combination of (a) and (b). We have with $\alpha = \sum_{k=0}^{m-1} a_k^2$ and

$$\begin{aligned}
Z &= \begin{bmatrix} 1 & b_1 & \cdots & b_{m-1} \\ 1 & 0 & \cdots & 0 \end{bmatrix}, p = 2, \\
\hat{\beta} &= \begin{bmatrix} \beta_1 \\ \beta_2 \end{bmatrix} + \begin{bmatrix} \alpha(\alpha - 1)\varepsilon_1 - \alpha^{-1} \left(\varepsilon_2 - \sum_{k=1}^{m-1} a_k \varepsilon_{2+k}\right) \\ -\varepsilon_1 + \alpha^{-1} \left(\varepsilon_2 - \sum_{k=1}^{m-1} a_k \varepsilon_{2+k}\right) \end{bmatrix}, \\
\mathrm{Cov}(\hat{\beta}) &= \alpha^2(\alpha - 1)^{-1} \begin{bmatrix} \alpha & -1 \\ -1 & 1 \end{bmatrix}.
\end{aligned}$$

Of course, more elaborate cases can be constructed by combining the previous examples in various ways as in (e).

8.5 INTERPOLATION AND INVERSE CORRELATIONS

In this section the problem of interpolation of a missing value X_t based on all the other values $\{X_s; s \neq t\}$ of a stationary process is studied. This *interpolation problem* was first introduced and solved by Kolmogorov (1941) in the spectral-domain; see Section 9.9.3. Using ideas from Section 8.2 we provide a time-domain solution and introduce the *inverse correlation function* as a yet different reparameterization of stationary processes.

Let $H_t' = \overline{sp}\{X_s; s \neq t\}$ be the space of all linear interpolators for X_t. The *best linear interpolator* of X_t, in symbol $\hat{X}_t'$, is simply the orthogonal projection of X_t onto H_t', its *interpolation error* is $\varepsilon_t' = X_t - \hat{X}_t'$ with the *interpolation error variance* $\sigma'^2 = \mathrm{Var}(X_t - \hat{X}_t')$. While the nondeterministic and purely nondeterministic processes were of interest in dealing with prediction problems, for interpolation problems the analogous concepts are *minimal* and *purely minimal* processes. A stationary process $\{X_t\}$ is said to be *minimal* if

$$X_t \notin H_t',$$

and *purely minimal* if

$$\cap_t H_t' = \{0\}.$$

Note that if $\{X_t\}$ is not minimal, then $\varepsilon_t' \equiv 0$ or $\sigma'^2 = 0$ so that error-free interpolation is possible. Such processes are not of much interest in the sequel.

For a minimal process $\{X_t\}$, its standardized interpolation error is

$$X_t' = \varepsilon_t' \, \sigma'^{-2}, \tag{8.37}$$

and the process $\{X_t'\}$ is called the *standardized two-sided innovation*, see Masani (1960) and Salehi (1979), or *inverse process* of $\{X_t\}$, Cleveland (1972). The above normalization is chosen so that the inverse process $\{X_t'\}$ has $\dfrac{1}{f(\lambda)}$ as its spectral density function.

The notion of minimality is more rigid than that of nondeterminism and unlike innovation processes which are orthogonal (white noise), the interpolation errors or inverse processes are not as the following lemma shows.

Lemma 8.9. (a) A minimal stationary process is nondeterministic.

(b) A minimal stationary process $\{X_t\}$ and its inverse process $\{X_t'\}$ are *biorthogonal*, that is

$$\mathrm{Cov}\,(X_s, X_t') = \delta_{s,t},\ s, t \in \mathbb{Z}.$$

Proof. (a) Since for all t, $H_{t-1} \subseteq H_t'$, the assertion is immediate.

(b) Since $X_t - \hat{X}_t'$ is orthogonal to all $X_s, s \neq t$, it is immediate that

$$\mathrm{Cov}\,(X_s, X_t') = 0.$$

For $s = t$, using orthogonality of $\hat{X}'_t$ and $X_t - \hat{X}'_t$ we obtain

$$\begin{aligned} \text{Cov}\,(X_t, X'_t) &= \text{Cov}\,(X_t - \hat{X}'_t + \hat{X}'_t, X_t - \hat{X}'_t)/\sigma'^2 \\ &= \text{Var}\,(X_t - \hat{X}'_t)/\sigma'^2 = 1. \end{aligned}$$

Q.E.D.

What are the necessary and sufficient conditions for a stationary process to be (purely) minimal? Since minimal processes are nondeterministic, these conditions are expected to be more stringent than the integrability of the logarithm of the spectral density of the process (Theorem VII (a)). In the spectral domain, Kolmogorov (1941) showed that the integrability of f^{-1} is the necessary and sufficient condition for the minimality of $\{X_t\}$. In recent years, the close connection between problems of prediction and interpolation is revealed through the missing value problems studied in Sections 2 and 3, see Nakazi (1984); Miamee and Pourahmadi (1988a); Cheng, Miamee and Pourahmadi (1998). The result in Theorem 10.3 (d), sometimes referred to as the Beurling theorem, plays a central role in proving and unifying these diverse approximation problems.

Theorem 8.10. Let $\{X_t\}$ be a stationary process with the spectral distribution F decomposed into its absolutely continuous and singular parts as

$$dF(\lambda) = f(\lambda)d\lambda + dF_s(\lambda).$$

Then,

(a) $\{X_t\}$ is *minimal*, if and only if f is positive a.e. (Leb.) and $f^{-1} \in L^1$. In this case,

$$\sigma'^2 = \left(\int \frac{d\lambda}{f(\lambda)} \right)^{-1}. \tag{8.38}$$

(b) $\{X_t\}$ is *purely minimal*, if and only if it is minimal and

$$dF_s \equiv 0.$$

(Makagon and Weron, 1976; Salehi, 1979).

In this section we provide a time-domain characterization of minimality and express the interpolator, the inverse process and the inverse correlation function in terms of the AR parameters and the one-sided innovation of $\{X_t\}$. For alternative derivations of some of these results in the spectral-domain, see Masani (1960).

Interpolation of X_0 based on $\{X_t; t \leq m,\ t \neq 0\}$ studied in Section 8.2, and the formulas for $\hat{X}'_{0,m}$ and σ'^2_m can be used to compute $\hat{X}'_0, \sigma'^2$ and the inverse process $\{X'_t\}$. In fact, since $H_{m,0} \to H'_0$ as $m \to \infty$, it follows that

$$\hat{X}'_0 = \lim_{m\to\infty} \hat{X}'_{0,m}, \quad \sigma'^2 = \lim_{m\to\infty} \text{Var}(X_0 - \hat{X}'_{0,m}). \tag{8.39}$$

Now, from (8.7)–(8.9) assuming that

$$\sum_{i=1}^{\infty} a_i^2 < \infty, \tag{8.40}$$

it follows that

$$\sigma'^2 = \sigma^2 \lim_{m\to\infty} \left(1+\sum_{i=1}^{m} a_i^2\right)^{-1} = \sigma^2 \left(1+\sum_{i=1}^{\infty} a_i^2\right)^{-1} \neq 0, \tag{8.41}$$

and, for $k = 1, 2, \cdots$,

$$\rho'_k = \lim_{m\to\infty} \beta_{k,m} = \left(1+\sum_{i=1}^{\infty} a_i^2\right)^{-1} \left(a_k - \sum_{i=1}^{\infty} a_i a_{i+k}\right), \tag{8.42}$$

This simple argument is the main ingredient for a time-domain characterization of minimal processes.

Theorem 8.11. Let $\{X_t\}$ be a nondeterministic stationary process with the AR parameters $\{a_k\}$. Then, $\{X_t\}$ is minimal, if and only if $\sum_{i=1}^{\infty} a_i^2 < \infty$.

Proof. Sufficiency follows from (8.41). To prove the necessity, suppose that $\{X_t\}$ is minimal and $\sum_{i=1}^{\infty} a_i^2 = \infty$, then arguments leading to (8.41) implies that $\sigma'^2 = 0$ and hence contradicting the minimality of $\{X_t\}$.

Q.E.D.

Next, we express X'_t in terms of $\varepsilon_t, \varepsilon_{t+1}, \cdots$. For notational simplicity, we take $\sigma^2 = 1$.

Theorem 8.12. Let $\{X_t\}$ be a minimal stationary process with $\{X'_t\}$ its inverse process. Then,

(a) $X'_t = \varepsilon_t - \sum_{k=1}^{\infty} a_k \varepsilon_{t+k}$, where $a_1, a_2, \cdots$ are the AR parameters of $\{X_t\}$ and $\{\varepsilon_t\}$, is its innovation.

(b) $\{X'_t\}$ is a purely nondeterministic process with the covariance function $\{\gamma'_k\}$ given by

$$\gamma'_0 = 1 + \sum_{i=1}^{\infty} a_i^2, \gamma'_k = -a_k + \sum_{i=1}^{\infty} a_i a_{i+k}, k = 1, 2, \cdots.$$

the correlation function $\{\rho'_k\}$ as in (8.42) and the spectral density function $\dfrac{1}{f}$.

(c) The best linear interpolator $\hat{X}_t'$ of X_t based on $X_s, s \neq t$. and the inverse process $\{X_t'\}$ have the following formal representations:

$$\begin{cases} \hat{X}_t' &= -\gamma_0'^{-1} \sum_{k\neq 0} \gamma_k' X_{t-k} = -\sum_{k=1}^{\infty} \rho_k' (X_{t-k} + X_{t+k}), \\ X_t' &= X_t + \sum_{k=1}^{\infty} \rho_k' (X_{t-k} + X_{t+k}). \end{cases}$$

(d) The covariance functions $\{\gamma_k\}$ and $\{\gamma_k'\}$ of $\{X_t\}$ and $\{X_t'\}$ satisfy the following infinite system of linear equations:

$$\sum_{j=-\infty}^{+\infty} \gamma_j' \gamma_{k-j} = \delta_{k_0}, \quad k \in \mathbb{Z},$$

provided that the series for $\hat{X}_t'$ in (c) is mean-convergent.

Proof. (a) follows from the Lemma 8.5 (i) and the definition of X_t'. (b) follows from (a), Corollary 5.13 and the definition of the spectral density. (c) and (d) follow from Lemma 8.9 (b).

Q.E.D.

The *inverse correlation function* $\{\rho_k'\}$ was introduced by Cleveland (1972) as the Fourier coefficients of the reciprocal of the spectral density of $\{X_t\}$; see Section 3.4.3. The motivation for introducing and using $\{\rho_k'\}$ in time series model identification is that if $\{X_t\}$ is a stationary ARMA(p,q) process, then its inverse process $\{X_t'\}$ is an ARMA(q,p) process obtained by interchanging the AR and MA coefficients. This follows from the special form of the spectral density of ARMA models and Theorem 8.12 (b); see Problems 8.9–8.13. Thus, if $\{X_t\}$ conforms to an AR(p) model then $\{X_t'\}$ and hence $\{\rho_k'\}$ behave precisely like the correlation function of an MA(p) model, that is ρ_k' will be zero for all lags greater than p. On the other hand, if $\{X_t\}$ conforms to an MA(q) model, then $\{\rho_k'\}$ behave precisely like the correlation function of an AR(q) model and decays gradually to zero. Chatfield (1979) provides practical guides on the use of inverse autocorrelations and the relative merits of partial and inverse correlations as aids to time series model identification.

8.6 CANONICAL CORRELATION BETWEEN PAST AND FUTURE

In this section we study the canonical correlation or the angle between the infinite past and the infinite future of a stationary process. Conceptually and from Section 8.1, it is evident that the canonical correlation between the past and future is closely related to the linear prediction of functionals of the future. For example, a stationary process with only a finite number of non-zero canonical correlations is particularly interesting, since then a finite number of predictors determines the best predictor of any linear functional of the future. It turns out that this property, namely the finiteness of the number of non-zero canonical correlations between the past and the future characterizes ARMA models.

Computation of canonical correlations and finding the spectral necessary and sufficient conditions for positivity of the angle between the past and future are discussed in this section. Applications of the latter concept to the characterization of stationary processes forming a basis for their spans and the convergence of the series representations of the predictor and interpolator are studied.

8.6.1 Canonical Correlations

Gelfand and Yaglom (1957) generalized the familiar method of computing canonical correlation of two random vectors in Section 7.4 to two infinite collections of random variables. Later, Helson and Szegö (1960), Yaglom (1963) and Jewell and Bloomfield (1983) studied the canonical correlations between the past and the future of a stationary process.

For a stationary process $\{X_t\}$, let us denote by $\mathcal{P}$ and $\mathcal{F}_k$ the subspaces generated by the past values $\{X_t; t \leq 0\}$ and the future values beyond time k, that is, $\{X_t; t \geq k\}$, respectively. The canonical (maximal) correlation between the past and future beyond k is defined by

$$\rho_1(k) = \sup \{|\operatorname{Corr}(X,Y)|; X \in \mathcal{P}, Y \in \mathcal{F}_k\}. \tag{8.43}$$

For the time being, since k is fixed we suppress the dependence of $\rho_1(k)$ on k and simply use ρ_1 for the quantity in (8.43). If $U_1 \in \mathcal{P}$ and $V_1 \in \mathcal{F}_k$ have correlation equal to ρ_1, they are called *the first canonical components* of $\{X_t\}$. By analogy with the definition of successive canonical correlations in Section 7.4, we inductively define the i-th canonical correlation $\rho_i, i = 2, 3, \cdots$, by

$$\begin{aligned}\rho_i(k) = \sup \{|\operatorname{Corr}(X,Y)|; X \in \mathcal{P}, X \perp U_1, \cdots, U_{i-1} \text{ and} \\ Y \in \mathcal{F}_k, y \perp V_1 \cdots, V_{i-1}\}.\end{aligned} \tag{8.44}$$

Similarly, the *i-th canonical components* $U_i \in \mathcal{P}$ and $V_i \in \mathcal{F}_k$ have correlation equal to ρ_i while $U_i \perp U_1, \cdots, U_{i-1}$ and $V_i \perp V_1, \cdots, V_{i-1}$. From the definition of successive canonical correlations it is evident that

$$1 \geq \rho_1 \geq \rho_2 \geq \cdots \geq 0.$$

Note that if $\mathcal{P}$ and $\mathcal{F}_k$ have non-empty intersection we can obtain a correlation of 1 between them. However, even if $\mathcal{P} \bigcap \mathcal{F}_k$ is empty, still it is possible to obtain correlations between past and future elements arbitrarily close to 1. To give a geometric flavor to this phenomenon, we introduce the notion of the angle θ between the two subspaces by setting $\cos\theta = \rho_1$, and say the $\mathcal{P}$ and $\mathcal{F}_k$ are at *positive angle* if $\theta > 0$ or $\rho_1 < 1$.

The following two patterns at the start and the end of the above nonincreasing sequence $\{\rho_i\}$ of canonical correlations are of particular interest:

(i) The number of non-zero canonical correlations is finite.

(ii) The first canonical correlation ρ_1 is strictly less than one or equivalently *the angle between the past and future is positive.*

The case (i) which turns out to be equivalent to $\{X_t\}$ being an ARMA(p, q) process is of practical interest and lies at the core of Akaike's (1975) approach to Markovian (state-space) representations using canonical components corresponding to non-zero canonical correlations. Such characterization of ARMA models is also used implicitly by Tsay and Tiao (1984) to identify the orders p and q of such models by scanning the sample canonical correlations.

Jewell and Bloomfield (1983) established explicitly the equivalence between processes with a finite number of non-zero canonical correlations and ARMA processes. They exploited the fact that for $\{X_t\}$ with the spectral density $f = |\varphi|^2, \varphi \in H^2$ outer (Chapter 10), the canonical correlations are, indeed, the spectrum of the Hankel operator on L^2 with the symbol $\frac{\varphi}{\bar{\varphi}}$, and the rank of this operator is finite when φ is rational or $\{X_t\}$ is an ARMA process. Alternative proofs and more general results were provided later by Shigeno (1989) and Vaninskii and Yaglom (1990) using more common tools from prediction theory. The following result is from Vaninskii and Yaglom (1990, Theorem 2).

Theorem 8.13. Let $\{X_t\}$ be a purely nondeterministic stationary process and let $k \geq 1$ and $M > 0$ be two integers. Then, the number of non-zero canonical correlations of $\{X_t\}$ defined by (8.43)–(8.44) is M, if and only if its spectral density $f(\lambda)$ is a rational function in $e^{i\lambda}$ so that

$$f(\lambda) = \frac{\sigma^2}{2\pi} \left| \frac{\beta(e^{i\lambda})}{\alpha(e^{i\lambda})} \right|^2,$$

where $\alpha(e^{i\lambda})$ and $\beta(e^{i\lambda})$ are polynomials of degrees at most M and $M + k - 1$, all the roots of the former are outside the unit circle and those of the latter are on or outside the unit circle.

Proof of this theorem and others in this section are rather long and require deeper results from the theory of Hardy spaces (Chapter 10) and functional analysis. We do not present these proofs here. The close connection between the canonical correlations of stationary processes and the Hankel operators on L^2 seems to have been discovered first by Grenander (1981). Peller and Khruschev (1982) used it extensively to develop a unified approach to characterize and/or obtain rates of convergence of some of the mixing conditions in Section 5.6.6. For the multivariate analogues of such results, see Peller (2000).

8.6.2 Angle Between Past and Future

The case (ii), that is positivity of the angle between past and future, is more of theoretical interest. It was studied first by Helson and Szegö (1960) for $k = 1$ and later by Helson and Sarason (1967) for $k > 1$. The latter work was in response to the

following Yaglom's (1963) question:
What are the (spectral) necessary and sufficient conditions for $\rho_1(k) \to 0$ as $k \to \infty$?

It is evident from (5.79) that $\alpha(k) \le \rho_1(k)$ regardless of the distribution of $\{X_t\}$. But for Gaussian processes, it was shown by Kolmogorov and Rozanov (1960) that $\rho_1(k) \le 2\pi\alpha(k)$, so that a positive answer to above question provides necessary and sufficient conditions for a Gaussian process to be strongly mixing, see Rosenblatt (1956) and Ibragimov and Rozanov (1978). The following theorems provide spectral characterizations of the aforementioned properties.

Theorem 8.14 Let $\{X_t\}$ be a stationary process with the spectral distribution F decomposed into its absolutely continuous and singular parts as

$$dF(\lambda) = f(\lambda)d\lambda + dF_s(\lambda).$$

Then, the following are equivalent:

(a) $\rho_1(1) < 1$,

(b) $dF_s(\lambda) \equiv 0$ and $f = e^{u+\tilde{v}}$, where u is a bounded real function, and $\tilde{v}$ is the conjugate of a bounded real function v with $\|v\|_\infty < \frac{\pi}{2}$. [For the definition of conjugate function, see Section 10.5.]

(c) $dF_v(\lambda) \equiv 0$ and there exists a positive real constant K such that for all subintervals I of the unit circle with length $|I|$,

$$\frac{1}{|I|}\int_I f(\lambda)d\lambda \cdot \frac{1}{|I|}\int_I f^{-1}(\lambda)d\lambda \le K. \tag{A$_2$}$$

(d) The exponentials $\{e^{ik\lambda}\}$ form a basis in $L^2(dF)$. Namely, every element $h \in L^2(dF)$ has a unique expansion

$$h(\lambda) \sim \sum_{j=-\infty}^{\infty} \hat{h}_j e^{ij\lambda},$$

which is convergent in the norm of $L^2(dF)$.

This theorem with the exception of part (c) was proved by Helson and Szegö (1960), part (c) was proved later by Hunt, Muckenhoupt and Wheeden (1973). The condition A_2 above is usually called the Muckenhoupt A_2-condition. It can be shown that the unbounded spectral density function $f(\lambda) = |1 - e^{i\lambda}|^{-2\alpha}$, $0 < \alpha < \frac{1}{2}$, satisfies the A_2-condition or has a representation as in the part (b) of the above theorem. For $k > 1$ there is a characterization similar to that in Theorem 8.14 for $\rho_1(k) < 1$.

Theorem 8.15 (Helson and Sarason, 1967). Let $\{X_t\}$ be a stationary process with the spectral distribution F decomposed as

$$dF(\lambda) = f(\lambda)d\lambda + dF_s(\lambda).$$

Then, the following are equivalent:
(a) $\rho_1(k) < 1$,
(b) $dF_s(\lambda) \equiv 0$ and

$$f = |P|^2 e^{u+\tilde{v}},$$

where u and v are as in Theorem 8.14 (b) and P is a polynomial of degree $k - 1$ in $e^{i\lambda}$ with all roots outside the unit circle.

From the representation of the spectral density function f in the above theorem it is immediate that any invertible MA$(k - 1)$ model enjoys the property $\rho_1(k) < 1$, that is, its past and future k units apart are at positive angle.

Finally, the following result shows that for $\rho_1(k) \to 0$ as $k \to \infty$, the spectral density function has to be much smoother than that in Theorem 8.15 (b).

Theorem 8.16. (Helson and Sarason, 1967; Sarason, 1972). Let $\{X_t\}$ be a stationary process with the spectral distribution

$$dF(\lambda) = f(\lambda)d\lambda + dF_s(\lambda).$$

Then, the following are equivalent:

(a) $\lim_{k\to\infty} \rho_1(k) = 0$,

(b) $dF_s(\lambda) \equiv 0$ and

$$f = |P|^2 e^{u+\tilde{v}},$$

where u and v are continuous real functions with $||v||_\infty < \frac{\pi}{2}$ and P is a polynomial with roots on the unit circle.

(c) $dF_s(\lambda) \equiv 0$ and

$$f = |P|^2 e^g,$$

where P is as before and g is a function of vanishing mean oscillation (VMO), that is,

$$\sup_I \frac{1}{|I|} \int_I |g - g_I| d\lambda \to 0 \text{ as } |I| \to 0,$$

with I a subarc of the unit circle, $|I|$ its length and $g_I = \frac{1}{|I|} \int_I g d\lambda$ is the average of g over I.

For more information on the connection among the three classes of spectral densities appearing in the above theorems; see Garnnet (1981). For a generalization of Theorem 8.16 to the setup of multivariate stationary processes, see Treil and Volberg (1999) and references therein.

8.6.3 Abstract Past and Future

The concept of canonical correlation between the past and future of a stochastic process can be generalized to any two subspaces M and N of an abstract Hilbert space by

$$\rho = \sup \{|(X,Y)|; X \in M, Y \in N, ||X|| = ||Y|| = 1\}.$$

The angle $\theta = \theta(M,N)$ between M and N is defined by $\cos\theta = \rho$, to ensure unique solution θ is restricted to the range $[0, \frac{\pi}{2}]$. The angle between M and N is said to be *positive* if $\theta > 0$ or equivalently $\rho < 1$.

In the rest of this section we study additional properties and consequences of the positivity of the angle which are useful either in proving some of the earlier theorems or are of independent interest. The next lemma provides equivalent geometric conditions for the positivity of the angle between two subspaces.

Theorem 8.17 Let M and N be any two subspaces of a Hilbert space H. Then the following are equivalent:

(a) $\rho = \rho(M,N) < 1$.

(b) $d(M,N) = inf\{||X - Y||; X \in M, Y \in N, ||X|| = ||Y|| = 1\} > 0$, that is, the unit spheres of M and N are at a positive distance.

(c) There is a constant K such that

$$||X|| \le K||X + Y||, \text{ for all } X \in M, Y \in N,$$

that is, the *truncation operator* $T : X + Y \to X$, is bounded.

(d) The vector sum $M \oplus N$ is a (closed) subspace of H; see Section 9.6.2.

Proof. Equivalence of (a) and (b) follows from the identity

$$||X - Y||^2 = ||X||^2 + ||Y||^2 - (X,Y) - (Y,X) = 2[1 - Re(X,Y)],$$

for all $X \in M$, $Y \in N$ with $||X|| = ||Y|| = 1$. (c) clearly implies (b) with $d(M,N) \ge K^{-1}$. Next, we prove (b) $\Rightarrow$ (c) by showing that not (c) implies not (b). Assume that (c) is not satisfied. Then there are $X_n \in M, Y_n \in N$ such that

$$0 < n||X_n - Y_n|| \le ||X_n||,$$

from which it follows that $||Y_n|| > 0$ and

$$\begin{aligned} \left\| \frac{X_n}{||X_n||} - \frac{Y_n}{||Y_n||} \right\| &= \left\| \frac{X_n}{||X_n||} - \frac{Y_n}{||X_n||} + \frac{Y_n}{||X_n||} - \frac{Y_n}{||Y_n||} \right\| \\ &\le \frac{||X_n - Y_n||}{||X_n||} + \frac{||Y_n - X_n||}{||X_n||} \\ &= \frac{2}{||X_n||}||X_n - Y_n|| < \frac{2}{n} \to 0, \end{aligned}$$

which contradicts (b).

(c) $\Longrightarrow$ (d). Suppose a sequence $\{Z_n\}$ with $Z_n \in M \oplus N$ converges to Z, we must show that $Z \in M \oplus N$. Let $Z_n = X_n + Y_n$ with $X_n \in M, Y_n \in N$. Then from (c), we have

$$\begin{aligned} \|X_m - X_n\| &\leq K\|Z_m - Z_n\|, \\ \|Y_m - Y_n\| &\leq K\|Z_m - Z_n\|. \end{aligned}$$

Since $\{Z_n\}$ is convergent and hence a Cauchy sequence, it follows that $\{X_n\}$ and $\{Y_n\}$ are both Cauchy sequences and therefore converge to some $X \in M$ and $Y \in N$, respectively. Thus, $Z = X + Y \in M \oplus N$, that is, the latter is closed.

(d) $\Longrightarrow$ (c). The operator $T : M \oplus N \to M$ defined by $T(X + Y) = X$ is linear and closed, hence by the Closed Graph Theorem (Royden, 1968) is bounded, this gives (c).

Q.E.D.

8.6.4 Basicity and Positivity of the Angle

We show that positivity of the angle between past and future of a stationary process is equivalent to $\{X_t\}$ forming a basis for its span $H(X)$. We say that a sequence $\{X_t\}$ forms a *Schauder basis* for its span $H(X)$, if for each $Y \in H(X)$, there is a unique sequence of scalars $\{c_k\}$ so that

$$Y = \sum_{-\infty}^{\infty} c_t X_t,$$

and the infinite series converges in the norm of $H(X)$. The following theorem due to Nikolskii, see Lacey (1974), provides a characterization of $\{X_t\}$ forming a basis for its span.

Theorem 8.18. *(Nikolskii).* A sequence $\{X_t\}$ is a Schauder basis for its span, if and only if there exists a real constant $K > 0$ such that

$$\left\| \sum_{k}^{l} c_t X_t \right\| \leq K \left\| \sum_{-\infty}^{\infty} c_t X_t \right\|,$$

for all $l \geq k$ and scalars $\{c_t\}$ with $\sum_{-\infty}^{\infty} c_t X_t \in H(X)$. [That is, the truncation is a bounded linear operator on $H(X)$, see Theorem 8.17 (c).]

With the help of this theorem and Theorem 8.17 we prove the main result of this section connecting positivity of the angle and the process forming a basis for its span.

Theorem 8.19. The past and future of a stationary process $\{X_t\}$ are at a positive angle, if and only if $\{X_t\}$ forms a Schauder basis for its span $H(X)$.

Proof. Suppose $\{X_t\}$ forms a basis for its span. Then from Theorem 8.18 with $l = 0, k = -\infty$, or $M = P$ and $N = F$, and Theorem 8.17 (a) and (c) the result follows. Proof of the converse is similar.

Q.E.D.

As a corollary of Theorem 8.18 we show that when $\{X_t\}$ forms a basis for its span, then $\{X_t\}$ is purely minimal.

Corollary 8.20. A process $\{X_t\}$ forming a Schauder basis for its span is purely minimal, i.e. $\cap_t H_t' = \{0\}$ where $H_t' = \overline{sp}\{X_s;\ s \neq t\}$.

Proof. Suppose that $0 \neq Y \in \cap_t H_t'$. Since $\{X_t\}$ forms a basis for $H(X)$, any $Y \in H(X)$ can be written uniquely as $Y = \sum_{-\infty}^{\infty} c_t X_t$. But $Y \in H_t'$ for each t, implies that $c_t = 0$, for $all\ t$, and therefore $Y = 0$.

Q.E.D.

Another consequence of the positivity of the angle or $\{X_t\}$ forming a basis for $H(X)$ is that every element of the time-domain has a mean-convergent series representation in terms of elements of $\{X_t\}$. In particular, this holds for $\hat{X}_t$ and $\hat{X}_t'$, the best linear predictor and interpolator, respectively.

8.7 COMPUTATION OF CANONICAL CORRELATIONS

The standard method of computing canonical correlations as developed in Section 7.4.2 leads to complicated equations involving inverse of infinite matrices when $m = n = \infty$. Jewell, Bloomfield and Bartmann (1983) extend this finite-dimensional approach to compute the canonical correlations of the past and future of a stationary process by replacing the correlation matrix by a bounded operator acting on infinite-dimensional subspaces. They develop a computational scheme for estimating the canonical correlations and components based on a convergent sequence of approximations using the power method of computing eigenvalues and eigenvectors of the Hankel operator with symbol $\frac{\varphi}{\bar{\varphi}}$. In this section, we use the following property of the orthogonal projection to develop a method for approximating ρ using only finite matrices: For N a subspace of H and $X \in H$ with P_N^X its orthogonal projection, we have

$$(X, P_N^X) = ||X||^2 - ||X - P_N^X||^2. \tag{8.45}$$

The next lemma which is an extension of the above identity reveals the deeper connection between prediction and correlation.

Lemma 8.21. Let M and N be two subspaces of H. Then,

$$\sup\{|\text{Corr}(X, Y)|; X \in M, Y \in N\} = \sqrt{1 - \inf_{\substack{X \in M \\ ||X||=1}} ||X - P_N^X||^2},$$

where P_N^X stands for the orthogonal projection of X onto N.

Proof. It is obvious that

$$\{\text{Corr}(X, P_N^X); X \in M\} \subseteq \{\text{Corr}(X, Y); X \in M, Y \in N\},$$

and therefore

$$\sup\{|\text{Corr}(X, P_N^X)|; X \in M\} \leq \sup\{|\text{Corr}(X, Y)|; X \in M, Y \in N\}. \tag{8.46}$$

For any $X \in M$, we have from (8.43) that

$$|\text{Corr}(X, Y)| \leq |\text{Corr}(X, P_N^X)|, \text{ for } all\ Y \in N,$$

which implies

$$\sup\{|\text{Corr}(X, Y)|; X \in M, Y \in N\} \leq \sup\{|\text{Corr}(X, P_N^X)|; X \in M\}. \tag{8.47}$$

It follows from (8.46) and (8.47) that

$$\sup\{|\text{Corr}(X, Y)|; X \in M, Y \in N\} = \sup\{|\text{Corr}(X, P_N^X)|; X \in M\}. \tag{8.48}$$

Also for any $X \in M$,

$$\begin{aligned}
\text{Corr}(X, P_N^X) &= \frac{(X, P_N^X)}{||X||||P_N^X||} = \frac{||P_N^X||^2}{||X||||P_N^X||} = \frac{||P_N^X||}{||X||} \\
&= \frac{\sqrt{||X||^2 - ||X - P_N^X||^2||X||}}{||X||} \\
&= \sqrt{1 - \left\|\frac{X}{||X||} - P_N \frac{X}{||X||}\right\|^2} \\
&= \sqrt{1 - ||Y - P_N^Y||^2},
\end{aligned}$$

where $Y = \dfrac{X}{||X||} \in M$.

But from (8.45) for any $Y \in M$ with $||Y|| = 1$ we have

$$\text{Corr}(Y, P_N^Y) = \sqrt{1 - ||Y - P_N^Y||^2},$$

it follows that

$$\{|\text{Corr}(X, P_N^X)|; X \in M\} = \left\{\sqrt{1 - ||Y - P_N^Y||^2};\ Y \in M,\ ||Y|| = 1\right\},$$

which means that

$$\sup\{|\text{Corr}(X, P_N^X)|; X \in M\} = \sqrt{1 - \inf_{\substack{X \in M \\ \|X\|=1}} \|X - P_N^X\|^2}. \tag{8.49}$$

Now, the desired result follows from (8.48) and (8.49).

Q.E.D.

For an integer $m \geq 1$, let

$$F_m = sp\{X_t; 1 \leq t \leq m\}, \rho_m = \rho(P, F_m).$$

Theorem 8.22. For $\{X_t\}$ a stationary process and notation as above we have

$$\rho = \rho(P, F) = \lim_{m \to \infty} \rho_m.$$

Proof. It is clear that the sequence $\{\rho_m\}$ is bounded and nondecreasing, therefore it is convergent and

$$\lim_{m \to \infty} \rho_m = \sup_m \rho_m. \tag{8.50}$$

Since $F_m \subseteq F$, it follows that

$$\rho_m \leq \rho, \text{ for } all\ m \geq 1,$$

and therefore from this and (8.50),

$$\lim_{m \to \infty} \rho_m \leq \rho. \tag{8.51}$$

To establish equality in (8.51), we note that for any two finite linear combinations $X = \sum_{k=0}^{n} a_k X_{-k}$ in P and $Y = \sum_{k=1}^{m} b_k X_k$ in F, we have

$$|\text{Corr}(X, Y)| \leq \rho_m \leq \sup_m \rho_m. \tag{8.52}$$

By taking the supremum of the left-hand side of (8.52) over all X and Y as above, we arrive at

$$\rho \leq \sup_m \rho_m. \tag{8.53}$$

Now, the desired result follows from (8.50), (8.51), and (8.53).

Q.E.D.

Note that ρ_m is the canonical correlation between $(\cdots, X_{-2}, X_{-1}, X_0)$ and $(X_1, \cdots, X_m)$, and Theorem 8.12 makes it possible to compute ρ_m using finite $m \times m$ matrices with no matrix inversion; see Theorem 8.1, Examples 8.1 and 8.2. However, to compute G_m as in (8.2)–(8.3) we need to factorize the covariance matrix Γ or the spectral density function of the process, see Miamee and Pourahmadi (1992).

Problems

8.1 (a) Find the best predictable linear functions of the future of the process in Example 8.1 for $m = 3, 4$.

(b) Prove (8.5), and compute D_m as a function of λ.

8.2 Find the best predictable linear functions of the future two values ($m = 2$) of the process

$$X_t = .6X_{t-1} + \varepsilon_t + .4\varepsilon_{t-1}, \operatorname{Var}(\varepsilon_t) = 1.$$

8.3 Prove Lemma 8.5 (iv) and provide details of the proof of other parts of the Lemma.

8.4 Derive the formula for $\hat{X}_{0,m}$ in Example 8.3.

8.5 Express the elements of the matrices A and B in (8.13) and (8.14), in terms of the AR parameters of the process $\{X_t\}$. For AR(p) and MA(1) processes compute the matrices D, C and $\operatorname{Cov}(X_M - \hat{X}'_M)$. Prove Theorem 8.7.

8.6 Let $\{X_t\}$ be a nondeterministic stationary process and $m \geq 1$ be an integer.

(a) Compute the linear least squares predictor of X_{m+1} based on $\{X_t; t \leq m, t \neq 0, 1\}$
and its prediction error variance.

(b) Compute the linear least squares predictor of X_0 based on

$$\{X_t; t = m, -1, -2, \cdots\}$$

and its prediction error variance. What happens to the prediction error variance as m gets larger?

(c) Discuss the results in (a) and (b) when $\{X_t\}$ is an invertible MA(1) process. Do the same for a stationary AR(1) process.

8.7 Show that the process $\{X_t\}$ of Example 8.3 (ii) is not minimal when $|b| = 1$.

8.8 Is the process $X_t = \varepsilon_t + 2\varepsilon_{t-1} + \varepsilon_{t-2}$ minimal? How about $X_t = \varepsilon_t + 3\varepsilon_{t-1} + 3\varepsilon_{t-2} + \varepsilon_{t-3}$?

8.9 Show that any stationary AR(p) process is minimal. Find its inverse process $\{X'_t\}$ and covariance function $\{\gamma'_k\}$. Compare this covariance function with that of an MA(p) process.

8.10 (a) Prove that a stationary process $\{X_t\}$ is an AR(p) process, if and only if

$$\gamma'_k = 0, \text{ for } |k| > p, \text{ and } \gamma'_p \neq 0.$$

(b) For a real-valued stationary AR(p) process with covariance function $\{\gamma_k\}$ and inverse covariance function $\{\gamma'_k\}$, define

$$\gamma = (\gamma_1, \cdots, \gamma_p) \quad , \quad \tilde{\gamma} = (\gamma'_1, \cdots, \gamma'_p)$$
$$\Gamma_p = (\gamma_{i-j})_{i,j=1}^{p} \quad , \quad H_p = (\gamma_{i+j})_{i,j=1}^{p}.$$

Show that the linear interpolator of X_t based on $\{X_s; s \neq t\}$ is given by

$$\hat{X}_t' = -\sum_{i=1}^{p} \gamma_i'(X_{t-i} + X_{t+i}),$$

and hence its coefficient vector $\tilde{\gamma}$ satisfies the normal equation

$$\tilde{\gamma} = -(\Gamma_p + H_p)^{-1}\gamma.$$

(Kanto, 1984).

(c) Use the result in (b) to verify the calculations in Example 8.3 (i) for $p = 1$ and 2.

8.11 Let $\{X_t\}$ be a minimal stationary process and k be a fixed positive integer.

(a) Compute $\hat{X}_0$ and $\hat{X}_k$, the linear least squares predictor of X_0 and X_k based on $\{X_t; t \neq 0, k\}$.

(b) Let $\eta_k = X_k - \hat{X}_k$ be the interpolation error. Compute $\text{Var}(\eta_k)$, $\text{Var}(\eta_0)$ and $\text{Cov}(\eta_0, \eta_k)$.

(c) Show that,

$$\text{Corr}\,(\eta_0, \eta_k) = -\rho_k'.$$

Thus, the inverse correlation ρ_k' is, indeed, the partial correlation between X_0 and X_k after eliminating the effect of the remaining random variables $\{X_t; t \neq 0, k\}$, Kanto (1984).

8.12 Let $\{X_t\}$ be a minimal stationary process. Show that for any $Y \in H(X)$ and integer $m \geq 0$, there are unique scalars c_k, $|k| \leq m$, and random variable $e_m \in \overline{sp}\{X_t; |t| > m\}$ such that

$$Y = \sum_{k=-m}^{m} c_k X_k + e_m.$$

Hint: Use the Regression Lemma of Chapter 6.

8.13 Let $\{X_t\}$ be a minimal stationary process and $\{\rho_k'\}$ be its inverse correlation function. Show that for any integer $m > 0$, we have

(a) $\hat{X}_t' = \sum_{|k|>0}^{m} \rho_k' X_{t-k} + e_{t,m}$, where $e_{t,m} \in \overline{sp}\{X_s; |s| > m\}$.

(b) $\hat{X}_t' = \sum_{k=1}^{m} \rho_k'(X_{t-k} + X_{t+k}) + e_{t,m}$.

(c) Find $\hat{X}_t'$ for an AR(p) process when $m \geq p$.

(d) Find $\hat{X}_t'$ for the MA(1) process $X_t = \varepsilon_t + b\varepsilon_{t-1}$, $|b| \leq 1$.

8.14 Let $\{X_t\}$ be a minimal stationary process. Prove that

$$\cap_t H_t' = \cap_{n=0}^{\infty} \overline{sp}\{X_t; |t| > n\}.$$

9

Stationary Sequences in Hilbert Spaces

The use of Hilbert space framework for solving the linear least squares prediction problems of stationary sequences was initiated by Kolmogorov (1941). This framework has proved indispensable for the study of similar problems in the more general setups of multivariate stationary sequences pursued by Wiener and Masani (1957, 1958); Rozanov (1967) and multiparameter stationary sequences (random fields) pursued by Helson and Lowdenslager (1958). Practical and mathematical ramifications of some results in this framework are discussed and reviewed more thoroughly in Robinson (1954); Parzen (1967); Mandrekar and Salehi (1983); various discussants of Norbert Wiener's collected work edited by Masani (1981); Helson (1964) and Rosenblum and Rovnyak (1985).

The first goal of this chapter is to provide some of the background and introductory material on Hilbert spaces needed in prediction theory of stochastic processes and more generally in probability and statistics. We introduce the basics of inner product spaces, complete normed spaces, orthogonal projections and bounded operators in the first five sections. The second goal is to rely on more advanced results like the spectral theory of unitary operators to establish most of the "spectral-domain" results for stationary sequences. The algebra of projection operators, the role of the angle between subspaces when projecting on the sum of subspaces and the ubiquitous von Neumann alternating projections algorithm are discussed in Section 9.6. Section 9.7 is concerned with the Herglotz representation of positive-definite functions as the Fourier transform of positive measures, the general theory of positive-definite functions and the reproducing kernel Hilbert spaces. These are used in Section 9.8 to prove the spectral theorem for unitary operators by noting that the function $\gamma_k = (U^k x, x)$ is positive-definite on the set of integers for any x from a Hilbert space and

U a unitary operator. Having proved the spectral theorem for unitary operators, one is within the easy reach of the proof of the spectral representation of stationary processes; this is elaborated in Section 9.9. The latter representation is then used extensively in solving the linear filtering and the subordination problem of stationary processes. This topic is rather general and includes the prediction and interpolation problems as special cases. The abstract Wold decomposition of a Hilbert space induced by an isometry from which most of the "time-domain" and "spectral-domain" results follow is covered in Chapter 10.

We should note that there are two apparently distinct, though similar, approaches to proving the spectral representation theorem for a stationary sequence. The first using the notion of stochastic integrals was developed by Cramer (1940) and employed by Doob (1944, 1953); Grenander and Rosenblatt (1957); Anderson (1971); Brillinger (1981); Lambert and Poskitt (1983); Brockwell and Davis (1991) and Fuller (1996), among others. The second approach exploiting the spectral theorem of unitary operators was advocated by Kolmogorov (1941) and used subsequently by Robinson (1954); Wiener and Masani (1957, 1958); Rozanov (1967) and Grenander (1981). Needless to say our approach here follows the latter approach. A mathematically astute reader interested in the theory of stationary sequences can go directly to Section 9.5.3 for the connection between stationary sequences and unitary operators.

9.1 INNER PRODUCT SPACES

In this section we show that the intuition gained from finite-dimensional spaces carries over to infinite-dimensional spaces and is the source of inspiration for some important concepts unifying and linking various apparently distinct areas of physical and mathematical sciences. The catalyst in this enterprise is the concept of *inner product* which is important in defining the notions of length, distance, angle, energy, variance and covariance, just to name a few. Most of the results on inner product spaces and operators introduced in the sequel can be found in Halmos (1957); Akhiezer and Glazman (1963), and Young (1989).

9.1.1 Definition and Examples

A complex vector space V is said to be an *inner product space* if for each $x, y \in V$ there can be defined a complex number (x, y), called the *inner product* of x and y, with the following properties:

(i) *Linearity:* $(\alpha x + \beta y, z) = \alpha(x, z) + \beta(y, z)$, for all $x, y, z \in V$ and complex scalars α, β.

(ii) *Conjugate Symmetry:* $(x, y) = \overline{(y, x)}$.

(iii) *Positivity:* $(x, x) \geq 0$, and $(x, x) = 0$ if and only if $x = 0$.

Hereafter, for a complex number $z = a + ib$, $\bar{z}$ stands for its complex conjugate, that is, $\bar{z} = a - ib$.

Example 9.1. Important examples of inner product spaces are listed below. When V is a space of functions on a set T, we use $f, g, h, \cdots$, to denote vectors in V.

(a) The complex Euclidean space $\mathbb{C}^n = \{x; x = (x_1, \cdots, x_n)', x_i \in \mathbb{C}, i = 1, \cdots, n\}$ with (x, y) defined by

$$(x, y) = \sum_{i=1}^{n} x_i \bar{y}_i, x, y \in \mathbb{C}^n,$$

is an inner product space.

(b) The set $\ell^2 = \left\{ x; x = (x_1, x_2, \cdots), x_i \in \mathbb{C}, i = 1, 2, \cdots, \sum_{i=1}^{\infty} |x_i|^2 < \infty \right\}$ with componentwise addition and scalar multiplication defined by

$$\begin{aligned} x + y &= (x_1 + y_1, x_2 + y_2, \cdots), \\ \lambda x &= (\lambda x_1, \lambda x_2, \cdots), \end{aligned}$$

for all $x, y \in \ell^2, \lambda \in \mathbb{C}$, is a vector space; see Problem 9.4. The complex number (x, y) defined by

$$(x, y) = \sum_{i=1}^{\infty} x_i \bar{y}_i, x, y \in \ell^2,$$

can be shown to be an inner product and hence ℓ^2 is an inner product space.

(c) Let $C[a, b]$ stand for the space of all continuous complex-valued functions on the interval $[a, b], a < b$, with the usual pointwise addition and scalar multiplication. For $f, g \in C[a, b]$, the complex number

$$(f, g) = \int_a^b f(t)\overline{g(t)}dt,$$

defines an inner product on $C[a, b]$.

For an integer $n \geq 0$, let $\mathcal{P}_n$ be the set of all polynomials on [a, b] of degrees at most n of the form $p(t) = \sum_{k=0}^{n} p_k t^k$ where p_k's are complex scalars. Of course, $\mathcal{P}_n$ is a subspace of $C[a, b]$ and hence an inner product space with inner product

$$(p, q) = \int_a^b p(t)\overline{q(t)}dt = \sum_{k=0}^{n} \sum_{\ell=0}^{n} p_k \overline{q}_\ell \int_a^b t^{k+\ell} dt, p, q \in \mathcal{P}_n.$$

More generally, $\mathcal{P}$ the space of all polynomials on [a, b] of arbitrary but finite degrees is a subspace of $C[a, b]$ and hence an inner product space.

(d) Let $W[a,b]$ stand for the space of all continuously differentiable functions on $[a,b]$. For $f \in W[a,b]$, let f' denote its derivative. Then, for $f, g \in W[a,b]$, the complex number $(f,g)_W$ given by

$$(f,g)_W = \int_a^b [f(t)\overline{g(t)} + f'(t)\overline{g'(t)}]dt,$$

defines an inner product on the space $W[a,b]$.

(e) Let Ω be an arbitrary set, $\mathcal{F}$ a σ-algebra of subsets of Ω and μ a nonnegative measure on $\mathcal{F}$. Define $L^2(\Omega, \mathcal{F}, \mu) = \{f; f : \Omega \to \mathbb{C}$ is a measurable function with $\int_\Omega |f|^2 d\mu < \infty\}$. For $f, g \in L^2(\Omega, \mathcal{F}, \mu)$, the complex number

$$(f,g) = \int_\Omega f\bar{g}d\mu,$$

defines an inner product on $L^2(\Omega, \mathcal{F}, \mu)$. Specific choice of Ω and μ gives rise to specific inner product spaces encountered in various areas of applications. For example, when $\Omega = \{1, 2, \cdots, n\}$ or $\{1, 2, \cdots, \}$ and $\mathcal{F}$ is all subsets of Ω with μ the counting measure on $\mathcal{F}$, then $L^2(\Omega, \mathcal{F}, \mu)$ reduces to $\mathbb{C}^n$ and ℓ^2, respectively. For $\Omega = (-\pi, \pi], \mathcal{F}$ Borel subsets of Ω and μ a finite measure, we write $L^2(\mu)$ for the space of all square integrable functions on $(-\pi, \pi]$ or the unit circle in the complex plane. For μ the normalized Lebesgue measure on the unit circle we simply write L^2 for $L^2(\mu)$. When μ is a probability measure, that is, $\mu(\Omega) = 1$ for an arbitrary set Ω, $L^2(\Omega, \mathcal{F}, \mu)$ is composed (thought) of all random variables with finite second moments. For more information on topics from real analysis, particularly the L^p spaces, see Royden (1968).

There is a price to be paid for the broad generality of $L^2(\Omega, \mathcal{F}, \mu)$. Unfortunately, the elements of $L^2(\Omega, \mathcal{F}, \mu)$ are not in general bona fide functions, rather they are equivalent classes of functions on Ω which differ on sets of measure zero. Thus for an $f \in L^2(\Omega, \mathcal{F}, \mu)$ and $\omega \in \Omega$ we are unable to ascertain what $f(\omega)$ is. Some examples of inner product spaces of (genuine) functions are given next.

(f) Let $\mathbb{D} = \{z \in \mathbb{C}; |z| < 1\}$ be the open unit disc in the complex plane. Then,

$$\tilde{H}^2 = \left\{ F; F : \mathbb{D} \to \mathbb{C}, F(z) = \sum_{k=0}^{\infty} c_k z^k, \sum_{k=0}^{\infty} |c_k|^2 < \infty \right\},$$

is a vector space with the usual pointwise addition and multiplication of functions. Since $c_k \to 0$, functions in $\tilde{H}^2$ are analytic in $\mathbb{D}$and hence infinitely differentiable. For $F, G \in \tilde{H}^2$ with $F(z) = \sum_{k=0}^{\infty} c_k z^k$ and $G(z) = \sum_{k=0}^{\infty} d_k z^k$, the complex number

$$(F,G)_{\tilde{H}^2} = \sum_{k=0}^{\infty} c_k \bar{d}_k,$$

defines an inner product on $\tilde{H}^2$ which is closely related to ℓ^2. For more information on Fourier series, analytic and harmonic functions; see Chapter 10 and Hoffman (1962).

Let $\partial\mathbb{D}$ denote the unit circle (the boundary of $\mathbb{D}$) in the complex plane and $e_n(\lambda) = e^{in\lambda}, \lambda \in (-\pi, \pi], n = 0, \pm 1, \pm 2, \cdots$. Define

$$H^2 = \left\{ f; f : \partial\mathbb{D} \to \mathbb{C}, f(e^{i\lambda}) \sim \sum_{k=0}^{\infty} c_k e_k(\lambda), \sum_{k=0}^{\infty} |c_k|^2 < \infty \right\}.$$

For $f, g \in H^2$, the complex number

$$(f, g) = \sum_{k=0}^{\infty} c_k \bar{d}_k, g(e^{i\lambda}) \sim \sum_{k=0}^{\infty} d_k e_k(\lambda),$$

defines an inner product on H^2. There is close relationship among elements and inner products in ℓ^2, H^2 and $\tilde{H}^2$. These function spaces play fundamental roles in prediction theory of stationary processes as illustrated in Chapter 10. The same is true of the next example which subsumes $\tilde{H}^2$.

(g) Let T be an arbitrary set and $K : T \times T \to \mathbb{C}$ be a complex-valued function. For a fixed $s \in T$, let $K_s(\cdot) : T \to \mathbb{C}$ be the function $K_s(t) = K(s,t), t \in T$. Define $V(K)$ as the vector space of all finite linear combinations of the form $\sum_{i=1}^{n} c_i K_{s_i}(\cdot)$, for arbitrary integer $n \geq 1$, scalars $c_1, \cdots, c_n$ and $s_1, \cdots, s_n \in T$.

One can show that for $f, g \in V(K)$ with $f = \sum_{i=1}^{n} c_i K_{s_i}(\cdot)$ and $g = \sum_{j=1}^{m} d_j K_{t_j}(\cdot)$, the complex number

$$(f, g)_K = \sum_{i=1}^{n} \sum_{j=1}^{m} c_i \bar{d}_j K(s_i, t_j),$$

defines an inner product on $V(K)$ if and only if K is *positive-definite*, that is,

$$\sum_{i=1}^{n} \sum_{j=1}^{n} c_i \bar{c}_j K(t_i, t_j) \geq 0,$$

for all integers $n \geq 1$, scalars $c_1, \cdots, c_n$ and $t_1, \cdots, t_n \in T$; see Problems 9.1 and 9.2.

The *length* or *norm* of a vector x, in symbols $||x||$, in an inner product space is the nonnegative number $\sqrt{(x, x)}$. Some standard examples of norms are as follows:

$$x \in \mathbb{C}^n, \quad ||x|| = (|x_1|^2 + \cdots + |x_n|^2)^{1/2},$$

$$f \in L^2(\Omega, \mathcal{F}, \mu), ||f|| = \left(\int_\Omega |f|^2 d\mu \right)^{1/2},$$

$$f \in V(K), ||f||_K^2 = \sum_{i=1}^{n} \sum_{j=1}^{n} c_i \bar{c}_j K(t_i, t_j).$$

9.1.2 Properties of Inner Products

Relationships between norm, inner product and positive-definite function (Example 9.1 (g)) can be seen through the positivity of the norm of a linear combination $x = \sum_{i=1}^{n} c_i x_i$:

$$||x||^2 = \left(\sum_{i=1}^{n} c_i x_i, \sum_{j=1}^{n} c_j x_j \right) = \sum_{i=1}^{n} \sum_{j=1}^{n} c_i \bar{c}_j (x_i, x_j) \geq 0. \tag{9.1}$$

For ease of reference we single out few frequently used properties of norm and inner product generalizing those from the plane geometry. Their proofs follow from (9.1) with the appropriate choice of c_i's and x_i's and hence are omitted.

Theorem 9.1. For x, y in an inner product space we have,

(a) *Cauchy-Schwarz Inequality:*

$$|(x, y)| \leq ||x|| ||y||,$$

with equality if and only if $y = cx$ for a scalar c.

(b) *Triangle Inequality:*

$$||x + y|| \leq ||x|| + ||y||,$$

with equality if and only if $y = cx$ for a positive scalar c.

(c) *Pythagorian Theorem:*

$$||x + y||^2 = ||x||^2 + ||y||^2,$$

if and only if x and y are *orthogonal*, that is, $(x, y) = 0$.

(d) *The Parallelogram Law:*

$$||x + y||^2 + ||x - y||^2 = 2||x||^2 + 2||y||^2.$$

(e) *Polarization Identity:*

$$\begin{aligned} 4(x, y) &= ||x + y||^2 - ||x - y||^2 + i||x + iy||^2 - i||x - iy||^2 \\ &= \sum_{k=1}^{4} i^k ||x + i^k y||. \end{aligned}$$

In view of the Pythagorian Theorem, the notion of *orthogonality* in inner product spaces is crucial for carrying out the analogy with the plane geometry. A set $\{x_t\}$ of vectors in V is said to be an *orthogonal set* if x_k is orthogonal to x_ℓ whenever $k \neq \ell$. A vector x is *normalized* or is a *unit vector* if $||x|| = 1$. A set $\{x_k\}$ of vectors in V is said to be an *orthonormal set* if it is an orthogonal set and each vector x_k is normalized, or more precisely, $(x_k, x_\ell) = \delta_{k\ell}$ for all k, ℓ, where $\delta_{k\ell}$ is the Kronecker's delta. For a finite orthogonal set a generalized Pythagorian Theorem holds:

$$\left\|\sum_{i=1}^{n} x_i\right\|^2 = \sum_{i=1}^{n} ||x_i||^2. \tag{9.2}$$

Example 9.2. (a) The set $\{x_k; k = 1, 2, \cdots\}$ with $x_k = (0, \cdots, 1, 0, \cdots)$ having 1 in the kth place and zeros elsewhere is an orthonormal set in ℓ^2.
(b) The sets $\{e_k; k = 0, \pm 1, \pm 2, \cdots\}$ and $\{e_k; k = 1, 2, \cdots\}$, where $e_k(\lambda) = e^{ik\lambda}$, $\lambda \in (-\pi, \pi]$, are orthonormal sets in L^2 and H^2, respectively, since

$$(e_k, e_\ell) = \int_{-\pi}^{\pi} e^{i(k-\ell)\lambda} d\lambda = \delta_{k\ell}.$$

But these are not orthogonal in $L^2(\mu)$ when μ is different from the Lebesgue measure.

A sequence $\{x_n\}$ of elements of an inner product space V is said to be *norm-convergent* to $x \in V$ if $||x_n - x|| \to 0$ as $n \to \infty$. This concept of convergence implies that an inner product is a continuous function on $V \times V$.

Theorem 9.2. (Continuity of Inner Products) . Let $\{x_n\}$ and $\{y_n\}$ be sequences of elements of V that are norm-convergent to $x, y \in V$, respectively. Then,

(a) $||x_n|| \to ||x||$,

(b) $(x_n, y_n) \to (x, y)$.

Proof. (a) Using the triangle inequality (Theorem 9.1 (b)) we have

$$\begin{aligned} ||x_n|| &= ||x_n - x + x|| \le ||x_n - x|| + ||x||, \\ ||x|| &= ||x - x_n + x_n|| \le ||x_n - x|| + ||x_n||, \end{aligned}$$

and hence

$$|||x_n|| - ||x||| \le ||x_n - x||,$$

from which the result follows.
(b) Using the linearity of inner products and the Cauchy-Schwarz inequality we have,

$$\begin{aligned} |(x_n, y_n) - (x, y)| &= |(x_n, y_n - y) + (x_n - x, y)| \\ &\le |(x_n, y_n - y)| + |(x_n - x, y)| \\ &\le ||x_n|| ||y_n - y|| + ||x_n - x|| ||y||, \end{aligned}$$

from which the desired result follows, since according to (a), $\{||x_n||\}$ being a convergent sequence is bounded.

Q.E.D.

A sequence in an inner product space could converge in norm to a vector outside V. For instance, the sequence $\{f_n\}$ in $C[0, 1]$ given by

$$f_n(t) = \max\{0, \min\{1, n(t - .5)\}\}, \tag{9.3}$$

can be seen (plot f_n) to converge pointwise to the *discontinuous* function

$$f(t) = \begin{cases} 0 & 0 \leq t < 1/2 \\ 1 & \frac{1}{2} \leq t \leq 1. \end{cases}$$

Such a vector space is *incomplete* or *not closed* with respect to its norm. The *completion* or *closure* of an inner product space V, in symbols $\bar{V}$, is obtained by adjoining limits of all convergent sequences to V. A complete inner product space is called a Hilbert space. More general complete spaces are defined next.

9.2 COMPLETE NORMED (BANACH) SPACES

In this section we review vector spaces whose *norms* are not necessarily defined via inner products. Since norm defines a natural metric (distance) on a vector space, standard notions of metric spaces will be used to study convergence, completeness and completion of such spaces.

9.2.1 Definition and Examples

From the definition of inner product and Theorem 9.1, it is evident that the norm in an inner product space do satisfy the following three properties of norm in any vector space. Let V be a vector space. A *norm* on V is a function $|| \cdot || : V \to R$ satisfying:

(a) *Positivity:* $||x|| > 0$, if $x \neq 0$;

(b) *Homogeniety:* $||\lambda x|| = |\lambda| ||x||$, for all scalars λ and vectors $x \in V$;

(c) *Triangle Inequality:* $||x + y|| \leq ||x|| + ||y||$, for all $x, y \in V$.

Example 9.3. Important examples of normed spaces are listed, in a slightly different order than those in Example 9.1. While it is easy to check properties (a) and (b) above, proof of (c) is more challenging for most of these examples.

(a) The space $C[a, b]$ of continuous functions on [a, b] with *supremum norm*, in symbols $|| \cdot ||_\infty$, defined by

$$||f||_\infty = \sup_{a \leq x \leq b} |f(x)|,$$

is a normed space. Note that the norm or distance between two functions f and g is the maximum modulus of $f(x)-g(x)$ for all $x \in [a,b]$. In the sequel $C(-\pi,\pi]$ will be denoted by C if there is no danger of confusion.

(b) For $1 \leq p \leq \infty$, the space $\ell^p = \left\{ x; x=(x_1,x_2,\cdots), x_i \in \mathbb{C}, \sum_{i=1}^{\infty} |x_i|^p < \infty \right\}$ with $||x||_p = \left(\sum_{i=1}^{\infty} |x_i|^p \right)^{1/p}$ is a normed space. For $p=\infty$, $||x||_\infty$ is defined as the supremum norm, that is $||x||_\infty = \sup_i |x_i|$. An important subspace of ℓ^∞ is the space c_0 of all $x=(x_1,x_2,\cdots) \in \ell^\infty$ so that $x_n \to 0$ as $n \to \infty$.

(c) Let Ω be an arbitrary set, $\mathcal{F}$ a σ-algebra of subsets of Ω and μ a nonnegative measure on $\mathcal{F}$. For $1 \leq p \leq \infty$, the space

$$L^p(\Omega,\mathcal{F},\mu) = \left\{ f; f:\Omega \to \mathbb{C} \text{ a measurable function with } \int |f|^p d\mu < \infty \right\}$$

with $||f||_p = (\int |f|^p d\mu)^{1/p}$ is a normed space, called the *Lebesgue space* of p-integrable functions with respect to μ. For $p=\infty$, the space $L^\infty(\Omega,\mathcal{F},\mu)$ is simply the space of all bounded measurable functions with the μ-essential supremum norm, that is

$$||f||_\infty = ess_\mu \sup_x |f(x)|,$$

which means the infimum of $\sup_x |g(x)|$ as g ranges over all bounded measurable functions which agree with f almost everywhere with respect to μ. As in Example 9.1 (e) different choice of Ω and μ gives rise to different normed spaces, accordingly for $\Omega = (-\pi,\pi]$ we shall use the notation $L^p(\mu)$ and L^p as before. For μ finite, the $L^p(\mu)$ spaces are nested, that is $L^p(\mu) \subseteq L^{p'}(\mu)$ for $p \geq p'$. In particular,

$$C \subset L^\infty \subset L^2 \subset L^1. \tag{9.4}$$

(d) Let $\mathbb{D}$ be the open unit disc in the complex plane. The space

$$\tilde{H}^p = \left\{ F; F \text{ analytic in } \mathbb{D} \text{ with } \sup_{0 \leq r \leq 1} \int |F(re^{i\lambda})|^p d\lambda < \infty \right\}$$

with norm $||F||_p = \sup_{0 \leq r \leq 1} \left(\int |F(re^{i\lambda})|^p d\lambda \right)^{1/p}$ is a normed space, called the *Hardy space* of analytic functions. Similarly, the space

$$H^p = \left\{ f \in L^p; f \sim \sum_{k=0}^{\infty} \hat{f}_k e_k \right\}$$

with $\hat{f}_k$ the kth Fourier coefficient of f and $||f||_p = (\int |f|^p d\lambda)^{1/p}$ is a normed space.

9.2.2 Complete Normed Spaces

The function

$$d(x, y) = ||x - y||,$$

defines a metric on the normed space $(V, ||\cdot||)$. Thus, every normed space is a metric space and the following notions of topology of metric spaces can be used in the study of convergence of sequences in a normed space:

(a) A sequence $\{x_n\}$ in the normed space $(V, ||\cdot||)$ is a *Cauchy sequence*, if for every $\varepsilon > 0$, there exists an integer n_0 such that $m, n \geq n_0$ implies that $||x_n - x_m|| < \varepsilon$.

(b) $(V, ||\cdot||)$ is said to be a *complete normed (Banach) space* if every Cauchy sequence in V converges to a limit in V.

(c) A *Hilbert Space* is an inner product space which is complete with respect to the norm induced by its inner product.

While every Hilbert space is a Banach space, the converse is not true. For example, $C[a, b]$ with the inner product given as the integral in Example 9.1 (c) is not complete, and therefore is not a Hilbert space. But, $C[a, b]$ with the supremum norm is a Banach space whose norm is not induced by any inner product. All other normed spaces introduced in Example 9.3 can be shown to be complete (Hoffman, 1962).

For many applications $C[a, b]$ with the integral inner product is more natural, in spite of the fact that it is incomplete. Its completion or closure with respect to the norm induced by the inner product is the larger space $L^2([a, b], \mathcal{F}, \lambda)$ which contains many discontinuous functions, see (9.3). In some approaches to measure theory and for our purpose it is more convenient to define $L^p(\mu)$ as the closure in the p-norm of $C(-\pi, \pi]$, and this is how we are going to view $L^p(\mu)$ from here on.

A non-empty subset M of a Banach space is called a (linear) manifold, if for all $x, y \in M$ and all scalars α, β, we have $\alpha x + \beta y \in M$. M is called a *subspace*, if it is a closed manifold, that is it contains the limit of every Cauchy sequence in M. For a non-empty subset M, the smallest manifold generated by M is called the *span* of M and is denoted by

$$sp\ M = \left\{ \sum_{k=1}^{n} c_k x_k;\ \text{for } n \geq 1,\ \text{scalars } c_i;\ \text{and} x_i, \in M;\ i = 1, \cdots, n \right\};$$

the subspace generated by M is called the *closed linear span* or *span closure* of M and is denoted by $\overline{sp}M$ or $\overline{sp}\{x; x \in M\}$. For two or more subsets $M_i, i = 1, 2, \cdots$ of H we use $sp\{M_i; i = 1, 2, \cdots\}$ for $sp \bigcup_{i=1}^{\infty} M_i$, for the manifold or span of all these subsets. Similar notations are used for their closed linear span. For example, in the sequel we use repeatedly the fact that $L^p = \overline{sp}\{e^{in\lambda}; n = 0, \pm 1, \cdots\}$ and $H^p = \overline{sp}\{e^{in\lambda}; n = 0, 1, 2, \cdots\}$ with respect to the p-norm defined in Example 9.3 (c).

9.3 ORTHOGONAL PROJECTION

The *minimum distance property* and the *Projection Theorem* are probably the main reasons for popularity of the Hilbert space setup for solving the least squares and other optimization problems. These are developed in this section with a view to compute projections as they arise from various areas of applications.

9.3.1 Convexity and Minimum Distance Property

A subset M of a Hilbert space is said to be *convex* if for each pair $x, y \in M$ and all $0 < \alpha < 1$,

$$\alpha x + (1 - \alpha)y \in M,$$

that is, the line-segment connecting x and y remains in M.

The *minimum distance property* states that for any non-empty closed convex set M in a Hilbert space H and any $y \in H$ there is a unique vector $\hat{y}$ in M which is *closest* to y, that is,

$$||y - \hat{y}|| = \inf_{x \in M} ||y - x|| = dist(y, M). \tag{9.5}$$

When M is a subspace $\hat{y}$ is called the *orthogonal projection* of y onto M. In the following theorem without loss of generality we take $y = 0$.

Theorem 9.3. Let M be a nonempty closed convex subset in a Hilbert space H. Then there is a *unique* vector in M with the *smallest* norm.

Proof. Let $d = \inf\{||x||; x \in M\}$. Then, there exists a sequence $\{x_n\}$ of vectors in M such that $||x_n|| \to d$. Since M is convex, we have $\frac{x_m + x_n}{2} \in M$ so that $\left\|\frac{x_m + x_n}{2}\right\| \geq d$ or $||x_m + x_n|| \geq 2d$, then from the parallelogram law (Theorem 9.1(d))

$$\begin{aligned} ||x_m - x_n||^2 &= 2||x_m||^2 + 2||x_m||^2 - ||x_m + x_n||^2 \\ &\leq 2||x_m||^2 + 2||x_n||^2 - 4d^2. \end{aligned}$$

Note that the right-hand side converges to $2d^2 + 2d^2 - 4d^2 = 0$, as $m, n \to \infty$, so that $\{x_n\}$ is a Cauchy sequence and hence $x_n \to x$, with x in M. Clearly $||x|| = \lim_n ||x_n|| = d$ and x is a vector in M with the smallest norm. To show that x is unique, suppose $x' \neq x$ is another vector in M with norm d. Then $\frac{x + x'}{2}$ is also in M, using parallelogram once again yields

$$\begin{aligned} ||\frac{x + x'}{2}||^2 &= \frac{1}{2}||x||^2 + \frac{1}{2}||x'||^2 - ||\frac{x - x'}{2}||^2 \\ &< \frac{1}{2}||x||^2 + \frac{1}{2}||x'||^2 = d^2, \end{aligned}$$

contradicting the definition of d.

Q.E.D.

9.3.2 Orthogonal Complement

The most important relation among vectors in a Hilbert space is orthogonality. We say that x is *orthogonal* to y, in symbols $x \perp y$, if $(x, y) = 0$. A vector x is *orthogonal* to a subset M of H, in symbol $x \perp M$, if $x \perp y$ for all y in M. More generally, a set N is said to be orthogonal to M, in symbol $N \perp M$, if $x \perp y$ for all x in N and all y in M.

The *orthogonal complement* of a subset M of H, in symbols $M^{\perp}$, is the set of all vectors x such that $x \perp M$. If M and N are subspaces such that $M \subset N$, the orthogonal complement of M in N is the set $N \bigcap M^{\perp}$. Next, we summarize properties of the orthogonal complement of a set.

Theorem 9.4. If M is a subset of H, then

(a) $M^{\perp}$ is a subspace and $M \bigcap M^{\perp} \subseteq \{0\}$.

(b) $M \subset M^{\perp\perp}$.

(c) If M and N are subsets of H with $M \subset N$, then $M^{\perp} \supset N^{\perp}$.

(d) If M is a subset of H, then $M^{\perp} = M^{\perp\perp\perp}$. Furthermore, if M is a subspace, then $M = M^{\perp\perp}$.

(e) If $\{M_i; i\}$ is a sequence of subspaces, then $(\overline{sp}\{M_i; i\})^{\perp} = \bigcap_i M_i^{\perp}$ and $(\bigcap_i M_i)^{\perp} = \overline{sp}\{M_i^{\perp}; i\}$.

Proof. (a) If x is in M and if y_1, y_2 are in $M^{\perp}$, then for any pair of complex numbers α, β,

$$(x, \alpha y_1 + \beta y_2) = \bar{\alpha}(x, y_1) + \bar{\beta}(x, y_2) = 0,$$

hence $\alpha y_1 + \beta y_2$ is in $M^{\perp}$ and $M^{\perp}$ is a manifold. $M^{\perp}$ is closed, since if x is in M and $\{y_n\}$ a sequence in $M^{\perp}$ converging to y, using continuity of the inner product, from $(x, y_n) = 0$, for all n, it follows that $(x, y) = 0$ so that $y \in M^{\perp}$. If $x \in M \bigcap M^{\perp}$, then $x \perp x$ and $x = 0$, that is, $M \bigcap M^{\perp} \subset \{0\}$.

(b) For any $x \in M$ and $y \in M^{\perp}, x \perp y$ so that $x \perp M^{\perp}$ or $x \in M^{\perp\perp}$. Proofs of (c)–(e) are left to the readers.

Q.E.D.

9.3.3 Orthogonal Vector Sums

To take full advantage of the concept of orthogonal complement we need the concept of *vector sum* of two subspaces M and N, in symbols $M + N$, defined to be the set of all vectors of the form $x + y$, with $x \in M$ and $y \in N$. It is evident that $M + N$ is a linear manifold and a subset of the subspace generated by M and N. We show here that a sufficient condition for $M+N$ to be a closed is that M and N be orthogonal. A necessary and sufficient condition for this in terms of positivity of the angle between M and N is given in Section 9.6.2.

Theorem 9.5. If M and N are orthogonal subspaces of a Hilbert space, then

(a) $M + N$ is closed.

(b) If M is a subspace, then $M + M^{\perp} = H$.

Proof. (a) Suppose that $\{z_n\}$ is a sequence of vectors in $M+N$ convergent to z in H, then for each n, $z_n = x_n + y_n$ with $x_n \in M$ and $y_n \in N$. Thus, for every m, n

$$||z_n - z_m||^2 = ||x_n - x_m||^2 + ||y_n - y_m||^2,$$

and hence $\{x_n\}$ and $\{y_n\}$ are Cauchy sequences. If $x_n \to x$ and $y_n \to y$, then $x \in M$ and $y \in N$, and $z_n \to x + y \in M + N$, that is, $z = x + y \in M + N$.

(b) If $M + M^{\perp} = N$, then by (a), N is a subspace. Since $M \subset N$ and $M^{\perp} \subset N$, it follows from Theorem 9.4 (c) that $N^{\perp} \subset M^{\perp}$ and $N^{\perp} \subset M^{\perp\perp}$ and hence $N^{\perp} = \{0\}$. Consequently, $N = N^{\perp\perp} = \{0\}^{\perp} = H$ as desired.

Q.E.D.

The *vector sum* of a sequence $\{M_i\}$ of subspaces, in symbols $\sum_i M_i$, is the set of all vectors of the form $\sum_i x_i$ with $x_i \in M_i$ for all i. It is evident that $\sum_i M_i$ is a linear manifold. The close connection between vector sums and spans is revealed in the next theorem.

Theorem 9.6. If $\{M_i\}$ is a sequence of subspaces and $M = \sum_i M_i$, then $\overline{sp}\{M_i, i\} = \overline{sp}\{M\}$.

Proof. Since $\bigcup_i M_i \subset \overline{sp}\{M\}$ and $\overline{sp}\{M\}$ is a subspace it follows that $\overline{sp}\{M_i, i\} \subset \overline{sp}\{M\}$. For the other inclusion, consider the set F of all vectors of the form $\sum_i x_i$ with finite terms and $x_i \in M_i$. Evidently F is dense in M and since $\overline{sp}\{M_i; i\}$ is closed, it follows that $M \subset \overline{sp}\{M_i; i\}$ and therefore $\overline{sp}\{M\} \subset \overline{sp}\{M_i; i\}$ as desired.

Q.E.D.

This theorem simply says that the vector sum of subspaces can fail to be closed (Example 9.6). More definite results can be obtained when the subspaces are orthogonal.

A sequence $\{M_i; i\}$ of subspaces is said to be *orthogonal* if $M_i \perp M_j$ for all $i \neq j$. The vector sums of an orthogonal sequence is called an *orthogonal sum*, and the symbol $\sum_i \oplus M_i$ is sometimes used to emphasize the orthogonality of M_i's.

Theorem 9.7. If $\{M_i; i\}$ is an orthogonal sequences of subspaces, then $\overline{sp}\{M_i; i\} = \sum_i M_i$. Furthermore, the representation of an element of $\sum_i M_i$ in the form $\sum_i x_i$ with $x_i \in M_i$ for all i, is unique.

Proof. Follows from Theorem 9.5 (a) and induction.

9.3.4 Projection Theorem

Now, we state and prove the fundamental *projection theorem*. The same name is sometimes used for the result in Theorem 9.5 (b). The following version is more explicit and suited for applications.

Theorem 9.8. (Projection Theorem) Suppose M is a subspace of the Hilbert space H. Then,

(a) For $x \in H$ there is a unique element $\hat{x} \in M$ such that

$$||x - \hat{x}|| = \inf\{||x - y||; y \in M\}. \tag{9.6}$$

(b) An $\hat{x} \in M$ satisfies (9.6) if and only if

$$x - \hat{x} \in M^{\perp}. \tag{9.7}$$

Proof. (a) The set $C = x - M$ is a non-empty closed convex subset of the Hilbert space H, hence by Theorem 9.3 there exists a unique vector $\hat{x}$ in M satisfying (9.6).

(b) If $\hat{x} \in M$ and $x - \hat{x} \in M^{\perp}$, then for any $y \in M$, $\; x - \hat{x} \perp \hat{x} - y$ and

$$\begin{aligned} ||x - y||^2 &= ||x - \hat{x} + \hat{x} - y||^2 \\ &= ||x - \hat{x}||^2 + ||\hat{x} - y||^2 \geq ||x - \hat{x}||^2, \end{aligned}$$

that is, $\hat{x}$ is the unique element of M satisfying (9.6). Conversely, if $\hat{x} \in M$ and $x - \hat{x} \notin M^{\perp}$, we show that $\hat{x}$ can not satisfy (9.6). Indeed, for any $y \in M$ with $(x - \hat{x}, y) \neq 0$, consider the vector

$$\tilde{x} = \hat{x} + (x - \hat{x}, y)\frac{y}{||y||^2} \in M,$$

some computation shows that

$$||x - \tilde{x}||^2 < ||x - \hat{x}||^2,$$

that is, $\tilde{x}$ is closer to x than $\hat{x}$.

Q.E.D.

The *orthogonality principle* (9.7) is useful in computing $\hat{x}$ when an *orthogonal basis* or a *generator* for M is known. The unique element $\hat{x}$ is called the *orthogonal projection* of x onto M. Other more informative notation for $\hat{x}$, that we use from time to time, are $P_M x$ or $(x|M)$.

Example 9.4. Let $x_1, \cdots x_n$ be n vectors in a Hilbert space H and $M = sp\{x_1, \cdots, x_n\}$. For any $x \in H$, the unique element $\hat{x}$ in M closest to x is of the form

$$\hat{x} = \beta_1 x_1 + \cdots + \beta_n x_n \tag{9.8}$$

for some scalars β_i's to be found. Using (9.7) we must have $x - \hat{x} \perp x_j$, or $\left(x - \sum_{i=1}^{n} \beta_i x_i, x_j\right) = 0, j = 1, 2, \cdots, n$, which leads to the system of linear equations

$$\sum_{i=1}^{n} (x_i, x_j)\beta_i = (x, x_j), j = 1, 2, \cdots, n.$$

Writing this in the matrix form

$$\begin{bmatrix} (x_1, x_1) & \cdots & (x_1, x_n) \\ (x_2, x_1) & \cdots & (x_2, x_n) \\ \vdots & & \vdots \\ (x_n, x_1) & \cdots & (x_n x_n) \end{bmatrix} \begin{bmatrix} \beta_1 \\ \beta_2 \\ \vdots \\ \beta_n \end{bmatrix} = \begin{bmatrix} (x, x_1) \\ \vdots \\ \vdots \\ (x, x_n) \end{bmatrix}, \tag{9.9}$$

brings in the $n \times n$ matrix $G = [(x_i, x_j)]_{i,j=1}^{n}$, called the *Gram matrix* or *Grammian* of $\{x_1, \cdots, x_n\}$. It is evident that G is invertible if and only if $x_1, \cdots, x_n$ are linearly independent. The minimum distance from x to M, in symbols dist(x, M), can be expressed in terms of β_j's and G:

$$\begin{aligned} \text{dist}(x, M) = \|x - \hat{x}\|^2 &= (x - \hat{x}, x - \hat{x}) = (x, x - \hat{x}) \qquad (9.10) \\ &= (x, x) - \sum_{i=1}^{n} \bar{\beta}_i (x, x_i) = \|x\|^2 - \boldsymbol{\beta}'^* G^{-1} \boldsymbol{\beta}, \end{aligned}$$

where $\boldsymbol{\beta} = (\beta_1, \cdots, \beta_n)$ and $\boldsymbol{\beta}^* = (\bar{\beta}_1, \cdots, \bar{\beta}_n)$. For $n = 1$, (9.8)-(9.10) reduces to

$$\begin{cases} \hat{x} = \dfrac{(x, x_1)}{(x_1, x_1)} x_1, \\ \|x - \hat{x}\|^2 = \|x\|^2 - \dfrac{|(x, x_1)|^2}{\|x_1\|^2}. \end{cases}$$

In applications, often n grows to infinity and the norm-convergence of the infinite series in (9.8) then needs to be addressed. For $x_1, \cdots, x_n$ orthogonal vectors, G is a diagonal matrix and solving (9.9) is a trivial matter; see (9.12)–(9.14) below. Otherwise, it is common to rely on the Gram-Schmidt process to orthogonalize $x_1, x_2, \ldots, x_n, \ldots$.

9.3.5 Gram-Schmidt Process

Given any sequence $\{x_t; t \geq 1\}$ of vectors from H, one could use the *Gram-Schmidt orthogonalization process* to produce an orthonormal sequence. This process is inductive and at its kth stage replaces x_k by a vector e_k such that,

$$\begin{aligned}&(1)\ e_k \text{ is a linear combination of } x_1, \cdots, x_k,\\ &(2) \text{ the sequence } \{e_1, e_2, \cdots\} \text{ is orthonormal,}\\ &(3)\ sp\{e_1, \cdots, e_n\} = sp\{x_1, \cdots, x_n\}, n \geq 1.\end{aligned} \tag{9.11}$$

It starts with $e_1 = \dfrac{x_1}{||x_1||}$, after $e_1, \cdots, e_k$ have been constructed, e_{k+1} is obtained by normalizing the vector $x_{k+1} - \sum_{i=1}^{k} (x_{k+1}, e_i) e_i$, which is orthogonal to $sp\{x_1, \cdots, x_k\} = sp\{e_1, \cdots, e_k\}$.

The importance of this orthogonalization process in applied mathematics, probability and statistics is well-known, for more information see Szegö and Grenander (1958). Two important examples of the use of the Gram-Schmidt process in the theory of orthogonal polynomials are:

(a) *Szegö's orthogonal polynomials* obtained by applying the Gram-Schmidt process to the sequence $\{1, e^{i\lambda}, e^{2i\lambda}, \cdots\}$ in $L^2(\mu)$; see Chapter 7.

(b) The *normalized Hermite polynomials* obtained by applying the Gram-Schmidt process to $\{x^n; n = 0, 1, 2, \cdots\}$ in $L^2(\Omega, \mathcal{F}, \mu)$ where Ω is the real line and μ is a measure with density $e^{-x^2/2}$ with respect to the measure on the line.

9.4 ORTHONORMAL BASES AND EXPANSION

In R^3 a vector $x = (x_1, x_2, x_3)$ is visualized and understood through its coordinates in the familiar (orthogonal) coordinate system. The notions of *orthogonal* basis and expansion attempt to do the same for abstract Hilbert spaces.

To appreciate the importance of orthonormal basis in computing projections, for the moment we return to the Example 9.4 with $e_1, \cdots, e_n$ orthonormal and $M = sp\{e_1, \cdots, e_n\}$. Then from (9.8)–(9.10) for any $x \in H$, $\beta_i = (x, e_i), i = 1, \cdots, n$, and

$$\hat{x} = \sum_{i=1}^{n} (x, e_i) e_i, \tag{9.12}$$

$$||\hat{x}||^2 = \sum_{i=1}^{n} |(x, e_i)|^2, \tag{9.13}$$

$$||x - \hat{x}||^2 = ||x||^2 - \sum_{i=1}^{n} |(x, e_i)|^2. \tag{9.14}$$

The numbers (x, e_i) are called the *coordinates* or *Fourier coefficients* of x relative to the orthonormal set $\{e_1, \cdots, e_n\}$ and (9.12) is called the *Fourier expansion* of $\hat{x}$ as an element of M. The *Bessel's inequality* follows from nonnegativity of the left-hand side of (9.14):

$$\sum_{i=1}^{n} |(x, e_i)|^2 \leq ||x||^2. \tag{9.15}$$

For $n = \infty$, (9.12)–(9.15) hold, when the series in (9.12) is norm-convergent in the sense defined below.

We say a series $\sum_{k=1}^{\infty} x_k$ *converges in norm* to x, and write $\sum_{k=1}^{\infty} x_k = x$, if its sequence of partial sums $\left\{\sum_{k=1}^{n} x_k\right\}_n$ converges to x in norm. Finding conditions for norm-convergence of a series is particularly simple when x_k's are orthogonal, as the following theorem shows.

Theorem 9.9. *(Riesz-Fischer Theorem).* Let $\{e_k; k = 1, 2, \cdots,\}$ be an orthonormal sequence in H and $\{c_k\}$ be a sequence of scalars. Then, $\sum_{k=1}^{\infty} c_k e_k$ converges in norm, if and only if

$$\sum_{k=1}^{\infty} |c_k|^2 < \infty.$$

Proof. Suppose the series is norm-convergent and let $\sum_{k=1}^{\infty} c_k e_k = x$, then for $n \geq m$,

$$\left(\sum_{k=1}^{n} c_k e_k, e_m\right) = \sum_{k=1}^{n} c_k (e_k, e_m) = c_m,$$

hence using the continuity of the inner product (Theorem 9.2 (b)), as $n \to \infty$, we get

$$(x, e_m) = c_m. \tag{9.16}$$

From Bessel's inequality (9.15) and (9.16) it follows that for all n

$$\sum_{m=1}^{n} |c_m|^2 = \sum_{m=1}^{n} |(x, e_m)|^2 \leq ||x||^2 < \infty,$$

and therefore $\{c_k\}$ is square-summable.

Conversely, suppose that $\sum_{k=1}^{\infty} |c_k|^2 < \infty$. Let

$$S_n = \sum_{k=1}^{n} c_k e_k,$$

then for $n \geq m$,

$$||S_n - S_m||^2 = \left\| \sum_{m+1}^{n} c_k e_k \right\|^2 = \sum_{m+1}^{n} |c_k|^2 \to 0,$$

as $m, n \to \infty$. Thus, $\{S_n\}$ is a Cauchy sequence and converges in norm to an element x of H.

Q.E.D.

A *basis* of a Hilbert space H is a maximal orthonormal subset in H. It is known that every Hilbert space has a basis (proof of this requires using Zorn's lemma or its equivalents). We say that H is *separable* if its basis is a countable subset of H, and can be viewed as a sequence $\{e_i\}_{i=1}^{\infty}$. Some properties of basis are stated and proved in the next theorem. All Hilbert spaces encountered in this book are separable.

Theorem 9.10. Let H be a separable Hilbert space and $\{e_i\}$ be an orthonormal sequence. Then, the following are equivalent:

(a) $\{e_i\}$ is a basis of H.

(b) If $x \in H$ and $x \perp e_i$ for all i, then $x = 0$.

(c) $H = \bar{sp}\{e_i, i = 1, 2, \cdots\}$.

(d) If $x \in H$, then $x = \sum_{i=1}^{\infty} (x, e_i) e_i$. *(Fourier expansion).*

(e) If $x, y \in H$, then $(x, y) = \sum_{i=1}^{\infty} (x, e_i) \overline{(y, e_i)}$. *(Parseval's identity).*

(f) If $x \in H$, then $||x||^2 = \sum_{i=1}^{\infty} |(x, e_i)|^2$.

Proof. (a) $\Rightarrow$ (b). If $0 \neq x \in H$ and $x \perp e_i$ for all i, then $x/||x||$ can be added to the family $\{e_i\}$, contradicting the maximality of $\{e_i\}$.

(b) $\Rightarrow$ (c). If $\overline{sp}\{e_i\} \subseteq H$ but not equal, then there exists a non-zero $x \in H$ such $x \perp e_i$ for all i, contradicting (b).

(c) $\Rightarrow$ (d). Every vector $x \in H$ has the form $\sum_i \alpha_i e_i$ for some α_i's. In fact, using continuity of inner products as in the proof of Theorem 9.9 we have $(x, e_j) = \lim_n \left(\sum_{i=1}^{n} \alpha_i e_i, e_j \right) = \alpha_j$ for every j as desired.

(d) $\Rightarrow$ (e) and (e) $\Rightarrow$ (f) are immediate from the proof of (c) $\Rightarrow$ (d).

(f) $\Rightarrow$ (a). If $\{e_i\}$ is not maximal and remains orthonormal after adjoining an x to it, then x can not satisfy (f).

Q.E.D.

While it is evident that the set $\{x_k; k = 1, 2, \cdots\}$ with $x_k = (0, \cdots, 0, 1, 0, \cdots)$ having 1 in the kth position and zeros elsewhere is a basis of ℓ^2, showing that the set of exponentials $\{e_k; k = 0, \pm 1, \cdots\}$ is a basis of L^2 relies on the deep Weierstrass approximation theorem. Meanwhile, *Fourier coefficients and expansion* of a function $f \in L^2$ with respect to the basis $\{e_k\}$ are

$$\begin{cases} \hat{f}_k = (f, e_k) = \int_{-\pi}^{\pi} f(\lambda) e^{-ik\lambda} d\lambda, k = 0, \pm 1, \cdots \\ f = \sum_{k=-\infty}^{\infty} \hat{f}_k e_k. \end{cases} \tag{9.17}$$

and the Fourier series for f converges to in L^2, see Theorem 9.9 and Theorem 9.10 (d). The norm of f given by

$$||f||^2 = \sum_{k=-\infty}^{\infty} |\hat{f}_k|^2, \tag{9.18}$$

is the same as the ℓ^2-norm of the sequence $\{\hat{f}_k\}$. This is a concrete realization of the fact that separable Hilbert spaces are isomorphic to ℓ^2 as proved next.

Theorem 9.11. A separable Hilbert space H is isomorphic either to $\mathbb{C}^n$ for some n or to ℓ^2.

Proof. If H has a finite basis $\{e_1, \cdots, e_n\}$, then for any $x \in H$ from Theorem 9.8 (d), (f) we get

$$x = \sum_{j=1}^{n} (x, e_j) e_j, \; ||x|| = \sum_{j=1}^{n} |(x, e_j)|^2.$$

Consider the transformation $V : H \to \mathbb{C}^n$ defined by

$$Vx = V \sum_{j=1}^{n} (x, e_j) e_j = ((x, e_1), \cdots, (x, e_n)).$$

It is easy to show that V is a linear one-to-one onto transformation with $||Vx|| = ||x||$ for all $x \in H$. Thus, H is isomorphic to $\mathbb{C}^n$. Virtually the same argument works when H has a countable infinite basis $\{e_1, e_2, \cdots\}$.

Q.E.D.

Geometrically, all separable Hilbert spaces look either like $\mathbb{C}^n$ or ℓ^2. A Hilbert space can be made more interesting by adding some external structure to its elements. For example, the spaces H^2 and $\tilde{H}^2$ of analytic functions are more interesting than ℓ^2. The nature of their zeros, factorizations and pointwise convergence of their Fourier series do not quite fit the Hilbert space paradigm, and are the source of some new exciting and challenging problems studied in Chapter 10.

9.5 BOUNDED OPERATORS

Study of linear functionals and operators occupies a major portion of theory of Hilbert spaces. Let H and K be two Hilbert spaces. A transformation $A : H \to K$ is said to be *linear* if

$$A(\alpha x + \beta y) = \alpha Ax + \beta Ay, \tag{9.19}$$

for all $x, y \in H$ and scalars α, β. A is said to be *bounded* if there exists a real constant C such that

$$||Ax|| \leq C||x||, \text{ for all } x \in H, \tag{9.20}$$

where the norm on the left is that of K. The *norm* of the bounded linear transformation A, in symbol $||A||$, is defined as $\sup_{x \neq 0} \frac{||Ax||}{||x||}$. It can be shown that $||A||$ is the smallest constant C satisfying (9.20) and that a linear transformation A is bounded if and only if it is *continuous*, in the sense that if $x_n \to x$ in the norm of H, then $Ax_n \to Ax$ in the norm of K. We use $\mathcal{B}(H, K)$ to denote the collection of all bounded linear transformations between H and K, when $H = K$ then $\mathcal{B}(H)$ is used in place of $\mathcal{B}(H, H)$ and elements of $\mathcal{B}(H)$ are simply called operators on H.

Examples 9.5. Some simple examples of linear operators on familiar Hilbert spaces are introduced next.

(a) Define S on ℓ^2 by

$$S(x_1, x_2, \cdots) = (0, x_1, x_2, \cdots),$$

clearly S is a linear operator with $||Sx|| = ||x||$ for all $x \in \ell^2$ so that S is bounded and $||S|| = 1$. Such a norm-preserving operator is called an *isometry*. One could also define the operator S^* on ℓ^2 by

$$S^*(x_1, x_2, x_3, \cdots) = (x_2, x_3, \cdots),$$

which is linear and bounded with $||S^*|| = 1$. The operators S and S^* on ℓ^2 are usually called the *shift and backward shift* on ℓ^2, respectively.

(b) Define M on $L^2[a, b]$ by

$$(Mf)(t) = h(t)f(t),$$

where $h \in L^\infty[a, b]$. Clearly M is linear and bounded, since for any f,

$$||Mf||^2 = \int_a^b |h(t)f(t)|^2 dt \leq ||h||_\infty^2 ||f||^2,$$

so that $||M|| \leq ||h||_\infty$. M is called the *multiplication operator*. For any subset A of $[a, b]$ the operator of multiplication by χ_A, the indicator function of A, is an orthogonal projection operator.

(c) Define K on $L^2[a, b]$ by

$$(Kf)(t) = \int_a^b k(s,t)f(s)ds,$$

where $k : [a, b]^2 \to \mathbb{C}$ is continuous. Clearly K is linear, using the Cauchy-Schwarz inequality we get

$$|(Kf)(t)|^2 \leq \int_a^b |k(s,t)|^2 ds \int_a^b |f(t)|^2 dt,$$

and hence

$$||Kf||^2 \leq ||f||^2 \int_a^b \int_a^b |k(s,t)|^2 dsdt,$$

so that K is bounded with

$$||K|| \leq \left(\int_a^b \int_a^b |k(s,t)|^2 dsdt \right)^{1/2}.$$

The operator K is called the *integral operator with kernel* $k(\cdot, \cdot)$.

9.5.1 Bounded Linear Functionals

The above definitions and properties hold in the important special case when K is the one-dimensional Hilbert space C of complex numbers. In this case a bounded linear transformation is called a *bounded linear functional.* The next theorem gives an elegant and extremely useful characterization of boundedness in terms of the inner product.

Theorem 9.12. *(The Riesz Representation Theorem.)* A linear functional $\varphi : H \to C$ is bounded if and only if there exists a *unique* vector y such that $\varphi(x) = (x, y)$ for all x.

Proof. If $\varphi(x) = (x, y)$ for all x, then from the Cauchy-Schwarz inequality we obtain $|\varphi(x)| = |(x, y)| \leq ||x|| ||y||$, so that φ is bounded with $||\varphi|| = \sup_{x \neq 0} \dfrac{||\varphi(x)||}{||x||} = ||y||$.

To prove the converse, suppose that φ is a bounded linear functional. Then, $M = \{x; \varphi(x) = 0\}$ is a subspace and either $M = H$ or $M \neq H$. In the first case it is evident that $\varphi(x) = 0 = (x, y)$ with $y = 0$. If $M \neq H$, then there is a non-zero vector $z \in H$ which is orthogonal to M, our aim is to show that a suitable multiple cz is the candidate for y. We note that for all $x \in M$ and for $y = cz$, since

$$\varphi(x) = 0 = (x, y)$$

the value of c is irrelevant. But if $x = dz$ for some complex number d, then $(x, y) = (dz, cz) = d\bar{c}||z||^2$ and a necessary and sufficient condition for the validity of the representation $\varphi(dz) = (dz, y)$ is that $c = \dfrac{\overline{\varphi(z)}}{||z||^2}$. Thus, with this choice of c it is true that $\varphi(x) = (x, y)$ with $y = cz$, if either $x \in M$ or x is a multiple of z. Since $\varphi(z) \neq 0$ and for an arbitrary vector x in H, $x - \beta z \in M$ if $\beta = \dfrac{\varphi(x)}{\varphi(z)}$, it follows that

$$\varphi(x) = \varphi(x - \beta z) + \varphi(\beta z) = (x - \beta z, y) + (\beta z, y) = (x, y),$$

as desired. Now, it is rather easy to establish the uniqueness of y.

Q.E.D.

Even though it is not eminently obvious, the Riesz representation theorem can be used to compute the orthogonal projection of any $y \in H$ onto a subspace M. For any $y \in H$, consider the bounded linear functional $\varphi_y(x) = (x, y)$. Since M is a subspace and hence a Hilbert space on its own right, the restriction of φ_y to M is a bounded linear functional, by the Riesz representation theorem there exists a unique $\hat{y} \in M$ such that

$$\varphi_y(x) = (x, y) = (x, \hat{y}), \text{ for all } x \in M,$$

or

$$(x, y - \hat{y}) = 0, \text{ for all } x \in M, \tag{9.21}$$

that is, $y - \hat{y} \perp M$, so that $\hat{y}$ is the orthogonal projection of y onto M.

9.5.2 Bounded Bilinear Functionals

A mapping $\phi : H \times H \to \mathbb{C}$ is called a *bilinear functional* if it is linear in the first argument and conjugate linear in the second. More precisely, with obvious notation, $\phi(.,.)$ satisfies the following two properties:

(a) $\phi(\alpha_1 x_1 + \alpha_2 x_2, y) = \alpha_1 \phi(x_1, y) + \alpha_2 \phi(x_2, y)$,

(b) $\phi(x, \beta_1 y_1 + \beta_2 y_2) = \bar{\beta}_1 \phi(x, y_1) + \bar{\beta}_2 \phi(x, y_2)$.

An inner product is the simplest example of a bilinear functional. The number $\sup_{||x||,||y|| \leq 1} |\phi(x, y)|$ is called the *norm* of ϕ and is denoted by $||\phi||$. A bilinear functional is said to be *bounded* if $||\phi|| < \infty$, evidently

$$||\phi|| = \sup_{||x||,||y|| \leq 1} |\phi(x, y)| = \sup_{x,y} \frac{|\phi(x, y)|}{||x||||y||}.$$

Therefore,

$$|\phi(x, y)| \leq ||\phi||||x||||y||, \; x, y \in H,$$

and a bounded bilinear functional is a continuous function of each of its arguments.

The fact that bounded linear functionals are inner products (Riesz representation theorem), is used to establish a one-to-one correspondence between bounded bilinear functionals and bounded operators on H.

Theorem 9.13. A bilinear functional ϕ on H is bounded, if and only if there exists a *unique* bounded linear operator A on H such that

$$\phi(x, y) = (Ax, y), \ x, y \in H.$$

Furthermore,

$$||\phi|| = ||A||.$$

Proof. Suppose ϕ is a bounded bilinear functional on H. For fixed $x \in H$, the mapping $\overline{\phi(x, .)}$ is a bounded linear functional on H, so by Riesz representation theorem, there exists a unique element R_x for which

$$\overline{\phi(x, y)} = (y, R_x) \text{ or } \phi(x, y) = (R_x, y), y \in H.$$

Let us define a mapping $A : H \to H$ by $Ax = R_x, x \in H$. Then,

$$\phi(x, y) = (Ax, y),$$

since

$$\phi(\alpha_1 x_1 + \alpha_2 x_2, y) = \alpha_1 \phi(x_1, y) + \alpha_2 \phi(x_2, y),$$

we get

$$(A(\alpha_1 x_1 + \alpha_2 x_2) - \alpha_1 A x_1 - \alpha_2 A_2 x_2, y) = 0, \ y \in H.$$

Hence,

$$A(\alpha_1 x_2 + \alpha_2 x_2) = \alpha_1 A x_1 + \alpha_2 A x_2,$$

so that A is a linear operator. Furthermore, A is bounded, since

$$||\phi|| = \sup \frac{|(Ax, y)|}{||x|| ||y||} \geq \sup \frac{(Ax, Ax)}{||x|| ||Ax||} = \sup \frac{||Ax||}{||x||}.$$

On the other hand, since $|(Ax, y)| \leq ||Ax|| ||y||$, we get

$$||\phi|| = \sup \frac{|\phi(x, y)|}{||x|| ||y||} = \sup \frac{|(Ax, y)|}{||x|| ||y||} \leq \sup \frac{||Ax||}{||x||},$$

so that $||A|| = ||\phi||$. That A is uniquely determined by $\phi(\cdot, \cdot)$ follows from the following argument. Suppose there is another operator A' such that, for all $x, y \in H$,

$$\phi(x, y) = (Ax, y) = (A'x, y).$$

Then,

$$(Ax - A'x, y) = 0, \ \text{for all } x, y \in H,$$

hence $Ax = A'x$, for all x, that is, $A = A'$.

Q.E.D.

The above characterization of bounded bilinear functionals is useful in the classification of operators. Let A be an operator on H. Then, $\phi(x, y) = (x, Ay)$ defines a bounded bilinear functional on H with norm $||A||$. From Theorem 9.13, there exists a unique operator denoted by A^* on H with norm $||A^*|| = ||A||$ such that

$$(x, Ay) = (A^*x, y), x, y \in H.$$

This operator A^* is called the *adjoint* of A. One can check that $(A^*)^* = A^{**} = A$.

If $A = A^*$, then A is said to be *self-adjoint (Hermitian)*. An operator A on H, is said to be *normal* if it commutes with its adjoint, that is, $AA^* = A^*A$. We say an operator A is *unitary*, if $AA^* = A^*A = I$ or $A^* = A^{-1}$. Unitary operators and their spectral decomposition are of particular interest in the study of stationary processes.

Bilinear functionals and adjoint of an operator are extensions of familiar ideas from finite-dimensional spaces and matrix theory. The useful notion of *quadratic form* can also be extended to the infinite-dimensional set up in a natural way: The *quadratic form* $\phi(\cdot)$ induced by a bilinear functional $\phi(\cdot,\cdot)$ is the function defined for each $x \in H$ by $\phi(x) = \phi(x, x)$. A routine computation yields the following useful analogue of the *polarization identity* relating inner products and norms.

Theorem 9.14. If $\phi(\cdot)$ is the quadratic form induced by a bilinear functional $\phi(\cdot,\cdot)$, then for any $x, y \in H$,

$$\phi(x,y) = \phi\left(\frac{x+y}{2}\right) - \phi\left(\frac{x-y}{2}\right) + i\phi\left(\frac{x+iy}{2}\right) - i\phi\left(\frac{x-iy}{2}\right).$$

Furthermore, a bilinear functional is uniquely determined by its quadratic form.

9.5.3 Unitary Operators and Stationarity

Intuitively, unitary operators are like generalized complex numbers of absolute value one. Their actions amount to rotation about the origin and they preserve inner products of vectors in H. As such they are closely related to stationary processes.

A sequence (two-sided sequence) $\{x_t; t \in \mathbb{Z}\}$ with $x_t \in H$ is said to be *stationary* if for all $t, k \in \mathbb{Z}$, the inner product (x_{t+k}, x_t) depends only on k. The function $\gamma(\cdot) : \mathbb{Z} \to \mathbb{C}$ defined by $\gamma(k) = (x_{t+k}, x_t)$ is called the covariance function of $\{x_t\}$. The subspace generated by a sequence is called its *time-domain*, in symbols $H(x) = \overline{sp}\{x_t; t \in \mathbb{Z}\}$.

The next lemma shows that the mapping U defined on the set $M = \{x_t; t \in \mathbb{Z}\}$ by $Ux_t = x_{t-1}$, extends to an isometry on $H(x)$.

Lemma 9.15. Let M be subset of a Hilbert space on which there is defined an *isometry* U, that is $(Ux, Uy) = (x, y)$, for all $x, y \in M$. Then U can be extended to an isometry on $\overline{sp}M$.

Proof. First we show that the isometry U is linear on M. Suppose that $x_1, \cdots, x_m$ and the element x defined by

$$x = \sum_{i=1}^{m} c_i x_i, \tag{1}$$

belong to M. Then, for any $y \in M$,

$$(Ux, Uy) = (x, y) = \sum_{i=1}^{m} c_i(x_i, y) = \sum_{i=1}^{m} c_i(Ux_i, Uy) = \left(\sum_{i=1}^{m} c_i U x_i, Uy\right),$$

and so

$$(Ux - \sum_{i=1}^{m} c_i U x_i, Uy) = 0. \tag{2}$$

In particular, for $y = x, x_1, \cdots, x_m$, from (2) we obtain

$$(Ux - \sum_{i=1}^{m} c_i\, Ux_i,\; Ux - \sum_{i=1}^{m} c_i\, Ux) = 0,$$

or

$$Ux = \sum_{i=1}^{m} c_i U x_i. \tag{3}$$

We define U on every element x of the form (1) by (3). Since every element $x \in \overline{sp} M$ is the limit of elements $x_n, n = 1, 2, \cdots$ of the form (1), we have,

$$\lim_{m,n} ||x_m - x_n|| = \lim_{m,n} ||Ux_m - Ux_n|| = 0,$$

and therefore $\lim_{n \to \infty} Ux_n$ exists. Setting

$$Ux = \lim_{n} Ux_n, \tag{4}$$

it is easy to check that the operator U defined on all of $\overline{sp} M$ by (3) and (4) is an isometry.

Q.E.D.

The one-to-one correspondence between stationary sequences and unitary operators is established in the next theorem.

Theorem 9.16. Let $\{x_t; t \in \mathbb{Z}\}$ be a sequence in a Hilbert space H. Then, the following are equivalent:

(a) $\{x_t; t \in \mathbb{Z}\}$ is stationary.

(b) There exists a unique unitary operator $U : H(x) \to H(x)$ such that

$$x_t = U^t x_0, t \in \mathbb{Z}.$$

Proof. (a) $\Rightarrow$ (b). By Lemma 9.15 the operator U defined by $Ux_t = x_{t-1}, t \in \mathbb{Z}$, extends to an isometry on $H(x)$. We need to show that this extension is unitary, that is, it maps $H(x)$ onto $H(x)$, or for any $y \in H(x)$ there is a $y' \in H(x)$ such that $Uy' = y$. Since every such y is the limit of elements $y_n, n = 1, 2, \cdots$ of the form

$$y_n = \sum_{m} c_{m,n} x_{t_m},$$

for the element y' we can take, for example,

$$y' = \lim_{n\to\infty} y'_n,$$

where

$$y'_n = \sum_m c_{m,n} x_{t_m+1}.$$

(b) $\Rightarrow$ (a) is easy to check.

Q.E.D.

Since two consecutive elements of a stationary sequence are related by $x_t = Ux_{t-1}$, the operator U is called the *(forward) shift* of the sequence; compare this with the operator S^* in Example 9.5 (a).

9.6 PROJECTION OPERATORS AND ANGLE BETWEEN SUBSPACES

Computing specific projections and the algebra of projection operators are of central importance in prediction theory and the spectral representation of unitary operators and stationary sequences. Some key results are covered in this section.

9.6.1 Orthogonal Projection Operators

The *orthogonal projection* operator or simply *projection* onto a subspace M is the operator $P_M = P$ defined for every $x \in H$ of the form $x_1 + x_2$, with $x_1 \in M$ and $x_2 \in M^\perp$, by $Px = x_1$, that is, to every $x \in H$ it assigns its closest element in M. This operator has the following useful properties.

Theorem 9.17. Let M and N be two subspaces of a Hilbert space.

(a) The projection P onto a subspace M is an *idempotent* ($P^2 = P$) and *Hermitian* ($P^* = P$) operator with range $\{x; Px = x\} = M$. If $M \neq \{0\}$, then $||P|| = 1$.

(b) If P is an idempotent and Hermitian operator and if M is the subspace $\{x; Px = x\}$, then P is the projection onto M.

(c) If P is the projection onto a subspace M, then $I - P$ is the projection onto the subspace $M^\perp$ and $M^\perp = \{x; Px = 0\}$.

(d) If M and N are subspaces with projections P and Q, then $M \perp N$ if and only if $PQ = 0$.

Proof. (a) From Theorem 9.5 (b) any $x \in H$ has a unique decomposition as $x = x_1 + x_2$, with $x_1 \in M$ and $x_2 \in M^\perp$, defining $Px = x_1$ it follows that P is a linear operator from H into H and

$$||Px||^2 = ||x_1||^2 \leq ||x_1||^2 + ||x_2||^2 = ||x||^2,$$

so that P is bounded with $||P|| \leq 1$. If M contains a non-zero vector x, then $Px = x$ implies that $||P|| = 1$. Since $P^2x = Px_1 = x_1 = Px$, it follows that $P^2 = P$, that is, P is idempotent. That P is Hermitian follows from the following computation with $y = y_1 + y_2, y_1 \in M, y_2 \in M^\perp$,

$$(Px, y) = (x_1, y) = (x_1, y_1) = (x, y_1) = (x, Py).$$

(b) Since $P^* = P_2^* P_1^* = P_2 P_1$, P is Hermitian if and only if P_1 and P_2 commute, in which case P is idempotent. Also, M the range of $P_1 P_2 (P_2 P_1)$ is contained in $M_1 (M_2)$ the range of $P_1 (P_2)$, thus $M \subset M_1 \bigcap M_2$. Conversely, if $x \in M_1 \bigcap M_2$, then $P_1 x = P_2 x = x$ and hence $Px = x$, so that $M_1 \subset M_2 \subset M$ as desired.

As for Q, since $Q = P_1 + (I - P_1)P_2$, it is a projection onto $M = \overline{sp}\{M_1, M_1^\perp \bigcap M_2\}$. Similarly, $Q = P_1(I - P_2) + P_2$ is a projection onto $M = \overline{sp}\{M_2, M_2^\perp \bigcap M_1\}$. Thus, from $M_1 \subseteq M \subseteq \overline{sp}\{M_1, M_2\}$ and $M_2 \subseteq M \subseteq \overline{sp}\{M_1, M_2\}$ it follows that $M = \overline{sp}\{M_1, M_2\}$.

(c) If $M_1 \subset M_2$, then for any vector x, $P_1 x \in M_2$ and $P_2 P_1 x = P_1 x$. Conversely, if $P_2 P_1 = P_1$ taking adjoint of both sides it follows that $P_1 P_2 = P_1$ and then for all x, $(P_1 x, x) = ||P_1 x||^2 = ||P_1 P_2 x||^2 \leq ||P_2 x||^2 = (P_2 x, x)$.

Q.E.D.

In view of (a)–(d), there is a natural one-to-one correspondence between subspaces and orthogonal projection operators. This observation is useful and suggests expressing geometric properties of subspaces in terms of the algebraic properties of their projections. Motivated by (d), we say two projections P and Q are *orthogonal*, in symbols $P \perp Q$, if $PQ = 0$.

Theorem 9.18. (a) If $P_i, i = 1, 2, \cdots, n$ are projections and $P = \sum_{i=1}^{n} P_i$, then P is a projection if and only if $P_i \perp P_j$ for $i \neq j$. If this condition is satisfied and M_i is the range of P_i and M the range of P, then $M = \sum_i M_i = \overline{sp}\{M_i; i\}$.

(b) The product $P = P_1 P_2$ of two projection operators P_1 and P_2 is a projection operator if and only if P_1 and P_2 *commute*, that is, $P_1 P_2 = P_2 P_1$. If this condition is satisfied and ranges of P, P_1 and P_2 are M, M_1 and M_2, respectively, then $M = M_1 \bigcap M_2$, and $Q = P_1 + P_2 - P_1 P_2$ is the projection onto $N = \overline{sp}\{M_1, M_2\}$.

(c) Suppose M_1 and M_2 are subspaces with projections P_1 and P_2, respectively, then $M_1 \subset M_2$ if and only if $P_2 P_1 = P_1$. In this case we say P_1 is *smaller* than P_2, in symbols $P_1 \prec P_2$, and the difference $P = P_2 - P_1$ is a projection onto $M = M_2 - M_1 = M_2 \bigcap M_1^\perp$.

(d) Let $\{P_i\}$ be an increasing sequence of projections $P_1 \prec P_2 \prec \cdots$ and $x \in H$. Then the sequence $\{P_n x\}$ converges to $\hat{x} \in H$ and the map $P : x \to \hat{x}$ is the projection onto $M = \overline{sp}\{\bigcup_i M_i\}$, where M_i is the range of $P_i, i = 1, 2, \cdots$.

Proof. Proofs of (a)–(c) are left to the reader.
(d) First, we show that $\{P_n x\}$ is a Cauchy sequence in H. Since $\{P_n\}$ is increasing by part (c) for $m < n$, $P_n - P_m$ is a projection and for $x \in H$,

$$||P_n x - P_m x||^2 = ||(P_n - P_m)x||^2 = ((P_n - P_m)x, x) = ||P_n x||^2 - ||P_m x||^2,$$

so that $||P_n x||^2$ increases with n but is bounded by $||x||^2$ and has a finite limit. Therefore $\{P_n x\}$ is a Cauchy sequence.

It is clear that P defined by $Px = \lim_{n\to\infty} P_n x$ is linear and bounded with $||P|| \leq 1$. Since for any $x, y \in H, (P_n x, P_n y) = (P_n x, y) = (x, P_n y)$, a passage to the limit yields

$$(Px, Py) = (Px, y) = (x, Py),$$

so that P is idempotent and Hermitian and thus a projection. The range of P includes each M_n, since $x \in M_n$ implies $Px = x$. If x is orthogonal to all M_n's then $Px = 0$, thus $M = \overline{sp}\{M_i; i\}$.

Q.E.D.

9.6.2 Angle Between Subspaces

In this section we study further the concept of *vector sum* from Section 9.3.3 and give a necessary and sufficient condition in terms of the angle between the two spaces for their vector sum to be closed. This is needed because of the following unpleasant fact that the vector sum of two subspaces can fail to be a subspace (Halmos, 1957); see Section 8.6.

Example 9.6. *(A Non-closed Vector Sum).* Here we construct two subspaces M and N for which

$$M + N \neq \overline{sp}\{M, N\}. \tag{9.22}$$

Since according to Theorem 9.5 equality holds when M and N are orthogonal subspaces, we move away from orthogonal M and N as far as possible and construct two subspaces which make an angle of zero degrees, in the sense defined in this section. More precisely, M and N are so that $M \bigcap N = \{0\}$ and for normalized vectors $x \in M,\ y \in N$, the inner product (x, y) is arbitrarily close to 1.

Let $\{e_n\}$ and $\{e'_n\}$ be two infinite orthonormal sequences such that $e_n \perp e'_m$ for all m and n, define $x_n = \alpha_n e_n + \beta_n e'_n$ with coefficients α_n and β_n chosen so that $\{x_n\}$ is orthonormal with some additional properties. For $\{x_n\}$ to be orthonormal, we need to find α_n, β_n such that $||x_n||^2 = |\alpha_n|^2 + |\beta_n|^2 = 1$. For simplicity and a technical reason we require that α_n, β_n to be strictly positive with $\sum_{n=1}^{\infty} \beta_n^2 < \infty$, so that $\beta_n \to 0$ and $\alpha_n \to 1$. The latter is particularly helpful for our goal since $(e_n, x_n) = \alpha_n$, for all n. For example, the sequences $\{\cos n^{-1}\}$ and $\{\sin n^{-1}\}$ have all the desired properties of $\{\alpha_n\}$ and $\{\beta_n\}$.

Let us set $M = \overline{sp}\{e_n\}, N = \overline{sp}\{x_n\}$. To establish (9.22), we show that the vector $y = \sum_n \beta_n e'_n$ is in $\overline{sp}\{M, N\}$ but not in $M + N$. Since $\sum_n \beta_n^2 < \infty$, the series defining y is norm-convergent. Furthermore, α_n being positive implies that $e'_n \in M + N$, for all n and therefore $y \in \overline{sp}\{M, N\}$. Next, using proof by contradiction we show that y can not be in $M + N$. Suppose $y \in M + N$, so that $y = x + z$ with $x \in M$ and $z \in N$, then

$$\begin{aligned}\beta_n = (y, e'_n) = (x + z, e'_n) &= (z, e'_n) = \left(\sum_n (z, x_n)x_n, e'_n\right) \\ &= (z, x_n)(x_n, e'_n) = (z, x_n)\beta_n,\end{aligned}$$

for every n. Since $\beta_n > 0$, it follows that $(z, x_n) = 1$ for every n which is impossible since (z, x_n) is the nth Fourier coefficient of z with respect to $\{x_n\}$.

Example 9.7 and Theorem 9.5 represent two extreme situations regarding the behavior of vector sums. To see what happens in other situations the notion of *angle between two subspaces* in a Hilbert space defined in Section 8.6.3 is needed.

The next theorem, provides the most satisfying result regarding closedness of vector sums. For a proof see Theorem 8.17.

Theorem 9.19. For two subspaces M and N, their vector sum is closed if and only if they are at positive angle.

9.6.3 Alternating Projections

Some least squares and approximation problems can be reduced to projecting on the sum of two subspaces M_1 and M_2. When $M_1 \perp M_2$, then $M_1 \bigcap M_2 = \{0\}$ and $M_1 + M_2$ is closed and projection onto the sum space $\overline{sp}\{M_1, M_2\}$, is simply the sum of the projections onto M_1 and M_2. We now consider projections onto $M = M_1 \bigcap M_2$ and $N = \overline{sp}\{M_1, M_2\}$ without requiring that $M_1 \perp M_2$. Let P_1, P_2, P and Q stand for the projections onto M_1, M_2, M and N, respectively. For P_1 and P_2 commuting Theorem 9.18 expresses P and Q in terms of $P_i, i = 1, 2$, in a simple manner. Otherwise, we show that they can be expressed in terms of limits of $(P_1P_2)^m$ and $(P_2P_1)^m$ as $m \to \infty$, giving rise to an iterative algorithm known as the *Von Neumann alternating projections* which was proved in J. Von Neumann's lectures on operator theory at Princeton in 1933–1934 and published later in 1950. For more information on the history and applications of this algorithm in probability and statistics, see Bickel et al. (1993, Appendix A.4) and Pourahmadi (1992); for engineering applications, see Kayalar and Weinert (1988).

Theorem 9.20. *(Von Neumann Alternating Projections).* With notation as above for any $x \in H$

(a) $Px = \lim_{m \to \infty} (P_2P_1)^m x = \lim_{m \to \infty} (P_1P_2)^m x.$

(b) $Qx = \lim_{m \to \infty} [I - (I - P_1)^m (I - P_2)^m]x.$

The following remarks and examples shed more light on the applications of this result. The proof is given at the end of this section. Since

$$\overline{sp}\{M_1, M_2\} = (M_1^{\perp} \bigcap M_2^{\perp})^{\perp},$$

(b) is an immediate consequence of (a). Therefore, we focus on understanding, proving and generalizing (a). The following simple example sheds some light on (a).

Example 9.7. Let H be a Hilbert space with $x, x_1, x_2 \in H, ||x_i|| = 1, \alpha_i = (x, x_i)$, $i = 1, 2, \rho = (x_1, x_2)$. Consider the one-dimensional subspaces $M_i = sp\{x_i\}, i = 1, 2$. Since for any $y \in H$,

$$P_i y = (y, x_i)x_i, i = 1, 2$$

we obtain

$$P_1 x = (x, x_1)x_1 = \alpha_1 x_1, P_2 P_1 x = \alpha_1 P_2 x_1 = \alpha_1 \rho x_2,$$

$$P_1 P_2 P_1 x = \alpha_1 |\rho|^2 x_1, (P_2 P_1)^2 = \alpha_1 |\rho|^2 \rho x_2.$$

Finally,

$$(P_2 P_1)^m x = \alpha_1 |\rho|^{2m-2} \rho x_2, \text{ for } m \geq 1. \tag{9.23}$$

Since ρ is the measure of angle between M_1 and M_2, and $|\rho| < 1$, if and only if $M_1 \bigcap M_2 = \{0\}$, it follows that when $|\rho| < 1$, then

$$0 = Px = \lim_{m \to \infty} (P_2 P_1)^m x,$$

and, in fact,

$$||Px - (P_2 P_1)^m x|| = \alpha_1 |\rho|^{2m-1}, \tag{9.24}$$

converges to zero exponentially fast.

Expanding $(I - P_i)^m$ in the formula for Q in Theorem 9.20 (b), we obtain the following representation for the action of Q on any $x \in H$:

$$\begin{aligned} Qx &= Px + \sum_{k=1}^{\infty} [P_1 (P_2 P_1)^{k-1} + P_2 (P_1 P_2)^{k-1} - (P_2 P_1)^k \\ &\quad -(P_1 P_2)^k] x. \end{aligned} \tag{9.25}$$

This is useful in computing the projection of an $x \in H$ onto $N = \overline{sp}\{M_1, M_2\}$, particularly when M_1 or M_2 is finite-dimensional with $M_1 \bigcap M_2 = \{0\}$; see Sections 8.2–8.3 and Problem 9.8 (c)–(e).

Corollary 9.21. Let M_1 and M_2 be two subspaces of a Hilbert space H where $M_2 = sp\{x_1, \cdots, x_m\}, 1 \leq m < \infty$ and $M_1 \bigcap M_2 = \{0\}$. Then, for any $x \in H$

$$Qx = P_1 x + \sum_{k=1}^{m} \beta_k (x_k - P_1 x_k), \tag{9.26}$$

where $\beta = (\beta_1, \cdots, \beta_m)$ satisfies

$$A\beta = b, \tag{9.27}$$

with $A = [(x_k - P_1x_k, x_\ell - P_1x_\ell)]_{k,\ell=1}^m, b = ((x_k - P_1x_k, x))_{k=1}^m$.

Proof. It is easy to verify that

$$N = \overline{sp}\{M_1, M_2\} = M_1 \oplus sp\{x_k - P_1x_k, 1 \le k \le m\}.$$

Then, for any $x \in H$,

$$Qx = P_1x + P_{sp\{x_k - P_1x_k, 1 \le k \le m\}}x,$$

and the desired result follows from Example 9.4.

Q.E.D.

To prove Theorem 9.20, use the notion of *positive operator*. An operator $T : H \to H$ is said to be *positive* if $(Tx, x) \ge 0$, for all $x \in H$. In general, for two projection operators P_1, P_2, the operator P_2P_1 is not a projection (Theorem 9.18 (b)), but fortunately $T = P_2P_1 : M_2 \to M_2$ is a positive (and hence hermitian operator).

Proof of Theorem 9.20. First we show that the operator $T = P_2P_1$ in the space M_2 is a positive operator with norm not greater than one. Indeed, for any $y \in M_2$, we have

$$\begin{aligned}(P_2P_1y, P_2P_1y) &= \|P_2P_1y\|^2 \le \|P_1y\|^2 = (P_1y, P_1y) = (P_1y, y) \\ &= (P_2P_1y, y) = \|P_1y\|^2 \le \|y\|^2,\end{aligned}$$

so that

$$0 \le (Ty, Ty) \le (Ty, y) \le (y, y). \tag{1}$$

Next, for any $x \in H$, we show that the sequence $\{T^kx\}$ converges in the norm of H. For $k \ge 1, T^kx = P_2P_1T^{k-1}x \in M_2$, putting $y = T^kx$ in (1) we get

$$0 \le (T^{k+1}x, T^{k+1}x) \le (T^{k+1}x, T^kx) \le (T^kx, T^kx). \tag{2}$$

Using the fact that T is positive and hence Hermitian, from (2) we obtain

$$0 \le (T^{2k+1}x, Tx) \le (T^{2k}x, Tx) \le (T^{2k-1}x, Tx).$$

Consequently, the sequence $\{(T^nx, Tx)\}$ is a decreasing sequence of positive numbers and therefore it is convergent. Since,

$$\begin{aligned}\|T^mx - T^nx\|^2 &= (T^mx, T^nx) - (T^mx, T^nx) - (T^nx, T^mx) + (T^nx, T^nx) \\ &= (T^{2m-1}x, Tx) - 2(T^{m+n-1}x, Tx) + (T^{2n-1}x, Tx),\end{aligned}$$

and it converges to zero as $m, n \to \infty$, it follows that $\{T^m x\}$ is a Cauchy sequence in H and hence convergent. This means that T^m converges strongly to some bounded operator P. Further, $T^{m+1}x = TT^m x = T^m Tx$, which as $m \to \infty$, gives

$$TPx = Px = PTx. \tag{3}$$

From (3), $T^m Px = Px$, and hence $P^2 x = \lim_{m\to\infty} T^m Px = Px$, that is, P is idempotent. As a limit of Hermitian operators, P is Hermitian and in M_2 the operator P is an orthogonal projection on the space of all $\{Px; x \in H\}$. From (3) we get

$$||P_2 P_1 Px|| \leq ||P_1 Px|| \leq ||Px|| = ||P_2 P_1 Px||.$$

Hence $||P_2P_1Px|| = ||P_1Px|| = ||Px||, P_2P_1Px = P_1Px = Px$, and $Px \in M_1 \bigcap M_2$. Conversely, if $x \in M_1 \bigcap M_2$, then $Tx = P_2P_1x = x$ and $Px = \lim_{n\to\infty} T^m x = x$, that is, in M_2 the operator P is the orthogonal projection onto $M_1 \bigcap M_2$. Then, for any $x \in H$, we have from (3),

$$Px = PTx = PP_2P_1x = \text{ Projection of } x \text{ onto } M_1 \cap M_2.$$

Q.E.D.

There is a close connection between the alternating projections and the theory of reproducing kernel Hilbert spaces introduced in the next section; see Aronszajn (1950); Parzen (1967) and Wahba (1990).

9.7 POSITIVE-DEFINITE FUNCTIONS

Positive-definition functions, see Example 9.1 (g), studied in this section often appear in diverse areas of applied mathematics. The close affinity among positive-definite functions, Hilbert spaces, moment problems and stochastic processes is best described by the *Kolmogorov's theorem*, a special case of which was stated in Theorem 5.1:

> a complex-valued function $K(s,t)$, $s,t \in T$, is positive-definite, if and only if, there exists a Hilbert space H and a family of vectors $\{X_t; t \in T\}$ with $X_t \in H$ such that

$$K(s,t) = (X_s, X_t). \tag{9.28}$$

Depending on the group structure of T and the nature of the function $K(.,.)$, we could construct H and $\{X_t; t \in T\}$ explicitly. For example, when $T = \mathbb{Z}$ and $K(s,t)$ depends only on $s - t$, then one may choose $H = L^2(\mu)$, $X_t = e^{it\lambda}, t \in \mathbb{Z}$, and (9.28) is essentially the Herglotz theorem representing a positive-definite function as the Fourier coefficients of a measure. We start the section with this simple and prototypical case of the *trigonometric moment problem*, then introduce the notion of reproducing kernel Hilbert spaces and use it to prove the Kolmogorov's theorem.

9.7.1 Herglotz Theorem

A univariate function $K : \mathbb{Z} \to \mathbb{C}$ is said to be *positive-definite* if

$$\sum_{s,t=1}^{n} c_s \bar{c}_t K(s-t) \geq 0, \tag{9.29}$$

for all integers $n \geq 1$, and scalars $c_1, \cdots, c_n$. Even though (9.29) is an algebraic property, the Herglotz theorem characterizes positive-definite functions as the Fourier coefficients of a distribution function. Its proof relies on the *Helly's theorem*: For any sequence of distribution functions $\{F_n\}$, there is a subsequence $\{F_{n_k}\}$ and a function $F : R \to [0,1]$ such that F is right-continuous and nondecreasing function with $F(-\infty) = 0, F(\infty) \leq 1$ and $F_{n_k}(t) \to F(t)$ at all continuity points of F.

Theorem 9.22. (Herglotz). A function $K(\cdot)$ on the set of integers is positive-definite, if and only if

$$K(t) = \int_{-\pi}^{\pi} e^{it\lambda} dF(\lambda), t \in \mathbb{Z}, \tag{9.30}$$

where $F(\cdot)$ is a unique right-continuous and non-decreasing function with $F(-\pi) = 0$.

Proof. If $K(\cdot)$ has the representation (9.30), then

$$\sum_{s,t=1}^{n} c_s \bar{c}_t K(s-t) = \int \sum_{s,t=1}^{n} c_s \bar{c}_t e^{i(s-t)\lambda} dF(\lambda) = \int \left| \sum_{t=1}^{n} c_t e^{it\lambda} \right|^2 dF(\lambda) \geq 0,$$

that is, it is a positive-definite function.

Conversely, suppose $K(\cdot)$ is a positive-definite function. Replacing c_t's in (9.29) by $e^{it\lambda}/\sqrt{n}, \lambda \in (-\pi, \pi]$, we obtain

$$f_n(\lambda) = \frac{1}{n} \sum_{s,t=1}^{n} K(s-t) e^{i(s-t)\lambda} = \sum_{|t|<n} \left(1 - \frac{|t|}{n}\right) K(t) e^{it\lambda} \geq 0.$$

Let us define a sequence $\{F_n(\cdot)\}_{n\geq 1}$ of distribution functions by

$$F_n(\lambda) = \int_{-\pi}^{\lambda} f_n(u) du, \ -\pi \leq \lambda \leq \pi.$$

It is evident that

$$\int e^{it\lambda} dF_n(\lambda) = \begin{cases} (1 - \frac{|t|}{n}) K(t), & |t| < n \\ 0 & \text{elsewhere.} \end{cases} \tag{9.31}$$

Since $0 \le F_n(\lambda) = \int_{-\pi}^{\lambda} dF_n(u) \le \int_{-\pi}^{\pi} dF_n(u) = K(0) < \infty$, for all n, it follows from Helly's theorem that there is a distribution function F and a subsequence $\{F_{n_k}\}$ of $\{F_n\}$ such that

$$\int e^{it\lambda} dF_{n_k}(\lambda) \to \int e^{it\lambda} dF(\lambda), \text{ as } k \to \infty. \tag{9.32}$$

Replacing n by n_k in (9.31) and letting $k \to \infty$, the desired result follows from (9.31) and (9.32).

To address the question of uniqueness of F, let us assume that two distinct distributions F_1 and F_2 satisfy (9.31). Then, their difference

$$\alpha(\lambda) = F_1(\lambda) - F_2(\lambda),$$

is a function of bounded variation for which

$$\int e^{it\lambda} d\alpha(\lambda) = 0, \text{ for all integers } t.$$

From the uniqueness of the Fourier coefficients, it follows that $\alpha(\lambda) =$ constant at all points of continuity. Since $F_i(-\pi) = 0, i = 1, 2$, we get $F_1 = F_2$.

Q.E.D.

9.7.2 Reproducing Kernel Hilbert Spaces (RKHS)

Since elements of the Hilbert space $L^2(T, \mathcal{F}, \mu)$ are equivalent classes of functions which differ on sets of measure zero, for given f and fixed $t \in T$, its value $f(t)$ is not well-defined. To get around this problem, we introduce the notion of *functional Hilbert space* or *reproducing kernel Hilbert spaces* (Aronszajn, 1950; Parzen, 1967; Wahba, 1990) as a space of complex-valued functions on T with the following properties:

(a) If $f, g \in H$ and α, β are scalars, then for any $t \in T$, $(\alpha f + \beta g)(t) = \alpha f(t) + \beta g(t)$, that is, the *point evaluation functionals* $f \to f(t)$ are defined and are linear.

(b) For each $t \in T$, there exists a positive constant M_t such that

$$|f(t)| \le M_t \|f\|, \text{ for all } f \in H, \tag{9.33}$$

that is, all point evaluation functionals are bounded.

An important example of such a space consisting of polynomials of degree n were discussed in Chapter 7. Next, we show how starting with a positive function K, one may construct such a Hilbert space. This also gives a proof of the Kolmogorov's theorem stated earlier in this section and in Section 5.1.

Given a positive-definite function $K(\cdot,\cdot)$ as in Example 9.1 (g) we construct a Hilbert space $H(K)$ of functions on T as follows. For each fixed $s \in T$, let $K_s(\cdot)$ stand for the univariate function

$$K_s(\cdot) = K(s, \cdot) : T \to \mathbb{C},$$

so that $K_s(t) = K(s,t)$, for all $t \in T$. Let $V(K)$ be the manifold of all finite linear combinations of the form

$$f = \sum_{i=1}^{n} c_i K_{s_i}(\cdot),$$

for $n \geq 1$, scalars $c_1, \cdots, c_n$, and $s_1, \cdots, s_n \in T$. Since $K(\cdot,\cdot)$ is a positive-definite function, the expression

$$\left(\sum_{i=1}^{n} c_i K_{s_i}, \sum_{j=1}^{m} d_j K_{t_j} \right) = \sum_{ij} c_i \bar{d}_j (K_{s_i}, K_{t_j}) = \sum_{i,j} c_i \bar{d}_j K(s_i, t_j),$$

defines an inner product on $V(K)$. It has the additional *reproducing property* that for any f of the above form,

$$(f, K_t) = f(t), \text{ for all } t \in T. \tag{9.34}$$

For any sequence $\{f_n\}$ in $V(K)$, we have from (9.33) that

$$|f_m(t) - f_n(t)| = |(f_m - f_n, K_t)| \leq \|K_t\| \|f_m - f_n\|,$$

so that norm-convergence implies pointwise-convergence at every point of T.

Let $H(K)$ be the completion of $V(K)$. Defining pointwise limit of a Cauchy sequence $\{f_n\}$ by

$$f(t) = \lim_{n \to \infty} f_n(t),$$

it follows that the reproducing property (9.34) is preserved for all $f \in H(K)$. Moreover, $H(K)$ is a functional or a reproducing kernel Hilbert space (RKHS).

Conversely, suppose H is a RKHS. Since for fixed $t \in T$ the point evaluation map $f(t)$ is a bounded linear functional on H, the Riesz representation theorem guarantees the existence of a unique function $K_t(\cdot) \in H$ satisfying (9.34) for any $f \in H$. In particular, choosing $f = K_s(\cdot),\ s \in T$, gives

$$(K_s(\cdot), K_t(\cdot)) = K_s(t) = K(s,t).$$

So that the function $K(\cdot,\cdot) : T \times T \to \mathbb{C}$ defined as the inner product of vectors $K_t(\cdot), t \in T$ in H, is positive-definite. It is called the *kernel function* or the *reproducing kernel* of H.

Theorem 9.23. If $\{e_j\}$ is an orthonormal basis for a RKHS H, then its kernel function K is given by

$$K(s,t) = \sum_{j} \overline{e_j(s)} e_j(t).$$

Proof. From (9.34), the Fourier expansion of $K_t(\cdot)$ in terms of $\{e_j\}$ is of the form (Theorem 9.10 (d))

$$K_t = \sum_j (K_t, e_j)e_j = \sum_j \overline{e_j(t)}e_j,$$

and from the Parseval's identity

$$K(s,t) = (K_s, K_t) = \sum_j \overline{e_j(s)}e_j(t).$$

Q.E.D.

The space $\tilde{H}^2$ of analytic functions on the unit disc, see Examples 9.1 (f), is a RKHS because each $f \in H^2$ with $f = \sum_{k=0}^{\infty} c_k e_k$ corresponds to $F \in \tilde{H}^2$ where for $|z| < 1$,

$$F(z) = \sum_{k=0}^{\infty} c_k z^k = (F, K_z), K_z = \sum_{k=0}^{\infty} \bar{z}^k e_n.$$

This proves at once that $F \to F(z)$ is a bounded linear functional. By Theorem 9.23 its kernel is given by

$$K(z,w) = \sum_{k=0}^{\infty} \bar{z}^k w^k = \frac{1}{1 - w\bar{z}}, |z|, |w| < 1. \tag{9.35}$$

9.8 THE SPECTRAL THEOREM FOR UNITARY OPERATOR

In this section we prove the spectral theorem for unitary operators using the fact that for any $x \in H$ and unitary operator U, the function $\gamma_k = (U^k x, x)$ is positive-definite on the set of integers. The proof reveals the close affinity of the result with the spectral representation of a stationary process, its covariance function, trigonometric moment problem and a host of other problems from various areas of applications (Landau, 1987b; Akhiezer and Glazman, 1963).

9.8.1 Spectral Measures

We start with the notion of *spectral measure* which is a projection-valued measure defined on $\mathcal{F}$, the σ-algebra of Borel subsets of $\Omega = (-\pi, \pi]$. Our presentation follows mostly Halmos (1957).

Let H be a Hilbert space. A function $E(\cdot)$ on $\mathcal{F}$ whose values are orthogonal projection operators onto subspaces of H is called a *spectral measure* if

(i) $E(\Omega) = I$, the identity operator, and

(ii) $E(\cdot)$ is *countably additive*, that is for a sequence $\{A_n\}$ of disjoint sets in $\mathcal{F}$,

$$E\left(\cup_{n=1}^{\infty} A_n\right) = \sum_{n=1}^{\infty} E(A_n).$$

The most typical spectral measure is obtained by taking $H = L^2(\mu)$ and defining $E(\cdot)$ by $E(A) = \chi_A$ for $A \in \mathcal{F}$, where χ_A stands for the indicator function of the set A, $E(A)$ is the operator projecting $L^2(\mu)$ onto the subspace of all functions vanishing outside A. Properties of a spectral measure are similar to those of a probability measure and can be proved in virtually the same manner. But, one must be mindful that the first measure is operator-valued and the second takes scalars as its values.

Theorem 9.24. Suppose $E(\cdot)$ is an spectral measure on $\mathcal{F}$. Then
(a) $E(\cdot)$ is *monotone* and *subtractive*, that is, if $A, B \in \mathcal{F}$ and $A \subseteq B$, we have $E(A) \prec E(B)$ and $E(B - A) = E(B) - E(A)$.

(b) $E(\cdot)$ is *modular* and *multiplicative*, that is, for $A, B \in \mathcal{F}$,

$$E(A \cup B) = E(A) + E(B) - E(A \cap B),$$

and

$$E(A \cap B) = E(A)E(B).$$

The next result shows the equivalence between countable additivity of spectral and scalar measures. This theorem and its proof set the tone for reducing problems regarding operator-valued measures first to vector measures and then to those for the more familiar scalar measures.

Theorem 9.25. A projection-valued function $E(\cdot)$ on $\mathcal{F}$ is a spectral measure if and only if,

(a) $E(\Omega) = I$, and

(b) For each pair of $x, y \in H$, the complex-valued set function $\mu_{x,y}$ defined for A in $\mathcal{F}$ by $\mu_{x,y}(A) = (E(A)x, y)$ is countably additive.

Proof. Suppose $E(\cdot)$ is a spectral measure, then (a) holds by definition, and (b) follows from the continuity of inner products (Theorem 9.2). Conversely, suppose that (a) and (b) holds. If $A, B \in \mathcal{F}$ are disjoint, then the identity

$$(E(A \cup B)x, y) = (E(A)x, y) + (E(B)x, y) = ((E(A) + E(B))x, y), x, y \in H,$$

shows that $E(A \cup B) = E(A) + E(B)$ holds, that is, $E(\cdot)$ is finitely additive. Similarly, if $\{A_n\}$ is a sequence of disjoint sets in $\mathcal{F}$ with $A = \bigcup_{n=1}^{\infty} A_n$, as above we argue that

$$(E(A)x, y) = \sum_n (E(A_n)x, y) = \left(\left(\sum_n E(A_n)\right) x, y\right), x, y \in H,$$

and hence $E(A) = \sum_n E(A_n)$. The trouble with this argument is that $\sum_n E(A_n)$ may not make sense. Next, we rely on the orthogonality of $\{E(A_n)\}$ and show that it does. For every vector $x \in H$, it follows from Theorem 9.24 (b), that $\{E(A_n)x\}$ is an orthogonal sequence of vectors. Thus,

$$\sum_n ||E(A_n)x||^2 = \sum_n (E(A_n)x, x) = (E(A), x, x) = ||E(A)x||^2 < \infty$$

and it follows that $\sum_{n=1}^{\infty} E(A_n)x$ is norm-convergent in H (Theorem 9.9). Let us define a mapping T by $Tx = \sum_{n=1}^{\infty} E(A_n)x$. It is evident that T is linear and bounded with $||T|| \leq 1$, so that $\sum_n E(A_n)$ makes sense as an operator.

Q.E.D.

Note that $\mu_{x,x}(\cdot)$, denoted hereafter by $\mu_x(\cdot)$, is a finite positive measure associated to the pair (E, x), where E is a spectral measure and $x \in H$ is a fixed vector. There is also a vector measure associated to the pair (E, x) which is of fundamental importance in the theory of stationary sequences. This is an H-valued *countably additive* measure defined by

$$Z_x(A) = E(A)x, A \in \mathcal{F}, \tag{9.36}$$

with the decisive property that

$$A, B \in \mathcal{F}, \text{ and } A, B \text{ disjoint } \Leftrightarrow Z_x(A) \perp Z_x(B). \tag{9.37}$$

Countable additivity of $Z_x(\cdot)$ is implicit in the proof of Theorem 9.25, so are the close relationship between μ_x and Z_x, namely, for $A \in \mathcal{F}$

$$\mu_x(A) = (E(A)x, E(A)x) = ||Z_x(A)||^2, \tag{9.38}$$

and the fact that the stochastic process $\{\tilde{Z}_x(\lambda); -\pi < \lambda \leq \pi\}$ defined by

$$\tilde{Z}_x(\lambda) = Z_x(-\pi, \lambda], -\pi < \lambda \leq \pi, \tag{9.39}$$

has *orthogonal increments*, namely, for any $-\pi < \lambda_1 < \lambda_2 < \lambda_3 < \lambda_4 \leq \pi$,

$$\tilde{Z}_x(\lambda_2) - \tilde{Z}_x(\lambda_1) \perp \tilde{Z}_x(\lambda_4) - \tilde{Z}_x(\lambda_3), \tag{9.40}$$

The Fourier transform of the H-valued measure Z_x, that is,

$$\int e^{it\lambda} dZ_x(\lambda), t = 0, \pm 1, \cdots \tag{9.41}$$

is simply the revered spectral representation of stationary sequences. An astute reader will appreciate the sense in which the integral (9.41) is defined and discover some of its properties while studying the proof of the spectral theorem for unitary operators.

9.8.2 The Spectral Theorem

It is interesting to note that the following proof of the spectral theorem for unitary operators relies only on integration with respect to scalar measures (Akhiezer and Glazman, 1963).

Theorem 9.26. If U is a unitary operator on a Hilbert space H, then there exists a unique spectral measure E on $(\Omega, \mathcal{F})$ such that

$$(Ux, y) = \int_{-\pi}^{\pi} e^{i\lambda} d(E(\lambda)x, y), \text{ all } x, y \in H. \tag{9.42}$$

The dependence of U on E is expressed symbolically by

$$U = \int_{-\pi}^{\pi} e^{i\lambda} dE(\lambda). \tag{9.43}$$

Proof. For convenience we divide up the proof into three steps.

Step I. *Integral representation of* $(U^k x, y)$: For a unitary operator U and fixed $x \in H$, the sequence $\{\gamma_k\}$ defined by

$$\gamma_k = (U^k x, x), \ k = 0, \pm 1, \cdots,$$

is positive-definite on the set of integers, since

$$\sum_{k,\ell=1}^{n} c_k \bar{c}_\ell \gamma_{k-\ell} = \sum_{k,\ell} c_k \bar{c}_\ell (U^k x, U^\ell_x) = \left(\sum_k c_k U^k x, \sum_\ell c_\ell U^\ell x \right) \geq 0.$$

Hence by the Herglotz theorem, there exists a unique finite measure μ_x on $(-\pi, \pi]$ such that

$$(U^k x, x) = \int e^{ik\lambda} d\mu_x(\lambda), k \in \mathbb{Z}. \tag{1}$$

Substituting x in (1) by $x + y, x - y, x + iy$ and $x - iy$, respectively, we obtain from Theorem 9.14 that

$$(U^k x, y) = \frac{1}{4} \sum_{\ell=1}^{4} i^\ell (U^k (x + i^\ell y), x + i^\ell y) = \int e^{ik\lambda} d\mu_{x,y}(\lambda), \tag{2}$$

where $\mu_{x,y}$, a corresponding linear combination of $\mu_{x \pm y}$ and $\mu_{x \pm iy}$, is a unique finite complex-valued measure on $(-\pi, \pi]$.

Step II. $\mu_{x,y}$ *is a bounded bilinear functional:* Using the fact that the left-hand side of (2) is a bounded bilinear functional for each k, and uniqueness of the Fourier coefficients it follows that $\mu_{x,y}(\lambda)$ is a bilinear functional with norm not exceeding one. Furthermore,

$$\mu_{x,y}(\lambda) = \overline{\mu_{y,x}(\lambda)}. \tag{3}$$

Therefore, by Theorem 9.13, there exists a unique family of bounded operators $\{E(\lambda); \lambda \in (-\pi, \pi]\}$ such that

$$\mu_{x,y}(\lambda) = (E(\lambda)x, y). \tag{4}$$

Step III. $E(\cdot)$ *is a spectral measure:* We show that $E(\cdot)$ is a projection-valued measure with $E(\Omega) = I$. Since $\int dE(\lambda) = E(\Omega)$, setting $k = 0$ in (5) below we obtain for all $x, y \in H$ that $(x, y) = (E(\Omega)x, y)$, and hence $E(\Omega) = I$.

From (3) and (4), we obtain

$$(E(\lambda)x, y) = \overline{(E(\lambda)y, x)} = (x, E(\lambda)y), \text{ all } x, y \in H,$$

so that $E(\lambda)$ is a Hermitian operator. To show that it is idempotent, combine (2) and (4) to obtain

$$(U^k x, y) = \int e^{ik\lambda} d(E(\lambda)x, y), \tag{5}$$

and replace y by $y = U^{-\ell}z, \ell = 0, \pm 1, \cdots$, to get

$$(U^k x, U^{-\ell}z) = \int e^{ik\lambda} d(E(\lambda)x, U^{-\ell}z),$$

or

$$(U^{k+\ell}x, z) = \int e^{ik\lambda} d(U^\ell E(\lambda)x, z). \tag{6}$$

But for λ fixed from (5) we have

$$(U^\ell E(\lambda)x, z) = \int e^{i\ell\lambda'} d(E(\lambda')E(\lambda)x, z),$$

substituting into (6), we obtain

$$(U^{k+\ell}x, z) = \int e^{ik\lambda} d\left\{\int e^{i\ell\lambda'} d(E(\lambda')E(\lambda)x, z)\right\}.$$

On the other hand,

$$(U^{k+\ell}x, z) = \int e^{i(k+\ell)\lambda} d(E(\lambda)x, z) = \int e^{ik\lambda} d\left\{\int_{-\pi}^{\lambda} e^{\ell\lambda'} d(E(\lambda')x, z)\right\}. \tag{7}$$

Thus, from (6) and (7),

$$\int_{-\pi}^{\pi} e^{i\ell\lambda'} d(E(\lambda')E(\lambda)x, z) = \int_{-\pi}^{\pi} e^{i\ell\lambda'} \chi_{(-\pi,\lambda]}(\lambda') d(E(\lambda')x, z), \ \ell = 0, \pm 1, \cdots$$

and the uniqueness of Fourier coefficients implies that for $\lambda' \leq \lambda$

$$(E(\lambda')E(\lambda)x, z) = (E(\lambda')x, z), \ x, z \in H.$$

Consequently,

$$E(\lambda')E(\lambda) = E(\lambda'), \text{ for } \lambda' \leq \lambda,$$

which implies that

$$E(\lambda)^2 = E(\lambda),$$

that is, $E(\lambda)$ is a projection operator, as desired.

Q.E.D.

9.8.3 The Spectral Integral

As noted earlier the proof of the spectral theorem for unitary operators only relies on integration with respect to a scalar measure. However, applications centered around (9.41) and (9.43) involve integration with respect to vector and operator measures which are of independent interest.

Fortunately, for any $x \in H$ and integer k the integral representation

$$U^k x = \int_{-\pi}^{\pi} e^{ik\lambda} dE(\lambda)x, \tag{9.44}$$

has a valid interpretation (definition) on its own right, in addition to being symbolic representations of (9.41)–(9.42). To elaborate, consider a subdivision of the interval $(-\pi, \pi]$:

$$-\pi = \lambda_0 < \lambda_1 < \lambda_2 < \cdots < \lambda_n = \pi,$$

and construct an operator

$$A_n = e^{ik\lambda_1} E(\Delta_1) + e^{ik\lambda_2} E(\Delta_2) + \cdots + e^{ik\lambda_n} E(\Delta_n),$$

where $\Delta_j = (\lambda_{j-1}, \lambda_j]$. Our goal is to show that A_n and U^k are close in the sense that for each $x \in H$,

$$||A_n x - U^k x|| \to 0 \tag{9.45}$$

when $\max_{1 \le j \le n} |\lambda_j - \lambda_{j-1}| \to 0$. That is, the sequence of operators A_n converges *strongly* to the operator U^k. Then, the limit of the sequence $\{A_n\}$ of operators is taken as the definition of the *spectral integral* (9.43).

To prove (9.45), it suffices (Problem 9.6) to show that as $n \to \infty$,

(a) for any $x, y \in H$

$$(A_n x, y) \to (U^k x, y),$$

(b) $||A_n x|| \to ||U^k x|| = ||x||$.

But, the validity of (a) is obvious, since

$$(A_n x, y) = \sum_{j=1}^{n} e^{ik\lambda j} (E(\Delta_j)x, y) \to \int_{-\pi}^{\pi} e^{ik\lambda} d\mu_{x,y}(\lambda) = (U^k x, y).$$

As for (b) we note that

$$||A_n x||^2 = (A_n x, A_n x) = \sum_{j,\ell=1}^{n} e^{ik(\lambda_j - \lambda_\ell)} (E(\Delta_j)x, E(\Delta_\ell)x)$$

$$= \sum_{j,\ell=1}^{n} e^{ik(\lambda_j - \lambda_\ell)} (E(\Delta_\ell)E(\Delta_j)x, x) = \sum_{j=1}^{n} (E(\Delta_j)x, x) = (x, x),$$

since $\sum_{j=1}^{n} E(\Delta_j) = I$, so that in (b) there is no need to take limit.

Replacing $e^{ik\lambda}$ in (9.44) by more general functions $g \in L^2(\mu_x)$ requires defining the integral

$$\int_{-\pi}^{\pi} g(\lambda)dE(\lambda)x,$$

so that its value is an element of H. For this, first choose a sequence $\{g_n\}$ of simple functions converging to g in the norm of $L^2(\mu_x)$, then set

$$\begin{cases} g_n(\lambda) = \sum_{j=1}^{n} c_{n_j} \chi_{\Delta_{nj}}, \\ y_n = \int g_n dE(\lambda)x = \sum_{j=1}^{n} c_{n_j} E(\Delta_{nj})x, n = 1, 2, \cdots, \end{cases}$$

where $\Delta_{n1}, \cdots, \Delta_{nn}$ is a partition of $(-\pi, \pi]$. Since for simple functions we have

$$||y_m - y_n||_H = \int |g_m - g_n|^2 d\mu_x = ||g_m - g_n||_{L^2(\mu_x)},$$

it follows that the sequence $\{y_n\}$ is Cauchy in H and converges to some element $y \in H$. This y is taken as the value of the above integral, that is,

$$\int g(\lambda)dE(\lambda)x = y. \tag{9.46}$$

The procedure used to define this integral is precisely the same as that for defining a stochastic integral in the spectral representation of stationary sequences; see (9.48)–(9.49) and Cramer (1940); Doob (1953) and Masani (1995).

9.9 THE SPECTRAL REPRESENTATION

The spectral representation of stationary sequences is now immediate from Theorem 9.26. Many consequences of this powerful result related to linear operations on stationary sequences are discussed in this section.

Theorem 9.27. For $\{x_t\}$ a stationary sequence in H there exists an orthogonal increment process $\{Z(\lambda); -\pi < \lambda \leq \pi\}$ such that

$$x_t = \int_{-\pi}^{\pi} e^{it\lambda} dZ(\lambda), t = 0, \pm 1, \cdots.$$

Proof. Let U be the unique shift operator of $\{x_t\}$ (Theorem 9.16) and $(E(\lambda); -\pi < \lambda \leq \pi)$ be its unique spectral measure (Theorem 9.26). Since

$$x_t = U^t x_0,$$

it follows from Theorem 9.26 that

$$x_t = \left(\int e^{it\lambda} dE(\lambda) \right) x_0 = \int e^{it\lambda} dE(\lambda) x_0 = \int e^{it\lambda} dZ_{x_0}(\lambda),$$

where $Z_{x_0} = E(A)x_0, A \in \mathcal{F}$, is described in (9.36)–(9.41).

Q.E.D.

The measure $\mu_{x_0}(A) = ||Z_{x_0}(A)||^2$ or equivalently the function F_{x_0} defined by

$$F_{x_0}(\lambda) = \mu_{x_0}(-\pi, \lambda], \ -\pi < \lambda \leq \pi, \tag{9.47}$$

is called the *spectral distribution* of $\{x_t\}$. When there is no danger of confusion we shall drop the subscript x_0. For more discussion and concrete examples of the spectral representation of stationary sequences, see Section 5.3.

9.9.1 Kolmogorov's Isomorphism

The Kolmogorov's isomorphism between the time-domain and the spectral-domain of a stationary sequence is immediate from its spectral representation.

For a function $g \in L^2(\mu)$, define the *stochastic integral* $\int g(\lambda)dZ(\lambda)$ where the integration is with respect to the H-valued measure $Z(\cdot)$ so that the equality

$$\left(\sum_j a_j x_j, \sum_k b_k x_k \right)_{H(x)} = \left(\sum_j a_j e_j, \sum_k b_k e_k \right)_{L^2(\mu)}, \tag{9.48}$$

which holds for all finite sums and scalars a_j's and b_j's, will continue to hold for limits of such linear combinations and trigonometric polynomials.

To meet this requirement, choose a sequence $\{g_n\}$ of trigonometric polynomials converging to g in the norm of $L^2(\mu)$, then follow the arguments leading to (9.46) and set

$$\int g dZ = y,$$

where y is the limit of $y_n = \int g_n dZ$. This, in particular shows that the mapping

$$g \to \int g dZ \tag{9.49}$$

is an isomorphism of the spectral-domain $L^2(\mu)$ onto the time-domain $H(x)$, called the *Kolmogorov's isomorphism*. Furthermore, the equality (9.48) holds when limits are taken in $H(x)$ and $L^2(\mu)$ norms. The function $g \in L^2(\mu)$ is called the *isomorph* of $y \in H(x)$. Finding the isomorphs of certain elements of $H(x)$ is of particular interest in prediction theory. The most notable examples are those corresponding to the innovation and predictors which depend on the optimal factorization of the spectral density and the requisite knowledge of the Hardy space H^2 covered in Chapter 10.

9.9.2 Linear Filtering and Subordination

The following basic results are the cornerstone of the idea of *linear filtering and subordination* of stochastic processes which is concerned with expressing one process in terms of others.

We say that two sequences $\{x_t\}$ and $\{y_t\}$, with $x_t = U^t x, y_t = U^t y$, are *stationarily correlated* if

$$(x_{t+k}, y_t) = (U^{t+k}x, U^t y) = (U^k x, y) = \gamma_{x,y}(k), \tag{9.50}$$

depends on k alone. The function $\{\gamma_{x,y}(k), k \in \mathbb{Z}\}$ is the *cross-covariance* of $\{x_t\}$ and $\{y_t\}$; note that when $x = y$ it reduces to the covariance function of a single stationary sequence $\{x_t\}$. Fortunately, the cross-covariance function also admits a spectral representation, see (9.42), as

$$\gamma_{x,y}(k) = \int_{-\pi}^{\pi} e^{ik\lambda} d\mu_{x,y}(\lambda), \tag{9.51}$$

where

$$\mu_{x,y}(A) = (E(A)x, E(A)y), A \in \mathcal{F}. \tag{9.52}$$

The function

$$F_{x,y}(\lambda) = \mu_{x,y}(0, \lambda], \quad -\pi < \lambda \leq \pi, \tag{9.53}$$

is called the *cross-spectral distribution* of $\{x_t\}$ and $\{y_t\}$. Note that $\mu_{x,y}$ is conjugate symmetric and from the Cauchy-Schwartz inequality we obtain

$$\begin{aligned} |\mu_{x,y}(A)| &= |(E(A)x, E(A)y)| \leq \|E(A)x\|^2 \|E(A)y\|^2 \\ &= \mu_x(A)\mu_y(A), A \in \mathcal{F}, \end{aligned} \tag{9.54}$$

so that it is absolutely continuous, for example, with respect to μ_x, μ_y and $\mu_x + \mu_y$.

It turns out that for U a unitary operator on H with the spectral measure E and fixed $x, y \in H$, the complex-valued measure $\mu_{x,y}$ associated to the ordered triplet (E, x, y) is of great interest in the theory of linear filtering and subordination of the pair of stationary sequences $\{U^t x\}$ and $\{U^t y\}$.

Theorem 9.28. Suppose $\{x_t\}$ is a stationary sequence in H with the shift operator U and $y \in H$ is arbitrary. Let $\hat{y}$ be the orthogonal projection of y onto $H(x)$ and $g \in L^2(\mu_x)$ be the isomorph of $\hat{y}$, that is,

$$\hat{y} = \int_{-\pi}^{\pi} g(\lambda) dE(\lambda) x.$$

Then, for any $A \in \mathcal{F}$, and $\mu_{x,y}$ defined as in (9.52) we have

(a) $\mu_{y,x}(A) = \mu_{\hat{y},x}(A) = \int_A g(\lambda) d\mu_x(\lambda)$.

(b) $\mu_{\hat{y}}(A) = \int_A |g(\lambda)|^2 d\mu_x(\lambda) = \int_A g(\lambda) d\mu_{x,\hat{y}}(\lambda)$.

Proof. Since $E(A)x \in H(x)$ and $y - \hat{y} \perp H(x)$, it follows that

$$\begin{aligned}\mu_{y,x}(A) &= (E(A)y, x) = (y - \hat{y} + \hat{y}, E(A)x) \\ &= (\hat{y}, E(A)x) = (E(A)\hat{y}, x) = \mu_{\hat{y},x}(A).\end{aligned}$$

As for the second equality in (a), let $\{g_n\}$ be a sequence of simple functions, as in Section 9.8.3, converging to g in $L^2(\mu_x)$. Then,

$$E(A)\int g_n(\lambda)dE(\lambda)x = \sum_{j=1}^{n} c_{n_j} E(\Delta_{nj} \cap A)x.$$

Using continuity of the operator $E(A)$ and definition of the integral involved, we obtain

$$\begin{aligned}E(A)\hat{y} &= E(A)\int g(\lambda)dE(\lambda)x = E(A)\left[\lim_{n\to\infty}\sum_{j=1}^{n} c_{n_j} E(\Delta_{n_j})x\right] \\ &= \lim_{n\to\infty}\sum_{j=1}^{n} c_{n_j} E(\Delta_{n_j} \cap A)x = \int_A g(\lambda)dE(\lambda)x,\end{aligned}$$

from which the integral representations of $\mu_{\hat{y},x}(A)$ and $\mu_{\hat{y}}$ in (a) and (b) are evident.
Q.E.D.

Let $\{x_t\}$ be a stationary sequence in H with the spectral representation $x_t = \int e^{it\lambda} dE(\lambda)x$. A stationary sequence $\{y_t\}$ is said to be obtained by a *linear filter (transformation)* of $\{x_t\}$ if

$$y_t = \int e^{it\lambda} g(\lambda)dE(\lambda)x, \tag{9.55}$$

for a function $g \in L^2(\mu_x)$. The function g is called the *spectral characteristic of the filter* or the isomorph of the vector y_0 in the time-domain. Since, evidently, $y_t \in H(x)$ for all t, the issue of *subordination* (inclusion of the information) of $\{y_t\}$ in $\{x_t\}$ is of interest. Historically, the concept of subordination and its characterization have played a major role in the development of prediction theory of stationary sequences; see Kolmogorov (1941); Rozanov (1967) and Rosenberg (1969).

Suppose $\{x_t\}$ and $\{y_t\}$ are any two sequence in H. We say,

(a) $\{y_t\}$ is *subordinate* to $\{x_t\}$ if $H(y) \subseteq H(x)$. (9.56)

(b) $\{y_t\}$ is *strongly subordinate* to $\{x_t\}$ if $H_s(y) \subseteq H_s(x)$, for all s.

(c) $\{y_t\}$ and $\{x_t\}$ are *mutually subordinate* or *equivalent* if $H(y) = H(x)$.

(d) $\{y_t\}$ and $\{x_t\}$ are *strongly equivalent* if $H_s(y) = H_s(x)$, for all s.

Next, we give several analytic characterizations of these geometric properties; these follow immediately from Theorem 9.28 by taking $y = y_0$.

Corollary 9.29. Let $\{x_t\}$ and $\{y_t\}$ be two stationary and stationarily correlated sequences in H. Then, the following are equivalent:

(a) $\{y_t\}$ is subordinate to $\{x_t\}$, that is, (9.56) holds.

(b) There exists $g \in L^2(\mu_x)$ such that $y_t = \int e^{it\lambda} g(\lambda) dE(\lambda)x$, that is, $\{y_t\}$ is obtained from $\{x_t\}$ by a linear filter.

(c) There exists $g \in L^2(\mu_x)$ such that for any $A \in \mathcal{F}$

$$\begin{cases} \mu_{yx}(A) = \int_A g d\mu_x, \\ \mu_y(A) = \int_A |g|^2 d\mu_x. \end{cases} \tag{9.57}$$

When (a) holds, the function g in (b) and (c) are equal a.e. (μ_x).

For a measure μ on $\Omega = (-\pi, \pi]$, set $H^2(\mu) = \overline{sp}\{e^{ik\lambda}; k \geq 0\}$ in $L^2(\mu)$. Note that for μ the Lebesgue measure, $H^2(\lambda)$ is the Hardy space H^2 discussed in Chapter 10.

Corollary 9.30. Let $\{x_t\}$ and $\{y_t\}$ be two stationary and stationarily correlated sequences in H. Then,

(a) $\{y_t\}$ and $\{x_t\}$ are mutually equivalent, if and only if there exists a g satisfying (9.57) such that $g \neq 0$ a.e. (μ_x) and $g^{-1} \in L^2(\mu_x)$.

(b) $\{y_t\}$ is strongly subordinate to $\{x_t\}$, if and only if there exists a $g \in H^2(\mu_x)$ satisfying (9.57).

The identities in (9.57) state that μ_{yx} and μ_y are absolutely continuous with respect to the measure μ_x with the Radon-Nykodym derivatives g and $|g|^2$, respectively. We note that these results provide analytic characterizations for the geometric notions of subordination. The next example establishes the close connection between linear filtering and linear processes defined in Section 5.2.

Example 9.8. Let $\{e_t\}$ be an orthonormal sequence in H with the shift operator U and the spectral measure E. Since

$$(e_k, e_0) = \delta_{k,0} = \int e^{ik\lambda} d\mu_e, \ k \in \mathbb{Z},$$

it follows from the uniqueness of Fourier coefficients that μ_e the spectral distribution of $\{e_t\}$ is precisely the normalized Lebesgue measure λ on $(-\pi, \pi]$. Thus, for any $g \in L^2$ with the Fourier coefficients $\{c_k\}$, the process $\{y_t\}$ defined by

$$y_t = \int e^{it\lambda} g(\lambda) dE(\lambda) e_0, \ t \in \mathbb{Z}$$

is subordinate to $\{e_t\}$ with the spectral density and linear representation,

$$\begin{cases} f_y = |g|^2, \\ y_t = \int e^{it\lambda} \sum_k c_k e^{-ik\lambda} dE(\lambda) e_0 = \sum_k c_k e_{t-k}. \end{cases} \tag{9.58}$$

This turns out to be the *general linear process* of Example 5.3.

As another application of Theorem 9.28, we obtain conditions for stationarity, invertibility and the formula for the spectral density of ARMA models. These were discussed also in Chapter 4.

Corollary 9.31. Let $\{e_t\}$ be an orthonormal sequence in H. Then, the sequence $\{x_t\}$ defined by

$$x_t - \sum_{k=1}^{p} \alpha_k x_{t-k} = e_t + \sum_{j=1}^{q} \beta_j e_{t-j}, \tag{9.59}$$

is stationary, if and only if $\alpha(z) = 1 - \alpha_1 z - \cdots - \alpha_p z^p \neq 0,\ |z| = 1$. [It is assumed that the polynomials $\alpha(z)$ and $\beta(z) = 1 + \beta_1 z + \cdots + \beta_q z^q$ have no common roots.] In this case the spectral density of $\{x_t\}$ is $f_x(\lambda) = \left|\dfrac{\beta(\lambda)}{\alpha(\lambda)}\right|^2$.

Proof. Suppose $\alpha(z) \neq 0$ on the unit circle. Then, the function $g(\lambda) = \dfrac{\beta(\lambda)}{\alpha(\lambda)}$ is in L^2 and the sequence

$$x_t = \int e^{it\lambda} g(\lambda) dE(\lambda) e_0$$

where $E(\cdot)$ is the spectral measure of the stationary sequence $\{e_t\}$ is stationary and its spectral density satisfies (9.58). Conversely, suppose (9.59) has a stationary solution $\{x_t\}$ with the spectral representation

$$x_t = \int e^{it\lambda} E_1(\lambda) x, \tag{9.60}$$

whose spectral measure $E_1(\cdot)$ is possibly different from that of $\{e_t\}$ since as yet it is not known whether $\{x_t\}$ and $\{e_t\}$ are stationarily correlated. Substituting (9.60) and the spectral representation of $\{e_t\}$ in (9.59) gives

$$\int e^{it\lambda} \alpha(\lambda) dE_1(\lambda) x = \int e^{it\lambda} \beta(\lambda) dE(\lambda) e, \text{ for all integers } t.$$

Consequently, from the uniqueness of the Fourier coefficients, we arrive at

$$\alpha(\lambda) dE_1(\lambda) x = \beta(\lambda) dE(\lambda) e, \ a.e.(\lambda).$$

Since $\alpha(\lambda)$ is a polynomial of degree p with at most a finite number of zeros on the unit circle, we have

$$dE_1(\lambda) x = \frac{\beta(\lambda)}{\alpha(\lambda)} dE(\lambda) e, a.e.(\lambda).$$

Now, from stationarity of $\{x_t\}$ we obtain

$$\mu_x(\Omega) = ||E_1(\Omega) x||^2 = \int \left|\frac{\beta(\lambda)}{\alpha(\lambda)}\right|^2 d\lambda < \infty,$$

so that $\alpha(z) \neq 0$ on the unit circle, and the formula for the spectral density function of $\{x_t\}$ follows.

Q.E.D.

Note that from (9.59), $\{e_t\}$ can also be obtained from $\{x_t\}$ by means of a linear filter, if $g \neq 0$ a.e. (μ_x) and $g^{-1} \in L^2(\mu_x)$. This amounts to $\beta(z) \neq 0$ on $|z| = 1$ which allows $\{e_t\}$ to be expressed in terms of $\{x_t\}$ in a possibly noncausal manner. Makagon and Salehi (1992) provide results similar to those in Corollary 9.31 for the solutions of infinite order analogues of (9.59) to be stationary, harmonizable and bounded harmonizable. For more details on the spectral representations of certain classes of nonstationary processes, see Priestley (1988).

9.9.3 Linear Interpolation

Linear interpolation and prediction problems are two of the most important special cases of the linear filtering problem. The former, which is easier to solve, is discussed in this section. The spectral-domain solution of the latter given in Chapter 10 is harder; it hinges on the optimal factorization of the spectral density function and requires some knowledge of function theory in the Hardy space H^2.

Let $\{x_t\}$ be a stationary sequence so that the entire sequence is known except at the time point $t = 0$. We would like to approximate x_0 by that linear combination of the observed values $\{x_t;\ t \neq 0\}$ which is best in the sense of smallest norm for the error. Evidently $\hat{x}_0$, the orthogonal projection of x_0 onto $\overline{sp}\{x_t;\ t = 0\}$, is the optimal interpolator. Thus, our problem is solved as soon as one finds the isomorph $g(\lambda)$ of $\hat{x}_0$ in $L^2(f)$. Starting from the spectral representation of $\{x_t\}$ in Theorem 9.27 (see also Theorem 9.28), we have

$$\hat{x}_0 = \int g(\lambda) dZ(\lambda).$$

Since $x_0 - \hat{x}_0$ is orthogonal to $\overline{sp}\{x_t;\ t \neq 0\}$, using the Kolmogorov's isomorphism we arrive at

$$(x_0 - \hat{x}_0, x_t) = \int (1 - g(\lambda)) f(\lambda) e^{-it\lambda} d\lambda = 0,\ t \neq 0. \tag{9.61}$$

It is clear that $g(\lambda)$ satisfies the above conditions, if and only if it is of the form

$$g(\lambda) = 1 - cf^{-1}(\lambda), \tag{9.62}$$

for a constant c. Since the interpolation error on the one hand is given by

$$\|x_0 - \hat{x}_0\|^2 = \int |1 - g(\lambda)|^2 f(\lambda) d\lambda = |c|^2 \int f^{-1}(\lambda) d\lambda,$$

and on the other by

$$\|x_0 - \hat{x}_0\|^2 = (x_0 - \hat{x}_0, x_0) = \int (1 - g(\lambda)) f(\lambda) d\lambda = c, \tag{9.63}$$

it follows that

$$\|x_0 - \hat{x}_0\|^2 = c = \left(\int f^{-1}(\lambda) d\lambda \right)^{-1}.$$

The interpolation error is evidently positive when f^{-1} is integrable, otherwise it can be shown to be zero. For more information on the coefficients of $\hat{x}_0$, see Section 8.5 and Problem 9.8 (b).

A slightly more general interpolation problem is that of interpolating several missing values $x_t,\ t = 0, 1, \ldots, m$, based on the known values $\{x_s;\ s \neq 0, 1, \ldots, m\}$; see Problem 9.8 (c)–(f). For simplicity in notation we focus on x_0. An argument analogous to those leading to (9.61)–(9.63) reveals that the isomorph $g(\lambda)$ of $\hat{x}_0$ satisfies

$$1 - g(\lambda) = \sigma'^2 \sum_{j=0}^{m} c_j e^{ij\lambda} f^{-1}(\lambda), \tag{9.64}$$

where $c_0 = 1$ and $c_1, \ldots, c_m$ are scalars minimizing the expression in the denominator of

$$\sigma'^2 = \|x_0 - \hat{x}_0\|^2 = \left(\min \int \left| \sum_{j=0}^{m} c_j e^{ij\lambda} \right|^2 f^{-1}(\lambda) d\lambda \right)^{-1}. \tag{9.65}$$

Thus, interestingly the problem of interpolation of x_0 based on infinite values of the sequence $\{x_t\}$ with the spectral density f is reduced to a finite prediction problem of a dual (inverse) process $\{x'_t\}$ with the spectral density f^{-1}. Aspects of this duality were presented in Chapter 8 and can be traced to the work of Yaglom (1962); Box and Tiao (1975); Salehi (1979); Nakazi (1984) and Miamee and Pourahmadi (1998).

Problems

9.1 Let $A = (a_{ij})$ be an $n \times n$ real symmetric matrix.

(a) When does the expression

$$(x, y)_A = x'Ay = \sum_{i=1}^{n} \sum_{j=1}^{n} a_{ij} x_i y_j,\ x, y \in R^n,$$

define an inner product on R^n? Check your answer for $n = 2$ and

$$A = \begin{pmatrix} 1 & -1 \\ -1 & 1 \end{pmatrix}.$$

Is the set $\{x \in R^2; x'Ax = 0\}$ a subspace of R^2?

(b) Suppose $(.,.)_A$ is an inner product on R^n, is there any relationship between the norms of $(R^n, (.,.)_A)$ and the Euclidean space R^n?

(c) For a finite positive integer n, let $T = \{1, 2, \ldots, n\}$ and note that a function $K : T \times T \to \mathbb{C}$ can be viewed as the $n \times n$ matrix $K = (K(i, j))_{i,j=1}^{n}$. For fixed $s, s_1, \ldots, s_m \in T$, $m \leq n$, interpret $K_s(\cdot)$ and $\sum_{i=1}^{m} c_i K_{s_i}(\cdot)$, defined in Example 9.1 (g), in terms of rows or columns of the $n \times n$ matrix K.

(d) Apply the Kolmogorov's theorem to the function K in (c) when K is positive-definite. What is the appropriate Hilbert space H in this situation? How does X_t in (9.28) relate to the matrix K or its square root?

9.2 Let $\mathcal{M}_{m,n}$ be the set of all $m \times n$ real matrices. Show that for $A, B \in \mathcal{M}_{m,n}$, the expression

$$(A, B) = tr AB',$$

defines an inner product on $\mathcal{M}_{m,n}$.

9.3 Let $\Gamma_n = (\gamma_{i-j})_{i,j=1}^{n}$, $n \geq 1$, be a sequence of invertible positive-definite Toeplitz matrices with $\gamma_{-k} = \bar{\gamma}_k$, and $\mathcal{P}_n$ be the space of trigonometric polynomials of degree n of the form

$$p(z) = p_0 + p_1 z + \cdots + p_n z^n, \; n \geq 0, |z| = 1.$$

For $p, q \in \mathcal{P}_n$ show that

$$(p, q)_n = \sum_{i=0}^{n} \sum_{j=0}^{n} p_i \bar{q}_j \gamma_{i-j},$$

defines an inner product satisfying

$$(zp, zq)_{n+1} = (p, q)_n.$$

9.4 Use the fact that for scalars $\alpha, \beta, 2|\alpha\beta| \leq |\alpha|^2 + |\beta|^2$, to show that ℓ^2 and $L^2(\Omega, \mathcal{F}, \mu)$ are linear spaces. Verify that the spaces in Example 9.1 are inner product spaces.

9.5 Prove Theorem 9.1.

9.6 Show that if $x_m, x \in H$ are such that,

$$(x_n, y) \to (x, y), \text{ for all } y \in H,$$

and

$$||x_n|| \to ||x||,$$

then, $||x_n - x|| \to 0$.

9.7 Suppose that $\{K_n\}$ is a sequence of positive-definite functions on $T \times T$. Show that

(a) $\alpha K_1 + \beta K_2$ is a positive-definite function for all scalars $\alpha, \beta \geq 0$. Is $K_1 - K_2$ a positive-definite function?

(b) $\lim_{n\to\infty} K_n$, when exists, is a positive-definite function.

(c) $\exp\{K_1\}$ is a positive-definite function.

9.8 Suppose that $\{x_t\}$ is a stationary sequence with the spectral density f satisfying $f^{-1} \in L^1$. Let $\hat{x}_0$ be the best linear interpolator of x_0 based on the rest, namely, the orthogonal projection of x_0 onto $\overline{sp}\{x_t;\ t \neq 0\}$.

(a) Show that the isomorph of $\hat{x}_0$ in $L^2(f)$ is the function $1 - \sigma'^2 f^{-1}$ where σ'^2 is given in (8.38). Express the interpolation error $||x_0 - \hat{x}_0||^2$ in terms of f.

(b) Consider the formal series

$$\hat{x}_0 \sim \sum_{k\neq 0} c_k x_k.$$

Express c_k in terms of f and state conditions on f so that the infinite series is norm-convergent.

(c) For a nonnegative integer m, let $\hat{x}_{0,m}$ be the orthogonal projection of x_0 onto $\overline{sp}\{x_t;\ t \neq 0, 1, \ldots, m\}$. Repeat (a) and (b) for $\hat{x}_{0,m}$. What happens when $m = 0$ or $m \to \infty$?

(d) For a nonnegative integer m, let $\tilde{x}_{0,m}$ be the orthogonal projection of x_0 onto $\overline{sp}\{x_t;\ t \neq 0, m\}$. Repeat (a) and (b) for $\tilde{x}_{0,m}$. What happens when $m = 0$ or $m \to \infty$?

(e) Repeat (a)–(d) for the orthogonal projection of x_0 onto $\overline{sp}\{x_t;\ t < -m\}$ for a fixed nonnegative integer m.

(f) Compare the size of the prediction errors and the analytical techniques employed to compute the projections in (c)–(e). For more information on this family of problems, see Nakazi (1984); Miamee and Pourahmadi (1988); Cheng, Miamee and Pourahmadi (1998) and Pourahmadi and Soofi (2000).

9.9 Consider the function $f = |\varphi|^2$ where $\varphi \in H^2$ is an outer function; see Chapter 10. Show that the maps

$$V_i\ :\ L^2(f) \to L^2,\ i = 1, 2$$

defined by $V_1 g = \varphi g$ and $V_2 g = \overline{\varphi} g$ are isometrics.

10

Stationarity and Hardy Spaces

The quintessential example of stationarity is the sequence of exponentials $\{e^{-ik\lambda};\ k = 0, \pm 1, \ldots\}$ in $L^2(\mu)$. Its "past and present" subspace $\overline{sp}\{e^{-ik\lambda};\ k \leq 0\}$ is precisely the (weighted) Hardy space $H^2(\mu)$; this reduces to the Hardy space H^2 when μ is the normalized Lebesgue measure on the unit circle. In this environment, one can re-interpret various orthogonalization techniques of prediction theory as attempts to connect the less familiar functions in $H^2(\mu)$ and their properties to the more familiar ones in H^2. Surprisingly, several central theorems about the Hardy space H^2 and the prediction theory are closely related. This close connection is brought about by the notions of shift operator for stationary sequences on the one hand, and that of shift-invariant subspaces in H^2, on the other. The thread holding and almost unifying the two subjects is the abstract Wold decomposition of a Hilbert space through the range of an isometry.

In Section 10.1, we begin with the abstract von Neumann-Wold-Halmos decomposition of a Hilbert space induced by an isometry and use it to obtain yet another proof of the Wold decomposition of a stationary sequence, and the Lebesgue decomposition of its spectral distribution. This naturally brings in the notion of *generating function* of a nondeterministic stationary sequence which happens to be an *outer function* in H^2. In Section 10.2, the Szegö's distance theorem so fundamental to prediction theory and function theory is proved and its applications are outlined. We use the abstract decomposition of a Hilbert space one more time in Section 10.3 to characterize invariant subspaces (Beurling Theorem) of H^2, to obtain inner-outer factorization of function in H^2 and to study the magnitude of modulus of such functions. The more traditional development and proofs of some of these results using analytic function theory can be found in Hoffman (1962) and Helson (1983).

Some background material on Fourier series, conjugate series, analytic and harmonic functions is covered in Sections 10.4 and 10.5.

10.1 THE WOLD DECOMPOSITION AND THE SPACE H^2

Can the Gram-Schmidt process be used to orthogonalize the sequence $\{e^{-it\lambda}; t \in \mathbb{Z}\}$? Although in principle the answer is in the affirmative, in practice there is a difficulty with getting the process off the ground, since there is no first element like x_1 to start the recursion (9.11). Notwithstanding this, for a sequence $\{x_t;\ t \in \mathbb{Z}\}$ an analogue of the recursion can be developed for the monotone subspaces

$$H_t(x) = \overline{sp}\{x_s; s \leq t\}, t = 0, \pm 1, \cdots.$$

Let $\hat{x}_t$ be the orthogonal projection of x_t onto $H_{t-1}(x)$ and ε_t be its *coprojection*, that is,

$$\varepsilon_t = x_t - \hat{x}_t, t = 0, \pm 1, \cdots$$

Then $\{\varepsilon_t; t \in \mathbb{Z}\}$ is an orthogonal set with possibly some zero elements, since $\varepsilon_t \in H_t(x)$ we have $H_t(\varepsilon) \subseteq H_t(x)$. Of course, computing $\hat{x}_t$, deciding precisely when $\varepsilon_t \neq 0$ and $H_t(\varepsilon) = H_t(x)$ are some of the problems that should be addressed.

For stationary sequences, however, these questions can be resolved with the help of the following abstract decomposition of a Hilbert space and its many unexpected consequences.

10.1.1 The Abstract Wold Decomposition

The celebrated Wold decomposition theorem, an elegant and powerful orthogonalization tool, was proved in 1938 for stationary sequences. The following is an abstract rendition of the result. It uses the range of an isometry to decompose a Hilbert space into direct sum of orthogonal subspaces, Halmos (1961); Masani (1962).

Theorem 10.1. *(von Neumann-Wold-Halmos).* Let V be an isometry on a Hilbert space H and $R = (VH)^{\perp}$ be the orthogonal complement of the range of V, then

(a) $V^m R \perp V^n R \perp \bigcap_{k=0}^{\infty} V^k H,$ for $m > n \geq 0,$

(b) $H = \sum_{k=0}^{\infty} V^k R + \bigcap_{k=0}^{\infty} V^k H.$

Proof. (a) For any $x, y \in R$, it follows that

$$(V^m x, V^n y) = (V^{*n} V^m x, y) = (V^{m-n} x, y) = 0,$$

since for an isometry $V^*V = I$, and $V^{m-n}x \in VH$. The second orthogonality statement follows from $R \perp VH$ and V being an isometry.

(b) Since $\{V^kR\}_{k\geq 0}$ is an orthogonal sequence of subspaces, $\sum_{k=0}^{\infty} V^kR$ is a closed subspace in H (Theorem 9.7) and by Theorem 9.4 (e) we have,

$$\left(\sum_{k=0}^{\infty} V^kR\right)^{\perp} = \bigcap_{k=0}^{\infty}(V^kR)^{\perp}.$$

Thus, the proof amounts to showing that the right-hand side is equal to $\bigcap_{k=0}^{\infty} V^kH$.

Assume that $x \in \bigcap_{k=0}^{\infty}(V^kR)^{\perp}$, we prove by induction that $x \in V^kH$ for all k. For $k = 0$, this is obvious. Thus, assume that $x \in V^kH$. Then $x = V^ky$ for some y. Since $x \in (V^kR)^{\perp}$, that is, $V^ky \perp V^kR$, it follows that $y \perp R$. This in turn implies that $y \in VH$, and hence that $x = V^ky \in V^{k+1}H$ as desired. Therefore

$$\sum_{k=0}^{\infty}(V^kR)^{\perp} \subseteq \bigcap_{k=0}^{\infty} V^kH.$$

Proof of the other inclusion uses a geometric fact that for any subspace M, $VM^{\perp} \subseteq (VM)^{\perp}$. Indeed, choose an $x \in M^{\perp}$, so that Vx is a typical element of $VM^{\perp}$, and choose a $y \in M$, so that Vy is a typical element of VM, then $Vx \perp Vy$ follows from $x \perp y$ and V being an isometry. The desired result, now follows from the computation

$$V^{k+1}H = V^k(VH) = V^kR^{\perp} \subseteq (V^kR)^{\perp}, k \geq 0.$$

Q.E.D.

As the first application of Theorem 10.1 we obtain the Wold's decomposition of nondeterministic stationary sequences by choosing V appropriately (Masani, 1962).

Theorem 10.2. *(Wold's Decomposition Theorem).* Let $\{x_t\}$ be a nondeterministic stationary sequence in a Hilbert space H, that is, the ε_t is non-zero for some t. Then,

(a) $\{\varepsilon_t\}$ is an orthogonal sequence, that is,

$$(\varepsilon_s, \varepsilon_t) = \delta_{st}\sigma^2, \ \ \sigma^2 = ||\varepsilon_0||^2 > 0.$$

(b) $H_t = \sum_{k=0}^{\infty} sp\{\varepsilon_{t-k}\} + H_{-\infty}$, $\sum_{k=0}^{\infty} sp\{\varepsilon_{t-k}\} \perp H_{-\infty}$.

(c) $x_t = \sum_{k=0}^{\infty} b_k\varepsilon_{t-k} + v_t$, $b_0 = 1, \sum_{k=0}^{\infty} |b_k|^2 < \infty$,

$$b_k\sigma^2 = (x_0, \varepsilon_{-k}).$$

The sequence $\{b_k\}_{k\geq 0} \in \ell^2$ is unique, $\{v_t\}$ is stationary with $v_t = P^{x_t}_{H_{-\infty}}, \varepsilon_s \perp v_s$, for all s, t, and $H_t(v) = H_{-\infty}(x)$ for all t.

Proof. (a) Follows from the definition of $\{\varepsilon_t\}$ and $\{x_t\}$ being nondeterministic. (Observe that $\varepsilon_{-1} \in H_{-1}$ and $\varepsilon_0 \perp H_{-1}$.)

(b) Let U be the unique unitary operator of the stationary sequence $\{x_t\}$, see Theorem 9.16. Since $UH_0 = H_{-1} \subset H_0$, the restriction V of U to H_0 is an isometry on the Hilbert space H_0. Hence, applying Theorem 10.1 with $R = H_0 \ominus H_{-1} = H_0 \bigcap H_{-1}^{\perp}$, (b) follows since

$$\bigcap_{k=0}^{\infty} V^k H_0 = \bigcap_{k=0}^{\infty} H_{-k} = H_{-\infty},$$

and

$$V^k R = V^k H_0 \cap H_{-1}^{\perp} = H_{-k} \cap H_{-k-1}^{\perp} = sp\{\varepsilon_{-k}\}.$$

The last equality holds because,

$$H_{-k} = \overline{sp}\{H_{-k-1}, x_{-k}\} = \overline{sp}\{H_{-k-1}, \varepsilon_{-k}\} \text{ and } H_{-k-1} \perp \varepsilon_{-k}.$$

(c) Follows from (b) and Theorems 9.7 and 9.10 (d).

Q.E.D.

The orthogonal sequence $\{\varepsilon_t\}$ is called the *innovation process*, σ^2 the *one-step prediction error variance*, the unique sequence $(1, b_1, b_2, \cdots) \in \ell^2$ the *moving average parameters* and the function $\varphi = \sigma \sum_{k=0}^{\infty} b_k e^{ik\lambda} \in H^2$ the *generating function* of the nondeterministic sequence $\{x_t\}$. It is of great interest to relate these characteristics of $\{x_t\}$ to its spectral distribution μ.

10.1.2 Lebesgue Decomposition of the Spectral Distribution

We show that the geometric decomposition of $\{x_t\}$ given above leads to the Lebesgue decomposition of its spectral measure μ. Fortunately, this paves the way for more analytic results which eventually connects φ to the (spectral) density of the measure μ or the process $\{x_t\}$.

Writing $\{x_t\}$ in Theorem 10.2 (c) as the sum of two *orthogonal* stationary sequences $\{u_t\}$ and $\{v_t\}$ with $u_t = \sum_{k=0}^{\infty} b_k \varepsilon_{t-k}$, splits virtually every relevant characteristics of $\{x_t\}$ into "orthogonal" sums. This is made more clear with the help of some examples. The first of which gives the Lebesgue decomposition of the spectral distribution μ_x into its absolutely continuous and singular parts with respect to the Lebesgue measure on the unit circle. Let μ_u and μ_v be the spectral measures of $\{u_t\}$ and $\{v_t\}$, respectively, then from Theorem 10.2 (c),

$$(x_k, x_0) = (u_k, u_0) + (v_k, v_0), k \in \mathbb{Z},$$

and from the spectral representation of these covariances and uniqueness of Fourier coefficients, it follows that

$$\mu_x = \mu_u + \mu_v. \tag{10.1}$$

Although μ_u is absolutely continuous with respect to the Lebesgue measure on $(-\pi, \pi]$, as yet we do not know much about the nature of μ_v. This is revealed next with the help of the spectral representation of $\{v_t\}$.

The Wold decomposition is also useful in expressing the spectral measure of $\{x_t\}$ in terms of the spectral measures of the stationary sequences $\{\varepsilon_t\}$, $\{u_t\}$ and $\{v_t\}$. Consider the spectral representations of these processes, that is, with obvious notation let

$$\begin{cases} \varepsilon_t = & \int e^{-it\lambda} dZ_\varepsilon(\lambda), \\ u_t = & \sum_{k=0}^{\infty} b_k \varepsilon_{t-k} = \int e^{-it\lambda} \left(\sum_{k=0}^{\infty} b_k e^{ik\lambda} \right) dZ_\varepsilon(\lambda), \\ v_t = & \int e^{-it\lambda} dZ_v(\lambda). \end{cases} \tag{10.2}$$

Then,

$$x_t = u_t + v_t = \int e^{-it\lambda} \left[\sum_{k=0}^{\infty} b_k e^{ik\lambda} dZ_\varepsilon(\lambda) + dZ_v(\lambda) \right],$$

so that the spectral measure of $\{x_t\}$ is

$$Z_x(\lambda) = \int_{-\pi}^{\lambda} \left[\sum_{k=0}^{\infty} b_k e^{ik\lambda} dZ_\varepsilon(\lambda) + dZ(\lambda) \right]. \tag{10.3}$$

We use (10.2) to find the isomorphs of ε_t, u_t and v_t in $L^2(\mu_x)$. Let $g \in L^2(\mu_x)$ be the isomorph of ε_0, then using Corollary 9.29 we have

$$\begin{aligned} \varepsilon_t &= \int e^{-it\lambda} g(\lambda) dZ_x(\lambda) \\ &= \int e^{-it\lambda} g(\lambda) \left[\sum_{k=0}^{\infty} b_k e^{ik\lambda} dZ_\varepsilon(\lambda) + dZ_v(\lambda) \right] \\ &= \int e^{-it\lambda} dZ_\varepsilon(\lambda). \end{aligned} \tag{10.4}$$

On comparing integrands, we find that

$$\begin{cases} g(\lambda) \sum_{k=0}^{\infty} b_k e^{ik\lambda} = 1, \text{ a.e. } (\lambda), \\ g(\lambda) = 0, \ \ a.e.(\mu_v). \end{cases} \tag{10.5}$$

These two conditions are incompatible unless there is a set S of Lebesgue measure zero such that

$$\int_S d\mu_v(\lambda) = \mu_v(-\pi, \pi], \tag{10.6}$$

that is unless μ_v is a singular measure with respect to the Lebesgue measure on $(-\pi, \pi]$. Since the expression for u_t shows that the $\{u_t\}$ process has an absolutely continuous spectral distribution with density

$$w(\lambda) = \left|\sigma \sum_{k=0}^{\infty} b_k e^{ik\lambda}\right|^2 = |\varphi(\lambda)|^2, \tag{10.7}$$

we conclude that the orthogonal decomposition of $\{x_t\}$ into $\{u_t\}$ and $\{v_t\}$ corresponds to decomposition of μ_x into the sum of its absolutely continuous and singular components as desired.

10.1.3 Decomposition of the Predictor

Theorem 10.2 (c) suggests that any finite approximating sum $\sum c_i x_{t-i}$, useful in prediction theory, splits into orthogonal sums:

$$\sum c_i x_{t-i} = \sum c_i u_{t-i} + \sum c_i v_{t-i}, \tag{10.8}$$

one involving only the $\{u_t\}$ process, the other only the $\{v_t\}$. This shows that a single linear operation might serve simultaneously as the predictor of $\{x_t\}$, $\{u_t\}$ and $\{v_t\}$ processes.

Since, for any integer $\ell \geq 1$,

$$x_0 = \sum_{k=0}^{\infty} b_k \varepsilon_{-k} + v_0 = \sum_{k=0}^{\ell-1} b_k \varepsilon_{-k} + \sum_{k=\ell}^{\infty} b_k \varepsilon_{-k} + v_0, \tag{10.9}$$

with $\sum_{k=0}^{\ell-1} b_k \varepsilon_{-k} \perp H_{-\ell}(x)$, $\sum_{k=\ell}^{\infty} b_k \varepsilon_{-k} \in H_{-\ell}(x)$ and $v_0 \in H_{-\infty}(x) \subset H_{-\ell}(x)$, it follows that the prediction (projection) of x_0 based on $x_{-\ell}, x_{-\ell-1}, \cdots$, in symbols $\hat{x}_{0,\ell}$, and its error term are

$$\hat{x}_{0,\ell} = \sum_{k=\ell}^{\infty} b_k \varepsilon_{-k} + v_0, x_0 - \hat{x}_{0,\ell} = \sum_{k=0}^{\ell-1} b_k \varepsilon_{-k}. \tag{10.10}$$

The same reasoning shows that the infinite sum on the right in (10.9) is also the prediction of u_0 based on $u_{-\ell}, u_{-\ell-1}, \cdots$. Therefore,

$$||x_0 - \hat{x}_{0,\ell}||^2 = ||u_0 - \hat{u}_{0,\ell}||^2 = \left\|\sum_{k=0}^{\ell-1} b_k \varepsilon_{-k}\right\|^2 = \sigma^2 \sum_{k=0}^{\ell-1} |b_k|^2. \tag{10.11}$$

Let h_ℓ be the isomorph of $\hat{x}_{0,\ell}$ in $L^2(\mu_x)$, using (10.1), we get

$$\begin{aligned} \|x_0 - \hat{x}_{0,\ell}\|^2 &= \int |1 - h_\ell|^2 d\mu_x \\ &= \int |1 - h_\ell|^2 d\mu_u + \int |1 - h_\ell|^2 d\mu_v \\ &= \|u_0 - \hat{u}_{0,\ell}\|^2 + \|v_0 - \hat{v}_{0,\ell}\|^2. \end{aligned} \tag{10.12}$$

From (10.11)–(10.12) it follows that

$$\|v_0 - \hat{v}_{0,\ell}\|^2 = 0, \tag{10.13}$$

and that h_ℓ is simultaneously the isomorph of the predictors of $\{x_t\}$, $\{u_t\}$ and $\{v_t\}$.

How does one compute this common isomorph h_ℓ? It turns out that knowing the isomorph of ε_0 is the key, but from (10.5) the isomorph of ε_0 is simply $\sigma\varphi^{-1} = \sigma/\varphi$ and hence that of $\{\varepsilon_t\}$ is $\sigma\{e^{-it\lambda}\ \varphi^{-1}\}$. Furthermore, it can be shown (Problem 10.2) that h_ℓ the isomorph of $\hat{x}_{0,\ell}$, is given by

$$h_\ell(\lambda) = e^{-i\ell\lambda}\left(1 - \sigma\varphi^{-1}\sum_{k=0}^{\ell-1} b_k e^{ik\lambda}\right). \tag{10.14}$$

This formula for the isomorph is usually the starting point for a serious study of the *autoregressive representation* of the predictors; see Akutowicz (1957); Wiener and Masani (1957); Pourahmadi (1984); Bloomfield (1985); Farshidi and Salehi (1992); Huang, Kerman and Weit (1997), Problems 10.2 and 10.3.

10.1.4 The Weighted Hardy Space $H^2(\mu)$

In this section, we specialize the setup of Theorem 10.2 by taking $H = L^2(\mu)$ and $\{x_t\} = \{e^{-ik\lambda};\ k \in \mathbb{Z}\}$ with μ as its spectral distribution and $H^2(\mu)$ as its "past and present" subspace. Some of the consequences of Theorems 10.1 and 10.2 are summarized below for ease of reference. Interestingly, these analytic and transparent consequences of the geometric Wold decomposition suggest ways to express σ^2 in terms of φ or the density w. Parts (c) and (d) are discussed and studied in more details in Sections 10.2 and 10.3, respectively. In the sequel, by $H_0^2(\mu)$ we mean the subspace $\overline{sp}\{e^{-ik\lambda};\ k \leq -1\}$ in $L^2(\mu)$.

Theorem 10.3. Suppose $1 \notin H_0^2(\mu)$. Then,

(a) $d\mu = wd\lambda + d\mu_s = |\varphi|^2 d\lambda + d\mu_s,\ \varphi \in H^2$

(b) $H^2(\mu) = H^2(w) \oplus L^2(\mu_s)$.

(c) Let h be the orthogonal projection of 1 onto $H_0^2(\mu)$ and set

$$\sigma^2 = \int |1 - h|^2 d\mu = \text{dist}(1,\ H_0^2(\mu))^2. \tag{10.15}$$

Then, σ^2 depends only on w and not on μ_s, that is,

$$\sigma^2 = \inf_{f \in H_0^2(\mu)} \int |1 - f|^2 d\mu = \inf_{f \in H_0^2(w)} \int |1 - f|^2 w d\lambda. \tag{10.16}$$

(d) $H^2 = \overline{sp}\{\varphi, e_1\varphi, \ldots\}$ in L^2.

Theorem 10.2 and its more abstract reincarnation Theorem 10.1 have played major roles in the development of prediction theory, factorizations of functions in H^2, characterization of shift operators and their invariant subspaces, just to name a few. The monographs by Hoffman (1962); Helson (1964); and Rosenblum and Rovnyak (1985) give penetrating accounts of the congenital role of the Wold decomposition in connecting the above and some apparently distinct and abstract topics from the theory of operators and operator-valued analytic functions.

The following simple and concrete example is the stepping stone for linking prediction theory with the function and operator theory. From here on, instead of the sequence of exponentials $\{e^{-ik\lambda};\ k \in \mathbb{Z}\}$ in $L^2(\mu)$, we shift attention to the more general sequence of functions $\{e^{-ik\lambda} f;\ k \in \mathbb{Z}\}$ with $f \in L^2$.

Example 10.1. For $f \in L^2$, the sequence $\{e_{-t} f, t \in \mathbb{Z}\}$ with $e_t = e^{it\lambda}$, is stationary in L^2 with the spectral density function $|f|^2$, since

$$(e_t f,\ e_{t+k} f) = \int e_k |f|^2 d\lambda, k \in \mathbb{Z}. \tag{10.17}$$

The subspace generated by its past and present

$$\overline{sp}\{e_{-t} f; t \le 0\} = \overline{sp}\{f, e_1 f, e_2 f, \cdots\},$$

denoted by M_f, is invariant under multiplication by e_1, that is,

$$e_1 \cdot M_f \subseteq M_f, \tag{10.18}$$

where $e_1 \cdot M_f = \{e_1 g;\ g \in M_f\}$. The sequence is *nondeterministic* if and only if

$$f \notin e_1 \cdot M_f. \tag{10.19}$$

For what functions f, is the sequence $\{e_{-t} f\}$ nondeterministic? What are the relations between f, the one-step prediction error variance σ^2 and the generating function φ?

Answering these questions requires deeper analysis of the modulus of the function f. The required tools for such analysis goes a bit beyond the basic Hilbert space framework. For example, we show that the sequence is nondeterministic if and only if $|f|$ equals the absolute value of an H^2 function, thus f cannot be zero too often on the unit circle. This sort of delicate analytical results are obtained using Szegö's distance theorem and properties of individual functions in the Hardy space H^2 discussed in the next two sections.

10.2 SZEGÖ'S DISTANCE THEOREM

For a given finite measure μ on the unit circle what is the distance of the constant function 1 to the subspace $H_0^2(\mu)$? Let $\mathcal{P}_0$ be the manifold generated by the exponentials $e^{ik\lambda}$, $k \geq 1$, note that $\mathcal{P}_0$ contains analytic trigonometric polynomials vanishing at zero or having zero as their leading Fourier coefficients. We wish to compute

$$\sigma^2 = \inf\left\{\int |1 - f|^2 d\mu; f \in \mathcal{P}_0, \int f d\lambda = 0\right\}, \tag{10.20}$$

and find the unique function h in $H_0^2(\mu)$ attaining the distance (Theorem 10.3 (c)). The classical theorem of Szegö in 1920 and Kolmogorov (1941) states that

$$\sigma^2 = \exp\left(\int \log w d\lambda\right), \tag{10.21}$$

so that the distance depends only on the absolutely continuous part of μ. In fact, Szegö proved the distance formula for absolutely continuous measures using orthogonal polynomials, and Kolmogorov extended it to the general case. That σ^2 does not depend on μ_s was already suggested by the Wold decomposition in Theorem 10.3 (c). So it remains to establish (10.21) for an absolutely continuous measure. From (10.21) it is evident that $1 \notin H_0^2(\mu)$ or a stationary sequence is nondeterministic, if and only if

$$\int \log w d\lambda > -\infty. \tag{10.22}$$

Since always $\log w < w$, the above condition is equivalent to the integrability of $\log\ w$ or $\log\ w \in L^1$. This condition essentially says that the function w cannot be zero on an interval or on a set of positive Lebesgue measure.

The distance formula expresses σ^2 as the geometric mean of the weight function w. Indeed, the familiar geometric-arithmetic mean inequality plays a role in its proof. Recall that if $a_1, \cdots, a_n$ are positive numbers, then

$$\exp\left(n^{-1}\sum_{k=1}^{n}\log a_k\right) \leq n^{-1}\sum_{k=1}^{n} a_k$$

and equality holds if and only if $a_1 = \cdots = a_n$. More generally, for a real-valued function $f \in L^1(\Omega, \mathcal{F}, \mu)$ where $\mu(\Omega) = 1$ the geometric-arithmetic mean inequality takes the form

$$\exp\left(\int_\Omega f d\mu\right) \leq \int_\Omega \exp(f) d\mu, \tag{10.23}$$

with equality, if and only if f is constant a.e. (μ). This is immediate from the simple inequality

$$1 + x \leq e^x, \text{ for all } x,$$

upon replacing x by $f - \int_\Omega f d\mu$, and integrating the resulting inequality with respect to μ.

The following outline of proof of Szegö's theorem from Helson (1983) relies on two characterizations of *outer functions*, one of which was encountered in Theorem 10.3 (d). More details on outer functions is given in the next section and in Hoffman (1962) and Helson (1983).

Theorem 10.4. (Szegö's Theorem). Let $w \in L^1$ be a nonnegative function. Then,

$$\inf_P \int |1+P|^2 w d\lambda = \exp\left(\int \log w d\lambda\right),$$

where P ranges over all analytic trigonometric polynomials vanishing at zero. The left hand side is 0 if $\log w \notin L^1$.

Proof. We prove the theorem only when $w = |\varphi|^2$ for some $\varphi \in H^2$, see Theorem 10.3 (a). Then, φ can be taken to be outer by Theorem 10.8 in the next section. We rely on the following observations and properties of outer functions;

(i) for $P = p_1 e^{i\lambda} + \ldots + p_n e^{in\lambda}$, $n \geq 1$, $\varphi \sim \hat{\varphi}_0 + \hat{\varphi}_1 e^{i\lambda} + \ldots$, the functions

$$(1+P)\varphi \sim \hat{\varphi}_0 + c_1 e^{i\lambda} + c_2 e^{i2\lambda} + \cdots, \tag{10.24}$$

share $\hat{\varphi}_0$ as their leading Fourier coefficient,

(ii) a function φ is outer if

$$\log|\hat{\varphi}_0| = \int \log|\varphi| d\lambda > -\infty, \tag{10.25}$$

(iii) for φ outer, the set of functions

$$\text{sp}\{\varphi, e_1\varphi, \cdots\} = \{(1+P)\varphi;\ P \in \mathcal{P}_0\} \tag{10.26}$$

is dense in H^2.

From (10.24) and (10.6) it follows that $\hat{\varphi}_0$ can be approximated by functions of the form $(1+P)\varphi$ in the norm of H^2. Thus,

$$\inf \int |1+P|^2 w d\lambda = \inf \int |(1+P)\varphi|^2 d\lambda \leq |\hat{\varphi}_0|^2.$$

The reverse inequality follows from the Parseval's identity (Theorem 9.10 (e)–(f)) applied to (10.24):

$$\int |(1+P)\varphi|^2 d\lambda \geq |\hat{\varphi}_0|^2.$$

Now, the result follows from (10.25) and noting that

$$|\hat{\varphi}_0|^2 = \exp\left(\int \log|\varphi|^2 d\lambda\right) = \exp\left(\int \log w d\lambda\right).$$

To show that the left side is 0 when $\log w \notin L^1$, we define $w_n = \max\left(w, \frac{1}{n}\right)$, for $n \geq 1$, and note that

$$\inf \int |1+P|^2 w d\lambda \leq \inf \int |1+P|^2 w_n d\lambda = \exp\left(\int \log w_n d\lambda\right),$$

for each n. As $n \to \infty$, the right-hand side tends to 0 by the dominated convergence theorem.

Q.E.D.

Szegö's distance (prediction) theorem is the starting point for obtaining many other prediction results. For example, showing that the condition (10.20) is necessary and sufficient for $\{x_t\}$ to be nondeterministic (Theorem VII) is immediate from Theorem 10.4. Similarly, many interpolation problems and characterization of minimality discussed in Chapter 8 and Section 9.9.3 are closely tied to the Szegö's theorem; see Helson and Szegö (1960); Helson (1964); Nakazi (1984); Miamee and Pourahmadi (1988); Cheng, Miamee and Pourahmadi (1998).

10.3 HARDY SPACES: THEOREMS OF BEURLING

Some properties of the Hardy space H^2 is studied in this section using Hilbert space methods; see Helson (1964, 1983). In particular, we show that functions in H^2 cannot be zero too often, in the sense that if $f \in H^2$ is zero on a set of positive measure on the boundary of $\mathbb{D}$, then $f = 0$ almost everywhere.

10.3.1 The Hardy Space H^2

H^2 is the subspace of L^2 spanned by the exponentials with nonnegative frequencies, that is, $H^2 = \overline{sp}\{e_k; k \geq 0\}$. Since Fourier expansions with respect to $\{e_k; k \in \mathbb{Z}\}$ are formally similar to Laurent expansions of analytic function theory, and elements of H^2 have Fourier expansions involving only nonnegative powers, they are called the *analytic* elements of L^2, even though they are defined only almost everywhere on the circle. In some situations we need to associate to each $f \in H^2$ an honest function $f \in \tilde{H}^2$ that is analytic in the open disc $\mathbb{D}$ in the complex plane. In spite of the differences between elements of H^2 and $\tilde{H}^2$, the functions in H^2 behave like analytic functions. For example, functions in H^2 cannot be zero too often. This function-theoretic result is proved next using the geometry of Hilbert spaces.

Theorem 10.5. (a) If $f \in H^2$ and is zero on a set of positive measure on $\partial\mathbb{D}$, then $f = 0$ almost everywhere.

(b) If $f \in H^2$ is real, then f is a constant.

Proof. (a) Suppose that f is zero on a set E of positive measure, but is not the zero function (proof by contradiction). If necessary we can divide f by powers of $e^{i\lambda}$ so that its leading Fourier coefficient $\hat{f}_0 \neq 0$. Let M be the closure of the convex set of functions of the form

$$(1 + c_1 e_1 + \cdots + c_n e_n) f, \tag{10.27}$$

in H^2, where $n \geq 1$ is arbitrary and $c_1, \cdots, c_n$ are complex scalars. Evidently, M is a closed convex subset of H^2, and each function in M vanishes on E and has $\hat{f}_0$ as its leading Fourier coefficient. The same is true for g, the element of M with the smallest norm, see Theorem 9.3. We arrive at a contradiction by proving that $|g|$ is a constant. For this, note that $g + u\ e_n g \in M$ for $n = 1, 2, \cdots$ and any complex number u. As a function of u,

$$||g + ue_n g||^2 = ||g||^2(1 + |u|^2) + 2\ Re\left(u \int |g|^2 e_n d\lambda\right)$$

has a minimum at $u = 0$, since g is the element with the smallest norm. This is only possible if the coefficient of the linear term is zero, that is,

$$\int |g|^2 e_n d\lambda = 0, \quad n = 1, 2, \cdots. \tag{10.28}$$

By taking complex conjugate, it follows that (10.28) holds for $n = -1, -2, \cdots$. Since the only nonzero Fourier coefficient of $|g|^2$ is the central one, it follows that $|g|$ is constant.

(b) For $f \in H^2$, it follows that

$$\int f e_n = 0,\ n = 1, 2, \cdots. \tag{10.29}$$

Since f is real, taking the complex conjugate of (10.29), shows that it holds for $n = -1, -2, \cdots$. Thus, f must be constant.

Q.E.D.

More quantitative information on the modulus of $f \in H^2$ must await deeper analysis of structure of functions in Hardy spaces.

10.3.2 Shift-Invariant Subspaces in H^2

The first theorem characterizing invariant subspaces of the shift operator on H^2 was obtained by Beurling using function - theoretic analysis of individual analytic functions. The proof given here is direct, geometric in nature and relies on Theorem 10.1 (Masani, 1962).

Theorem 10.6. (Beurling). Let M be a proper subspace of H^2 which is shift-invariant, namely,

$$e_1 \cdot M \subset M.$$

Then,

$$M = h \cdot H^2,$$

where $h \in H^\infty$, $|h| = 1$ a.e. and h is determined by M up to a constant factor.

Proof. Define $V : M \to M$ by $Vf = e_1 f$, it follows from $e_1 \cdot M \subset M$ that V is an isometry on the Hilbert space M. Since

$$\bigcap_{k=0}^{\infty} V^k M \subseteq \bigcap_{k=0}^{\infty} e_k \cdot H^2 = \{0\},$$

and from Theorem 10.1 (b)

$$M = \sum_{k=0}^{\infty} V^k R, \tag{10.30}$$

where $R = M \ominus e_1 \cdot M = M \bigcap (e_1 \cdot M)^\perp \subseteq H^2$, it only remains to show that $R = sp\{h\}$, for some $h \in H^\infty$, or that R is one-dimensional. For then,

$$V^k R = e_k \cdot sp\{h\},$$

and (10.30) will imply that for any $f \in M$, there are unique scalars c_k, with $\sum_{k=0}^{\infty} |c_k|^2 < \infty$, such that

$$f = \sum_{k=0}^{\infty} c_k e_k h = h \left(\sum_{k=0}^{\infty} c_k e_k \right) \in h \cdot H^2,$$

as desired.

Let $h \in R$, then from orthogonality of R and $V^k R$, it follows that $h \perp e_k h$ or

$$\int |h|^2 e_k = 0, \ k = 1, 2, \cdots. \tag{10.31}$$

Taking the complex conjugate of (10.31), shows that it holds for $k = -1, -2, \cdots$. Thus, $|h|^2$ must be constant and $h \in H^\infty$, since $h \in R \subset H^2$. Suppose there are two different functions $h_1, h_2 \in H^\infty$ such that $h_1 \cdot H^2 = h_2 \cdot H^2$. Then $\bar{h}_2 h_1 \cdot H^2 = H^2$ so that $\bar{h}_2 h_1$ is analytic. Similarly $h_2 \bar{h}_1$ is analytic, and $\bar{h}_2 h_1 = \dfrac{h_1}{h_2}$ is constant by Theorem 10.5 (b).

Q.E.D.

Analytic functions of modulus one called *inner functions* are important in the theory of Hardy spaces. Their function-theoretic properties are studied extensively in Hoffman (1962).

Recall that for a function $f \in L^2$, M_f stands for the subspace of L^2 generated by its shifts, that is, $M_f = \overline{sp}\{e_k f; k \geq 0\}$. A function $\varphi \in H^2$ is called *outer* if

$$M_\varphi = H^2. \tag{10.32}$$

These functions play crucial roles in function and prediction theory because of Theorem 10.3 (d) and the following result providing detailed information about the structure of functions in H^2.

Theorem 10.7. (Inner-Outer Factorization). Each function $f \in H^2$ has a factorization

$$f = h\varphi,$$

where h is inner or constant and φ is outer. The factors are unique up to constant factors of modulus one.

Proof. If f is outer, there is nothing to prove. Otherwise, M_f is a proper subspace of H^2 satisfying $e_1 \cdot M_f \subset M_f$, and must be of the form $h \cdot H^2$ for some inner function h, see Theorem 10.6. In particular, $f = h\varphi$ with $\varphi \in H^2$ and the desired factorization is obtained if we show that φ is outer. It is easy to show that

$$hH^2 = M_f = hM_\varphi.$$

Thus, $M_\varphi = H^2$, that is, φ is outer as desired. If also $f = h_1\varphi_1$ with h_1 inner and φ_1 outer, then $M_f = h_1 \cdot H^2 = h_2 \cdot H^2$, and h_1/h_2 is a constant by Theorem 10.5 (b).

Q.E.D.

An outer function is determined by its modulus up to a constant factor.

Theorem 10.8. (a) Suppose φ and $\psi \in H^2$ are outer functions with $|\varphi| = |\psi|$. Then,

$$\varphi = c\psi,$$

for a constant c. Furthermore, if $\hat{\varphi}_0 > 0$, then $c = 1$.

(b) Let w be a nonnegative integrable function such that $\log w \in L^1$. Then, the function

$$\varphi(z) = \exp\left(\int_{-\pi}^{\pi} \frac{e^{i\lambda} + z}{e^{i\lambda} - z} \log\sqrt{w(\lambda)}d\lambda\right),\ |z| < 1, \tag{10.33}$$

is in H^2 and outer. Moreover, on the unit circle

$$w = |\varphi|^2.$$

Proof. (a) Since $|\varphi| = |\psi|$, it follows that $\varphi = u \cdot \psi, |u| = 1$ a.e. Thus, $M_\varphi = u \cdot M_\psi$ or $H^2 = u \cdot H^2$, the latter follows from (10.32) since both φ and ψ are outer. Therefore, u is in H^2, and the same is true for $\bar{u}$, implying that u is an arbitrary constant of modulus one. If $\hat{\varphi}_0 > 0$, then it is evident that $c = 1$.

(b) It is evident that $\varphi(z)$ defined above is analytic in the open disc and

$$|\varphi| = e^u,$$

where u is the Poisson integral of $\log \sqrt{w}$. Thus, for $z = re^{i\lambda}$, we have

$$\int_{-\pi}^{\pi} |\varphi(re^{i\lambda})| d\lambda = \int_{-\pi}^{\pi} e^{u(re^{i\lambda})} d\lambda.$$

Since $d\mu = P_r(\lambda - t)dt$ is a probability measure, we have using the geometric-arithmetic mean inequality (10.23) that

$$\begin{aligned} e^{u(re^{i\lambda})} &= \exp\left(\int_{-\pi}^{\pi} P_r(\lambda - t) \log \sqrt{w(t)} dt\right) \\ &\leq \int_{-\pi}^{\pi} P_r(\lambda - t) \sqrt{w(t)} dt. \end{aligned}$$

Thus, for $0 \leq r < 1$,

$$\int_{-\pi}^{\pi} |\varphi(re^{i\lambda})|^2 d\lambda \leq \int_{-\pi}^{\pi} w(t) dt < \infty,$$

so that $\varphi \in H^2$ (Lemma 10.10 (b)) and

$$\lim_{r \to 1} |\varphi(re^{i\lambda})|^2 = \lim_{r \to 1} e^{u(re^{i\lambda})} = w(\lambda), \text{ a.e.}$$

The representation (10.33) is usually taken as the definition of outer functions, see Hoffman (1962, page 62), and its equivalence to (10.25)–(10.26) is well-known (Hoffman, 1962, p. 101).

Q.E.D.

10.4 FOURIER SERIES

We shall work with functions defined on $(-\pi, \pi]$ or the unit circle T in the complex plane. If f is such a function its value at λ will be denoted by $f(\lambda)$ or $f(e^{i\lambda})$ to indicate which domain is under consideration. $d\lambda$ stands for the normalized Lebesgue measure on $(-\pi, \pi]$, also the limits of integration in $\int_{-\pi}^{\pi} f\, d\lambda$ will be omitted often. We use C to denote the space $C(-\pi, \pi]$ of all continuous functions on $(-\pi, \pi]$. For more detailed discussion of results in this and next section, see Hoffman (1962) and Helson (1983).

For $f \in L^1$, its *Fourier coefficients* are the complex numbers

$$\hat{f}_k = \int_{-\pi}^{\pi} f(\lambda) e^{-ik\lambda} d\lambda, k = 0, \pm 1, \cdots \tag{10.34}$$

and its *Fourier series* is the formal series

$$f \sim \sum_{k=-\infty}^{\infty} \hat{f}_k e^{ik\lambda}. \tag{10.35}$$

The question of how to recapture f from its formal Fourier series is of fundamental importance in analysis and various areas of applications. The first natural step in recapturing f from (10.35) is to form the partial sums

$$s_n(f) = \sum_{k=-n}^{n} \hat{f}_k e_k,$$

and search for conditions on f which guarantee convergence of $s_n(f)$ to f in some reasonable sense. Note that

$$\begin{aligned} s_n(f) &= \sum_{k=-n}^{n} \hat{f}_k e_k = \sum_{k=-n}^{n} \left(\int f(\lambda) e^{ik\lambda} d\lambda \right) e^{ik\theta} \\ &= \int f(\lambda) \sum_{k=-n}^{n} e^{ik(\theta-\lambda)} d\lambda = \int f(\lambda) D_n(\theta - \lambda) d\lambda, \end{aligned} \tag{10.36}$$

where D_n is the partial sum of the Fourier series $\sum_{k=-\infty}^{\infty} e^{ik\lambda}$.

It was known as early as 1876 that there exists a continuous function f on $(-\pi, \pi]$ for which $s_n(f)$ does not converge to f at any point $\lambda \in (-\pi, \pi]$. However, Fejer in 1904 showed that for a continuous function f the averages

$$\bar{s}_n(f) = \frac{s_0(f) + \cdots + s_{n-1}(f)}{n}, n = 1, 2, \cdots \tag{10.37}$$

called the nth *Cesaro means* of the Fourier series of f converges uniformity to f. The integral representation of the Cesaro means of the Fourier series of any $f \in L^1$ is given by

$$\bar{s}_n(f) = \int f(\lambda) K_n(\theta - \lambda) d\lambda, \tag{10.38}$$

where $K_n(\lambda)$ is the nth Cesaro mean of the series $\sum_{k=-\infty}^{\infty} e^{ik\lambda}$, that is,

$$K_n(\lambda) = \frac{D_0(\lambda) + \cdots + D_{n-1}(\lambda)}{n}. \tag{10.39}$$

It satisfies the recursion

$$(n+1)K_{n+1}(\lambda) - nK_n(\lambda) = \sum_{k=-n}^{n} e^{ik\lambda} = \sum_{k=0}^{n} e^{ik\lambda} + \sum_{k=1}^{n} e^{-ik\lambda}$$

$$= \frac{1-e^{i(n+1)\lambda}}{1-e^{i\lambda}} + \frac{1-e^{-i(n+1)\lambda}}{1-e^{-i\lambda}} - 1 = \frac{\cos\, n\lambda - \cos\,(n+1)\lambda}{1-\cos\,\lambda}.$$

Since $K_1(\lambda) \equiv 1$, it follows that

$$K_n(\lambda) = \frac{1}{n}\cdot\frac{1-\cos\, n\lambda}{1-\cos\,\lambda} = \frac{1}{n}\left(\frac{\sin\dfrac{n}{2}\lambda}{\sin\dfrac{\lambda}{2}}\right)^2. \tag{10.40}$$

The sequence $\{K_n\}$ is called *Fejer's kernels* on $(-\pi, \pi]$. They are symmetric probability density functions with concentration at $\lambda = 0$, that is,

$$\begin{cases} \text{(i)} & K_n \geq 0, K_n(\lambda) = K_n(-\lambda), \\ \text{(ii)} & \int K_n(\lambda)d\lambda = 1, \\ \text{(iii)} & \text{If } I \text{ is any open interval about } \lambda = 0, \text{ then} \\ & \lim\limits_{n\to\infty} \sup\limits_{\lambda\notin I} K_n(\lambda) = 0, \ (|\lambda| \leq \pi). \end{cases} \tag{10.41}$$

Property (i) is evident from (10.40), (ii) follows from (10.39) and (iii) is a consequence of the following simple inequalities. If $0 < \delta < \pi$ and $\delta \leq |\lambda| \leq \pi$, then

$$\left(\sin\,\frac{\lambda}{2}\right)^2 \geq \left(\sin\,\frac{\delta}{2}\right)^2,$$

so that

$$K_n(\lambda) \leq \frac{1}{n(\sin\,\delta/2)^2}.$$

These properties of K_n are crucial for proving the following theorem.

Theorem 10.9. Let C be the space of continuous functions on $(-\pi, \pi]$.

(a) (*Fejer Theorem, 1904*). If $f \in C$ and $f(-\pi) = f(\pi)$, then the Cesaro means of the Fourier series of f converges uniformly to f, that is,

$$||\bar{s}_n(f) - f||_\infty \to 0 \text{ as } n \to \infty.$$

(b) (*Weierstrass Approximation Theorem*). For any $f \in C$ and any $\epsilon > 0$ there exists a trigonometric polynomial

$$p(\lambda) = \sum_{k=-n}^{n} c_k e^{ik\lambda},$$

such that for all $\lambda \in (-\pi, \pi]$,

$$|f(\lambda) - p(\lambda)| < \epsilon.$$

(c) (Lebesgue, 1905). For $1 \leq p < \infty,\ f \in L^p,\ ||\bar{s}_n(f) - f||_p \to 0$, as $n \to \infty$.

Proof. (a) By extending f to a periodic function on the real line with period 2π, (10.38) can be written as

$$\bar{s}_n(f)(\theta) = \int f(\theta - \lambda) K_n(\lambda) d\lambda.$$

The condition $f(-\pi) = f(\pi)$ makes this periodic extension of f continuous. Since $\int K_n = 1$, we have

$$\bar{s}_n(f)(\theta) - f(\theta) = \int [f(\theta - \lambda) - f(\theta)] K_n(\lambda) d\lambda = \int_{|\lambda|<\delta} + \int_{|\lambda|\geq\delta},$$

for any $0 < \delta < \pi$. Thus,

$$|\bar{s}_n(f)(\theta) - f(\theta)| \leq \sup_{|\lambda|<\delta} |f(\theta - \lambda) - f(\theta)| + 2||f||_\infty \sup_{|\lambda|\geq\delta} K_n(\lambda). \qquad (10.42)$$

Since a continuous f on the closed interval [a,b] is uniformly continuous, it follows from (10.41) (iii) and (10.42) that $\bar{s}_n(f)(\lambda)$ converges to f uniformly on $[-\pi, \pi]$.

(b) Follows from (a) with p being a certain $\bar{s}_n(f)$. The only concern is the possibility of $f(-\pi) \neq f(\pi)$. This can be resolved by choosing a complex number c so that $f(\lambda) + ce^{i\lambda}$ is 2π-periodic.

(c) Follows from (a) using a standard argument and using the fact that C is dense in L^p. For $f \in L^p$, there exists a $g \in C$ such that

$$||f - g||_p < \epsilon.$$

Then, for $n \geq 1$,

$$\begin{aligned} ||\bar{s}_n(f) - f||_p &= ||\bar{s}_n(f) - \bar{s}_n(g) + \bar{s}_n(g) + g - g - f||_p \\ &\leq ||f - g||_p + ||\bar{s}_n(g) - g||_p + ||\bar{s}_n(f) - \bar{s}_n(g)||_p. \end{aligned}$$

The result follows since the first two terms are small and it can be shown that

$$||\bar{s}_n(f) - \bar{s}_n(g)||_p \leq ||f - g||_p.$$

Q.E.D.

10.5 CONJUGATE SERIES AND ANALYTIC FUNCTIONS

The relation between L^2 and H^2 spaces can be understood better through the notion of *conjugate Fourier series* of the series (10.35) defined by

$$\tilde{f} \sim -i \sum_{k=-\infty}^{\infty} (sg\ k) \hat{f}_k e^{ik\lambda}, \tag{10.43}$$

where $sgk = 1$ if $k > 0, -1$ if $k < 0$, and $sg0 = 0$.

Formally,

$$f + i\tilde{f} \sim \hat{f}_0 + 2\sum_{k=1}^{\infty} \hat{f}_k e^{ik\lambda}, \tag{10.44}$$

is a series of *analytic type*. The series

$$F(z) = \sum_{k=-\infty}^{\infty} \hat{f}_k r^{|k|} e^{ik\lambda}, z = re^{i\lambda}, 0 \leq r < 1, \tag{10.45}$$

converges absolutely in the unit disc (since $|\hat{f}_k| \leq \|f\|_1$) and defines a *harmonic function*, whose conjugate is

$$\tilde{F}(z) = -i \sum_k (sg\ k) \hat{f}_k r^{|k|} e^{ik\lambda}, \tag{10.46}$$

so that $F(z) + i\tilde{F}(z)$ is a bona fide *analytic function* in the disc. Here $\tilde{F}$ is normalized so that $\tilde{F}(0) = 0$.

The following questions are of interest to us: Does $\tilde{f}$ in (10.43) define a function? If so, does it belong to L^2? How can $\tilde{f}$ be recaptured from the series in (10.43)?

Fortunately, when $f \in L^2$, then $\tilde{f} \in L^2$ and $f + i\tilde{f} \in H^2$. These are simple consequences of the Parseval's identity (Theorem 9.10 (e)). In this case the function $\tilde{f}$ is said to be *conjugate* to f. In recapturing f and $\tilde{f}$ from (10.45)–(10.46), the *Poisson's kernel* plays a role similar to Fejer's kernel and an analogue of Theorem 10.9 holds. To this end, note that

$$\begin{aligned} F(z) = \sum_k \hat{f}_k r^{|k|} e^{ik\lambda} &= \int f(\theta) \left(\sum_k r^{|k|} e^{ik(\lambda-\theta)} \right) d\theta \\ &= \int f(\theta) P_r(\lambda - \theta) d\theta, \end{aligned}$$

where

$$\begin{aligned} P_r(\theta) &= \sum_k r^{|k|} e^{ik\theta} = \sum_{k=0}^{\infty} r^{|k|} e^{ik\theta} + \overline{\sum_{k=0}^{\infty} r^{|k|} e^{ik\theta}} - 1 \\ &= C_r(\theta) + \overline{C_r(\theta)} - 1 = 2\, Re C_r(\theta) - 1 = Re[2C_r(\theta) - 1] \\ &= Re\left[\frac{1 + re^{i\theta}}{1 - re^{i\theta}}\right] = \frac{1 - r^2}{1 - 2r\cos\theta + r^2}. \end{aligned} \tag{10.47}$$

The family of functions $\{P_r; 0 \leq r < 1\}$ and $\{C_r; 0 \leq r < 1\}$ are called *Poisson's kernel* and *Cauchy's kernel*, respectively. The kernel

$$H_r(\theta) = 2C_r(\theta) - 1 = \frac{1 + re^{i\theta}}{1 - re^{i\theta}}, \tag{10.48}$$

appearing in (10.47) is also of interest, because first $P_r(\theta) = ReH_r(\theta)$ and with $Q_r(\theta) = ImH_r(\theta)$, called the *conjugate Poisson kernel*, we have

$$\tilde{F}(z) = \int f(\theta) Q_r(\lambda - \theta) d\theta. \tag{10.49}$$

Moreover,

$$\begin{aligned} F(z) + i\tilde{F}(z) &= \int f(\theta) H_r(\lambda - \theta) d\theta \\ &= \int \frac{e^{i\theta} + z}{e^{i\theta} - z} f(e^{i\theta}) d\theta \end{aligned} \tag{10.50}$$

is an analytic function in the unit disc $\mathbb{D}$. This shows how to the Fourier series (10.35) one can associate a harmonic function (10.45) and analytic function (10.50); see Problem 10.7.

For any $f \in H^2$ with the Fourier expansion $f = \sum_{k=0}^{\infty} \hat{f}_k e_k$, $\sum_{k=0}^{\infty} |\hat{f}_k|^2 < \infty$, the radius of convergence of the power series $\sum_{k=0}^{\infty} \hat{f}_k z^k$ is greater or equal to 1 since $\hat{f}_k \to 0$, therefore it defines an analytic function $F(z)$ in the open disc $\mathbb{D}$.

We show that the *extension mapping* $f \to F$ (from H^2 to $\tilde{H}^2$) is continuous in the sense defined below and shed more light on the relationship between analytic functions defined on the unit disc and circle.

Lemma 10.10. (a) (*Continuity of Extension Mapping*) If $f_n \to f$ in H^2, then $F_n(z) \to F(z)$ for $|z| < 1$. Moreover, the convergence is uniform on each disc $\{z; |z| \leq r\}, 0 < r < 1$.

(b) Let $g(z) = \sum_{k=0}^{\infty} c_k z^k$ be analytic in $\mathbb{D}$ and define

$$g_r(e^{i\lambda}) = g(re^{i\lambda}), 0 < r < 1.$$

Then $g_r \in H^2$ for each r and $\sum_{k=0}^{\infty} |c_k|^2 < \infty$ if and only if the norms $||g_r||$ are bounded.

(c) For $f \in H^2$, let $f_r(e^{i\lambda}) = f(re^{i\lambda}), 0 < r < 1$. Then $f_r \to f$ in L^2 as $r \to 1$.

Proof. (a) From the representation of $F(z)$ and K_z in (9.35) we obtain

$$|F_n(z) - F(z)| = |(f_n - f, K_z)| \leq ||f_n - f|| ||K_z||.$$

Since $||K_z||^2 = \sum_{k=0}^{\infty} |z|^{2k} = (1 - |z|^2)^{-1}$, it follows that if $|z| \leq r$, then

$$|F_n(z) - F(z)| \leq ||f_n - f||(1 - r^2)^{-1},$$

and hence the result.

(b) Since for each fixed $0 < r < 1$, the series $\sum_{k=0}^{\infty} c_k r^k z^k$ converges uniformly on the unit circle, it follows that the Fourier expansion of g_r in L^2 is $\sum_{k=0}^{\infty} c_k r^k e_k$ and hence $g_r \in H^2$. With g_r defined as above $||g_r||^2 = \sum_{k=0}^{\infty} |c_k|^2 r^{2k}$, let $\beta = \sum_{k=0}^{\infty} |c_k|^2$ and $\beta_r = \sum_{k=0}^{\infty} |c_k|^2 r^{2k}$. Since $\beta_r \leq \beta$ for $0 < r < 1$, it follows that if $\beta < \infty$, then the β_r's are bounded. Conversely, suppose that $\beta_r \leq C$ for all r. It follows that for each positive integer n,

$$\begin{aligned}
\sum_{k=0}^{n} |c_k|^2 &= \left(\sum_{k=0}^{n} |c_k|^2 - \sum_{k=0}^{n} |c_k|^2 r^{2k}\right) + \sum_{k=0}^{n} |c_k|^2 r^{2k} \\
&\leq \sum_{k=0}^{n} |c_k|^2 (1 - r^{2k}) + \beta_r \\
&\leq \sum_{k=0}^{n} |c_k|^2 (1 - r^{2k}) + C.
\end{aligned}$$

For fixed n, choose r so that $\sum_{k=0}^{n} |c_k|^2(1-r^{2k}) < 1$. This can be done since the finite sum is a polynomial in r that vanities at $r = 1$. Therefore, $\sum_{k=0}^{n} |c_k|^2 \leq 1 + C$ for all n, which implies that $\sum_{k=0}^{\infty} |c_k|^2 \leq 1 + C < \infty$.

(c) If $f = \sum_{k=0}^{\infty} c_k e_k$, then $f_r = \sum_{k=0}^{\infty} c_k r^k e_k$, and

$$\|f - f_r\|^2 = \sum_{k=0}^{\infty} |c_k|^2(1-r^k)^2 \leq \sum_{k=0}^{n} |c_k|^2(1-r^k)^2 + \sum_{k=n+1}^{\infty} |c_k|^2,$$

for any positive integer n. As in (b), choosing n large enough to make the second summand small, and then choosing r close to 1 so that the first summand is small, it follows that $\|f - f_r\|$ is small when r is close to 1.

Q.E.D.

Since convergence in L^2 implies the existence of a subsequence converging almost everywhere, it follows from Theorem 10.10 (c) that there exists a sequence $\{r_n\}, r_n \to 1$ such that $f_{r_n} \to f$ almost everywhere (Problem 10.7).

For a finite complex measure defined on $(-\pi, \pi]$, its Fourier coefficients

$$\hat{\mu}_k = \int e^{-ik\lambda} d\mu(\lambda), k = 0, \pm 1, \cdots$$

and the associated Fourier series can be defined as in (10.35), and problem of recapturing μ from its Fourier series be discussed in analogy with functions. In this direction, it is instructive to identify a function $f \in L^1$ with the measure $\int_A f d\lambda,\ A \in \mathcal{F}$. A measure μ on the circle is called *analytic* if its negative Fourier coefficients are zero, that is,

$$\int e^{ik\lambda} d\mu(\lambda) = 0,\ k = -1, -2, \cdots.$$

A theorem of F. and M. Riesz, Hoffman (1962, p. 47), says that an analytic measure is necessarily absolutely continuous with respect to the Lebesgue measure. In other words, analytic measures can be identified with H^1 functions or the subspace of functions in L^1 whose negative Fourier coefficients are zero. A geometric proof of this theorem, in the spirit of proofs of results in Section 10.3, is given by Rosenberg (1969).

An important example of a measure that is not absolutely continuous with respect to the Lebesgue measure is the *Dirac delta measure* δ or the point mass at zero. This measure assigns one to a Borel set containing the point 0 and otherwise its assigns the measure zero. The Fourier coefficients of δ are all equal to 1 and its Fourier series is

$$\delta \sim \sum_{k=-\infty}^{\infty} e^{ik\lambda}.$$

Of course, the partial sum s_n of this series is precisely the *Dirichelet kernel* D_n appearing in (10.36). Its nth Cesaro mean is the Fejer kernel K_n and its Abel mean

$$\sum_{k=-\infty}^{\infty} r^{|k|} e^{ik\lambda}, 0 \leq r < 1, \tag{10.51}$$

is the Poisson kernel $P_r(\lambda)$. This may explain partly why $K_n(\cdot)$ and $P_r(\cdot)$ approach the delta measure; see (10.41).

Problems

10.1 Show that $\{x_t\}$ and $\{\varepsilon_t\}$ in Theorem 10.2 are stationarily correlated. Prove the same for $\{v_t\}$ and $\{\varepsilon_t\}$.

10.2 **(a)** Show that h_ℓ given in (10.14) is the isomorph of $\hat{x}_{0,\ell}$.
Hint: Use the orthogonality principle (9.7).

(b) Obtain the coefficients of the formal series representation of $\hat{x}_{0,\ell}$ in terms of $x_{-\ell}, x_{-\ell-1}, \ldots$.

(c) Let $\varphi^{-1} \sim \sum_{k=0}^{\infty} d_k e^{ik\lambda}$ be the Fourier series of φ^{-1} and $\{S_n\}$ be the sequence of its partial sums. Show that the series for $\hat{x}_{0,1}$ converges in the mean, if and only if

$$\overline{\lim_n} \int_{-\pi}^{\pi} |S_n(e^{i\lambda})|^2 w(\lambda) d\lambda \leq 1, \tag{C}$$

(Farshidi and Salehi, 1992).

(d) Verify if $w(\lambda) = |1 + e^{i\lambda}|^2$ satisfies the condition (C).

(e) Show that if w satisfies the *boundedness* condition, that is, $0 < c \leq w \leq d < \infty$, then it satisfies (C). (Wiener and Masani, 1957).

(f) Show that if $w \in L^\infty$ and $w^{-1} \in L^1$, then w satisfies (C). (Masani, 1960).

(g) Show that if $w \in A_2$ (see Chapter 8), then it satisfies (C). (Helson and Szegö, 1960; Pourahmadi, 1985).

10.3 The series for $\hat{x}_{0,1}$ is said to be *Abel summable* if the series for φ^{-1} is Abel summable in the norm of $L^2(w)$, namely, for the function

$$(\varphi^{-1})(re^{i\lambda}) = \sum_{k=0}^{\infty} r^k d_k e^{ik\lambda}, 0 < r < 1,$$

we have

$$\lim_{r \to 1^-} \int |(\varphi^{-1})(re^{i\lambda}) - \varphi^{-1}(e^{i\lambda})|^2 w(\lambda) d\lambda = 0.$$

(a) Show that the series for the predictor $\hat{x}_{0,1}$ is Abel summable, if and only if

$$\lim_{r \to 1^-} \int \left| \frac{\varphi(e^{i\lambda})}{\varphi(re^{i\lambda})} \right|^2 d\lambda = 1. \tag{S}$$

(Huang, Kerman and Weit, 1997).

(b) Show that if $w \in \bigcup_{p>1} A_p$, then w satisfies (S).

(c) Let P be a polynomial with all its roots on the unit circle. Show that if $\frac{w}{|P|^2} \in A_2$ then it satisfies (S). (Pourahmadi, 1984).

10.4 Provide the details of applying Theorem 10.2 to the stationary sequence $\{e^{-ik\lambda}\,;\, k \in \mathbb{Z}\}$ in $L^2(\mu)$ and the proof of Theorem 10.3. Use part (b) and Szegö's theorem to show that a stationary sequence $\{x_t\}$ with the spectral distribution μ is *purely nondeterministic*, if and only if $\mu_s \equiv 0$ and w satisfies (10.22).

10.5 Prove the geometric-arithmetic mean inequality (10.23).

10.6 Let $w(\lambda) = \exp\{-|\lambda|^{-\alpha}\}$, $\alpha > 0$. For what values of α, (10.22) is satisfied?

10.7 **(a)** For $0 \le r < 1$, show that the Poisson kernel $P_r(\lambda)$ has the following properties:

(i) $P_r(\lambda) \ge 0, P_r(\lambda) = P_r(-\lambda)$, $P_r(\cdot)$ is continuous on the circle and $\int P_r(\lambda) d\lambda = 1$.

(ii) If I is any open interval about $\lambda = 0$, then

$$\lim_{r \to 1} \sup_{\lambda \notin I} P_r(\lambda) = 0, \ (|\lambda| \le \pi).$$

(b) Compute and compare the Fourier coefficients of $P_r(\cdot), C_r(\cdot)$ and $H_r(\cdot)$.

(c) For $f \in L^2$, compute and compare the inner products $(P_r, f), (C_r, f)$ and (H_r, f).

(d) For $f \in C$, define $f_r(\lambda) = \int f(\theta) P_r(\lambda - \theta) d\theta$. Show that f_r converges uniformly to f as $r \to 1$.

(e) For $f \in L^p$, define f_r as in (d). Show that as $r \to 1$, f_r converges to f in the p-norm for $1 \le p < \infty$.

(f) Compare the results in (d)-(e) with those in Theorem 10.9 (a), (c). Describe their roles in capturing f and $\tilde{f}$ from (10.45) and (10.46).

Appendix A Multivariate Distributions

The early developments of prediction theory by Wiener (1949) and others were centered around the Gaussian processes; see Dym and McKean (1976) and Ibragimov and Rozanov (1978). It was Kolmogorov (1941) who showed that the Gaussian assumption and, in fact, the whole probabilistic setup are dispensible by formulating a linear least squares prediction theory of stationary sequences in a Hilbert space framework. However, distributions of stochastic processes do play a central role in computing the best predictors which turns out to be the conditional expectation of future values given the past; see Section 5.4; Shepp, Slepian and Wyner (1980) and Tong (1990).

As concrete examples of finite-dimensional distributions of stochastic processes we introduce multivariate normal and t distributions and study their conditional distributions. These distributions are special members of the broader class of *elliptically contoured distributions* and share some properties of multivariate normal distributions that are of interest in the non-linear prediction of stochastic processes. With the exception of the multivariate normal distribution, the prediction error variances of other members of this class are nonconstant and depend on the values of the conditioning variables; see Lemma A.2 (c) and Lemma A.3 (b). For multivariate t autoregressive models these prediction error variances satisfy a generalized autoregressive conditionally heteroscedastic (GARCH) model; see Tarami and Pourahmadi (2000).

A.1 NORMAL RANDOM VECTORS

An n-dimensional random vector $X = (X_1, \ldots, X_n)$ is said to be *multivariate normal (Gaussian)* if and only if $Y = \sum_{i=1}^{n} c_i X_i$ is univariate normal for all nonrandom vectors $0 \neq (c_1, \ldots, c_n) \in R^n$. In words, all linear combinations of entries of a multivariate normal random vector is normally distributed. Let $E(X) = \mu$ and $\text{Cov}(X) = \Sigma$. Then, the first two moments μ and Σ completely determine the distribution of X; assuming that Σ is strictly positive-definite, its density at $x \in R^n$ is given by

$$f(x|\mu, \Sigma) = \frac{1}{(2\pi)^{n/2}|\Sigma|^{1/2}} \exp\left\{-\frac{1}{2}(x-\mu)'\Sigma^{-1}(x-\mu)\right\}.$$

The notation $X \sim N_n(\mu, \Sigma)$ with n as the subscript is used to emphasize the dimension of an n-dimensional multivariate normal with mean μ and covariance matrix Σ. We suppress n when it is not the focus of the discussion. The characteristic function of X at $t \in R^n$ is given by

$$\phi_X(t) = \exp\left\{it'\mu + \frac{1}{2}t'\Sigma t\right\}.$$

Using either the density function or the characteristic function, one can show that the subvectors of X are independent if and only if they are uncorrelated. Normal distributions enjoy closure properties under linear transformations, marginalization and conditioning as summarized in the following lemma.

Lemma A.1. (a) If $X \sim N(\mu, \Sigma)$, A and b are constant matrices of suitable dimensions, then the random vector $Y = AX + b \sim N(A\mu + b,\ A\Sigma A')$.

(b) If $X_i \sim N(\mu_i, \Sigma_i)$, $i = 1, \ldots, m$, are independent and the constant matrices $A_1, \ldots, A_m$ and b have suitable dimensions, then $Y = \sum_{i=1}^{m} A_i X_i + b \sim N\left(\sum_{i=1}^{m} A_i \mu_i, \sum_{i=1}^{m} A_i \Sigma_i A_i'\right)$.

(c) *(Marginal Distributions)* If $X \sim N(\mu, \Sigma)$ with the conformable partitions

$$X = \begin{pmatrix} X_1 \\ X_2 \end{pmatrix}, \ \mu = \begin{pmatrix} \mu_1 \\ \mu_2 \end{pmatrix} \text{ and } \Sigma = \begin{pmatrix} \Sigma_{11} & \Sigma_{12} \\ \Sigma_{21} & \Sigma_{22} \end{pmatrix}.$$

then

$$X_i \sim N(\mu_i, \Sigma_{ii}), \ i = 1, 2,$$

so that the marginal distributions are also normally distributed.

(d) *(Conditional Distributions)* If X is distributed and partitioned as in (c) and $(X_1|X_2)$ denotes the conditional distribution of X_1 given X_2, then

$$(X_1|X_2) \sim N(\mu_1(X_2), \Sigma_{11}(X_2)),$$

where $\mu_1(X_2)$ the MMSE prediction of X_1 based on X_2 is given by

$$\mu_1(X_2) = \mu_1 + \Sigma_{12}\Sigma_{22}^{-1}(X_2 - \mu_2),$$

which depends on X_2, but the covariance of the prediction error

$$\Sigma_{11}(X_2) = \Sigma_{11} - \Sigma_{12}\Sigma_{22}^{-1}\Sigma_{21},$$

does not depend on X_2. The matrix $\boldsymbol{\beta} = \Sigma_{12}\Sigma_{22}^{-1}$ is usually called the *regression matrix* of X_1 on X_2.

(e) For a bivariate normal $X = (X_1, X_2)$ with the covariance matrix

$$\Sigma = \begin{pmatrix} \sigma_1^2 & \rho\sigma_1\sigma_2 \\ \rho\sigma_1\sigma_2 & \sigma_2^2 \end{pmatrix},$$

where ρ is the correlation coefficient between X_1 and X_2, the regression matrix reduces to the scalar

$$\beta = \rho\frac{\sigma_1}{\sigma_2},$$

and the conditioinal variance of X_1 given X_2 reduces to the familiar expression

$$\Sigma_{11}(X_2) = (1 - \rho^2)\sigma_1^2,$$

which is the portion of variance of X_1 explained by X_2.

For a complex matrix $A = (a_{ij})$, that is, a matrix with the complex-valued entries, its conjugate transpose is denoted and defined by $A^* = (\overline{a}_{ji})$. Consider the $n \times 1$ complex vector $\mu = \mu_1 + i\mu_2$, the $n \times n$ complex matrix $\Sigma = \Sigma_1 + i\Sigma_2$ for which $\Sigma = \Sigma^*$ and $c^*\Sigma c \geq 0$ for all $c \in \mathbb{C}^n$. We say that the complex random vector $X = X_1 + iX_2$ has a complex multivariate normal distribution, in symbol $X \sim N_c(\mu, \Sigma)$, if

$$\begin{bmatrix} X_1 \\ X_2 \end{bmatrix} \sim N_{2n}\left(\begin{bmatrix} \mu_1 \\ \mu_2 \end{bmatrix}, \frac{1}{2}\begin{bmatrix} \Sigma_1 & -\Sigma_2 \\ \Sigma_2 & \Sigma_1 \end{bmatrix}\right).$$

For more information on multivariate normal distributions, see Anderson (1984) and Muirhead (1982).

A.2 MULTIVARIATE t-DISTRIBUTIONS

The density function of a multivariate t-distribution is quite different from that of the multivariate normal, yet the corresponding random vector enjoys the closure properties under marginalization and conditioning. The differences between the two distributions are subtle and revealing, see Lemma A.2 (c).

Let μ be an $n \times 1$ vector and Σ an $n \times n$ positive-definite matrix. Then, the random vector X is said to have a multivariate t *distribution* with mean μ, scaling matrix Σ and v degrees of freedom, in symbol $X \sim t_n(\mu, \Sigma, v)$, if its density at $x \in R^n$ is given by

$$f(x|\mu, \Sigma, v) = c_n |\Sigma|^{-1/2} \left[1 + v^{-1}(x-\mu)'\Sigma^{-1}(x-\mu)\right]^{-\frac{n+v}{2}}, \tag{A.1}$$

where c_n, the normalizing constant, is given by

$$c_n = (2\pi)^{-n/2}\Gamma\left(\frac{n+v}{2}\right)\Gamma^{-1}\left(\frac{v}{2}\right).$$

The subscript n emphasizing the dimension of X will be suppressed when there is no danger of confusion. Using (A.1) it can be shown that $E(X) = \mu$ and $\text{Cov}(X) = \dfrac{v}{v-2}\Sigma$ for $v > 2$. Small values of v correspond to heavy tailed distributions. In fact, $v = 1$ corresponds to the *multivariate Cauchy distribution*. For larger v, the density converges to that of the multivariate normal, because $\left(\frac{v}{2}\right)^{-\frac{n}{2}}\Gamma\left(\dfrac{n+v}{2}\right)\Gamma^{-1}\left(\frac{v}{2}\right)$ converges to 1 and the other factor converges to the $\exp\left\{-\dfrac{1}{2}(x-\mu)'\Sigma^{-1}(x-\mu)\right\}$.

In analogy with Lemma A.1 we summarize some basic properties of multivariate t distributions next.

Lemma A.2. (a) If $X \sim t(\mu, \Sigma, v)$, A and b are constant matrices of suitable dimensions. Then, the random vector $Y = AX + b \sim t(A\mu + b, A\Sigma A', v)$.

(b) *(Marginal Distributions)* If $X \sim t(\mu, \Sigma, v)$ with the conformable partitions

$$X = \begin{pmatrix} X_1 \\ X_2 \end{pmatrix}, \mu = \begin{pmatrix} \mu_1 \\ \mu_2 \end{pmatrix}, \Sigma = \begin{pmatrix} \Sigma_{11} & \Sigma_{12} \\ \Sigma_{21} & \Sigma_{22} \end{pmatrix},$$

then

$$X_i \sim t\left(\mu_i, \Sigma_{ii}, v\right), \ i = 1, 2,$$

so that the marginals are also distributed as multivariate t.

(c) *(Conditional Distributions)* If X is distributed and partitioned as in (b) and $(X_1|X_2)$ denotes the conditional distribution of X_1 given X_2, then

$$(X_1|X_2) \sim t(\mu_1(X_2), \Sigma_{11}(X_2), v + p_2),$$

where p_2 is the dimension of X_2 and

$$\mu_1(X_2) = \mu_1 + \Sigma_{12}\Sigma_{22}^{-1}(X_2 - \mu_2),$$

$$\Sigma_{11}(X_2) = g(X_2)(\Sigma_{11} - \Sigma_{12}\Sigma_{22}^{-1}\Sigma_{21}),$$

with

$$g(X_2) = \frac{v + p_2}{v - 2 + (X_2 - \mu_2)'\Sigma_{22}^{-1}(X_2 - \mu_2)}.$$

In stark contrast to the multivariate normal case, the conditional covariance of X_1 given X_2 does depend on X_2 through the scalar function $g(X_2)$. For a bivariate $X = (X_1, X_2)$ distributed as $t(\mu, \Sigma, v)$ with

$$\Sigma = \begin{pmatrix} \sigma^2 & \rho\sigma_1\sigma_2 \\ \rho\sigma_1\sigma_2 & \sigma_2^2 \end{pmatrix},$$

we have

$$\Sigma_{11}(X_2) = g(X_2)(1 - \rho^2)\sigma_1^2,$$

and

$$g(X_2) = \frac{v + 1}{v - 2 + \left(\frac{X_2 - \mu_2}{\sigma_2}\right)^2}.$$

Note that $g(X_2) = 1$, if and only if $X_2 = \mu_2 \pm \sqrt{3}\sigma_2$. Otherwise, $g(X_2) \neq 1$ and consequently $\Sigma_{11}(X_2)$ is either less or more than $\sigma_1^2(1 - \rho^2)$, the prediction error variance of X_1 based on X_2 in the Gaussian case. For more information on multivariate t distributions, see Box et al. (1994) and Muirhead (1982).

A.3 ELLIPTICALLY CONTOURED DISTRIBUTIONS

The multivariate normal and t distributions are two of the most distinguished members of the class of elliptically contoured distributions. There are many ways to define this class, in what follows we assume the existence of a density function.

An n-dimensional random vector X is said to have an *elliptically contoured* (EC) distribution with mean μ and covariance matrix Σ, in symbol $X \sim EC_n(\mu, \Sigma, h)$, if its density function is of the form

$$f(x|\mu, \Sigma, h) = c_n|\Sigma|^{-1/2}h\left((x - \mu)'\Sigma^{-1}(x - \mu)\right),$$

for some h where c_n is the normalizing constant not depending on μ and Σ. The choice of $h(\cdot)$ for the multivariate normal and t distributions are self-evident.

The marginal and conditional distributions of EC distributions are EC. As in the normal case the conditional expectation of a subvector of X given the remaining component is linear in fixed variables, but unlike the normal case the conditional covariance does depend on the fixed variables through a scalar function.

Lemma A.3. If $X \sim EC(\mu, \Sigma, h)$ with the conformable partitions

$$X = \begin{pmatrix} X_1 \\ X_2 \end{pmatrix}, \mu = \begin{pmatrix} \mu_1 \\ \mu_2 \end{pmatrix}, \Sigma = \begin{pmatrix} \Sigma_{11} & \Sigma_{12} \\ \Sigma_{21} & \Sigma_{22} \end{pmatrix}.$$

Then,

(a) *(Marginal distributions)* $X_i \sim EC(\mu_i, \Sigma_i, h),\ i = 1, 2.$

(b) *(Conditional Distributions)* If X is distributed and partitioned as above and $(X_1|X_2)$ denotes the conditional distribution of X_1 given X_2, then

$$(X_1|X_2) \sim EC(\mu_1(X_2), \Sigma_{11}(X_2), h),$$

where

$$\mu_1(X_2) = \mu_1 + \Sigma_{12}\Sigma_{22}^{-1}(X_2 - \mu_2),$$

and

$$\Sigma_{11}(X_2) = g(X_2)(\Sigma_{11} - \Sigma_{12}\Sigma_{22}^{-1}\Sigma_{21})$$

for some function g provided that $E(X_1|X_2)$ and $\text{Cov}(X_1|X_2)$ exist.

Fortunately, for the multivariate normal and t distributions the forms of $g(\cdot)$ are known. For other EC distributions finding explicit formulas for $g(\cdot)$ is rather challenging; see Cambanis, Fotopoulos and He (2000). For more information and examples of EC, see Muirhead (1982).

Appendix B
The Bayesian Forecasting

The Kolmogorov-Wiener theory of prediction of stationary processes is rather rigid so far as incorporation of additional and/or prior information is concerned; see West and Harrison (1997) and Pourahmadi and Soofi (2000). Moreover, the theory assumes the full knowledge of the covariance or the spectral density function. Two of the most common ways of getting rid of the unknown parameters are the plug-in and Bayesian forecasting systems. The former is based on the prevalent idea in classical statistics that any statistical model can be converted into a forecasting system by replacing the unknown parameters by their estimates. The latter is based on introducing a prior distribution for the unknown parameters with the focus on predictions involving observables rather than the parameters and validating the predictions using the data at hand. The goal is to avoid as much as possible the traditional and unrealistic references to hypothetical replications of the data generating mechanism, alternative data sequences or events that did not occur; see Kabaila (1999) and Skouras and Dawid (1998, 1999). In particular, no concept of *expected performance* will be used as measure of merit of a forecasting system.

In this appendix, following the recent important work of Skouras and Dawid (1998, 1999), we provide merely glimpses into the efficiency of plug-in and Bayesian forecasting systems. It is expected that this line of research will serve as a catalyst to bring even closer the cognate areas of statistical forecasting systems, model selection using

BIC and the theory of stochastic complexity (Rissanen, 1989), parameter estimation, model testing and robustness (Dawid, 1992).

The starting point of classical statistical inference is the distribution of the data Y given the parameters θ, denoted by $f_{Y|\theta}$. However, the Bayesian approach to inference resumes with specifying a prior distribution $\pi(\cdot)$ for the parameter θ before collecting or examining the data (Casella and Berger, 1990; West and Harrison, 1997). Then, the *marginal distribution* of Y is obtained by integrating out the parameter θ:

$$f_Y(y) = \int_\Theta f_{Y|\theta}(y|\theta)\pi(\theta)d\theta, \tag{B.1}$$

where Θ is the parameter space. The *posterior distribution* $(\theta|Y^t)$ of θ given Y^t obtained using Bayes's theorem is

$$\pi_t(\theta) = \pi(\theta|Y^t) = cf_{Y|\theta}(y|\theta)\pi(\theta), \tag{B.2}$$

where c is the normalizing constant.

Finally, the Bayesian forecasting system (BFS) based on the pair $\{f_{Y|\theta}, \pi(\theta); \theta \in \Theta\}$ at time t, issues the point prediction

$$m_{B,t} = \int_\Theta m_t(\theta)\pi_{t-1}(\theta)d\theta, \ t = 1, 2, \ldots \tag{B.3}$$

as the conditional mean of Y_t given Y^{t-1}, calculated under the marginal distribution f_Y given in (B.1). The predictive variance of $m_{B,t}$ defined by

$$P_{B,t} = E(|Y_t - m_{B,t}|^2|Y^{t-1}) \tag{B.4}$$

and computed also under f_Y, plays crucial roles in establishing the efficiency of a BFS. Some concrete examples of (B.1) - (B.4) in the context of normal linear regression are useful in illustrating the Bayesian machinery.

B.1 BAYESIAN REGRESSION MODELS

Consider a linear regression model and, for the moment, suppose that the vector of observations Y and the regression parameters β are related via

$$(Y|\beta) \sim N_n(X\beta, \Sigma). \tag{B.5}$$

This is the $f_{Y|\beta}$ needed in (B.1).

In the Bayesian framework the uncertainty about the unknown parameters β is expressed through a prior distribution $\pi(\beta)$. Take this prior to be

$$\beta \sim N(\beta_0, R_0), \tag{B.6}$$

where β_0 and R_0 are known matrices. Having this prior distribution, one can evaluate the integral in (B.1). In some situations there are alternatives to integration. For this, note that the two-stage specification of distributions of $(Y|\beta)$ and β is equivalent to the twin equations

$$\begin{cases} Y = X\beta + V, & V \sim N(0, \Sigma), \\ \beta = \beta_0 + W, & W \sim N(0, R_0), \end{cases} \tag{B.7}$$

where β_0, R_0 and Σ are assumed known, and V is independent of β and W. Note the resemblance of these equations with the state-space models (2.57).

Using the additivity and closure properties of multivariate normal random vectors (Lemma A.1), we construct the joint distribution of (Y, β), the marginal distribution of Y, and finally the conditional distribution $(\beta|Y)$ or the posterior of β given the data Y. In general, when the distributions in (B.5)–(B.6) are nonnormal, one must rely on (B.1)–(B.4) to compute numerically the distributions needed for implementation of a BFS; see West and Harrison (1997) for details. In most applications, the matrices Σ and R_0 are scalar multiples of the identity matrices of appropriate dimensions.

Theorem B.1. Suppose the distributions in the linear regression model (2.12) are specified as in (B.5)–(B.6). Then,

(a) *(Joint Distribution)*

$$\begin{pmatrix} Y \\ \beta \end{pmatrix} \sim N\left[\begin{pmatrix} X\beta_0 \\ \beta_0 \end{pmatrix}, \begin{pmatrix} XR_0X' + \Sigma & XR_0 \\ R_0X' & R_0 \end{pmatrix}\right].$$

(b) *(Marginal Distribution)* $Y \sim N(X\beta_0, XR_0X' + \Sigma)$.

(c) *(Posterior Distribution)* $(\beta|Y) \sim N(\beta_1, R_1)$, where

$$\beta_1 = \beta_0 + Ke,$$

with $e = Y - X\beta_0,\ K = R_0X'(XR_0X' + \Sigma)^{-1}$ and

$$R_1 = R_0 - R_0X'(XR_0X' + \Sigma)^{-1}XR_0.$$

(d) *(Predictive Distribution)* $(Y_t|Y^{t-1}) \sim N(\mu_t, \sigma_t^2),\ t = 2, \ldots, n$, where μ_t and σ_t^2 are certain functions of X, R_0, Σ and β_0.

Proof. (a) Using (B.7) since $Y = X\beta + V = X(\beta_0 + W) + V = X\beta_0 + XW + V$, and

$$\begin{pmatrix} Y \\ \beta \end{pmatrix} = \begin{pmatrix} X\beta_0 \\ \beta_0 \end{pmatrix} + \begin{pmatrix} I_n & X \\ 0 & I_p \end{pmatrix}\begin{pmatrix} V \\ W \end{pmatrix},$$

is a linear transformation of the $(n+p)$-dimensional normal random vector $\begin{pmatrix} V \\ W \end{pmatrix}$ with mean zero and covariance matrix diag(Σ, R_0), the result follows from Lemma A.1 (a). (b)–(c) follows by identifying Y with X_1 and β with X_2 in the partition of X in Lemma A.1 (c) and (d). (d) follows by appropriate partitioning of the random vector Y in (b) and using Lemma A.1 (d).

Q.E.D.

The mean and covariance matrix of the posterior distribution of β given the data Y clearly demonstrate the role of the data in updating the prior mean β_0, but the update of the prior covariance matrix R_0 does not depend on Y. This should not be surprising in view of Lemma A.1 (d). The above results are illustrated for the simple, yet non-trivial, case of the constant mean model.

Example B.1: For $p = 1,\ \beta = \mu,\ X = (1, \ldots, 1) = 1_n,\ R_0 = \sigma_0^2$ and $\Sigma = \sigma^2 I_n$, we have

$$e = Y - \mu_0 1_n,\ K = \sigma_0^2 1_n'(\sigma_0^2 1_n 1_n' + \sigma^2 I_n)^{-1},$$

$$R_1 = \sigma_0^2 - \sigma_0^4 1_n'(\sigma_0^2 1_n 1_n' + \sigma^2 I_n)^{-1} 1_n.$$

Since

$$(\sigma_0^2 1_n 1_n' + \sigma^2 I_n)^{-1} = \sigma^{-2}\left[I_n - \frac{\sigma_0^2}{n\sigma_0^2 + \sigma^2} 1_n 1_n'\right],$$

the updated prior mean

$$\begin{aligned} \mu_1 &= \mu_0 + \frac{n\sigma_0^2}{n\sigma_0^2 + \sigma^2}(\bar{Y} - \mu_0) \\ &= \frac{\sigma^2}{n\sigma_0^2 + \sigma^2}\mu_0 + \frac{n\sigma_0^2}{n\sigma_0^2 + \sigma^2}\bar{Y}, \end{aligned}$$

is a mixture of the prior mean μ_0 and the sample mean $\bar{Y}$ of the data, and the updated variance of μ_1 is

$$R_1 = \frac{\sigma_0^2 \sigma^2}{n\sigma_0^2 + \sigma^2}.$$

It is interesting and instructive to note that Theorem B.1 can be proved verbatim for the following dynamic (state-space) analogue of the twin equations (B.7):

$$\begin{cases} Y_t = X_t \beta_t + V_t & , \quad V_t \sim N(0, \Sigma_t), \\ \beta_t = \beta_{t-1} + W_t & , \quad W_t \sim N(0, R_t). \end{cases} \tag{B.8}$$

In this setup $\{\beta_t\}$ is viewed as the unobservable state of a system. Furthermore, $\{V_t\}$ and $\{W_t\}$ are white noises which are independent of each other. The similarity of (B.8) with the state-space equations (2.57) is self-evident.

Recursive methods of updating the state vector and computing the predictors in (B.8) are discussed in Chapter 4. More general dynamic models are discussed extensively by West and Harrison (1997).

Next, we present mild conditions under which the Bayesian and Plug-in SFSs are efficient in the sense defined in Section 2.5.5 (Skouras and Dawid, 1998, 1999).

B.2 EFFICIENCY OF BFS

We focus on the point prediction $m_{B,t}$ and its predictive variance $P_{B,t}$ for a positive prior $\pi(\cdot)$. It turns out that the efficiency of such a BFS is determined by the boundedness of the sequence of predictive variances $\{P_{B,t}\}$.

Theorem B.2 A BFS with a positive prior density for θ is efficient if

$$\sup_{t\geq 1} P_{B,t} < \infty. \tag{B.9}$$

This elegant and powerful result showing that a SFS based on the conditional means of a Bayesian marginal distribution f_Y, is efficient under the natural assumption that its predictive variances stay bounded with probability one for almost all θ, is proved in Skouras and Dawid (1998). In the next example, we give a condition for the Bayesian predictors to be efficient in the linear regression setup.

Example B.2: We continue with the regression setup of the Example B.1, but using the prior (B.6) with $\beta_0 = 0$ and $R_0 = \sigma_0^{-2} I_p$. The marginal distribution of the future observation Y_{t+1}, in this case, is a univariate normal with the mean and variance given, respectively, by

$$\begin{aligned} m_{B,t+1} &= \hat{\beta}_t' x_{t+1}, \\ P_{B,t+1} &= \sigma^2\{1 + x_{t+1}'(\sigma_0^2 I_p + X_t'X_t)^{-1} x_{t+1}\}, \end{aligned} \tag{B.10}$$

where $\hat{\beta}_t = (\sigma_0^2 I_p + X_t'X_t)^{-1} X_t'Y^t$. In view of Theorem B.2, the BFS $\{m_{B,t}\}$ in (B.10) is efficient if

$$\sup_{t\geq 1} x_{t+1}'(\sigma_0^2 I_p + X_t'X_t)^{-1} x_{t+1} < \infty,$$

or equivalently if

$$\sup_{t>t_0} x_{t+1}'(X_t'X_t)^{-1} x_{t+1} < \infty, \tag{B.11}$$

for some t_0, chosen so that $X_t'X_t$ is invertible for all $t > t_0$.

B.3 EFFICIENCY OF A PLUG-IN SFS

We present sufficient conditions for the efficiency of a plug-in SFS by comparing its point predictions with the predictions of an efficient BFS.

Consider the plug-in SFS $\{m_t(\hat{\theta}_t)\}$ of Section 2.5.5, based on an estimator sequence $\{\hat{\theta}_t\}$, where $\hat{\theta}_t$ is defined for all $t \geq 1$ with values in the parameter space Θ. It is natural to expect that the efficiency of such a SFS is very much dependent on the quality of the estimators $\{\hat{\theta}_t\}$. The following theorem due to Skouras and Dawid (1998) gives precise conditions for this heuristic result to hold.

Theorem B.3 The plug-in SFS $\{m_t(\hat{\theta}_t)\}$ based on the estimator sequence $\{\hat{\theta}_t\}$ is efficient provided that the following three conditions hold:

Condition 1. For a positive prior probability density $\pi(\theta)$ for θ, let $\tilde{\theta}_t$ denote the posterior mean of θ based on the data Y^t and let V_t denote the posterior expectation of $||\theta - \tilde{\theta}_t||^2$ and $P_{B,t}$ be as in (B.4). Assume that $\{P_{B,t}\}$ satisfies (B.9) and $\{V_t\}$ satisfies

$$\sum_{t=1}^{\infty} V_t^2 < \infty.$$

Condition 2. The sequence of estimators $\{\hat{\theta}_t\}$ satisfies

$$\sum_{t=1}^{\infty} ||\hat{\theta}_t - \tilde{\theta}_t||^2 < \infty.$$

Condition 3. The parameter set Θ is open and convex. The predictive means $m_t(\theta)$ are twice contiuously differentiable with respect to $\theta = (\theta_1, \ldots, \theta_p)$ with derivatives $\frac{\partial}{\partial \theta} m_t(\theta)$ and $\frac{\partial^2}{\partial \theta \partial \theta'} m_t(\theta)$ satisfying

$$\sup_t \sup_\theta \{d_{t,1}(\theta) + d_{t,2}(\theta)\} < \infty,$$

where

$$d_{t,1}(\theta) = \left\| \frac{\partial}{\partial \theta} m_t(\theta) \right\|,$$

and

$$d_{t,2}(\theta) = \sup \left\{ h' \frac{\partial^2}{\partial \theta \partial \theta'} m_t(\theta) h;\ h \in R^p,\ h'h = 1 \right\}.$$

We note that these conditions are less intuitive and more demanding than (B.9) needed for the efficiency of a BFS. The following remarks about these conditions sheds light on their nature and why they might be needed. The requirements in Condition 3 are smoothness requirements and force the first two derivatives to be bounded. Condition 2 requiring the square summability of the distances between $\hat{\theta}_t$ and the posterior mean $\tilde{\theta}_t$ is the only condition on the estimator $\hat{\theta}_t$. This is a weak condition for the MLE $\hat{\theta}_t$; see Skouras and Dawid (1998). Condition 1 guarantees the existence of an efficient BFS and it controls the behavior of the posterior variance.

B.4 PROBABILITY FORECASTING SYSTEMS

Consider the setup of Section 2.5.5 and a forecaster who wishes to associate with any observed set of data $y^t = (y_1, \ldots, y_t)$ a *predictive distribution* for the next observation Y_{t+1}, instead of a point prediction.

For example, if $\{Y_t\}$ is a sequence of i.i.d. $N(\mu, 1)$ random variables, based on y^t our forecaster may issue the predictive distribution $N(\bar{y}_t, 1)$ for the next observation Y_{t+1}. We refer to this as a *probability forecasting system* (PFS). Such PFSs evidently lead to natural point prediction rules through the mean, median or mode of the predictive distribution. How does one assess the performance of a PFS? Does reasonable notions of efficiency of a PFS imply the efficiency of the ensuing point prediction systems? In the non-time series context some of these questions are addressed by Geisser (1993). Here following Skouras and Dawid (1999), we introduce a measure of empirical performance for a PFS in terms of the observables rather than using the unobservable components of a model or its parameters. The surprisingly close link between the PFS, the statistical problems of model selection using BIC and the theory of stochastic complexity (Rissanen, 1989), model testing and robustness is discussed in Dawid (1992).

An empirical measure of performance and a notion of efficiency for PFSs can be developed paralell to those in Section 2.5.5, as follows. Let F denote a distribution or a PFS for $Y = \{Y_t\}$. The restriction of F to the first t observations $Y^t = (Y_1, \ldots, Y_t)$ is denoted by F^t and the predictive distribution $F(Y_{t+1}|y^t)$ by F_{t+1}. Lower-case letters f, f^t, and f_{t+1} are used to denote densities of the above distributions , respectively, when they exist. For a data-sequence $y^n = (y_1, \ldots, y_n)$, we assess the success of a PFS F^n by its empirical performance, namely, comparing the realized predictive distributions $\{F_t;\ t \le n\}$ with the realized outcomes y^t of Y^t, $t \le n$. In this framework, the performance of a PFS is evaluated using the data at hand, rather than using its *expected performance* by averaging over all possible realizations, which involve data and forecasts that were never observed or issued.

Instead of the running sum of the squared prediction errors in Section 2.5.5, one may use the logarithmic score to assess the performance of predictive densities issued by a PFS. Thus, the performance of the predictive density $f_t(\cdot)$ when the outcome of Y_t turns out to be y_t, is measured by $\log f_t(y_t)$. The *cumulative logarithmic score* for assessing a PFS F based on the data-sequence y^n is

$$L_n(F) = \sum_{t=1}^{n} \log f_t(y_t) = \log f^n(y^n).$$

This has the pleasing interpretation that the performance of F is measured using the overall predictive loglikelihood. Now, it is natural to compare any two PFSs F and G in the light of the data y^n, by the difference between their predictive loglikelihoods, namely $D_n(F, G) = L_n(F) - L_n(G)$. Since larger loglikelihoods are desirable, we prefer F over G if $D_n(F, G)$ is positive, and the reverse if it is negative. Due to the possible alternating signs of $D_n(F, G)$ as a function of the data y^n and n, we may end up prefering F over G for some n and the reverse for others. This complication can be resolved by resorting to asymptotics as $n \to \infty$, or relying on the infinite outcome sequence $(y_1, y_2, \ldots)$. There are three distinct cases to consider. If $D_n(F, G) \to \infty$ as $n \to \infty$, the F is ultimately preferable to G, and the opposite if $D_n(F, G) \to -\infty$. If the difference $D_n(F, G)$ stays bounded above and below, F and G can not be distinguished in which case they are considered to be equivalent.

Having the above empirical performance measure, it is desirable to construct a PFS which performs at least as well as any other PFS for all possible $\theta \in \Theta$ and data. This motivates the following definition of efficiency of a PFS. Given a parametric family $\{f_{Y|\theta}\}$ of distributions for a sequence Y, a PFS F is said to be *efficient* for $\{f_{Y|\theta}\}$, if for any other PFS G, we have

$$\lim_{n\to\infty} \sup D_n(F, G) < \infty,$$

for "almost all" data and $\theta \in \Theta$.

How does one construct (efficient) PFSs? It turns out that as in Section 2.5.6 and Appendix B.2, BFSs based on everywhere positive prior densities $\pi(\theta)$ on θ are efficient (Dawid, 1984). In this framework, the predictive distribution is simply the conditional distribution of Y_{t+1} given y^t calculated under the Bayesian marginal distribution F_Y. As before, one could use the alternative method of plug-in PFSs by computing the conditional distribution of Y_{t+1} given y^t using $f_{Y|\theta}$ and replacing the unknown parameter θ by an estimate $\hat{\theta}_t$ based on the current data y^t. This popular method not only obviates the need to specify a prior distribution for θ, but its predictive distributions are usually easier to compute than those of the BFSs. A major disadvantage of the plug-in method is that the uncertainty of the estimator of θ is not accounted for in the predictive distribution. This, in turn, may lead to underestimation of the uncertainty for the future observation Y_{t+1} and hence to the inefficiency of the ensuing PFS.

When is a plug-in PFS efficient? Using the fact that a BFS is efficient, one needs to show that the predictive distributions of a plug-in method are "close" to those of an efficient BFS. One may use, for example, the Kullback-Leibler distance between two distributions F and G as the metric on the space of distributions. Conditions for efficiency of plug-in PFSs using this metric and others are discussed in detail in Skouras and Dawid (1999).

References

[1] B. Abraham and J. Ledolter, *Statistical Methods for Forecasting*, Wiley, New York, 1983.

[2] R.J. Adler, R.E. Feldman and M.S. Taqqu, *A Practical Guide to Heavy Tails: Statistical Techniques and Applications*, Birkhäuser, 1998.

[3] H. Akaike, Fitting autoregressive models for prediction, *Annals of the Institute of Statistical Mathematics*, Tokyo, **21**, 243-247 (1969).

[4] H. Akaike, Markovian representation of stochastic processes by canonical variables, *SIAM J. Control*, **13**, 162-173 (1975).

[5] N.I. Akhiezer and I.M. Glazman, *Theory of Linear Operators in Hilbert Space*, **Vol I, II**, F. Ungar Publishing Company, New York, 1963.

[6] E.J. Akutowicz, On an explicit formula in least squares prediction, *Math. Scan.*, **5**, 261-266 (1957).

[7] H.Z. An, Z.G. Chen and E.J. Hannan, Autocorrelation, autoregression and autoregressive approximation. *Ann. Statist.*, **10**, 926-936 (1982)

[8] T.W. Anderson, A note on a vector-variate normal distribution and a stationary autoregressive process. *Journal of Multivariate Analysis*, **32**, 149-150 (2000).

[9] T.W. Anderson, *An Introduction to Multivariate Analysis*, 2nd ed., Wiley, New York, 1984.

[10] T.W. Anderson, Asymptotically efficient estimation of covariance matrices with linear structure. *Ann. Statist.* **1**, 135-141 (1973).

[11] T.W. Anderson, *The Statistical Analysis of Time Series*, Wiley, New York, 1971.

[12] C.F. Ansley, An algorithm for the exact likelihood of a mixed autoregressive-moving process. *Biometrika*, **66**, 59-65 (1979)

[13] M. Arato, On the sufficient statistics for stationary Gaussian random processes. *Theory of Probab. and Its Applications*, **6**, 199-201 (1961).

[14] N. Aronszajn, Theory of reproducing kernels. *Trans. Amer. Math. Soc.*, 337-404 (1950).

[15] J. Barnard, R. McCulloch and X.L. Meng, Modeling covariance matrices in terms of standard deviations and correlations, with application to shrinkage. *Statistica Sinica*, **10**, 1281-1311 (2000).

[16] M.S. Bartlett, Correlation or spectral analysis? *Statistician*, **27**, 147-158 (1978).

[17] B. Baxter, An asymptotic result for the finite predictor, *Math. Scand.*, **10**, 137-144 (1962).

[18] J. Beran, *Statistics for Long-Memory Processes*, Chapman & Hall, New York, 1994.

[19] K.N. Berk, Consistent autoregressive spectral estimates, *Ann. Statist.*, **2**, 489-502 (1974)

[20] R.J. Bhansalli, On a relationship between the inverse of a stationary covariance matrix and the linear interpolator, *J. of Applied Probability*, **27**, 156-170 (1990).

[21] P.J. Bickel, C.A.J. Klaassen, Y. Ritov and J.A. Wellner, *Efficient and Adaptive Estimation for Semiparametric Models*, Johns Hopkins University Press, 1993.

[22] P. Billingsley, The Lindeberg-Lèvy theorem for martingales. *Proc. Amer. Math. Soc.*, **12**, 788-792 (1961).

[23] P. Bloomfield, *Fourier Analysis of Time Series: An Introduction*, Wiley, New York, 1976.

[24] P. Bloomfield, On series representations for linear predictors. *Ann. Probab.*, **13**, 226-233 (1985).

[25] G.E.P. Box, G.M. Jenkins and G.C. Reinsel, *Time Series Analysis—Forecasting and Control*, (Revised 3rd Edition), Prentice Hall, NJ 1994.

[26] G.E.P. Box and D.A. Pierce, Distribution of residual autocorrelations in autoregressive integrated moving average time series models. *J. of Amer. Statist. Assoc.*, **65**, 1509-1526 (1970).

[27] G.E.P. Box and G.C. Tiao, Intervention analysis with applications to environmental and economic problems, *J. of the American Statistical Association*, **70**, 70-79 (1975).

[28] G.E.P. Box and G.C. Tiao, Comparison of forecast and actuality. *Applied Statistics*, **25**, 195-200 (1976).

[29] R.C. Bradley, Basic properties of strong mixing conditions, in *Dependence in Probability and Statistics* (E. Eberlein and M. Taqqu, eds.), Birkhäuser, 165-192, 1986.

[30] D.R. Brillinger, *Time Series: Data Analysis and Theory*, Holt, New York, 1981.

[31] P.J. Brockwell and R.A. Davis, *Time Series: Theory and Methods*, Springer-Verlag, New York, 1991.

[32] P.J. Brown, N.D. Le and J.V. Zidek, Inference for a covariance matrix. *In Aspects of Uncertainty*, eds. P.R. Freeman and A.F.M. Smith, Chichester, Wiley, UK, 77-90. (1994).

[33] J.P. Burg, Maximum entropy spectral analysis, in *Proc. 37th Meeting Soc. of Exploration Geophysicists*, Oklahoma City, OK, 1967.

[34] S. Cambanis and A.R. Soltani, Prediction of stable processes: Spectral and moving average representations. *Z. Wahrsch Verw. Geb.*, **66**, 593-612 (1984).

[35] S. Cambanis, C.D. Hardin and A. Weron, Innovations and Wold decompositions of stable processes. *Probab. Theory and Rel. Fields*, **79**, 1-27 (1988).

[36] S. Cambanis, S.B. Fotopoulos and L. He, On the conditional variance for scale mixtures of normal distributions, *J. of Multivariate Analysis*, **74**, 163-193 (2000).

[37] M.J. Campbell and A.M. Walker, A survey of statistical work on the Mackenzie River series of annual Canadian lynx trappings for the years 1821–1934, and a new analysis (with discussion). *J. Roy. Statist. Soc.*, **A**, 158, 419-466 (1977).

[38] G. Casella and R.L. Berger, Statistical Inference, Duxbury Press, Belmont, California, 1990.

[39] C. Chatfield, Inverse autocorrelations. *J.R. Statist. Soc* A, **142**, 363-377 (1979).

[40] R. Cheng and M. Pourahmadi, Baxter's inequality and convergence of finite predictors of multivariate stochastic processes. *Probab. Theory Relat. Fields*, **95**, 115-124 (1993).

[41] R. Cheng, A.G. Miamee and M. Pourahmadi, Some extremal problems in $L^p(w)$. *Proc. of the Amer. Math. Soc.*, **126**, 2333-2340 (1998).

[42] R. Cheng, A.G. Miamee and M. Pourahmadi, Regularity and minimality of infinite variance processes. *J. of Theoretical Probability*, **13**, 1115-1122 (2000).

[43] S. Chib and E. Greenberg, Hierarchical analysis of SUR with extensions to correlated serial errors and time-varying parameter models, *J. of Econometrics*, **68**, 339-360 (1995).

[44] B.S. Choi, *ARMA Model Identification*, Springer Series in Statistics, New York, 1992.

[45] W.S. Cleveland, The inverse autocorrelations of time series and their applications, *Technometrics*, **14**, 277-283 (1972).

[46] W.S. Cleveland, *Visualizing Data*, AT&T Bell Laboratories, New Jersey.

[47] T. Cover and S. Thomas, *Elements of Information Theory.* Wiley, New York, 1990.

[48] D.R. Cox and N. Wermuth, *Multivariate Dependencies: Models, Analysis and Interpretation*, London: Chapman and Hall, 1996.

[49] H. Cramer, On the theory of stationary random processes, *Ann. of Math.*, **41**, 214-230 (1940).

[50] B.R. Cullis and C.A. McGilchrist, A model for the analysis of growth data from designed experiments, *Biometrics*, **46**, 131-142 (1990).

[51] A.C. Davison and S. Sardy, The partial scatterplot matrix, *J. of Computational and Graphical Statistics*, **9**, 750-758 (2000).

[52] A.P. Dawid, Statistical theory. The prequential approach (with Discussion), *J.R. Statist. Soc. A*, **147**, 278-292 (1984).

[53] A.P. Dawid, Prequential analysis, stochastic complexity and Bayesian inference. *Bayesian Statistics 4*, J.M. Bernard, J.O. Berger, A.P. Dawid and A.F.M. Smith, (eds.), 109-125, Oxford University Press, 1992.

[54] K.S. Dawson, C. Gennings and W.H. Carter, Two graphical techniques useful in detecting correlation structure in repeated measures data, *American Statistician*, **51**, 275-283 (1997).

[55] S. Degerine, Partial autocorrelation function for a scalar stationary discrete-time process, *Proc. 3rd Franco-Belgian Meeting of Statisticians*, Bruxelles: Publications des Facultes Universitaires Sait-Louis, 79-94 (1982).

[56] S. Degerine, Canonical partial autocorrelation function of a multivariate time series. *Ann. of Statist.*, **18**, 961-971 (1990).

[57] S. Degerine, Sample partial autocorrelation function of a multivariate time series, *J. of Multivariate Anal.*, **50**, 294-313 (1994).

[58] B.W. Dickinson, Sufficient statistics for stationary discrete-time Gaussian random processes, *J. of Time Series Analysis*, **3**, 165-168 (1982).

[59] P.J. Diggle, Spatial and longitudinal data analysis: Two histories with a common future?, In *Modelling Longitudinal and Spatially Correlated Data, Methods, Applications, and Future Directions*, Springer Lecture Notes in Statistics, NO. 122, T.G. Gregoire, et al., Eds., New York, Springer-Verlag, 387-402, 1997.

[60] P.J. Diggle, K.Y. Liang and S.L. Zeger, *Analysis of Longitudinal Data*, Claredon Press, Oxford, 1994.

[61] P.J. Diggle, *Time Series: A Biostatistical Introduction*, Clarendon Press, Oxford, 1990.

[62] P.J. Diggle and A.P. Verbyla, Nonparametric estimation of covariance structure in longitudinal data, *Biometrics*, **54**, 401-415 (1998).

[63] J.L. Doob, The elementary Gaussian processes, *Ann. Math. Statist.*, **15**, 229-282 (1944).

[64] J.L. Doob, *Stochastic Processes*, Wiley, New York, 1953.

[65] P. Doukhan, *Mixing Properties and Examples*, Springer, 1994.

[66] J. Durbin, The state space approach to time series analysis and its potential for official statistics, *Australia and New Zealand J. of Statistics*, **42**, 1-23 (2000).

[67] J. Durbin, The fitting of time series models, *Rev. Int. Inst. Stat.*, **28**, 233-244 (1960).

[68] H. Dym and H.P. McKean, *Gaussian Processes, Function Theory, and the Inverse Spectral Problem*, Academic Press, 1976.

[69] E. Eberlein and M. Taqqu, *Dependence in Probability and Statistics*, Birkhäuser, 1986.

[70] R. Engle, W. Granger, J. Rice and A. Weiss, Semiparametric estimates of the relationship between weather and electricity sales, *J. of the American Statistical Association*, **81**, 310-320 (1986).

[71] J. Farshidi and H. Salehi, Autoregressive expansion of the linear predictor for stationary stochastic process, in *Nonstationary Processes and Their Applications*, (A.G. Miamee, eds.), 176-193 World Scientific, 1992.

[72] G. Forchini, The density of sufficient statistics for a Gaussian AR(1) model in terms of generalized functions, *Statistics and Probability Letters*, **50**, 237-243 (2000).

[73] J. Franke, ARMA processes have maximal entropy among time series with prescribed autocovariances and impulse responses. *Adv. Appl. Prob.*, **17**, 810-840 (1985).

[74] W. Fuller, *Introduction to Time Series Analysis*, Second ed., Wiley, New York, 1996.

[75] K.R. Gabriel, Antidependence analysis of an ordered set of variables, *Ann. of Mathematical Statistics*, **33**, 201-212 (1962).

[76] R. F. Galbraith and J.I. Galbraith, On the inverse of some patterned matrices arising in the theory of stationary time series. *J. Appl. Prob.*, **11**, 63-71 (1974).

[77] J. Gani, *The Making of Statisticians*, Springer-Verlag, New York, 1982.

[78] J.B. Garnet, *Bounded Analytic Functions*, Academic Press, New York, 1981.

[79] S. Geisser, *Predictive Inference: An Introduction*, Chapman and Hall, New York, 1993.

[80] I.M. Gelfand and A.M. Yaglom, Calculation of the amount of information about a random function contained in another such function, *Upsehi Mat. Nauk* **12**, 3-52 (1957) [English translation: American Math. Soc. Transl. (2), **12**, 199-246 (1959)].

[81] I.C. Gohberg, *I. Schur Methods in Operator Theory and Signal Processing*, Birkhäuser Verlag, Basel, 1986.

[82] G.H. Golub, and C.F. Van Loan, *Matrix Computations*, Second Ed., Johns Hopkins University Press, Baltimore, 1989.

[83] U. Grenander, *Abstract Inference*, Wiley, New York, 1981.

[84] U. Grenander and M. Rosenblatt, *Statistical Analysis of Stationary Time Series*, Wiley, New York, 1957.

[85] P. Hall and C.C. Heyde, *Martingale Limit Theory and Its Applications*, Academic Press, New York, 1980.

[86] P. Halmos, *Introduction to Hilbert Space*, Chelsea Publishing Company, New York, 1957.

[87] P. Halmos, Shifts on Hilbert spaces, *für Reine and Angewende Mat.*, **208**, 102-112 (1961).

[88] E.J. Hannan, *Time Series Analysis*, London, Methuen, 1960.

[89] E.J. Hannan, *Multiple Time Series*, Wiley, New York, 1970.

[90] E.J. Hannan, The uniform convergence of autocovariances. *Ann. Statist.*, **2**, 803-806 (1974).

[91] E.J. Hannan, Rational transfer function approximation, *Statist. Sci.*, **2**, 135-161 (1987).

[92] E.J. Hannan, and M. Deistler, *The Statistical Theory of Linear Systems*, Wiley, New York, 1988.

[93] E.J. Hannan and J. Rissanen, Recursive estimation of ARMA order, *Biometrika*, **69**, 81-94 (1982).

[94] A. Harvey, *Forecasting, Structural Time Series Models and the Kalman Filter*, Cambridge University Press, Cambridge, 1989.

[95] T.J. Hastie and R.J. Tibshirani, *Generalized Additive Models*, Chapman and Hall, New York, 1990.

[96] K. Hayes and J. Haslett, Simplifying general least squares, *American Statistician*, **53**, 376-381 (1999).

[97] H. Helson, *Harmonic Analysis*, Addison-Wesley, 1983.

[98] H. Helson, *Lectures on Invariant Subspaces*, Academic Press, New York, 1964.

[99] H. Helson and D. Sarason, Past and Future, *Math. Scand.*, **21**, 5-16 (1967).

[100] H. Helson and G. Szegö, A problem in prediction theory, *Ann. Math. Pura Appl.*, **51**, 107-138 (1960).

[101] H. Helson and D. Lowdenslager, Prediction theory and Fourier series in several variables, *Acta Mathematica*, **99**, 165-202 (1958).

[102] K. Hoffman, *Banach Spaces of Analytic Functions*, Prentice-Hall, Englewood Cliffs, NJ, 1962.

[103] A.J. Hoogstrate, F.C. Palm and G.A. Pfann, Pooling in dynamic panel-data models: An application to forecasting GDP growth rates, *J. of Business & Economic Statistics*, **18**, 274-283 (2000).

[104] D. Huang, On the maximal entropy property for ARMA processes and ARMA approximation, *Adv. Appl. Prob.*, **22**, 612-626 (1990).

[105] M.L. Huang, R.A. Kerman and Y. Weit, Abel summability of the autoregressive series for the best linear least squares predictors, *Illinois J. of Math.*, **41**, 577-588 (1997).

[106] R. Hunt, B. Muckenhoupt and R. Wheeden, Weighted norm inequalities for the conjugate function and Hilbert transform. *Trans. Amer. Math. Soc.*, **176**, 227-251 (1973).

[107] I.A. Ibragimov and Yu. A. Rozanov, *Gaussian Random Processes*, Springer-Verlag, New York, 1978.

[108] J.D. Isaacson and D.L. Zimmerman, Combining temporally correlated environmental data from two measurement systems, *J. of Agricultural, Biological and Environmental Statistics*, **5**, 398-416 (2000).

[109] N.P. Jewell and P. Bloomfield, Canonical correlations of past and future for time series: Definitions and theory, *Ann. Statist.*, **11**, 837-847 (1983).

[110] N.P. Jewell, P. Bloomfield and F.C. Bartmann, Canonical correlations of past and future for time series: Bounds and computation, *Ann. Statist.*, **11**, 848-855 (1983).

[111] M.C. Jones, Randomly choosing parameters from the stationarity and invertibility region of autoregressive-moving average models, *Appl. Statist.*, **36**, 134-138 (1987).

[112] R.H. Jones, Maximum likelihood fitting of ARMA models to time series with missing observations, *Technometrics*, **22**, 389-395 (1980).

[113] P. Kabaila, The relevance property for prediction intervals, *J. of Time Series Analysis*, **20**, 655-662 (1999).

[114] T. Kailath, A view of three decades of linear filtering, *IEEE Trans. Inform. Theory*, **20**, 145-181 (1974).

[115] T. Kailath, Signal processing applications of some moment problems, *Moments in Mathematics, H.J. Landau (ed.), Proceedings of Symposia in Applied Mathematics, American Math. Soc.*, **37**, 71-109 (1987)

[116] R.E. Kalman, A new approach to linear filtering and prediction problems, *Trans. Amer. Soc. Mech. Eng., J. Basic Engineering*, **82**, 35-45, 1960.

[117] R.E. Kalman and R.S. Bucy, New results in linear filtering and prediction theory, *J. of Basic Engineering, Transactions ASME Series D*, **83**, 95-108 (1961).

[118] A.J. Kanto, A characterization of the inverse autocorrelation function, *Commun. Statist.-Theor. Meth.*, **13**, 2503-2510 (1984).

[119] S. Kayalar and H.L. Weinert, Error bounds for the method of alternating projections, *Math. Control Signals Systems*, **1**, 43-59 (1988).

[120] M.G. Kenward, A method for comparing profiles of repeated measurements, *Applied Statistics*, **36**, 296-308 (1987).

[121] A.Y. Khinchin, Sulle successioni stazionarie di eventi, *Giornale Institute Italian Attuari*, **3**, 267-373 (1932).

[122] A.Y. Khinchin, Korrleationstheorie der Stationären Stochastischen Prozesse, *Mathematische Annalen*, **709**, 604-615 (1934).

[123] J.L. Klein, *Statistical Visions in Time: A History of Time Series Analysis, 1662–1938*, Cambridge University Press, Cambridge, 1997.

[124] R. Kohn and C.F. Ansley, Computing the likelihood and its derivative for a Gaussian ARMA model, *J. Statist. Comput. Simul.*, **22**, 229-63 (1985).

[125] A.N. Kolmogorov, Sur l'interpolation et extrapolation des suites stationnaires, *C.R. Acad. Sci. Paris*, **208**, 2043-2045 (1939).

[126] A.N. Kolmogorov, Stationary sequences in a Hilbert space, *Bull. Moscow State University*, **2**, 1-40 (1941).

[127] A.N. Kolmogorov and Yu.A. Rozanov, On strong mixing conditions for stationary Gaussian random processes, *Theory of Probability and Its Applications*, **5**, 204-208 (1960).

[128] L.H. Koopmans, *The Spectral Analysis of Time Series*, Academic Press, New York, 1974.

[129] M.G. Krien, On an extrapolation problem of A.N. Kolmogorov, *Dokl. Akad. Nauk SSSR*, **46**, 306-309 (1944).

[130] H.E. Lacey, *The Isometric Theory of Classical Banach Spaces*. Springer-Verlag, New York, 1974.

[131] N.M. Laird and J.H. Ware, Random-effects models for longitudinal data, *Biometrics*, **38**, 963-974 (1982).

[132] P.J. Lambert and D.S. Poskitt, *Stationary Processes in Time Series Analysis: The Mathematical Foundations*, Vandenhoeck and Ruprecht, Gottingen, 1983.

[133] H.J. Landau, Maximum entropy and the moment problem, *Bull. of the Amer. Math. Soc.*, **16**, 47-77 (1987a).

[134] H.J. Landau, Moments in Mathematics, *Proc. of Symposia in Applied Mathematics*, **37**, Amer. Math. Soc. (1987b).

[135] N. Levinson, A heuristic exposition of Wiener's mathematical theory of prediction and filtering. *J. of Mathematics and Physics*, **26**, 110-119 (1947).

[136] K.Y. Liang and S.L. Zeger, Longitudinal data analysis using generalized linear models, *Biometrika*, **73**, 13-22 (1986).

[137] T.C. Lin and M. Pourahmadi, Nonparametric and non-linear models and data mining in time series: a case-study on the Canadian lynx data. *Appl. Statist.*, **47**, 187-201 (1998).

[138] J.K. Lindsey, *Models for Repeated Measurements*. Second Ed., Oxford University Press, Oxford, 1999.

[139] G.M. Ljung and G.E.P. Box, The likelihood function of stationary autoregressive-moving average models. *Biometrika*, **66**, 265-270 (1979).

[140] A. Luceño, Fast optimization of the exact likelihood of AR and ARMA processes, *Comput. Stat. Data Anal.*, **17**, 51-63 (1994).

[141] A. Makagon and H. Salehi, On bounded and harmonizable solutions of infinite order ARMA systems, *New Directions in Time Series Analysis*, Brillinger et al. (eds), 193-213 (1992).

[142] A. Makagon and A. Weron, Wold-Cramer concordance theorems for interpolation of q-variate stationary processes over locally compact abelian group, *J. Multivariate Anal.*, **6**, 123-137 (1976).

[143] V. Mandrekar and H. Salehi, *Prediction Theory and Harmonic Analysis*. North-Holland, Amsterdam, 1983.

[144] J.M. Marriott and A.F.M. Smith, Reparameterization aspects of numerical Bayesian methodology for autoregressive moving-average models, *J. of Time Series Analysis*, **13**, 327-343 (1992).

[145] P.R. Masani, The prediction theory of multivariate stochastic processes, III, *Acta Math.*, **104**, 141-162 (1960).

[146] P.R. Masani, Shift invariant spaces and prediction theory, *Acta Math*, **107**, 275-290 (1962).

[147] P.R. Masani, *Commentary on the prediction-theoretic papers, Norbert Wiener: Collected Works III*, P.R. Masani (ed.), MIT Press, Cambridge, MA 276-306 (1981).

[148] P.R. Masani, The aftermath of Cramer's work on stochastic processes, *Scand. Actuarial J.*, **1**, 66-96 (1995).

[149] MathSoft, Inc., *S-PLUS version 3.4*, StatSci Division, WA, 1996.

[150] P. McCullagh and J.A. Nelder, *Generalized Linear Models*, Second Ed., Chapman & Hall, London, 1989.

[151] D.N. McClosky, The art of forecasting: from ancient to modern times, *Cato Journal*, **12**, 23-43 (1992).

[152] A.G. Miamee and M. Pourahmadi, Best approximation in $L^p(d\mu)$ and prediction problems of Szegö, Kolmogorov, Yaglom and Nakazi. *J. London Math. Soc.*, **38**, 133-145 (1988a).

[153] A.G. Miamee and M. Pourahmadi, Wold decomposition, prediction and parameterization of stationary processes with infinite variance, *Probab. Theory and Related Fields*, **79**, 145-164 (1988b).

[154] A.G. Miamee and M. Pourahmadi, Computation of canonical correlation between past and future of a time series, *J. of Time Series Analysis*, **13**, 345-351 (1992).

[155] J.W. Miller, Exact maximum likelihood estimation in autoregressive processes, *J. of Time Series Analysis*, **16**, 607-615 (1995).

[156] C. Min and A. Zellner, Bayesian and non-Bayesian methods for combining models and forecasts with applications to forecasting international growth rates, *J. of Econometrics*, **56**, 89-118 (1993).

[157] M. Minozzo and A. Azzalini, On the unimodality of the exact likelihood function for normal AR(2) series, *J. of Time Series Analysis*, **14**, 497-509 (1993).

[158] P.A.P. Moran, The statistical analysis of the sunspot and lynx cycles, *J. Animal Ecol.*, **1**, 163-173 (1953).

[159] R.J. Muirhead, *Aspects of Multivariate Statistical Theory*, Wiley, New York, 1983.

[160] A. Munoz, V. Carey, J.P. Schouten, M. Segal and B. Roser, A parametric family of correlation structures for the analysis of longitudinal data, *Biometrics*, **48**, 733-742 (1992).

[161] J.F. Muth, Optimal properties of exponentially weighted forecasts. *J. Amer. Statist. Assoc.*, **55**, 299-305 (1960).

[162] T. Nakazi, Two problems in prediction theory, *Studia Math.*, **78**, 7-14 (1984).

[163] P. Newbold, The exact likelihood function for a mixed autoregressive-moving average process. *Biometrika*, **61**, 423-426 (1974).

[164] H. Niemi, Theoretical properties of linear predictors for ARIMA models, *Scand. J. Statist.*, **II**, 113-122 (1984).

[165] V. Núñez-Antón and D.L. Zimmerman, Modeling nonstationary longitudinal data, *Biometrics*, **56**, 699-705 (2000).

[166] R.A. Olshen, Asymptotic properties of the periodogram of a discrete stationary process, *J. Appl. Probab.*, **4**, 508-528 (1967).

[167] D.R. Osborn, Maximum likelihood estimation of moving average processes, *Ann. Econ. Social Meas.*, **5**, 75-87 (1976).

[168] E. Parzen, *Time Series Analysis Papers*, Holden-Day, San Francisco, 1967.

[169] E. Parzen, Some recent advances in time series modeling, *IEEE Trans. on Automatic Control*, **AC-19**, 723-730 (1974).

[170] V.V. Peller, Regularity conditions for vectorial stationary process, *Operator Theory: Advances and Applications*, **113**, 287-301 (2000).

[171] V.V. Peller and S.V. Khruschev, Hankel operators, best approximation and stationary Gaussian processes, *Russian Math. Surveys*, **37**, 53-124 (1982).

[172] D.T. Pham and S. Degerines, Efficient computation of autoregressive estimates through a sufficient statistic, *IEEE Trans. on Acoust., Speech and Signal Processing*, **38**, 175-177 (1990).

[173] D. Piccolo and T.G. Wilson, A unified approach to ARMA model identification and preliminary estimation, *J. of Time Series Analysis*, **5**, 183-204 (1984).

[174] D.A. Pierce, A duality between autoregressive and moving average processes concerning their least squares estimates, *Ann. Math. Statist.*, **41**, 422-426 (1970).

[175] J.D. Pinheiro and D.M. Bates, Unconstrained parameterizations for variance-covariance matrices, *Statist. Comp.*, **6**, 289-296 (1996).

[176] D.S.G. Pollock, *A Handbook of Time-Series Analysis, Signal Processing and Dynamics*, Academic Press, San Diego, 1999.

[177] M. Pourahmadi, Taylor expansion of $\exp(\sum_{k=0}^{\infty} a_k z^k)$ and some applications, *Amer. Mathematical Monthly*, **91**, 303-307. (1984).

[178] M. Pourahmadi, A matrical extension of the Helson-Szego theorem and its applications in multivariate prediction, *J. of Multivariate Analysis*, **5**, 265-275 (1985).

[179] M. Pourahmadi, Autoregressive representations of multivariate stationary stochastic processes, *Probab. Theory and Rel. Fields*, **80**, 315-322 (1988).

[180] M. Pourahmadi, Estimation and interpolation of missing values of a stationary time series, *J. of Time Series Analysis*, **10**, 149-169 (1989).

[181] M. Pourahmadi, Alternating projections and interpolation of stationary processes. *J. Appl. Probab.*, **29**, 921-931 (1992).

[182] M. Pourahmadi, Fundamental roles of the idea of regression and Wold decomposition in time series. *New Directions in Time Series Analysis*, D. Brillinger et al. (eds.), Springer-Verlag, New York (1992).

[183] M. Pourahmadi, Joint mean-covariance models with applications to longitudinal data: unconstrained parameterisation. *Biometrika*, **86**, 677-690 (1999).

[184] M. Pourahmadi, Maximum likelihood estimation of generalized linear models for multivariate normal covariance matrix, *Biometrika*, **87**, 425-435 (2000).

[185] M. Pourahmadi and E.S. Soofi, Prediction variance and information worth of observations in time series. *J. of Time Series Analysis*, **21**, 413-434 (2000),

[186] M.B. Priestley, *Spectral Analysis and Time Series*, Vols. 1 and 2, Academic Press, London, 1981.

[187] M.B. Priestley, *Non-linear and Non-stationary Time Seris Analysis*, Academic Press, London, 1988.

[188] M.H. Quenouille, Approximate tests of correlation in time series, *J. of Roy. Statist. Soc.*, **13, 11**, 68-84 (1949).

[189] F.L. Ramsey, Characterization of the partial autocorrelation function, *Ann. of Statist.*, **2**, 1296-1301 (1974).

[190] C.R. Rao, *Linear Statistical Inference and Its Applications*, Wiley, New York, 1973.

[191] F. Riesz and B. Nagy, *Functional Analysis*, Frederick Ungar Co., New York, 1955.

[192] J. Rissanen, *Stochastic Complexity in Statistical Inquiry*, World Scientific Publishing, Singapore, 1989.

[193] E. Robinson, *Infinitely Many Variates*, Griffin's Statistical Monographs and Courses, London, 1954.

[194] M. Rosenberg, Mutual subordination of multivariate stationary processes over any locally compact Abelian group, *Zeit. Wahr*, **12**, 333-343 (1969).

[195] M. Rosenblatt, A central limit theorem and a strong mixing condition, *Proc. Nat. Acad. Sci. USA*, **42**, 43-47 (1956).

[196] M. Rosenblatt, *Stationary Sequences and Random Fields*, Birkhauser, Boston, 1985.

[197] M. Rosenblatt, *Gaussian and Non-Gaussian Linear Time Series and Random Fields*, Springer-Verlag, New York, 2000.

[198] M. Rosenblatt and U. Grenander, An extension of a theorem of G. Szegö and its application to the theory of stochastic processes. *Trans. Amer. Math. Soc.*, **76**, 112-126 (1954).

[199] M. Rosenblum and J. Rovnyak, *Hardy Classes and Operator Theory*, Oxford Mathematical Monographs, Oxford, 1985.

[200] H. Royden, *Real Analysis*, Macmillan, 1968.

[201] Yu. A. Rozanov, *Stationary Random Processes*, Holder-Day, San Francisco, 1967.

[202] H. Salehi, Algorithm for linear interpolator and interpolation error for minimal stationary stochastic processes. *Ann. Probab.*, **7**, 840-846 (1979).

[203] SAS Institute Inc. *SAS/STAT Software: Changes and Enhancements through Release* 6.12, Cary, NC: SAS Institute Inc, 1997.

[204] G. Samorodnitsky and M.S. Taqqu, *Stable Non-gaussian Random Processes: Stochastic Models with Infinite Variance*, Chapman and Hall, New York, 1994.

[205] D.E. Sarason, An addendum to "Past and future," *Math. Scand.*, **30**, 62-64 (1972).

[206] A. Schuster, On the investigation of hidden periodicities with application to a supposed 26 day period of meteorological phenomena, *Terr. Magn.*, **3**, 13-41 (1898).

[207] A. Schuster, On the Periodicities of Sunspots, *Philosophical Transactions*, **206A**, 69-100, 1906.

[208] G. Schwarz, Estimating the dimension of a model, *Ann. Statist.*, **6**, 461-464 (1978).

[209] S.R. Searle, G. Casella and C.E. McCulloch, *Variance Components*, Wiley, New York, 1992.

[210] W. Sharpe, Capital asset prices: A theory of market equilibrium under conditions of risk, *J. of Finance*, **19**, 425-442 (1964).

[211] L.A. Shepp, D. Slepian and H.D. Wyner, On prediction of moving-average processes. *Bell System Technical Journal*, **59**, 367-415 (1980).

[212] R. Shibata, Selection of the order of an autoregressive model by Akiake's information criterion. *Biometrika*, **63**, 117-126 (1976).

[213] H. Shigeno, On a characterization of autoregressive moving average process by the canonical structure between past and future spaces, *J. Japan Statist. Soc.*, **19**, 35-47 (1989).

[214] R.H. Shumway and D. Stoffer *Time Series Analysis and Its Applications*, Springer-Verlag, 2000.

[215] M.M. Siddiqui, On the inversion of the sample covariance matrix in a stationary autoregressive process,*Ann. Math. Statist.*, **29**, 585-588 (1958).

[216] K. Skouras and A.P. Dawid, On efficient point prediction systems, *J.R. Statist. Soc. B*, **60**, 765-780 (1998).

[217] K. Skouras and A.P. Dawid, On efficient probability forecasting systems, *Biometrika*, **86**, 765-784 (1999).

[218] E. Slutsky, The summation of random causes as the source of cyclic processes, (Russian) *Problem of Economic Conditions* **3**; Engl. transl. *Econometrica*, **5**, 105-106 (1937).

[219] C.G. Small and D.L. McLeish, *Hilbert Space Methods in Probability and Statistical Inference*, Wiley, New York, 1994.

[220] M. Smith, Modeling and short-term forecasting of New South Wales electricity system load, *J. of Business & Economics Statistics*, **18**, 465-478 (2000).

[221] M. Smith and R. Kohn, Bayesian parsimonious covariance matrix estimation, preprint (2000).

[222] N.C. Stenseth, W. Falck, K.S. Chan, O.N. Bjørnstad, M. O'Donoghue, H. Tong, R. Boonstra, S. Boutin, C.J. Krebs and N.G. Yaccuz, From patterns to processes: Phase and density dependencies in the Canadian lynx cycle, *Proc. Nat. Acad. Sci. USA*, **95**, 15430-15435 (1998).

[223] G. Szegö and U. Grenander, *Toeplitz Forms and Their Applications*, University of California Press, Berkeley, CA, 1958.

[224] M. Taniguchi and Y. Kakizawa, *Asymptotic Theory of Statistical Inference for Time Series*, Springer-Verlag, New York, 2000.

[225] B. Tarami and M. Pourahmadi, Multivariate t autoregressions: Innovations, prediction variances and exact likelihood equations, preprint (2000).

[226] H. Tong, *Nonlinear Time Series: A dynamic system approach*, Oxford Press, New York, 1990.

[227] S. Treil and A. Volberg, Completely regular multivariate stationary processes and the Muckenhoupt condition. *Pacific J. of Math.*, **190**, 361-382 (1999).

[228] R.S. Tsay, Outliers, level shifts, and variance changes in time series. *J. of Forecasting*, **7**, 1-20 (1988).

[229] R.S. Tsay and G.C. Tiao, Consistent estimates of autoregressive parameters and extended sample autocorrelations functions for stationary and non-stationary ARIMA models, *J. of Amer. Statist. Assoc.*, **79**, 84-96 (1984).

[230] R.S. Tsay and G.C. Tiao, Use of canonical analysis in time series model identification, *Biometrika*, **72**, 299-315 (1985).

[231] J.W. Tukey, Discussion emphasizing the connection between analysis of variance and spectrum analysis, *Technometrics*, **3**, 1-29 (1961).

[232] K.L. Vaninskii and A.M. Yaglom, Stationary processes with a finite number of non-zero canonical correlations between future and past, *J. of Time Series Analysis*, **11**, 361-375 (1990).

[233] B. Vidakovic, *Statistical Modeling by Wavelets*, Wiley, New York, 1999.

[234] G. Wahba, Spline Models for Observational Data, *SIAM, CBMS-NSF*, **59** (1990).

[235] G. Walker, On periodicity in series of related terms, *Proc. of the Royal Society*, **131**, 518-532 (1931).

[236] E.J. Wegman, Another look at Box-Jenkins forecasting procedures, *Comm. in Statist.*, **B 15**, 523-530 (1986).

[237] M. West and J. Harrison, *Bayesian Forecasting and Dynamic Models*, Springer Series in Statistics, New York, 1997.

[238] J.S. White, Asymptotic expansions for the mean and variance of the serial correlation coefficient, *Biometrika*, **48**, 85-94 (1961).

[239] J. Whittaker, *Graphical Models in Applied Multivariate Statistics*. Wiley & Sons, New York, 1990.

[240] P. Whittle, *Hypothesis Testing and Time Series Analysis*, Uppsala, Almquist and Wiksell, 1951.

[241] P. Whittle, Gaussian estimation in stationary times series, *Bull. I.S.I.*, **39**, 105-129 (1961).

[242] P. Whittle, *Prediction and Regulation*, Van Nostrand-Reinhold, Princeton, NJ, 1963.

[243] N. Wiener, *Time Series*, MIT Press, Cambridge, MA, 1949.

[244] N. Wiener and P.R. Masani, The prediction theory of multivariate stationary processes, I. *Acta Math.*, 111-150 (1957); II. *Acta Math. 99*, 93-137 (1958).

[245] J. Wise, The autocorrelation function and the spectral density function, *Biometrika*, **42**, 151-159 (1955).

[246] H. Wold, *A Study in the Analysis of Stationary Time Series*, Almquist and Wiksell, Stockholm, 1954.

[247] R.D. Wolfinger, Heterogeneous variance-covariance structures for repeated measures. *J. of Agricultural, Biological and Environmental Statistics*, **1**, 205-230 (1996).

[248] W.A. Woodward and H.L. Gray, New ARMA models for Wolfer's sunspot data, *Commu. in Statist., Simul. and Comput.*, **7**, 97-115 (1978).

[249] A.M. Yaglom, *An Introduction to the Theory of Stationary Random Functions*. Dover Publications, New York, 1962.

[250] A.M. Yaglom, Stationary Gaussian processes satisfying the strong mixing condition and best predictable functional, *Proc. Int. Research Seminar of the Statistical Laboratory*, University of California, Berkeley, 241-252 (1963).

[251] A.M. Yaglom, *Correlation Theory of Stationary and Related Random Functions*, Vols. I, II, Springer-Verlag, 1986.

[252] R. Yang and J.O. Berger, Estimation of a covariance matrix using a reference prior. *Ann. of Statist.*, **22**, 1195-1211 (1994).

[253] N. Young, *An Introduction to Hilbert Space*, Cambridge University Press, Cambridge, 1989.

[254] G.U. Yule, On the time correlation problem, *J.R. Statist. Soc.*, **84**, 497-526 (1921).

[255] G.U. Yule, Why do we sometimes get nonsense-correlations between times series? A study in sampling and the nature of time series. *J.R. Statist. Soc.*, **89**, 1-65 (1926).

[256] G.U. Yule, On a model of investigating periodicities in disturbed series with special reference to Wolfer's sunspot numbers, *Phil. Trans. A*, **226**, 267-298 (1927).

[257] A. Zellner, An error-component procedure (ECP) for introducing prior information about covariance matrices and analysis of multivariate regression models, *International Economic Review*, **20**, 201-214 (1979).

[258] A. Zellner, An efficient method of estimating seemingly unrelated regression equations and tests for aggregation bias, *J. of the American Statistical Association*, **57**, 348-368 (1962).

[259] L.P. Zhao and R.L. Prentice, Use of quadratic exponential model to generate estimating equations for means, variances and covariances. In *Estimating Functions*, Ed. V. P. Godambe, p. 103-112, Claredon Press, Oxford, 1991.

[260] D.L. Zimmerman & V. Núñez-Antón, Structured antedependence models for longitudinal data, in *Modeling Longitudinal and Spatially Correlated Data: Methods, Applications, and Future Directions. Springer Lecture Notes in Statistics*, NO.122, Ed. T.G. Gregoire et al., Springer-Verlag, New York, 63-76 (1997).

[261] D.L. Zimmerman, Viewing the correlation structure of longitudinal data through a PRISM, *Amer. Statistician*, **54**, 310-318 (2000).

Index

Author Index

WILEY SERIES IN PROBABILITY AND STATISTICS
ESTABLISHED BY WALTER A. SHEWHART AND SAMUEL S. WILKS

Probability and Statistics Section

*ANDERSON · The Statistical Analysis of Time Series
ARNOLD, BALAKRISHNAN, and NAGARAJA · A First Course in Order Statistics
ARNOLD, BALAKRISHNAN, and NAGARAJA · Records
BACCELLI, COHEN, OLSDER, and QUADRAT · Synchronization and Linearity: An Algebra for Discrete Event Systems
BARNETT · Comparative Statistical Inference, *Third Edition*
BASILEVSKY · Statistical Factor Analysis and Related Methods: Theory and Applications
BERNARDO and SMITH · Bayesian Statistical Concepts and Theory
BILLINGSLEY · Convergence of Probability Measures, *Second Edition*
BOROVKOV · Asymptotic Methods in Queuing Theory
BOROVKOV · Ergodicity and Stability of Stochastic Processes
BRANDT, FRANKEN, and LISEK · Stationary Stochastic Models
CAINES · Linear Stochastic Systems
CAIROLI and DALANG · Sequential Stochastic Optimization
CONSTANTINE · Combinatorial Theory and Statistical Design
COOK · Regression Graphics
COVER and THOMAS · Elements of Information Theory
CSÖRGŐ and HORVÁTH · Weighted Approximations in Probability Statistics
CSÖRGŐ and HORVÁTH · Limit Theorems in Change Point Analysis
*DANIEL · Fitting Equations to Data: Computer Analysis of Multifactor Data, *Second Edition*
DETTE and STUDDEN · The Theory of Canonical Moments with Applications in Statistics, Probability, and Analysis
DEY and MUKERJEE · Fractional Factorial Plans
*DOOB · Stochastic Processes
DRYDEN and MARDIA · Statistical Shape Analysis
DUPUIS and ELLIS · A Weak Convergence Approach to the Theory of Large Deviations
ETHIER and KURTZ · Markov Processes: Characterization and Convergence
FELLER · An Introduction to Probability Theory and Its Applications, Volume I, *Third Edition,* Revised; Volume II, *Second Edition*
FULLER · Introduction to Statistical Time Series, *Second Edition*
FULLER · Measurement Error Models
GHOSH, MUKHOPADHYAY, and SEN · Sequential Estimation
GIFI · Nonlinear Multivariate Analysis
GUTTORP · Statistical Inference for Branching Processes
HALL · Introduction to the Theory of Coverage Processes
HAMPEL · Robust Statistics: The Approach Based on Influence Functions
HANNAN and DEISTLER · The Statistical Theory of Linear Systems
HUBER · Robust Statistics

*Now available in a lower priced paperback edition in the Wiley Classics Library.

Probability and Statistics (Continued)

HUSKOVA, BERAN, and DUPAC · Collected Works of Jaroslav Hajek—with Commentary
IMAN and CONOVER · A Modern Approach to Statistics
JUREK and MASON · Operator-Limit Distributions in Probability Theory
KASS and VOS · Geometrical Foundations of Asymptotic Inference
KAUFMAN and ROUSSEEUW · Finding Groups in Data: An Introduction to Cluster Analysis
KELLY · Probability, Statistics, and Optimization
KENDALL, BARDEN, CARNE, and LE · Shape and Shape Theory
LINDVALL · Lectures on the Coupling Method
MANTON, WOODBURY, and TOLLEY · Statistical Applications Using Fuzzy Sets
MORGENTHALER and TUKEY · Configural Polysampling: A Route to Practical Robustness
MUIRHEAD · Aspects of Multivariate Statistical Theory
OLIVER and SMITH · Influence Diagrams, Belief Nets and Decision Analysis
*PARZEN · Modern Probability Theory and Its Applications
PEÑA, TIAO, and TSAY · A Course in Time Series Analysis
PRESS · Bayesian Statistics: Principles, Models, and Applications
PUKELSHEIM · Optimal Experimental Design
RAO · Asymptotic Theory of Statistical Inference
RAO · Linear Statistical Inference and Its Applications, *Second Edition*
RAO and SHANBHAG · Choquet-Deny Type Functional Equations with Applications to Stochastic Models
ROBERTSON, WRIGHT, and DYKSTRA · Order Restricted Statistical Inference
ROGERS and WILLIAMS · Diffusions, Markov Processes, and Martingales, Volume I: Foundations, *Second Edition;* Volume II: Îto Calculus
RUBINSTEIN and SHAPIRO · Discrete Event Systems: Sensitivity Analysis and Stochastic Optimization by the Score Function Method
RUZSA and SZEKELY · Algebraic Probability Theory
SCHEFFE · The Analysis of Variance
SEBER · Linear Regression Analysis
SEBER · Multivariate Observations
SEBER and WILD · Nonlinear Regression
SERFLING · Approximation Theorems of Mathematical Statistics
SHORACK and WELLNER · Empirical Processes with Applications to Statistics
SMALL and McLEISH · Hilbert Space Methods in Probability and Statistical Inference
STAPLETON · Linear Statistical Models
STAUDTE and SHEATHER · Robust Estimation and Testing
STOYANOV · Counterexamples in Probability
TANAKA · Time Series Analysis: Nonstationary and Noninvertible Distribution Theory
THOMPSON and SEBER · Adaptive Sampling
WELSH · Aspects of Statistical Inference
WHITTAKER · Graphical Models in Applied Multivariate Statistics
YANG · The Construction Theory of Denumerable Markov Processes

Applied Probability and Statistics Section

ABRAHAM and LEDOLTER · Statistical Methods for Forecasting
AGRESTI · Analysis of Ordinal Categorical Data
AGRESTI · Categorical Data Analysis

*Now available in a lower priced paperback edition in the Wiley Classics Library.

Applied Probability and Statistics (Continued)

ANDERSON, AUQUIER, HAUCK, OAKES, VANDAELE, and WEISBERG · Statistical Methods for Comparative Studies

*ARTHANARI and DODGE · Mathematical Programming in Statistics

ASMUSSEN · Applied Probability and Queues

*BAILEY · The Elements of Stochastic Processes with Applications to the Natural Sciences

BARNETT and LEWIS · Outliers in Statistical Data, *Third Edition*

BARTHOLOMEW, FORBES, and McLEAN · Statistical Techniques for Manpower Planning, *Second Edition*

BASU and RIGDON · Statistical Methods for the Reliability of Repairable Systems

BATES and WATTS · Nonlinear Regression Analysis and Its Applications

BECHHOFER, SANTNER, and GOLDSMAN · Design and Analysis of Experiments for Statistical Selection, Screening, and Multiple Comparisons

BELSLEY · Conditioning Diagnostics: Collinearity and Weak Data in Regression

BELSLEY, KUH, and WELSCH · Regression Diagnostics: Identifying Influential Data and Sources of Collinearity

BHAT · Elements of Applied Stochastic Processes, *Second Edition*

BHATTACHARYA and WAYMIRE · Stochastic Processes with Applications

BIRKES and DODGE · Alternative Methods of Regression

BLISCHKE AND MURTHY · Reliability: Modeling, Prediction, and Optimization

BLOOMFIELD · Fourier Analysis of Time Series: An Introduction, *Second Edition*

BOLLEN · Structural Equations with Latent Variables

BOULEAU · Numerical Methods for Stochastic Processes

BOX · Bayesian Inference in Statistical Analysis

BOX and DRAPER · Empirical Model-Building and Response Surfaces

*BOX and DRAPER · Evolutionary Operation: A Statistical Method for Process Improvement

BUCKLEW · Large Deviation Techniques in Decision, Simulation, and Estimation

BUNKE and BUNKE · Nonlinear Regression, Functional Relations and Robust Methods: Statistical Methods of Model Building

CHATTERJEE and HADI · Sensitivity Analysis in Linear Regression

CHERNICK · Bootstrap Methods: A Practitioner's Guide

CHILÈS and DELFINER · Geostatistics: Modeling Spatial Uncertainty

CLARKE and DISNEY · Probability and Random Processes: A First Course with Applications, *Second Edition*

*COCHRAN and COX · Experimental Designs, *Second Edition*

CONOVER · Practical Nonparametric Statistics, *Second Edition*

CORNELL · Experiments with Mixtures, Designs, Models, and the Analysis of Mixture Data, *Second Edition*

*COX · Planning of Experiments

CRESSIE · Statistics for Spatial Data, *Revised Edition*

DANIEL · Applications of Statistics to Industrial Experimentation

DAVID · Order Statistics, *Second Edition*

*DEGROOT, FIENBERG, and KADANE · Statistics and the Law

DODGE · Alternative Methods of Regression

DOWDY and WEARDEN · Statistics for Research, *Second Edition*

GALLANT · Nonlinear Statistical Models

GLASSERMAN and YAO · Monotone Structure in Discrete-Event Systems

GNANADESIKAN · Methods for Statistical Data Analysis of Multivariate Observations, *Second Edition*

GOLDSTEIN and LEWIS · Assessment: Problems, Development, and Statistical Issues

GREENWOOD and NIKULIN · A Guide to Chi-Squared Testing

*HAHN · Statistical Models in Engineering

*Now available in a lower priced paperback edition in the Wiley Classics Library.

Applied Probability and Statistics (Continued)

HAHN and MEEKER · Statistical Intervals: A Guide for Practitioners
HAND · Construction and Assessment of Classification Rules
HAND · Discrimination and Classification
HEIBERGER · Computation for the Analysis of Designed Experiments
HEDAYAT and SINHA · Design and Inference in Finite Population Sampling
HINKELMAN and KEMPTHORNE: · Design and Analysis of Experiments, Volume 1: Introduction to Experimental Design
HOAGLIN, MOSTELLER, and TUKEY · Exploratory Approach to Analysis of Variance
HOAGLIN, MOSTELLER, and TUKEY · Exploring Data Tables, Trends and Shapes
HOAGLIN, MOSTELLER, and TUKEY · Understanding Robust and Exploratory Data Analysis
HOCHBERG and TAMHANE · Multiple Comparison Procedures
HOCKING · Methods and Applications of Linear Models: Regression and the Analysis of Variables
HOGG and KLUGMAN · Loss Distributions
HOSMER and LEMESHOW · Applied Logistic Regression, *Second Edition*
HØYLAND and RAUSAND · System Reliability Theory: Models and Statistical Methods
HUBERTY · Applied Discriminant Analysis
JACKSON · A User's Guide to Principle Components
JOHN · Statistical Methods in Engineering and Quality Assurance
JOHNSON · Multivariate Statistical Simulation
JOHNSON and KOTZ · Distributions in Statistics
JOHNSON, KOTZ, and BALAKRISHNAN · Continuous Univariate Distributions, Volume 1, *Second Edition*
JOHNSON, KOTZ, and BALAKRISHNAN · Continuous Univariate Distributions, Volume 2, *Second Edition*
JOHNSON, KOTZ, and BALAKRISHNAN · Discrete Multivariate Distributions
JOHNSON, KOTZ, and KEMP · Univariate Discrete Distributions, *Second Edition*
JUREČKOVÁ and SEN · Robust Statistical Procedures: Aymptotics and Interrelations
KADANE AND SCHUM · A Probabilistic Analysis of the Sacco and Vanzetti Evidence
KELLY · Reversability and Stochastic Networks
KHURI, MATHEW, and SINHA · Statistical Tests for Mixed Linear Models
KLUGMAN, PANJER, and WILLMOT · Loss Models: From Data to Decisions
KLUGMAN, PANJER, and WILLMOT · Solutions Manual to Accompany Loss Models: From Data to Decisions
KOTZ, BALAKRISHNAN, and JOHNSON · Continuous Multivariate Distributions, Volume 1, *Second Edition*
KOVALENKO, KUZNETZOV, and PEGG · Mathematical Theory of Reliability of Time-Dependent Systems with Practical Applications
LAD · Operational Subjective Statistical Methods: A Mathematical, Philosophical, and Historical Introduction
LEPAGE and BILLARD · Exploring the Limits of Bootstrap
LINHART and ZUCCHINI · Model Selection
LITTLE and RUBIN · Statistical Analysis with Missing Data
LLOYD · The Statistical Analysis of Categorical Data
MAGNUS and NEUDECKER · Matrix Differential Calculus with Applications in Statistics and Econometrics, *Revised Edition*
MANN, SCHAFER, and SINGPURWALLA · Methods for Statistical Analysis of Reliability and Life Data
McLACHLAN · Discriminant Analysis and Statistical Pattern Recognition
McLACHLAN and KRISHNAN · The EM Algorithm and Extensions
McLACHLAN and PEEL · Finite Mixture Models
MEEKER and ESCOBAR · Statistical Methods for Reliability Data

*Now available in a lower priced paperback edition in the Wiley Classics Library.

Applied Probability and Statistics (Continued)

MYERS and MONTGOMERY · Response Surface Methodology: Process and Product in Optimization Using Designed Experiments
NELSON · Accelerated Testing, Statistical Models, Test Plans, and Data Analyses
NELSON · Applied Life Data Analysis
OCHI · Applied Probability and Stochastic Processes in Engineering and Physical Sciences
OKABE, BOOTS, and SUGIHARA · Spatial Tesselations: Concepts and Applications of Voronoi Diagrams
PANKRATZ · Forecasting with Dynamic Regression Models
PANKRATZ · Forecasting with Univariate Box-Jenkins Models: Concepts and Cases
PORT · Theoretical Probability for Applications
POURAHMADI · Foundations of Time Series Analysis and Prediction Theory
PUTERMAN · Markov Decision Processes: Discrete Stochastic Dynamic Programming
RACHEV · Probability Metrics and the Stability of Stochastic Models
RÉNYI · A Diary on Information Theory
RIPLEY · Spatial Statistics
RIPLEY · Stochastic Simulation
ROUSSEEUW and LEROY · Robust Regression and Outlier Detection
RUBIN · Multiple Imputation for Nonresponse in Surveys
RUBINSTEIN · Simulation and the Monte Carlo Method
RUBINSTEIN and MELAMED · Modern Simulation and Modeling
RYAN · Statistical Methods for Quality Improvement, *Second Edition*
SCHIMEK · Smoothing and Regression: Approaches, Computation, and Application
SCHUSS · Theory and Applications of Stochastic Differential Equations
SCOTT · Multivariate Density Estimation: Theory, Practice, and Visualization
*SEARLE · Linear Models
SEARLE · Linear Models for Unbalanced Data
SEARLE, CASELLA, and McCULLOCH · Variance Components
SENNOTT · Stochastic Dynamic Programming and the Control of Queueing Systems
STOYAN, KENDALL, and MECKE · Stochastic Geometry and Its Applications, *Second Edition*
STOYAN and STOYAN · Fractals, Random Shapes and Point Fields: Methods of Geometrical Statistics
THOMPSON · Empirical Model Building
THOMPSON · Sampling
THOMPSON · Simulation: A Modeler's Approach
TIJMS · Stochastic Modeling and Analysis: A Computational Approach
TIJMS · Stochastic Models: An Algorithmic Approach
TITTERINGTON, SMITH, and MAKOV · Statistical Analysis of Finite Mixture Distributions
UPTON and FINGLETON · Spatial Data Analysis by Example, Volume 1: Point Pattern and Quantitative Data
UPTON and FINGLETON · Spatial Data Analysis by Example, Volume II: Categorical and Directional Data
VAN RIJCKEVORSEL and DE LEEUW · Component and Correspondence Analysis
VIDAKOVIC · Statistical Modeling by Wavelets
WEISBERG · Applied Linear Regression, *Second Edition*
WESTFALL and YOUNG · Resampling-Based Multiple Testing: Examples and Methods for *p*-Value Adjustment
WHITTLE · Systems in Stochastic Equilibrium
*ZELLNER · An Introduction to Bayesian Inference in Econometrics

*Now available in a lower priced paperback edition in the Wiley Classics Library.

Biostatistics Section

ARMITAGE and DAVID (editors) · Advances in Biometry
BROWN and HOLLANDER · Statistics: A Biomedical Introduction
CHOW and LIU · Design and Analysis of Clinical Trials: Concepts and Methodologies
DUNN and CLARK · Applied Statistics: Analysis of Variance and Regression, *Second Edition*
*ELANDT-JOHNSON and JOHNSON · Survival Models and Data Analysis
*FLEISS · The Design and Analysis of Clinical Experiments
FLEISS · Statistical Methods for Rates and Proportions, *Second Edition*
FLEMING and HARRINGTON · Counting Processes and Survival Analysis
KADANE · Bayesian Methods and Ethics in a Clinical Trial Design
KALBFLEISCH and PRENTICE · The Statistical Analysis of Failure Time Data
LACHIN · Biostatistical Methods: The Assessment of Relative Risks
LANGE, RYAN, BILLARD, BRILLINGER, CONQUEST, and GREENHOUSE · Case Studies in Biometry
LAWLESS · Statistical Models and Methods for Lifetime Data
LEE · Statistical Methods for Survival Data Analysis, *Second Edition*
MALLER and ZHOU · Survival Analysis with Long Term Survivors
McNEIL · Epidemiological Research Methods
McFADDEN · Management of Data in Clinical Trials
*MILLER · Survival Analysis, *Second Edition*
PIANTADOSI · Clinical Trials: A Methodologic Perspective
WOODING · Planning Pharmaceutical Clinical Trials: Basic Statistical Principles
WOOLSON · Statistical Methods for the Analysis of Biomedical Data

Financial Engineering Section

HUNT and KENNEDY · Financial Derivatives in Theory and Practice
ROLSKI, SCHMIDLI, SCHMIDT, and TEUGELS · Stochastic Processes for Insurance and Finance
SHAFER and VOVK · Probability and Finance: It's Only a Game!

Texts, References, and Pocketbooks Section

AGRESTI · An Introduction to Categorical Data Analysis
ANDĚL · Mathematics of Chance
ANDERSON · An Introduction to Multivariate Statistical Analysis, *Second Edition*
ANDERSON and LOYNES · The Teaching of Practical Statistics
ARMITAGE and COLTON · Encyclopedia of Biostatistics: Volumes 1 to 6 with Index
BARTOSZYNSKI and NIEWIADOMSKA-BUGAJ · Probability and Statistical Inference
BENDAT and PIERSOL · Random Data: Analysis and Measurement Procedures, *Third Edition*
BERRY, CHALONER, and GEWEKE · Bayesian Analysis in Statistics and Econometrics: Essays in Honor of Arnold Zellner
BHATTACHARYA and JOHNSON · Statistical Concepts and Methods
BILLINGSLEY · Probability and Measure, *Second Edition*
BOX · R. A. Fisher, the Life of a Scientist
BOX, HUNTER, and HUNTER · Statistics for Experimenters: An Introduction to Design, Data Analysis, and Model Building

*Now available in a lower priced paperback edition in the Wiley Classics Library.

Texts, References, and Pocketbooks (Continued)

BOX and LUCEÑO · Statistical Control by Monitoring and Feedback Adjustment
CHATTERJEE and PRICE · Regression Analysis by Example, *Third Edition*
COOK and WEISBERG · Applied Regression Including Computing and Graphics
COOK and WEISBERG · An Introduction to Regression Graphics
COX · A Handbook of Introductory Statistical Methods
DANIEL · Biostatistics: A Foundation for Analysis in the Health Sciences, *Sixth Edition*
DILLON and GOLDSTEIN · Multivariate Analysis: Methods and Applications
*DODGE and ROMIG · Sampling Inspection Tables, *Second Edition*
DRAPER and SMITH · Applied Regression Analysis, *Third Edition*
DUDEWICZ and MISHRA · Modern Mathematical Statistics
DUNN and CLARK · Basic Statistics: A Primer for the Biomedical Sciences, *Third Edition*
EVANS, HASTINGS, and PEACOCK · Statistical Distributions, *Third Edition*
FISHER and VAN BELLE · Biostatistics: A Methodology for the Health Sciences
FREEMAN and SMITH · Aspects of Uncertainty: A Tribute to D. V. Lindley
GROSS and HARRIS · Fundamentals of Queueing Theory, *Third Edition*
HALD · A History of Probability and Statistics and their Applications Before 1750
HALD · A History of Mathematical Statistics from 1750 to 1930
HELLER · MACSYMA for Statisticians
HOEL · Introduction to Mathematical Statistics, *Fifth Edition*
HOLLANDER and WOLFE · Nonparametric Statistical Methods, *Second Edition*
HOSMER and LEMESHOW · Applied Logistic Regression, *Second Edition*
HOSMER and LEMESHOW · Applied Survival Analysis: Regression Modeling of Time to Event Data
JOHNSON and BALAKRISHNAN · Advances in the Theory and Practice of Statistics: A Volume in Honor of Samuel Kotz
JOHNSON and KOTZ (editors) · Leading Personalities in Statistical Sciences: From the Seventeenth Century to the Present
JUDGE, GRIFFITHS, HILL, LÜTKEPOHL, and LEE · The Theory and Practice of Econometrics, *Second Edition*
KHURI · Advanced Calculus with Applications in Statistics
KOTZ and JOHNSON (editors) · Encyclopedia of Statistical Sciences: Volumes 1 to 9 with Index
KOTZ and JOHNSON (editors) · Encyclopedia of Statistical Sciences: Supplement Volume
KOTZ, READ, and BANKS (editors) · Encyclopedia of Statistical Sciences: Update Volume 1
KOTZ, READ, and BANKS (editors) · Encyclopedia of Statistical Sciences: Update Volume 2
LAMPERTI · Probability: A Survey of the Mathematical Theory, *Second Edition*
LARSON · Introduction to Probability Theory and Statistical Inference, *Third Edition*
LE · Applied Categorical Data Analysis
LE · Applied Survival Analysis
MALLOWS · Design, Data, and Analysis by Some Friends of Cuthbert Daniel
MARDIA · The Art of Statistical Science: A Tribute to G. S. Watson
MASON, GUNST, and HESS · Statistical Design and Analysis of Experiments with Applications to Engineering and Science
McCULLOCH and SEARLE · Generalized, Linear, and Mixed Models
MONTGOMERY, PECK, and VINING · Introduction to Linear Regression Analysis, *Third Edition*
MURRAY · X-STAT 2.0 Statistical Experimentation, Design Data Analysis, and Nonlinear Optimization
PRESS and TANUR · The Subjectivity of Scientists and the Bayesian Approach

*Now available in a lower priced paperback edition in the Wiley Classics Library.

Texts, References, and Pocketbooks (Continued)

PURI, VILAPLANA, and WERTZ · New Perspectives in Theoretical and Applied Statistics
RENCHER · Linear Models in Statistics
RENCHER · Methods of Multivariate Analysis
RENCHER · Multivariate Statistical Inference with Applications
ROSS · Introduction to Probability and Statistics for Engineers and Scientists
ROHATGI and SALEH · An Introduction to Probability and Statistics, *Second Edition*
RYAN · Modern Regression Methods
SCHOTT · Matrix Analysis for Statistics
SEARLE · Matrix Algebra Useful for Statistics
STYAN · The Collected Papers of T. W. Anderson: 1943–1985
TIAO, BISGAARD, HILL, PEÑA, and STIGLER (editors) · Box on Quality and Discovery: with Design, Control, and Robustness
TIERNEY · LISP-STAT: An Object-Oriented Environment for Statistical Computing and Dynamic Graphics
WONNACOTT and WONNACOTT · Econometrics, *Second Edition*
WU and HAMADA · Experiments: Planning, Analysis, and Parameter Design Optimization

JWS/SAS Co-Publications Section

FREUND and LITTELL · SAS System for Regression, *Third Edition*
KHATTREE and NAIK · Applied Multivariate Statistics with SAS Software, *Second Edition*
KHATTREE and NAIK · Multivariate Data Reduction and Discrimination with SAS Software

WILEY SERIES IN PROBABILITY AND STATISTICS

ESTABLISHED BY WALTER A. SHEWHART AND SAMUEL S. WILKS

Editors
Robert M. Groves, Graham Kalton, J. N. K. Rao, Norbert Schwarz, Christopher Skinner

Survey Methodology Section

BIEMER, GROVES, LYBERG, MATHIOWETZ, and SUDMAN · Measurement Errors in Surveys
COCHRAN · Sampling Techniques, *Third Edition*
COUPER, BAKER, BETHLEHEM, CLARK, MARTIN, NICHOLLS, and O'REILLY (editors) · Computer Assisted Survey Information Collection
COX, BINDER, CHINNAPPA, CHRISTIANSON, COLLEDGE, and KOTT (editors) · Business Survey Methods
*DEMING · Sample Design in Business Research
DILLMAN · Mail and Telephone Surveys: The Total Design Method, *Second Edition*
DILLMAN · Mail and Internet Surveys: The Tailored Design Method
GROVES and COUPER · Nonresponse in Household Interview Surveys

*Now available in a lower priced paperback edition in the Wiley Classics Library.

Survey Methodology (Continued)

GROVES · Survey Errors and Survey Costs

GROVES, BIEMER, LYBERG, MASSEY, NICHOLLS, and WAKSBERG · Telephone Survey Methodology

*HANSEN, HURWITZ, and MADOW · Sample Survey Methods and Theory, Volume 1: Methods and Applications

*HANSEN, HURWITZ, and MADOW · Sample Survey Methods and Theory, Volume II: Theory

KISH · Statistical Design for Research

*KISH · Survey Sampling

KORN and GRAUBARD · Analysis of Health Surveys

LESSLER and KALSBEEK · Nonsampling Error in Surveys

LEVY and LEMESHOW · Sampling of Populations: Methods and Applications, *Third Edition*

LYBERG, BIEMER, COLLINS, de LEEUW, DIPPO, SCHWARZ, TREWIN (editors) · Survey Measurement and Process Quality

SIRKEN, HERRMANN, SCHECHTER, SCHWARZ, TANUR, and TOURANGEAU (editors) · Cognition and Survey Research

VALLIANT, DORFMAN, and ROYALL · Finite Population Sampling and Inference: A Prediction Approach